METHODS OF QUANTUM FIELD THEORY IN STATISTICAL PHYSICS

METHODS OF QUANTUM FIELD THEORY IN STATISTICAL PHYSICS

A. A. ABRIKOSOV

L. P. GORKOV

I. E. DZYALOSHINSKI

Institute for Physical Problems
Academy of Sciences, U.S.S.R.

Revised English Edition
Translated and Edited by

Richard A. Silverman

DOVER PUBLICATIONS, INC.
NEW YORK

Published in Canada by General Publishing Company, Ltd., 30 Lesmill Road, Don Mills, Toronto, Ontario.
Published in the United Kingdom by Constable and Company, Ltd.

This Dover edition, first published in 1975, is an unabridged republication, with slight corrections, of the work originally published by Prentice-Hall, Inc., Englewood Cliffs, New Jersey, in 1963.

International Standard Book Number: 0-486-63228-8
Library of Congress Catalog Card Number: 75-17174

Manufactured in the United States of America
Dover Publications, Inc.
180 Varick Street
New York, N.Y. 10014

AUTHORS' PREFACE TO THE RUSSIAN EDITION

In recent years, remarkable success has been achieved in statistical physics, due to the extensive use of methods borrowed from quantum field theory. The fruitfulness of these methods is associated with a new formulation of perturbation theory, primarily with the application of "Feynman diagrams." The basic advantage of the diagram technique lies in its intuitive character: Operating with one-particle concepts, we can use the technique to determine the structure of any approximation, and we can then write down the required expressions with the aid of correspondence rules. These new methods make it possible not only to solve a large number of problems which did not yield to the old formulation of the theory, but also to obtain many new relations of a general character. At present, these are the most powerful and effective methods available in quantum statistics.

There now exists an extensive and very scattered journal literature devoted to the formulation of field theory methods in quantum statistics and their application to specific problems. However, familiarity with these methods is not widespread among scientists working in statistical physics. Therefore, in our opinion, the time has come to present a connected account of this subject, which is both sufficiently complete and accessible to the general reader.

Some words are now in order concerning the material in this book. In the first place, we have always tried to exhibit the practical character of the new methods. Consequently, besides a detailed treatment of the relevant mathematical apparatus, the book contains a discussion of various special problems encountered in quantum statistics. Naturally, the topics dealt with here do not exhaust recent accomplishments in the field. In fact, our choice of subject matter is dictated both by the extent of its general physical interest and by its suitability as material illustrating the general method.

We have confined ourselves to just one of the possible formulations of quantum statistics in field theory language. For example, we do not say anything about the methods developed by Hugenholtz, and by Bloch and de Dominicis. From our point of view, the simplest and most convenient method is that based on the use of Green's functions, and it is this method which is taken as fundamental in the present book.

It is assumed that the reader is familiar with the elements of statistical physics and quantum mechanics. The method of second quantization, as well as all information needed to derive the field theory methods used here, can be found in Chapter 1. This chapter is of an introductory character, and contains a brief exposition of contemporary ideas on the nature of energy spectra, together with some simple examples.

Unless the contrary is explicitly stated, we use a system of units in which both Planck's constant $\hbar$ and the velocity of light c equal 1 (the latter is important in Chapters 6 and 7). Moreover, temperature is expressed in energy units, so that $k = 1$.

The authors would like to express their gratitude to L. P. Pitayevski and Academician L. D. Landau for their valuable advice on the material discussed here.

1961

A. A. A.

L. P. G.

I. E. D.

AUTHORS' PREFACE TO THE REVISED ENGLISH EDITION

We are very pleased that our book is now appearing in English and will therefore be accessible to a larger audience. In the time that has elapsed since the Russian version was written, a number of new results have been obtained in quantum statistics, mainly in the theory of superconductivity and the theory of transport phenomena in the Fermi liquid. However, since these topics are of a more specialized character and involve very formidable calculations, we would not have included them in any event.

Sections 21 and 22 of the present edition have been drastically changed, and are now written in a more modern form, based on the use of the field theory technique for finite temperatures. We have also made a variety of smaller additions and corrections in other sections. We would like to express our appreciation to Dr. R. A. Silverman who, despite the exigencies of the publication schedule, not only allowed us to introduce these changes, but also effectively gave us complete control over the new edition.

1963

A. A. A.
L. P. G.
I. E. D.

TRANSLATOR'S PREFACE

So far, the series "Selected Russian Publications in the Mathematical Sciences" has been devoted to works in the field of pure and applied mathematics. Theoretical physics is now represented by a unique volume, whose authors need no introduction to anyone interested in many-body theory and its ramifications.

The present edition is the product of the closest collaboration with the authors. Not only have they carefully studied and commented upon a photocopy of the original typescript, but they have also read through and corrected all the galley proof. In addition to the major changes described in the Authors' Preface to the Revised English Edition, substantive changes of all sorts (many in answer to my queries) have been incorporated at every stage of the project, even at the very last moment. It is my hope that this exceptional degree of control has resulted in an authoritative version, insofar as this is possible in such a rapidly developing field.

For the system of references, I have chosen "letter-number form," which is so suitable for making last-minute additions without introducing excessive perturbations. For example, A7 refers to the seventh paper (or book) whose (first) author's surname begins with the letter A, where the entire Bibliography is arranged in lexicographic order, and in chronological order as well, whenever there are several papers by the same author.

My task would have been immeasurably more difficult were it not for the indefatigable assistance of Dr. N. R. Werthamer of the Bell Telephone Laboratories, who gave the entire translation a careful reading and suggested numerous improvements. All of his suggestions were examined in due course by the authors, who occasionally made counterproposals. The present version is the result of this interplay, which was particularly successful in unearthing typographical errors in the Russian original. The availability of Dr. Werthamer, at short notice and at all hours, has been one of my greatest assets in bringing this project to a successful conclusion. I would also like to acknowledge helpful advice from Professor J. M. Luttinger of Columbia University and Professor B. Zumino of New York University, who were both given desk copies of the translation shortly before it became the subject of a five months' correspondence between New York and Moscow.

I feel a debt of a very special nature to the senior author of this book (the patient "A. A." of so many marginalia), whom I have come to regard as a friend as well as a literary colleague.

R. A. S.

CONTENTS

2 METHODS OF QUANTUM FIELD THEORY FOR $T = 0$ (*Continued*).

3 THE DIAGRAM TECHNIQUE FOR $T \neq 0$, Page 97.

4 THEORY OF THE FERMI LIQUID, Page 154.

1

GENERAL PROPERTIES OF MANY-PARTICLE SYSTEMS AT LOW TEMPERATURES

1. Elementary Excitations. The Energy Spectrum and Properties of Liquid He^4 at Low Temperatures

1.1. Introduction. Quasi-particles. Statistical physics studies the behavior of systems consisting of a very large number of particles. In the last analysis, the macroscopic properties of liquids, gases and solids are due to microscopic interactions between the particles making up the system. Obviously, a complete solution of the problem, involving determination of the behavior of each individual particle, is out of the question. Fortunately, however, the overall macroscopic characteristics are determined only by certain average properties of the system.

To be explicit, we now consider some thermodynamic properties. The macroscopic state of a system is specified by giving three independent thermodynamic variables, e.g., the pressure P, the temperature T, and the average number N of particles in the system. From a quantum-mechanical point of view, a closed system of N particles is characterized by its energy levels E_n. Suppose that from the system we single out a volume (subsystem) which can still be regarded as macroscopic. The number of particles in such a subsystem is still very large, whereas the interaction forces between particles act at distances whose order of magnitude is that of atomic dimensions. Therefore, apart from boundary effects, we can regard the subsystem itself as closed, and characterized by certain energy levels (for a given number of particles). Since the subsystem actually interacts with other parts of the closed system, it does not have a fixed energy and a fixed number of particles, and, in fact, it has a nonzero probability of occupying any energy state.

As is familiar from statistical physics (see e.g., L8),[1] the microscopic derivation of thermodynamic formulas is based on the *Gibbs distribution*, which gives the following probability of finding the subsystem in the energy state E_{nN}, with a number of particles equal to N:

$$w_{nN} = Z^{-1}e^{-(E_{nN}-\mu N)/T}\,. \tag{1.1}$$

In this formula, T denotes the absolute temperature, μ the chemical potential, and Z a normalization factor which is determined from the condition

$$\sum_{n,N} w_{nN} = 1. \tag{1.2}$$

According to (1.1), we have

$$Z = \sum_{n,N} e^{-(E_{nN}-\mu N)/T}. \tag{1.3}$$

The quantity Z is called the *grand partition function.* If the energy levels E_{nN} are known, the partition function can be calculated. This immediately determines the thermodynamic functions as well, since the formula

$$\Omega = -T\ln Z \tag{1.4}$$

relates the quantity Z to the thermodynamic potential Ω (involving the variables V, T and μ).

Obviously, the simplest use that can be made of these formulas is to calculate the thermodynamic functions of ideal gases, since in this case the energy is just the sum of the energies of the separate particles. However, in general, it is impossible to determine the energy levels of a system consisting of a large number of interacting particles. Therefore, so far, interactions between particles in quantum statistics have been successfully taken into account only when the interactions are sufficiently weak, and perturbation-theory calculations of thermodynamic quantities have been carried out only to the first or second approximations. In the majority of physical problems, where the interaction is far from small, an approach based on the direct use of formulas (1.1)–(1.4) is unrealistic.

The case of very low temperatures is somewhat exceptional. As $T \to 0$, the important energy levels in the partition function are the *weakly excited states*, whose energies differ only very little from the energy of the ground state. The character of the energy spectrum of the system in this region of energies can be ascertained in some detail, by using very general considerations which are valid regardless of the magnitude and specific features of the interaction between the particles.

---

[1] The reference scheme is explained in the Translator's Preface.

As an example illustrating the subsequent discussion, consider the excitation of lattice vibrations in a crystal. As long as the vibrations are small, we can regard the lattice as a set of coupled harmonic oscillators. Introducing normal coordinates, we obtain a system of $3N$ linear oscillators with characteristic frequencies ω_i (N is the number of atoms). According to quantum mechanics, the energy spectrum of such a system is given by the formula

$$E = \sum_{i=1}^{3N} \omega_i\left(n_i + \frac{1}{2}\right),$$

where the n_i are arbitrary nonnegative integers (including zero), and various sets of numbers n_i correspond to various energy levels of the system. The lattice vibrations can be described as a superposition of monochromatic plane waves, propagating in the crystal. Each wave is characterized by a wave vector and a frequency, and also by an index s specifying the type of wave involved. Because of the possibility that various types of waves can propagate in the crystal, the frequency ω, regarded as a function of the wave vector $\mathbf{k}$, is not a single-valued function, but rather consists of several branches $\omega_s(\mathbf{k})$, where the total number of branches equals $3r$ (r is the number of atoms belonging to one unit cell of the crystal). For small momenta, three of these branches (the *acoustic branches*) are characterized by the fact that the frequency depends linearly on the wave vector:

$$\omega_s(\mathbf{k}) = u_s(\theta, \varphi)|\mathbf{k}|.$$

For other momenta, the curve $\omega_s(\mathbf{k})$ begins with some finite value for $\mathbf{k} = 0$, and depends weakly on $\mathbf{k}$ in the region of small wave numbers.[2]

From a knowledge of the frequency spectrum, the energy levels and the matrix elements of the displacements of the atoms of the lattice (the coordinates of the oscillators), we can calculate completely, at least in principle, both the thermodynamic and the kinetic characteristics of the vibrating lattice. However, instead of the model of coupled oscillators, it turns out in practice to be very convenient to use another, equivalent model. This model can be obtained by applying the quantum-mechanical correspondence principle, which states that every plane wave corresponds to a set of moving "particles," with momentum[3] determined by the wave vector $\mathbf{k}$ and energy determined by the frequency $\omega_s(\mathbf{k})$.[4] Thus, an excited state of the lattice can be thought of as an aggregate of such "particles" (called *phonons*), moving freely in the volume occupied by the crystal. This leads to an expression for

---

[2] For more detailed information on the spectrum of lattice vibrations, see e.g., Peierls' book P1.

[3] Actually, $\mathbf{k}$ is not a momentum but a "quasi-momentum" (see P1), but here this distinction is not important.

[4] We recall that $\hbar = 1$ in the system of units used here. This means that the energy has dimensions $\sec^{-1}$ and the momentum has dimensions cm^{-1}. To convert to ordinary units, we have to multiply all energies and momenta by $\hbar$.

the energy levels of the system which is analogous to that for an ideal gas. In fact, n_i can be interpreted as the number of phonons in the state i, where $i = (\mathbf{k}, s)$, and the numbers n_i range over all nonnegative integers. It follows that phonons obey Bose statistics, even when the atoms making up the system have half-integral spins.

At very low temperatures, the most important role is played by phonons with small energies. According to what was just said about the branches of the frequency spectrum, the phonons with the smallest energies correspond to the acoustic branches, in the region of small momenta. In this case, the function $\omega(\mathbf{k})$ is linear, and this fact alone permits us to draw a number of qualitative conclusions, e.g., to deduce that the heat capacity of the lattice is proportional to T^3.

For quantitative calculations, the *isotropic Debye model* is often used instead of the spectrum of the actual lattice. In this model, it is assumed that the low-frequency part of the spectrum, instead of having three acoustic branches, is the same as that of an isotropic body, i.e., consists of longitudinal phonons with energy $\omega_l(k) = u_l k$, and transverse phonons with two possible polarizations and the same dependence $\omega_t(k) = u_t k$ of the energy on the momentum. Furthermore, it is assumed that the momenta of the phonons do not exceed a certain upper bound k_D, determined by a normalization involving the appropriate number of degrees of freedom. Then it is clear that

$$k_D \sim \frac{1}{a},$$

where a is the interatomic distance. This model leads to Debye's well-known interpolation formula for the heat capacity of solids. Later on, we shall use this model to study the interaction of electrons and phonons in a metal.

If we take into account the small anharmonic terms in the potential energy of the vibrating lattice, the expression for the energy (given above) is no longer exact, and transitions between states with different sets of numbers n_i are now possible. This fact can also be interpreted in phonon language, in terms of various interaction processes between phonons, leading to scattering of phonons by phonons and the creation of new phonons. In other words, in a rigorous analysis it is only an approximation to regard the phonons as freely moving particles. The role of the anharmonic terms becomes greater as the amplitude of the vibrations increases, i.e., as the temperature is raised. In the phonon model, the number of phonons increases as the temperature is raised, and this increases the importance of interactions between phonons. Therefore, the very concept of phonons as freely moving particles is applicable only in the region of temperatures that are not too high (considerably lower than the melting point).

We now consider the general case. By analogy with the example just considered, to construct a model of the energy spectrum for the weakly excited states of a system, we make the basic assumption that to a first approximation the structure of the energy levels obeys the same principle as that of the energy levels of an ideal gas. In other words, it is assumed that any energy level can be obtained as the sum of the energies of a certain number of "quasi-particles" or "elementary excitations", with momentum $\mathbf{p}$ and energy $\varepsilon(\mathbf{p})$, moving in the volume occupied by the system. [Generally speaking, the dispersion law $\varepsilon(\mathbf{p})$ for these excitations will be different from the expression $\varepsilon_0(\mathbf{p}) = p^2/2m$ for the energy of free particles.] It should be emphasized that the elementary excitations are the result of collective interactions of the particles of the system, and therefore pertain to the system as a whole and not to its separate particles. In particular, the number of elementary excitations is certainly not the same as the total number of particles in the system.

All energy spectra can be divided into two broad categories, spectra of the *Bose type* and spectra of the *Fermi type*. In the first case, the excitations have integral-valued intrinsic moments (spins) and obey Bose statistics. In the second case, the excitations have half-integral spins and obey Fermi statistics. According to quantum mechanics, the spin of any system can only change by an integer. It follows that Bose excitations can appear or disappear one at a time, whereas Fermi excitations always appear and disappear in pairs.

As already mentioned in the example given above (involving lattice vibrations), the statistics of the elementary excitations do not have to be the same as the statistics of the particles making up the system. It is obvious only that a Bose system cannot have excitations with half-integral spins.

The elementary excitations do not correspond to exact statistical states of the system, but instead represent wave packets, i.e., superpositions of large numbers of exact stationary states with a narrow spread in energy. As a result, transitions from one state to another have nonzero probability. This leads to spreading of the packet, i.e., *attenuation* (or *damping*) of the excitations. Therefore, a description of the system by using elementary excitations is possible only as long as the (energy) width of the packet determining its attenuation is small compared to the energy of the packet.

Spread of the packet and the concomitant attenuation of the elementary excitations can be regarded as the result of interactions between the "quasi-particles," during which the laws of energy and momentum conservation are satisfied. Clearly, all such processes can be divided into processes in which one excitation "decays" into several others, and processes in which excitations are "scattered" by each other. As we shall see below, decay of excitations can take place only at sufficiently high energies. Moreover, scattering processes are important only when the number of excitations is sufficiently large. Thus, at low temperatures, where low-energy excitations are important and there are few of them, both types of processes leading to the attenuation of excitations are unimportant. The weakness of the interactions between excitations at low temperatures allows us to regard them as an ideal gas of "quasi-particles."

At present, because of both experimental data and direct theoretical calculations, the ideas just presented, concerning the structure of energy spectra, are well-established facts. Of course, the energy spectra of different physical objects (e.g., liquids consisting of the isotopes He^3 and He^4, metals, dielectrics, etc.) are completely different. For example, the spectrum of liquid He^4 is of the Bose type, while the spectrum of liquid He^3 and the electronic spectra of metals are of the Fermi type.[5]

1.2. The spectrum of a Bose liquid.[6] An example of a system with a spectrum of the Bose type is the *Bose liquid*, i.e., a liquid consisting of atoms with integral-valued spins. In nature, there exists only one such liquid which does not solidify at absolute zero, namely, liquid helium (more precisely, the isotope He^4). Since the atoms of He^4 have spin zero, for all practical purposes we can confine ourselves to this case.

The dependence of the excitation energy of a Bose liquid on the momentum in the limit of small values of the momentum can be determined by very general considerations. The region of small values of p corresponds to long-wavelength oscillations of the liquid. But such oscillations are just ordinary sound waves. From this we conclude at once that for small p, the elementary excitations are identical with acoustic quanta (phonons), for which the relation between energy and momentum is well known. In fact, noting that the acoustic frequency ω is related to the wave vector by the formula $\omega = uk$, where u is the velocity of sound, we immediately find that the desired relation between ε and p is

$$\varepsilon = up. \tag{1.5}$$

Thus, for small momenta, the energy of an excitation in the Bose liquid depends linearly on its momentum, and the coefficient of proportionality is just the velocity of sound.

As the momentum increases, the function $\varepsilon(p)$ ceases to be linear, and the subsequent behavior of the curve $\varepsilon(p)$ cannot be determined quite so easily. Then we have recourse to the following argument,[7] which allows us to draw a variety of conclusions concerning the behavior of $\varepsilon(p)$ for arbitrary momenta: The energy of the liquid is a functional of its density $\rho(\mathbf{r})$ and its hydrodynamic velocity $\mathbf{v}(\mathbf{r})$, i.e.,

$$E(\rho, \mathbf{v}) = \frac{1}{2} \int \rho \mathbf{v}^2 \, d\mathbf{r} + E^{(1)}(\rho), \tag{1.6}$$

[5] To avoid misunderstanding, we observe at this point that henceforth we shall always be dealing exclusively with the isotropic model of a metal, which of course is far from representing the true state of affairs. Electronic spectra in actual metals are sharply anisotropic, and hence many of the results given in this book have only a qualitative character when applied to metals.

[6] The ideas given here concerning the spectrum of a Bose liquid were first proposed by Landau (L1, L2).

[7] The derivation given here is due to Pitayevski (P3).

where $E^{(1)}$ is the part of the energy which is independent of the velocity. If we consider small oscillations, then

$$\rho(\mathbf{r}) = \bar{\rho} + \delta\rho(\mathbf{r}),$$

where $\bar{\rho}$ is the equilibrium density, which is independent of the coordinates, and $\delta\rho(\mathbf{r})$, $\mathbf{v}(\mathbf{r})$ are small quantities describing the oscillations. It should be noted that by definition

$$\bar{\rho} = \frac{1}{V}\int \rho(\mathbf{r})\, d\mathbf{r}, \qquad \int \delta\rho(\mathbf{r})\, d\mathbf{r} = 0,$$

where V is the volume occupied by the liquid.

To within quantities of the second order in $\delta\rho$ and $\mathbf{v}$, we can replace the function $\rho(\mathbf{r})$ in the first term of the right-hand side of (1.6) by its average value $\bar{\rho}$. To the same accuracy, the expression for $E^{(1)}$ can be written in the form

$$E^{(1)}(\rho) = E^{(1)}(\bar{\rho}) + \int \psi(\mathbf{r})\delta\rho(\mathbf{r})\, d\mathbf{r} + \frac{1}{2}\int \varphi(\mathbf{r}, \mathbf{r}')\delta\rho(\mathbf{r})\delta\rho(\mathbf{r}')\, d\mathbf{r}\, d\mathbf{r}'.$$

The functions $\psi(\mathbf{r})$ and $\varphi(\mathbf{r}, \mathbf{r}')$ are determined only by the properties of the liquid when it is unperturbed by oscillations, i.e., when it is homogeneous and isotropic. As a result,

$$\psi(\mathbf{r}) = \text{const} = \psi,$$

while $\varphi(\mathbf{r}, \mathbf{r}')$ depends only on $|\mathbf{r} - \mathbf{r}'|$:

$$\varphi(\mathbf{r}, \mathbf{r}') = \varphi(|\mathbf{r} - \mathbf{r}'|).$$

Therefore, the first-order term in the expansion of $E^{(1)}$ is simply proportional to

$$\int \delta\rho(\mathbf{r})\, d\mathbf{r} = 0,$$

and we finally have

$$E = E^{(1)}(\bar{\rho}) + \frac{1}{2}\bar{\rho}\int \mathbf{v}^2\, d\mathbf{r} + \frac{1}{2}\int \varphi(|\mathbf{r} - \mathbf{r}'|)\delta\rho(\mathbf{r})\delta\rho(\mathbf{r}')\, d\mathbf{r}\, d\mathbf{r}'.$$

The velocity $\mathbf{v}$ is related to the density oscillations by the equation of continuity

$$\dot{\rho} + \operatorname{div}(\rho\mathbf{v}) = 0,$$

which can be written as

$$\dot{\rho} + \bar{\rho}\operatorname{div}\mathbf{v} = 0 \tag{1.7}$$

to within first-order terms in $\delta\rho$ and $\mathbf{v}$.

We now go over to Fourier components, writing

$$\delta\rho(\mathbf{r}) = \frac{1}{V}\sum_{\mathbf{p}} \rho_{\mathbf{p}} e^{i\mathbf{p}\cdot\mathbf{r}},$$

$$\mathbf{v}(\mathbf{r}) = \frac{1}{V}\sum_{\mathbf{p}} \mathbf{v}_{\mathbf{p}} e^{i\mathbf{p}\cdot\mathbf{r}},$$

$$\varphi(\mathbf{r}) = \frac{1}{V}\sum_{\mathbf{p}} \varphi_{\mathbf{p}} e^{i\mathbf{p}\cdot\mathbf{r}},$$

and we bear in mind that small oscillations of the liquid are always longitudinal, i.e., the velocity $\mathbf{v}_\mathbf{p}$ of the wave with wave vector $\mathbf{p}$ is always directed along $\mathbf{p}$:

$$\mathbf{v}_\mathbf{p} = a_\mathbf{p}\mathbf{p}.$$

Then it is an easy consequence of (1.7) that

$$\mathbf{v}_\mathbf{p} = i\frac{\dot{\rho}_\mathbf{p}}{\bar{\rho}}\frac{\mathbf{p}}{p^2}$$

and

$$E = E^{(1)}(\bar{\rho}) + \frac{1}{V}\sum_\mathbf{p}\left(\frac{|\dot{\rho}_\mathbf{p}|^2}{2\bar{\rho}p^2} + \frac{1}{2}\varphi_\mathbf{p}|\rho_\mathbf{p}|^2\right). \tag{1.8}$$

The first term in (1.8) represents the energy of the unperturbed liquid, while the second term decomposes into a sum of terms each of which is just the energy of a harmonic oscillator of frequency $\omega_\mathbf{p}$, where

$$\omega_\mathbf{p}^2 = \bar{\rho}p^2\varphi_\mathbf{p}. \tag{1.9}$$

Thus, we see that every small oscillation of the liquid decomposes into elementary oscillations or *elementary excitations*, described by the equations of suitable harmonic oscillators. In the quantum-mechanical case, the energy of each such harmonic oscillator is given by the formula

$$\varepsilon_\mathbf{p} = \omega_\mathbf{p}\left(n + \frac{1}{2}\right) \qquad (n = 0, 1, 2, \ldots).$$

The structure of the spectrum of the system corresponds completely to this model of the elementary excitations, i.e., the spectrum is the sum of the energies of different numbers of elementary excitations, and the dependence of the energy $\varepsilon(\mathbf{p})$ of an elementary excitation on the momentum is given by (1.9) and the obvious formula

$$\varepsilon(\mathbf{p}) = \omega_\mathbf{p}.$$

To complete the solution of the problem, we have to express $\varphi_\mathbf{p}$ in terms of characteristics of the system. To do this, we note that in the quantum-mechanical case, the energy of the ground state of the system is not $E^{(1)}(\bar{\rho})$, as in the classical case. Instead, we have to take into account the *zero-point energy* of the oscillators, which, as is well known, equals $\frac{1}{2}\omega_\mathbf{p}$ for each oscillator. Thus, the energy of the ground state of the Bose liquid equals

$$E_0 = E^{(1)}(\bar{\rho}) + \sum_\mathbf{p}\frac{1}{2}\omega_\mathbf{p},$$

where

$$\frac{1}{2}V\omega_\mathbf{p} = \frac{1}{2\bar{\rho}p^2}\overline{|\dot{\rho}_\mathbf{p}|^2} + \frac{1}{2}\varphi_\mathbf{p}\overline{|\rho_\mathbf{p}|^2} = \varphi_\mathbf{p}\overline{|\rho_\mathbf{p}|^2} \tag{1.10}$$

[see (1.8)]. From (1.9) and (1.10), we immediately obtain[8]

$$\varepsilon(\mathbf{p}) = \omega_\mathbf{p} = \frac{p^2}{2mS(\mathbf{p})}, \tag{1.11}$$

---

[8] Formula (1.11) was first obtained by Feynman (F1) by another method. His derivation is much more complicated, and in our opinion, no more general than that given here.

where

$$S(\mathbf{p}) = \frac{\overline{|\rho_{\mathbf{p}}|^2}}{Vm\bar{\rho}}$$

is the Fourier component of the *density correlation function*

$$S(\mathbf{r} - \mathbf{r}') = \frac{\overline{[n(\mathbf{r}) - \bar{n}][n(\mathbf{r}') - \bar{n}]}}{\bar{n}} \qquad \left(n(\mathbf{r}) = \frac{1}{m}\,\rho(\mathbf{r})\right) \tag{1.12}$$

[$n(\mathbf{r})$ is the number of particles per unit volume.] While it is impossible to calculate the quantity $S(\mathbf{p})$, formula (1.11) enables us to draw a variety of important conclusions about the form of $\varepsilon(\mathbf{p})$. On the other hand, from a knowledge of some general properties of the spectrum $\varepsilon(\mathbf{p})$, we can ascertain the behavior of $S(\mathbf{p})$, which determines the character of interaction processes between the liquid and various particles.[9]

As already mentioned, in the region of small momenta, the excitation energy is a linear function of momentum, i.e., $\varepsilon \approx up$. This implies that $S(\mathbf{p})$ also depends linearly on the momentum:

$$S \approx \frac{p}{2mu}.$$

In the region of small distances (or equivalently, the region of large momenta), it is well known (see e.g., L8, Sec. 115) that the function $S(\mathbf{r})$ has the form

$$S(\mathbf{r}) = \delta(\mathbf{r}) + \nu(\mathbf{r}), \tag{1.13}$$

where $\nu(\mathbf{r})$ has no singularities as $\mathbf{r} \to 0$. Therefore, in Fourier components, we have

$$S(\mathbf{p}) = 1 + \nu(\mathbf{p}),$$

where $\nu(\mathbf{p}) \to 0$ as $p \to \infty$. Thus, for large momenta, $S(\mathbf{p}) \approx 1$ and

$$\varepsilon(\mathbf{p}) \approx \frac{p^2}{2m},$$

i.e., the energy of the elementary excitations is the same as the energy of a free atom of the liquid (an atom of liquid He^4).

For intermediate values of the momentum, the function $S(\mathbf{p})$ can either increase monotonically from 0 to 1 as p increases, or else it can have a minimum for the value $p \sim 1/a$, where a is the interatomic distance. (This follows from a dimensionality argument, since in a problem involving a liquid, the interatomic distance is the only parameter with the dimension of length.) In the latter case, the spectrum of the elementary excitations can have the form shown in Fig. 1. The conjecture that the excitation spectrum of liquid He^4 has a minimum for $p \sim 1/a$ was first made by Landau (L1, L2).

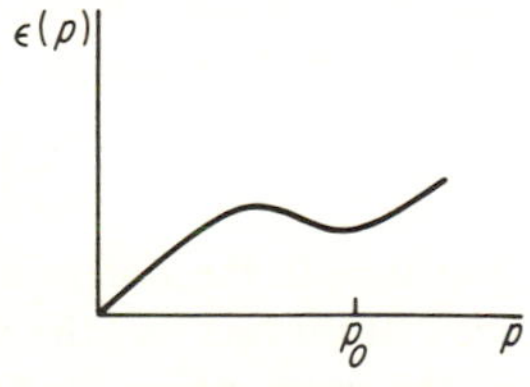

FIGURE 1

It should be noted that the derivation just given of formula (1.11) is based

[9] E.g., neutrons (see Chap. 3, Sec. 17).

on the hydrodynamical approximation, in which the liquid is regarded as a continuous medium. This approximation is no longer valid when distances of the order of interatomic distances (or momenta of order $1/a$) become important. Therefore, formula (1.11), which is valid for small momenta, has to be regarded as an interpolation between the region of small momenta and the limiting region of large momenta (where the particles are essentially free and the elementary excitations are the same as particles, i.e., have energy $p^2/2m$).

Of course, the spectrum of elementary excitations in liquid He^4 cannot be calculated in full detail. The most exact curves of the function $\varepsilon(p)$ have been obtained recently from experiments on the scattering of neutrons in He^4 (see Y1).

From a knowledge of the energy spectrum, we can calculate the thermodynamic functions of liquid He^4, or, more exactly, the difference between the values of these functions at a given temperature and at $T = 0$. Here, depending on the value of T, different parts of the spectrum (shown in Fig. 1) play the most important role. In the region of very low temperatures, the most important part of the spectrum is that corresponding to small p, i.e., the phonons. For higher temperatures, the excitations in the neighborhood of the minimum of $\varepsilon(p)$ [at $p = p_0$] become most important. Expanding $\varepsilon(p)$ in powers of $p - p_0$, we obtain[10]

$$\varepsilon(p) = \Delta + \frac{1}{2m^*}(p - p_0)^2. \tag{1.14}$$

In this part of the energy spectrum, the elementary excitations are called "rotons."

All the thermodynamic functions can be represented as sums of a "phonon" part and a "roton" part. To find the thermodynamic potential, it is sufficient to substitute (1.1) and (1.14) into the formula

$$\Omega = VT \int \ln\left[1 - e^{[\mu - \varepsilon(p)]/T}\right] \frac{d\mathbf{p}}{(2\pi)^3} \tag{1.15}$$

(see L8, Sec. 53). Here the following facts must be kept in mind: In the first place, the number of excitations is not fixed, but is itself determined by the equilibrium condition, which states that the free energy is a minimum with respect to changes in the number of particles, i.e.,

$$\left(\frac{\partial F}{\partial N}\right)_{V,T} = \mu = 0 \qquad (F = \Omega + \mu N). \tag{1.16}$$

For $\mu = 0$, the potential Ω coincides with the free energy F. Secondly, in view of the fact that the energy ε_{rot} of the rotons is always very large compared to the temperatures under discussion, the Bose distribution of the rotons can

[10] For He^4, the values of the constants appearing in (1.14) are

$$\Delta = 11.4 \times 10^{11}\ \text{sec}^{-1}, \quad p_0 = 1.92 \times 10^8\ \text{cm}^{-1}, \quad m^* = 0.16\ m_{He^4}$$

(see Y1).

be replaced by a Boltzmann distribution, since for $T \ll \varepsilon_{\rm rot}$, we can confine ourselves to the first-order term in the expansion of

$$\ln\left(1 - e^{-\varepsilon_{\rm rot}/T}\right)$$

with respect to the small quantity $e^{-\varepsilon_{\rm rot}/T}$. This leads to the Boltzmann formula

$$F_{\rm rot} = -VT\int e^{-\varepsilon_{\rm rot}/T}\frac{d\mathbf{p}}{(2\pi)^3}.$$

Taking these facts into account, we find that

$$F_{\rm phon} = -V\frac{\pi^2T^4}{90u^3}, \qquad F_{\rm rot} = -V\frac{2m^{*1/2}T^{3/2}p_0^2}{(2\pi)^{3/2}}\,e^{-\Delta/T}. \tag{1.17}$$

From (1.17) we can easily obtain all the other thermodynamic functions.

1.3. Superfluidity. The most interesting property of the Bose liquid is the property of "superfluidity," i.e., the possibility of flowing through capillary tubes without friction. Landau (L1) showed that this property is a consequence of the assumptions made about the form of the excitation spectrum.

Consider a Bose liquid at absolute zero, flowing with velocity $\mathbf{v}$ in a capillary. In the coordinate system fixed with respect to the liquid, the liquid is at rest and the capillary moves with velocity $-\mathbf{v}$. As a result of friction between the liquid and the wall of the capillary, the liquid begins to be "carried along" by the wall. This means that the liquid begins to have nonzero energy and momentum, which is possible only if elementary excitations appear in the liquid. As soon as a single such excitation appears, the liquid acquires momentum $\mathbf{p}$ and energy $\varepsilon(\mathbf{p})$.

Now, suppose we go back to the coordinate system fixed with respect to the capillary. In this system, the energy of the liquid equals

$$\varepsilon + \mathbf{p}\cdot\mathbf{v} + \tfrac{1}{2}Mv^2.$$

Thus, the appearance of an excitation changes the energy by an amount $\varepsilon + \mathbf{p}\cdot\mathbf{v}$. In order for such an excitation to appear, the change in energy must be negative, i.e.,

$$\varepsilon + \mathbf{p}\cdot\mathbf{v} < 0.$$

The quantity $\varepsilon + \mathbf{p}\cdot\mathbf{v}$ takes its minimum value when $\mathbf{p}$ and $\mathbf{v}$ have opposite directions. Thus, in any case, we must have

$$\varepsilon - pv < 0 \quad \text{or} \quad v > \frac{\varepsilon}{p}.$$

This means that in order for it to be possible for any excitations to appear in the fluid, the velocity must satisfy the condition

$$v > \left(\frac{\varepsilon}{p}\right)_{\rm min}. \tag{1.18}$$

The minimum value of ε/p corresponds to the point of the curve $\varepsilon(p)$ where

$$\frac{d\varepsilon}{dp} = \frac{\varepsilon}{p}, \tag{1.19}$$

i.e., the point where a line drawn from the origin of coordinates is tangent to the curve $\varepsilon(\mathbf{p})$. Thus, superfluid flow can occur only in the case where the velocity of the liquid is less than the velocity of the elementary excitation at the points satisfying the condition (1.19). (We recall that $d\varepsilon/dp$ is the velocity of the elementary excitation.) For every Bose liquid, there always exists at least one point where the condition (1.19) is satisfied, namely, the origin of coordinates $p = 0$. Since for values of p near zero, the excitations move with the velocity of sound, the superfluidity condition is certainly not satisfied for flow velocities exceeding the velocity of sound u

One more critical point is present in the spectrum of He^4. It is clear from the form of Fig. 1 that this point lies to the right of the roton minimum. Using (1.14), we easily find that the velocity of the superfluid flow must satisfy the inequality

$$v < \frac{1}{m^*}\left(\sqrt{p_0^2 + 2m^*\Delta} - p_0\right),$$

or

$$v < \frac{\Delta}{p_0},$$

if we take account of the numerical values of the constants (see footnote 10, p. 10), which show that $p_0^2 \gg 2m^*\Delta$. Therefore, we finally arrive at the conclusion that there is certainly no superfluid flow in He^4 at velocities exceeding Δ/p_0.

At nonzero temperatures, excitations appear in the Bose liquid. It is not hard to see that this does not change the argument just given concerning the appearance of new excitations during the flow. However, it is interesting to examine the effect on the liquid's motion of the excitations already present in the liquid. To do so, we imagine that a "gas of elementary excitations" moves in the liquid with some macroscopic velocity $\mathbf{v}$. In this case, the distribution function is obtained from the distribution function of a gas at rest by changing ε to $\varepsilon - \mathbf{p}\cdot\mathbf{v}$. The momentum of the gas per unit volume is given by the integral

$$\mathbf{P} = \int \mathbf{p}n(\varepsilon - \mathbf{p}\cdot\mathbf{v})\,\frac{d\mathbf{p}}{(2\pi)^3}. \tag{1.20}$$

For small velocities, $n(\varepsilon - \mathbf{p}\cdot\mathbf{v})$ can be expanded with respect to $\mathbf{p}\cdot\mathbf{v}$. As a result, we obtain

$$\mathbf{P} = -\int \mathbf{p}(\mathbf{p}\cdot\mathbf{v})\,\frac{\partial n}{\partial \varepsilon}\,\frac{d\mathbf{p}}{(2\pi)^3} = -\frac{\mathbf{v}}{3}\int p^2\,\frac{\partial n}{\partial \varepsilon}\,\frac{d\mathbf{p}}{(2\pi)^3}. \tag{1.21}$$

It follows from (1.21) that $\mathbf{P}$, the momentum of the moving "excitation gas," is proportional to $\mathbf{v}$, the velocity of motion; obviously, the coefficient of proportionality is the mass of a moving body. Thus, we see that motion of the excitation gas relative to the liquid is accompanied by transport of mass. Of course, the separate excitations can interact with the walls of the capillary and be scattered by them. This leads to an exchange of momentum between the excitation gas and the walls, which means that the motion of the excitation gas is viscous. Since, as we have just seen, the motion of the excitation gas is accompanied by transport of mass, we arrive at the conclusion that in a Bose liquid already containing excitations, viscous flow can occur with velocities which do not violate the superfluidity condition (1.18). However, it is important that the viscous motion be accompanied by a transport of mass. This mass is by no means the same as the mass of the whole liquid. In fact, it is determined by formula (1.21), and depends on the number of excitations (in particular, $\mathbf{P} = 0$ for $T = 0$).

Thus, we have the following general picture of the motion of a Bose liquid when the velocity is such that the superfluidity condition holds: First, we consider the temperature $T = 0$ (absolute zero). If the liquid is initially in the ground state, i.e., if it contains no elementary excitations, then no excitations can appear later and the motion is superfluid. For $T \neq 0$, the picture changes in an essential way. Now the fluid contains excitations whose number is determined by the appropriate statistical formulas. Although new excitations cannot appear, nothing, as noted above, can prevent the excitations already present from colliding with the walls, thereby exchanging momentum with the walls. Only a part of the mass of the liquid participates in this viscous motion, as described by (1.21). The remaining part of the mass of the liquid moves as before, with no friction between it and the walls or between it and the part of the fluid participating in the viscous flow. Thus, at $T \neq 0$, a Bose liquid represents a kind of mixture of two liquids, one of which is "superfluid" and the other "normal," moving with no friction between them.

Of course, in reality no such separation occurs, and there are simply two motions in the liquid, each of which has its own effective mass or density. The "normal" density is the coefficient of proportionality between the momentum and the velocity of a unit volume of the moving gas of excitations. Substituting into (1.21) first a Bose distribution with $\varepsilon = up$ and then a Boltzmann distribution with ε given by (1.14), we find the following formulas for the phonon and roton parts of the normal density:

$$\rho_{n,\,\mathrm{phon}} = \frac{2\pi^2 T^4}{45u^5}, \qquad \rho_{n,\,\mathrm{rot}} = \frac{2m^{*1/2}p_0^4 e^{-\Delta/T}}{3(2\pi)^{3/2}T^{1/2}}. \tag{1.22}$$

The remaining part of the density of the fluid, denoted by ρ_s, corresponds to superfluid motion, and hence

$$\rho = \rho_n + \rho_s. \tag{1.23}$$

Let $\mathbf{v}_n$ denote the macroscopic velocity of the gas of excitations, and let $\mathbf{v}_s$ denote the velocity of the superfluid liquid. Then the velocity $\mathbf{v}_s$ has the following basic property: If we put the Bose liquid in a cylinder and rotate the cylinder about its axis, the normal part is "carried along" by the walls of the cylinder and itself begins to rotate. On the other hand, the superfluid part remains at rest, and hence does not have to be taken into account. In other words, the rotation of the superfluid part is always irrotational, a fact which is expressed mathematically by the condition

$$\operatorname{curl} \mathbf{v}_s = 0. \tag{1.24}$$

The motion of the superfluid part of the liquid imposes certain conditions on the excitations. In fact, we note that it is precisely in the reference system fixed with respect to the superfluid part that the function $\varepsilon(p)$ has the form discussed above. In the rest system, we obviously have

$$\varepsilon' = \varepsilon(p) + \mathbf{p}\cdot\mathbf{v}_s, \tag{1.25}$$

where $\mathbf{p}$ is the momentum in the reference system fixed with respect to the superfluid liquid. This has to be taken into account in writing the *transport equation* for the excitations, which therefore takes the form

$$\frac{\partial n}{\partial t} + \nabla_{\mathbf{r}} n\cdot\nabla_{\mathbf{p}}\varepsilon' - \nabla_{\mathbf{p}} n\cdot\nabla_{\mathbf{r}}\varepsilon' = I(n), \tag{1.26}$$

where $I(n)$ is the *collision integral.*

The fact that a Bose liquid contains two types of motions with different velocities leads to a very distinctive kind of hydrodynamics, whose equations can be derived from the transport equation (1.26). This derivation, which we do not give here, has been carried out by Khalatnikov and is described in his review articles (K3, K4).[11] The "two-velocity hydrodynamics" of a Bose liquid differs from ordinary hydrodynamics in many ways. In particular, it turns out that two different kinds of oscillations can occur in a Bose liquid, with two different velocities of propagation. The oscillations of the first kind represent ordinary sound, or what is called "first sound," with velocity of propagation equal to u. In a sound wave of this kind, the liquid moves as a whole, i.e., the normal and superfluid parts do not separate. The oscillations of the second kind, the so-called "second sound," propagate with velocity

$$u_2 = \sqrt{\frac{\rho_s T S^2}{\rho_n \rho C}}, \tag{1.27}$$

where C and S are the heat capacity and entropy, respectively, per unit volume. In a wave of this kind, the oscillations of the normal and superfluid parts of the liquid have opposite phases, and hence the total flow vector of the liquid is

$$\mathbf{j} = \rho_n \mathbf{v}_n + \rho_s \mathbf{v}_s \approx 0.$$

We shall not spend any more time on problems relating to the hydrodynamics of a superfluid liquid. The propagation of sound in liquid He^4, as well as interaction processes between the excitations (leading to various dissipation phenomena such as viscosity, heat conduction, etc.), are analyzed in many special papers, and are discussed in detail in the surveys by Lifshitz (L12) and Khalatnikov (K3, K4), to which we refer the reader.

[11] Hydrodynamic equations for superfluid He^4, valid for velocities that are not too large, were first obtained by Landau (L1).

Let us now consider what can be said about the behavior of a Bose liquid at higher temperatures, when the number of excitations in the liquid becomes large. In this case, interaction between excitations can no longer be neglected, and our picture of the excitations of a gas of free particles ceases to correspond to reality. As a result, the formulas (1.17) for the thermodynamic quantities F_{phon} and F_{rot}, calculated for the gas model, lose their meaning. This applies equally well to the formula (1.22) for the normal density. However, the picture of two kinds of motion in the Bose liquid, occurring with appropriate effective densities, is not directly connected with the model of the excited state considered above, and it can be assumed that this picture is preserved for relatively high temperatures. The same applies to the hydrodynamical equations, since they are actually consequences only of conservation laws, from which they can be derived (see K3, K4). As the temperature increases, the normal density ρ_n increases until it reaches a value equal to ρ. At this point, called the *λ-point*, a phase transition occurs in helium. Below the transition point, superfluid motion is possible. However, above the transition point, superfluid motion is no longer possible, and the hydrodynamics of the Bose liquid do not differ from ordinary hydrodynamics.

In principle, the transition from $\rho_n \neq \rho$ to $\rho_n = \rho$ might take place either continuously or discontinuously. It follows from experiment that the phase transition in liquid helium is a transition of the second kind, and is not accompanied by release or absorption of any latent heat (see K1, p. 224). This implies that the normal density ρ_n grows continuously as the temperature increases and becomes equal to ρ at the λ-point.

Considerably above the λ-point, helium has no peculiarities of behavior as compared with an ordinary liquid. As for the neighborhood of the λ-point, there is good reason to expect a number of essentially new properties. The problem of the behavior of various characteristics of systems, especially their thermodynamic properties, in the neighborhood of a point where a phase transition of the second kind occurs, remains unsolved at present, and represents one of the most interesting problems of the physics of matter in the condensed state.

2. The Fermi Liquid

2.1. Excitations in a Fermi liquid. We now consider a system of interacting particles obeying Fermi statistics, restricting ourselves to the case where the spin of the particles equals $\frac{1}{2}$, since for all practical purposes, we are concerned only with liquid He^3, electrons in metals or nuclear matter. A system of interacting Fermi particles with spin $\frac{1}{2}$ will be called a *Fermi liquid.*

A theory of the weakly excited states of a Fermi liquid has been constructed by Landau (L3, L4). The basis for this theory is the assumption that the excitation spectrum of the Fermi liquid is formed by the same principle as the spectrum of an ideal Fermi gas. Therefore, before discussing the Fermi liquid, it makes sense to establish the connection between the familiar picture of an excited state of a Fermi gas and the notion of elementary excitations.

As is well known, in the ground state of an ideal Fermi gas at $T = 0$, the particles occupy all quantum states with momenta less than some limiting value p_0, and all states with larger momenta are unoccupied. In momentum space, the occupied states form a sphere of radius p_0, called the *Fermi sphere.* The quantity p_0, called the *Fermi momentum*, is determined from the condition that the number of states with $p_0 < p$ should equal the number of particles, i.e.,

$$p_0 = \left(\frac{3\pi^2 N}{V}\right)^{1/3}, \tag{2.1}$$

where N/V is the density of the particles.

In an excited state, the momentum distribution of the particles is different. It is not hard to see that every such state can be constructed from the ground state by moving a number of particles one after another from the interior to the exterior of the Fermi sphere. Each such elementary act leads to a state differing from the original state by the presence of a particle with $p > p_0$ and of a "hole" with $p < p_0$. Obviously, it is just these particles with $p > p_0$ and these holes with $p < p_0$ which play the role of the elementary excitations in an ideal Fermi gas. They have spin $\frac{1}{2}$, can appear or disappear only in pairs, and have momenta in the neighborhood of p_0 for weakly excited states. It is convenient to measure the energy of these elementary excitations from the surface of the Fermi sphere (i.e., from $p_0^2/2m$). Thus, the energy of excitations of the particle type is

$$\xi = \frac{p^2}{2m} - \frac{p_0^2}{2m} \approx v(p - p_0),$$

where $v = p_0/m$, and the energy of excitations of the hole type is

$$-\xi = \frac{p_0^2}{2m} - \frac{p^2}{2m} \approx v(p_0 - p).$$

In Landau's theory, it is assumed that a weakly excited state of a Fermi liquid greatly resembles a weakly excited state of a Fermi gas, and can be described by using a set of elementary excitations with spin $\frac{1}{2}$ and momenta in the neighborhood of p_0. Basic to Landau's theory is the assumption that the quantity p_0 is related to the particle density of the liquid by the same formula (2.1) as in the case of an ideal Fermi gas (this will be proved in Chap. 4). Just as in the gas, the excitations in the liquid are of two types, "particles" with momenta greater than p_0, and "holes" with momenta less than p_0, which can appear and disappear only in pairs. It follows that the number of "particles" must equal the number of "holes."

Despite the great resemblance between excitations in a Fermi liquid and excitations in an ideal Fermi gas, there are also important differences between them, due to the fact that excitations in the liquid interact with each other. The clearest manifestation of this interaction is the existence of superfluid Fermi liquids (or superconducting Fermi liquids, if we are talking about electrons in a metal). It is not hard to see that the spectrum of a Fermi gas (described above) does not lead to superfluidity. In fact, an arbitrarily small energy is sufficient to excite a Fermi gas, i.e., to form a "particle" with $p > p_0$ and a "hole" with $p < p_0$, while at the same time, the total momentum of the pair can have the value $2p_0$. Therefore

$$\left(\frac{\varepsilon}{p}\right)_{\min} = 0,$$

and according to (1.18), this implies that the critical velocity is zero, so that superfluidity is absent. The appearance of superfluidity is connected with the fact that a certain type of interaction of quasi-particles leads to a radically different structure of the spectrum. In particular, it turns out that excitation of such a Fermi liquid requires an expenditure of energy which cannot be less than a certain amount. In such cases, it is customary to say that there is a *gap* in the excitation spectrum.

For the time being, we shall not discuss superfluid Fermi liquids (Chap. 7 is devoted to such liquids). Instead, we now consider the properties of the excitations of normal Fermi liquids.

As a result of interaction between excitations, the very idea of elementary excitations makes sense only near the Fermi momentum p_0. As already mentioned on p. 5, one can only talk about an elementary excitation in the case where its attenuation is small compared to its energy. The amount of attenuation is determined either by processes in which one excitation decays into several others, or by collisions of excitations with each other. If the energy of an excitation is large compared to the temperature of the liquid, then decay processes play the chief role, and the amount of attenuation is proportional to the probability of these processes. It is not hard to see[12]

[12] To estimate this probability, it is convenient to use the analogy with a Fermi gas. Consider the following process: A particle with momentum $\mathbf{p}_1$ ($p_1 > p_0$) interacts with one of the particles inside the Fermi sphere with momentum $\mathbf{p}_2$ ($p_2 < p_0$). As a result, two particles appear with momenta $\mathbf{p}_3$ and $\mathbf{p}_4 = \mathbf{p}_1 + \mathbf{p}_2 - \mathbf{p}_3$, where $p_3 > p_0, p_4 > p_0$. Thus, the particle with momentum $\mathbf{p}_1$ "has decayed" into particles with momenta $\mathbf{p}_3$, $\mathbf{p}_4$ and a hole with momentum $\mathbf{p}_2$. The total probability of such a process is proportional to

$$\int \delta(\varepsilon_1 + \varepsilon_2 - \varepsilon_3 - \varepsilon_4)\, d\mathbf{p}_2\, d\mathbf{p}_3,$$

$$p_2 < p_0, \quad p_3 > p_0, \quad p_4 = |\mathbf{p}_1 + \mathbf{p}_2 - \mathbf{p}_3| > p_0.$$

It is easy to see that for $p_1 - p_0 \ll p_0$, the permissible regions of variation for the magnitudes of the vectors $\mathbf{p}_2$ and $\mathbf{p}_3$ are such that

$$p_0 < p_3 < p_1 + p_2 - p_0, \qquad 2p_0 - p_1 < p_2 < p_0.$$

The angle between $\mathbf{p}_1$ and $\mathbf{p}_2$ can be arbitrary. Then the angle between $\mathbf{p}_3$ and $\mathbf{p}_1 + \mathbf{p}_2$ is determined by the condition that energy be conserved, and integration over this angle cancels the delta function. The remaining integral over $d\mathbf{p}_2\, d\mathbf{p}_3$ is carried out near $p_2 \approx p_3 \approx p_0$, and gives the factor $(p_1 - p_0)^2$.

by taking account of the laws of energy and momentum conservation, and also the condition that the number of "particles" should equal the number of "holes", that the probability of decay is proportional to $(p - p_0)^2$. On the other hand, the energy of the excitation is proportional to $p - p_0$. Thus, it is clear that the attenuation is relatively small only for excitations with momenta in the neighborhood of p_0.

In the case of an equilibrium Fermi liquid at finite temperatures, the average energy of the "particles" and "holes" is of order T. Since the excitations obey Fermi statistics, the number of excitations is also proportional to T. It is easily shown that for such excitations, the probabilities of decay and scattering are of the same order,[13] and are in fact proportional to T^2. It follows that the description of a Fermi liquid in terms of elementary excitations applies only at sufficiently low temperatures.

The properties of the energy spectrum of a Fermi liquid can be made more intuitive by using a model constructed by analogy with a Fermi gas: Suppose the ground state of the system corresponds to a set of quasi-particles filling a Fermi sphere with limiting momentum p_0. Then formula (2.1) can be interpreted as the statement that the number of quasi-particles equals the number of particles of the liquid. In this model, the excitations are in complete correspondence with the concept of "particles" and "holes," and in particular, the fact that the number of "particles" equals the number of "holes" expresses the conservation of the number of quasi-particles. If we denote the distribution function of the quasi-particles by $n(\mathbf{p})$, then changes in $n(\mathbf{p})$ are subject to the condition

$$\int \delta n \, d\mathbf{p} = 0. \tag{2.2}$$

This gas model is convenient for further study of the properties of the Fermi liquid. However, it must be borne in mind that the very concept of quasi-particles has meaning only near the surface of the Fermi sphere, and hence all properties of the gas model for which quasi-particles far from the surface play an important role have no counterparts in a real Fermi liquid.

2.2. The energy of the quasi-particles. In addition to the assumptions just made about the character of the elementary excitations, Landau's theory is based on one more assumption concerning the interaction of the quasi-particles, i.e., that the interaction can be described by using a self-consistent field acting on one quasi-particle due to the other quasi-particles surrounding it. Then the energy of the system is no longer equal to the sum of the

[13] For an almost-ideal Fermi gas, these processes represent essentially the same phenomenon, and the corresponding probability is proportional to

$$\int \delta(\varepsilon_1 + \varepsilon_2 - \varepsilon_3 - \varepsilon_4) n(\varepsilon_2)[1 - n(\varepsilon_3)][1 - n(\varepsilon_4)] \, d\mathbf{p}_2 \, d\mathbf{p}_3.$$

Formally, one can talk in terms of scattering for $|\mathbf{p}_2| > p_0$ and decay for $|\mathbf{p}_2| < p_0$. For $\varepsilon_1 - \mu \sim T$, the integral is in both cases proportional to T^2.

energies of the separate quasi-particles, but rather is a functional of their distribution functions. It is natural to define the energy of an individual quasi-particle as the variational derivative of the total energy E with respect to the distribution function:

$$\delta E = 2V \int \varepsilon \delta n \frac{d\mathbf{p}}{(2\pi)^3}. \tag{2.3}$$

(The factor 2 comes about as a result of summing over the projections of the spin.) In fact, it is clear from this formula that ε is just the change in the energy of the system due to one extra quasi-particle with momentum $\mathbf{p}$.[14]

In formulas (2.2) and (2.3), it is assumed that the distribution of quasi-particles is spatially homogeneous. In practice, this restriction reduces to the fact that spatial inhomogeneity can only take place at distances greatly exceeding the wavelengths of the quasi-particles. Since we are considering only excitations near the Fermi surface, i.e., with momenta near p_0, it follows from formula (2.1) that the corresponding wavelength is of the order of interatomic distances. Thus, the requirement of spatial homogeneity essentially introduces no restrictions at all.

In the presence of a magnetic field, and also in the case of a ferromagnetic system, the distribution function must be regarded as an operator acting on the spin indices, i.e., as a density matrix $n_{\alpha\beta}$. Moreover, the energy of the quasi-particles is also an operator $\varepsilon_{\alpha\beta}$. (If no magnetic field is present and if the system is not ferromagnetic, the operators $n_{\alpha\beta}$ and $\varepsilon_{\alpha\beta}$ are proportional to the unit matrix.) Therefore, in the general case, formula (2.3) should be written as

$$\delta\left(\frac{E}{V}\right) = \sum_{\alpha,\beta} \int \varepsilon_{\alpha\beta} \delta n_{\beta\alpha} \frac{d\mathbf{p}}{(2\pi)^3}.$$

It is convenient to write this expression in the abbreviated form

$$\delta\left(\frac{E}{V}\right) = \mathrm{Sp}_\sigma \int \varepsilon \delta n \frac{d\mathbf{p}}{(2\pi)^3}, \tag{2.4}$$

where ε and n are understood to be the appropriate matrices, and the symbol Sp_σ (from *Spur = trace*) calls for the operation of summing the diagonal elements of the matrix product $\varepsilon \delta n$.

The definition (2.4) of the energy of the quasi-particles implies that their equilibrium distribution function is actually a Fermi distribution. To prove this, it is most convenient to use the familiar expression

$$\frac{S}{V} = -\mathrm{Sp}_\sigma \int [n \ln n + (1 - n) \ln (1 - n)] \frac{d\mathbf{p}}{(2\pi)^3} \tag{2.5}$$

---

[14] Recall that $n(\mathbf{p})$ is the momentum distribution of the quasi-particles, i.e.,

$$2 \int n(\mathbf{p}) \frac{d\mathbf{p}}{(2\pi)^3}$$

is the number of quasi-particles per unit volume.

for the entropy.[15] This formula stems from purely combinatorial considerations, and its applicability to the Fermi liquid is due to the fact that by hypothesis the energy levels of the quasi-particles are classified in the same way as the energy levels of the particles in an ideal gas.

From the condition that the entropy should be a maximum when the number of particles and the energy are conserved, i.e., when

$$\delta N = 0, \qquad \delta E = 0,$$

we can find the distribution function

$$n(\varepsilon) = n_F(\varepsilon) = \frac{1}{e^{(\varepsilon - \mu)/T} + 1}, \tag{2.6}$$

by subjecting n to a variation δn. Here, the energy ε is a functional of n, so that formula (2.6) is actually a very complicated implicit definition of $n(\varepsilon)$.

Being a functional of n, ε also depends on the temperature. This dependence can be represented in the following form. If we use $\varepsilon^{(0)}(\mathbf{p})$ to denote the equilibrium energy of the quasi-particles for $T = 0$, then ε is given by the formula

$$\begin{aligned}\varepsilon(\mathbf{p}, \boldsymbol{\sigma}) &= \varepsilon^{(0)}(\mathbf{p}, \boldsymbol{\sigma}) + \delta\varepsilon(\mathbf{p}, \boldsymbol{\sigma}) \\ &= \varepsilon^{(0)}(\mathbf{p}, \boldsymbol{\sigma}) + \mathrm{Sp}_{\sigma'} \int f(\mathbf{p}, \boldsymbol{\sigma}; \mathbf{p}', \boldsymbol{\sigma}')\delta n(\mathbf{p}', \boldsymbol{\sigma}') \frac{d\mathbf{p}'}{(2\pi)^3},\end{aligned} \tag{2.7}$$

for small deviations from equilibrium or for small temperatures. Here $\delta n = n - n_F\ (T = 0)$, while f is an operator depending on the momenta and spin operators of two quasi-particles. In formula (2.7) we use a notation which indicates the matrix character of the quantities involved. As already noted, ε and n are matrices in the spin variables. To emphasize this fact, we have written them in the form $\varepsilon(\mathbf{p}, \boldsymbol{\sigma})$ and $n(\mathbf{p}, \boldsymbol{\sigma})$, where σ_x, σ_y, σ_z are the familiar Pauli matrices, connected with the spin operator $\mathbf{s}$ of the quasi-particles by the relation

$$\mathbf{s} = \frac{1}{2}\boldsymbol{\sigma}.$$

The operator f is a matrix both with respect to the spin variables appearing in the left-hand side of (2.7), and with respect to the spin variables of the operator δn appearing in the integrand in the right-hand side of (2.7). A more detailed version of formula (2.7) is

$$\varepsilon_{\alpha\beta}(\mathbf{p}) = \varepsilon^{(0)}_{\alpha\beta}(\mathbf{p}) + \int f_{\alpha\beta,\,\gamma\delta}(\mathbf{p}, \mathbf{p}')\delta n_{\delta\gamma}(\mathbf{p}') \frac{d\mathbf{p}}{(2\pi)^3},$$

which explains the meaning of the expression $f(\mathbf{p}, \boldsymbol{\sigma}; \mathbf{p}', \boldsymbol{\sigma}')$. The function f defined in this way is the second variational derivative with respect to δn of the energy per unit volume [cf. (2.7) and (2.4)], and hence f is symmetric under the permutation $\mathbf{p}, \boldsymbol{\sigma} \rightleftarrows \mathbf{p}', \boldsymbol{\sigma}'$. The function f is a very important

[15] As usual, by $\mathrm{Sp}_\sigma f(n)$ we mean the sum of the eigenvalues of the matrix $f(n)$.

characteristic of the Fermi liquid. As we shall see below (Chap. 4), f is connected with the forward scattering amplitude of two quasi-particles. The dependence of f on the spin variables can be written in the general form

$$f(\mathbf{p}, \boldsymbol{\sigma}; \mathbf{p}', \boldsymbol{\sigma}') = \varphi(\mathbf{p}, \mathbf{p}') + \sigma_i \sigma'_k \zeta_{ik}(\mathbf{p}, \mathbf{p}'). \tag{2.8}$$

If the spin interactions are due to exchange,[16] the second term in the right-hand side of (2.8) has the form $(\boldsymbol{\sigma} \cdot \boldsymbol{\sigma}')\zeta(\mathbf{p}, \mathbf{p}')$.

In the absence of a magnetic field, the energy ε of the quasi-particles does not depend on the spin. Thus, the function $\varepsilon^{(0)}$ in (2.7) depends only on p and can be expanded in a series

$$\xi(p) = \varepsilon^{(0)}(p) - \mu(0) = v(p - p_0) \tag{2.9}$$

in $p - p_0$, where $\mu(0)$ is the chemical potential at $T = 0$ and v is a constant. The quantity v, which represents the velocity of the excitations at the Fermi surface, can be written in the form

$$v = \frac{p_0}{m^*}, \tag{2.10}$$

where m^* is the *effective mass*. As shown by Landau (L3), there is a definite connection between m^* and f. The argument goes as follows:

First, we write the formula expressing the simple fact that the momentum of a unit volume of the liquid is the same as a flow of mass. Obviously, the momentum of a unit volume of the Fermi liquid is the same as the momentum of the quasi-particles, i.e., equals

$$2 \int \mathbf{p} n \frac{d\mathbf{p}}{(2\pi)^3}.$$

On the other hand, because of the assumption that the number of particles in the Fermi liquid equals the number of quasi-particles, the current of particles of the fluid coincides with the current of quasi-particles, and equals

$$2 \int \mathbf{v} n \frac{d\mathbf{p}}{(2\pi)^3},$$

where $\mathbf{v}$ is the velocity of the quasi-particles. The density of the mass current is obtained from this expression by simply multiplying it by the mass m of an atom of the liquid. Noting that $\mathbf{v} = \nabla_{\mathbf{p}} \varepsilon$, by definition, we can write the

---

[16] Ordinarily, one distinguishes several kinds of spin-dependent interactions between particles: (1) exchange interaction, involving the possibility of permutation of identical particles; (2) spin-orbit interaction, caused by the relativistic interaction of electric fields with a moving magnetic moment; (3) direct magnetic interaction of moments. Usually, the exchange interaction greatly exceeds all other forms of interaction. A characteristic feature of exchange interaction is its invariance with respect to spatial rotations of the total angular momentum of the system of particles. The scalar product $\boldsymbol{\sigma} \cdot \boldsymbol{\sigma}'$ has this property.

condition for equality of the momentum and the corresponding flow of mass in the form

$$\int \mathbf{p} n \frac{d\mathbf{p}}{(2\pi)^3} = m \int (\nabla_{\mathbf{p}} \varepsilon)\, n \frac{d\mathbf{p}}{(2\pi)^3}. \tag{2.11}$$

We now vary (2.11) with respect to n, bearing in mind that the accompanying change in the energy ε is connected with the quantity δn by formula (2.7), which can be written in the form

$$\delta\varepsilon = \frac{1}{2}\, \mathrm{Sp}_\sigma \mathrm{Sp}_{\sigma'} \int f(\mathbf{p}, \boldsymbol{\sigma}; \mathbf{p}', \boldsymbol{\sigma}')\delta n' \frac{d\mathbf{p}'}{(2\pi)^3}$$

when there is no magnetic field, i.e., when n and ε do not depend on the spin. It follows that

$$\int \frac{\mathbf{p}}{m} \delta n \frac{d\mathbf{p}}{(2\pi)^3} = \int (\nabla_{\mathbf{p}}\varepsilon)\, \delta n \frac{d\mathbf{p}}{(2\pi)^3} + \frac{1}{2}\, \mathrm{Sp}_\sigma \mathrm{Sp}_{\sigma'} \int n \delta n' \nabla_{\mathbf{p}} f(\mathbf{p}, \boldsymbol{\sigma}; \mathbf{p}', \boldsymbol{\sigma}') \frac{d\mathbf{p}\, d\mathbf{p}'}{(2\pi)^6}.$$

In the second integral, we integrate by parts with respect to $\mathbf{p}$ and permute the variables $\mathbf{p}$, $\boldsymbol{\sigma}$ with $\mathbf{p}'$, $\boldsymbol{\sigma}'$, obtaining

$$\int \frac{\mathbf{p}}{m} \delta n \frac{d\mathbf{p}}{(2\pi)^3} = \int (\nabla_{\mathbf{p}}\varepsilon)\, \delta n \frac{d\mathbf{p}}{(2\pi)^3} - \frac{1}{2}\, \mathrm{Sp}_\sigma \mathrm{Sp}_{\sigma'} \int \delta n\, f(\mathbf{p}, \boldsymbol{\sigma}; \mathbf{p}', \boldsymbol{\sigma}') \nabla_{\mathbf{p}'} n' \frac{d\mathbf{p}\, d\mathbf{p}'}{(2\pi^6)}.$$

Since δn is arbitrary, this implies at once that

$$\frac{\mathbf{p}}{m} = \nabla_{\mathbf{p}}\varepsilon - \frac{1}{2}\, \mathrm{Sp}_\sigma \mathrm{Sp}_{\sigma'} \int f(\mathbf{p}, \boldsymbol{\sigma}; \mathbf{p}', \boldsymbol{\sigma}') \nabla_{\mathbf{p}'} n' \frac{d\mathbf{p}'}{(2\pi)^3}.$$

At $T = 0$, the energy ε near the Fermi surface has the form (2.9), while

$$\nabla_{\mathbf{p}'} n' \approx -\frac{\mathbf{p}'}{p'} \delta(p' - p_0).$$

Observing that f depends only on the angle χ between $\mathbf{p}$ and $\mathbf{p}'$, because of the isotropy of the liquid, we find that

$$\frac{1}{m^*} = \frac{1}{m} - \frac{p_0}{2(2\pi)^3}\, \mathrm{Sp}_\sigma \mathrm{Sp}_{\sigma'} \int f(\chi) \cos\chi\, d\Omega, \tag{2.12}$$

where $f(\chi)$ is the value of f for $|\mathbf{p}| = |\mathbf{p}'| = p_0$. The integral in (2.12) is over all directions of the vector $\mathbf{p}'$ ($d\Omega$ is an element of solid angle). This formula relates the mass of an atom of the liquid to the effective mass of the quasi-particles, and it remains valid, except for small corrections, at sufficiently low temperatures.

The heat capacity of a Fermi liquid can be expressed in terms of m^* by using the ordinary formula for a Fermi gas. In fact, according to (2.3), the heat capacity per unit volume is

$$C_V = \left(\frac{\partial(E/V)}{\partial T}\right)_{N,V} = 2 \int \varepsilon \left(\frac{\partial n}{\partial T}\right)_N \frac{d\mathbf{p}}{(2\pi)^3}. \tag{2.13}$$

It is not hard to show that replacing ε by $\varepsilon^{(0)}$ in the integrand of (2.13) gives a relative error of order $[T/\mu(0)]^3$. Thus, to the approximation linear in T, we find that the heat capacity is given by the same formula

$$C_V = \frac{1}{3} m^* p_0 T \tag{2.14}$$

as for an ordinary gas. The same is also true of the entropy at low temperatures.[17]

2.3. Sound. The propagation of sound in a Fermi liquid has certain features which differ from those of a Bose liquid: At temperatures that are not too low, sound propagates according to the laws of ordinary hydrodynamics, with an attenuation proportional to the time τ between collisions of excitations. As the temperature is lowered, the probability of collisions decreases like the square of the "spread" of the Fermi distribution, and hence the collision time increases like T^{-2}. At temperatures for which τ is of the order ω^{-1}, sound ceases to propagate at all. However, it turns out that as the temperature is lowered further, sound is again able to propagate, but in general with a different velocity. Moreover, sound can no longer be thought of as a simple wave of compression and rarefaction. This phenomenon was predicted by Landau (L4), who called it "zero sound." Since it is only the relation between ω and τ which determines the nature of sound, these two kinds of sound can be characterized as *low-frequency sound* ($\omega\tau \ll 1$) and *high-frequency sound* ($\omega\tau \gg 1$). When the temperatures are not too low, i.e., when the condition $\omega\tau \ll 1$ is satisfied, the velocity of sound u is determined by the compressibility, in the usual manner. As we now show, u depends on the function f in an essential way (L3).

It is convenient to express the compressibility in terms of $\partial\mu/\partial N$, the derivative of the chemical potential with respect to the number of particles. Using the fact that the chemical potential depends only on N/V, we find that

$$\frac{\partial\mu}{\partial N} = -\frac{V^2}{N^2}\frac{\partial P}{\partial V} = \frac{1}{N}\frac{\partial P}{\partial(N/V)}, \tag{2.15}$$

where P is the pressure. This immediately implies the relation

$$u^2 = \frac{\partial P}{\partial \rho} = \frac{\partial P}{\partial(mN/V)} = \frac{1}{m} N \frac{\partial\mu}{\partial N} \tag{2.16}$$

between $\partial\mu/\partial N$ and u^2.

The derivative $\partial\mu/\partial N$ can be calculated as follows: Since $\mu \approx \varepsilon(p_0)$, changes in μ are due both to changes in p_0 and changes in the shape of the function $\varepsilon(p)$:

$$\delta\mu = \frac{1}{2} \operatorname{Sp}_\sigma \operatorname{Sp}_{\sigma'} \int f \, \delta n' \frac{d\mathbf{p}'}{(2\pi)^3} + \frac{\partial\varepsilon^{(0)}(p_0)}{\partial p_0} \delta p_0. \tag{2.17}$$

---

[17] Formula (2.13) can be used to determine m^* from experimental data on the heat capacity. According to (2.1), the momentum p_0 is determined by the density of the liquid. Thus, for liquid He^3, we find that

$$p_0 = 0.76 \times 10^8 \text{ cm}^{-1}, \qquad m^* = 2m_{He^3}$$

(see B9, K2).

(It is assumed that no magnetic field is present.) According to (2.1), the changes δN and δp_0 are connected by the relation

$$\delta N = \frac{1}{\pi^2} p_0^2 \delta p_0 V.$$

Since only changes $\delta n'$ near the Fermi surface are important in the integrand in formula (2.17), we can carry out the integration over the magnitude of the momentum, obtaining

$$\int f\, \delta n' \frac{d\mathbf{p}'}{(2\pi)^3} = \frac{\delta N}{8\pi V} \int f\, d\Omega.$$

It follows that

$$\frac{\partial \mu}{\partial N} = \frac{1}{16\pi V} \mathrm{Sp}_\sigma \mathrm{Sp}_{\sigma'} \int f\, d\Omega + \frac{\pi^2}{p_0 m^* V}. \tag{2.18}$$

Using formula (2.12) for the effective mass and formula (2.1), we find that

$$u^2 = \frac{p_0^2}{3m^2} + \frac{1}{6m}\left(\frac{p_0}{2\pi}\right)^3 \mathrm{Sp}_\sigma \mathrm{Sp}_{\sigma'} \int f(\chi)\,(1 - \cos\chi)\, d\Omega. \tag{2.19}$$

This formula gives the velocity of sound in the frequency range $\omega\tau \ll 1$. In the absence of interaction, we have just

$$u^2 = \frac{p_0^2}{3m^2}.$$

To study the propagation of sound in the frequency range $\omega\tau \gg 1$, we apply the usual transport equation

$$\frac{\partial n}{\partial t} + \nabla_{\mathbf{r}} n \cdot \nabla_{\mathbf{p}} \varepsilon - \nabla_{\mathbf{p}} n \cdot \nabla_{\mathbf{r}} \varepsilon = I(n), \tag{2.20}$$

where $I(n)$ is the collision integral. For small deviations from equilibrium, the distribution function can be written in the form

$$n = n_F + \delta n,$$

where n_F is the equilibrium distribution, and δn is a small increment, which is a periodic function of time:

$$\delta n \sim e^{i(\mathbf{k}\cdot\mathbf{r} - \omega t)}.$$

Since the collision integral $I(n)$ is of order $\delta n/\tau$, it can be neglected compared to the term $\partial n/\partial t$. In linearizing equation (2.20), it must be kept in mind that ε is a functional of n, and hence $\nabla_{\mathbf{r}}\varepsilon$ does not equal zero. According to (2.7), we have

$$\nabla_{\mathbf{r}} \varepsilon = \mathrm{Sp}_{\sigma'} \int f \nabla_{\mathbf{r}} \delta n' \frac{d\mathbf{p}'}{(2\pi)^3}.$$

In view of the remark just made, we obtain

$$(\mathbf{v}\cdot\mathbf{k} - \omega)\delta n - (\mathbf{v}\cdot\mathbf{k}) \frac{\partial n_F}{\partial \varepsilon} \mathrm{Sp}_{\sigma'} \int f\, \delta n' \frac{d\mathbf{p}'}{(2\pi)^3} = 0. \tag{2.21}$$

It follows from the form of this equation that δn is proportional to

$$\frac{\partial n_F}{\partial \varepsilon} \approx -\,\delta(\varepsilon - \mu).$$

Setting

$$\delta n = \frac{\partial n_F}{\partial \varepsilon}\,\nu,$$

we find that

$$(\mathbf{v}\cdot\mathbf{k} - \omega)\nu + (\mathbf{v}\cdot\mathbf{k})\,\frac{1}{2}\,\mathrm{Sp}_{\sigma'}\int F\nu'\,\frac{d\Omega'}{4\pi} = 0, \tag{2.22}$$

where

$$F(\chi) = f(\chi)\,\frac{p_0 m^*}{\pi^2}. \tag{2.23}$$

If we choose $\mathbf{k}$ as the polar axis, and if we denote the velocity of wave propagation by $\tilde{u} = \omega/k$, then equation (2.22) becomes

$$(s - \cos\theta)\,\nu(\theta, \varphi, \boldsymbol{\sigma}) = \cos\theta\,\frac{1}{2}\,\mathrm{Sp}_{\sigma'}\int F(\chi)\,\nu(\theta', \varphi', \boldsymbol{\sigma}')\,\frac{d\Omega}{4\pi}, \tag{2.24}$$

where $s = \tilde{u}/v$.

Formula (2.24) makes clear the basic distinction between ordinary sound and sound propagating in a Fermi liquid for $\omega\tau \gg 1$. In the first case, the distribution function remains isotropic in the reference system in which the liquid as a whole is at rest. This means that the radius of the Fermi sphere changes, and, moreover, its center oscillates about the point $\mathbf{p} = 0$. In the second case, the distribution function changes in a more complicated way, and the Fermi surface does not remain spherical. In fact, the change in the Fermi surface is determined by the function ν.

First, we examine the spin-independent solution of equation (2.24). In this case, of the whole function $F(\chi)$ all that remains is the part $\Phi(\chi)$ connected with the function φ in (2.8). We begin by considering the simplest case of all, i.e.,

$$\Phi = \Phi_0 = \text{const.}$$

Then from (2.24) we obtain

$$\nu = \text{const}\cdot\frac{\cos\theta}{s - \cos\theta}\,e^{i(\mathbf{k}\cdot\mathbf{r} - \omega t)}. \tag{2.25}$$

As we shall soon see, s must be larger than unity, which means that the Fermi surface is stretched in the direction of motion.

Substituting (2.25) into (2.24), with $F = \Phi_0$, we find an equation for s. After carrying out the integration, we obtain

$$\frac{s}{2}\ln\frac{s+1}{s-1} - 1 = \frac{1}{\Phi_0}. \tag{2.26}$$

It is clear from (2.26) that if s is real (corresponding to undamped waves), then s must be larger than unity, i.e.,

$$\tilde{u} > v. \tag{2.27}$$

Formula (2.24) shows that this condition remains valid for any function Φ. Moreover, since the left-hand side of equation (2.26) is always positive, it is clear that the condition for the existence of "zero sound" is that Φ_0 be positive.

If the function Φ_0 is large, s is also large. From equation (2.26), we find that

$$s \to \sqrt{\frac{\Phi_0}{3}}$$

as $\Phi_0 \to \infty$. On the other hand, as $\Phi_0 \to 0$, we have $s \to 1$, i.e., $\tilde{u} \to v$. This is the case of an almost-ideal Fermi gas. It is not hard to see that the conclusion that $s \to 1$ as $\Phi \to 0$ does not depend on the form of Φ. In fact, it follows from (2.24) that as $\Phi \to 0$, we have $s \to 1$ while ν differs from zero only for small values of θ. According to (2.19), in a "weakly non-ideal" Fermi gas,

$$u^2 = \frac{p_0^2}{3m^2}, \quad \text{i.e.,} \quad u \approx \frac{v}{\sqrt{3}} \approx \frac{\tilde{u}}{\sqrt{3}}.$$

Thus, the velocity of zero sound can exceed the velocity of ordinary sound by a factor of $\sqrt{3}$. It should be noted that in the limit of an almost-ideal Fermi gas, τ increases greatly, and as a result, the frequency range corresponding to zero sound becomes wider; on the other hand, ordinary sound exists only in the region of very low frequencies.

In the general case of an arbitrary function $\Phi(\chi)$, equation (2.24) can no longer be solved so simply. If we expand $\nu(\theta, \varphi)$ and $\Phi(\chi)$ in series of spherical harmonics, we can separate the equations for the amplitudes corresponding to spherical harmonics with different azimuthal numbers m (i.e., with different factors $e^{im\varphi}$). Furthermore, the number m cannot exceed the largest of the numbers l in the expansion

$$\Phi(\chi) = \sum_l \Phi_l P_l(\cos\chi)$$

of the function $\Phi(\chi)$ in Legendre polynomials. Thus, we arrive at the conclusion that in the general case, several "zero sounds" can occur, with anisotropic distribution functions in the plane perpendicular to the direction of propagation $\mathbf{k}$. Just as in the simplest case, whether or not such oscillations appear depends on the form of the function Φ. For example, if

$$\Phi = \Phi_0 + \Phi_1 \cos\chi,$$

the condition for appearance of oscillations with $\nu \sim e^{i\varphi}$ is that $\Phi_1 > 6$. We observe that compression and rarefaction of the liquid does not occur in waves of this kind.

In the case where the function f depends on the spins of the particles, special waves, which might be called *spin waves*, can propagate in the liquid. In fact, suppose, for example, that the function $F(\chi)$ has the form

$$F(\chi) = \Phi(\chi) + Z(\chi)(\boldsymbol{\sigma}\cdot\boldsymbol{\sigma}'), \tag{2.28}$$

corresponding to exchange interaction of the spins. Then, in addition to spin-independent solutions, equation (2.24) is satisfied by a function of the form

$$\nu = \boldsymbol{\nu}\cdot\boldsymbol{\sigma}, \tag{2.29}$$

where $\boldsymbol{\nu}$ is an unknown vector. For the function $\boldsymbol{\nu}$, we obtain the equation

$$(s - \cos\theta)\,\boldsymbol{\nu} = \cos\theta \int Z\boldsymbol{\nu}' \frac{d\Omega'}{4\pi}. \tag{2.30}$$

The equations for the components of the vector $\boldsymbol{\nu}$ differ from the equation for a spin-independent $\boldsymbol{\nu}$ only by the presence of Z instead of Φ. Therefore, the entire subsequent argument carries through for spin waves. It can be shown (see L3) that the zeroth-order term in the expansion of Z in spherical harmonics determines the expression for the magnetic susceptibility of the Fermi liquid. For liquid He^3, this term turns out to be negative, which, in all likelihood, means that it is impossible for spin waves to propagate in this liquid.

The case of electrons in a metal is somewhat special. It is obvious that in a metal, oscillations cannot occur which are accompanied just by variations of the electron density alone, with no changes in the crystal lattice. Such oscillations would lead to the appearance of uncompensated electric charge, and hence excitation of such oscillations would require very large energies. In all probability, this means that in the case of Coulomb forces, the function f contains an infinite constant, which is independent of the angle (see also Sec. 22). According to (2.26), this implies that $s = \infty$. However, this argument pertains only to density oscillations. Under certain conditions, higher "sounds" with $\nu \sim e^{im\varphi}$ (where $m \neq 0$) can propagate in an electron liquid, as well as spin waves which do not involve density changes.

The possibility of having sound waves propagate at $T = 0$ means that the excitation spectrum of the liquid contains Bose phonon branches, with the energy depending linearly on the momentum, i.e., $\varepsilon_i = u_i p$. However, the corrections to the thermodynamic quantities due to phonons contain higher powers of T (the heat capacity is $\sim T^3$), which are not taken into account in the approximation considered here.

Later on (see Chap. 4), we shall show how the basic results of the theory just presented can be derived from a microscopic analysis of a system of fermions with arbitrary short-range interaction forces.

Landau's theory, in the form given here, pertains primarily to the properties of liquid He^3 at low temperatures. The presence of Coulomb interactions between the particles leads to various special features, some of which will be exhibited in Sec. 22, with the help of a simple model. Superfluid (or superconducting) Fermi systems differ even more basically from an ordinary Fermi liquid, and Chap. 7 is devoted to a study of the properties of superconductors. Finally, we call attention to ferromagnetic Fermi systems, which also differ from the model considered here. The properties of Fermi liquids of this kind are studied in the paper by Abrikosov and Dzyaloshinski (A2).

3. Second Quantization

In a certain sense, the theory of Bose and Fermi liquids presented above is of a phenomenological character, and is based on definite assumptions about the spectrum of the temperature excitations. In what follows, we shall be concerned with a microscopic justification of this theory. The present section contains an exposition of some of the necessary mathematical tools, commonly known as the *method of second quantization.*[18]

Suppose we have a system of N noninteracting particles, which can occupy certain states with wave functions $\varphi_1(\xi), \varphi_2(\xi), \ldots$, forming a complete orthonormal set; here, ξ denotes any set of variables characterizing the state of a particle (usually, its coordinates and spin projections). Instead of giving the complete wave function of the system of particles, we can obviously describe the system by specifying the numbers of particles to be found in the states $\varphi_1, \varphi_2, \ldots$. This means that we go over to a new representation, called the *second-quantized representation*, where the occupation numbers $N_1, N_2, \ldots$ act as new variables. Consider first the case of particles obeying Bose statistics. As is well known, the complete wave function of a system of Bose particles is symmetric with respect to permutations of the variables corresponding to different particles. It is not hard to verify that the wave function corresponding to the occupation numbers $N_1, N_2, \ldots$ has the form

$$\Phi_{N_1N_2}\cdots = \left(\frac{N_1!N_2!\cdots}{N!}\right)^{1/2} \sum_{\text{P}} \varphi_{p_1}(\xi_1)\varphi_{p_2}(\xi_2)\cdots\varphi_{p_N}(\xi_N), \tag{3.1}$$

where the p_i index the different states, and the sum is taken over all possible permutations of different numbers p_i. The factor in front of the sum in (3.1) is introduced for normalization purposes, so that

$$\int |\Phi|^2 \prod_i d\xi_i = 1.$$

We shall regard $\Phi_{N_1N_2\cdots}$ as a function of the variables N_i.

Now let $F^{(1)}$ be some operator which is symmetric in all the particles and has the form

$$F^{(1)} = \sum_a f_a^{(1)}, \tag{3.2}$$

where f_a is an operator acting only on functions of ξ_a. It is not hard to see that if such an operator acts on the function $\Phi_{N_1N_2\cdots}$, it either carries it into the same function or into another function in which the state of one particle has been changed. As a result, the matrix elements of $F^{(1)}$ with respect to the functions (3.1) have the form[19]

---

[18] In view of the fact that the method of second quantization (see e.g., L7, Chap. 9) is basic in the formalism developed later, we think it advisable to give a brief sketch of the method here.

[19] The notation is essentially that used in Sec. 62 of the book L7.

$$\overline{F^{(1)}} = \sum_i f_{ii}^{(1)} N_i \qquad \text{for diagonal elements,}$$

$$(F^{(1)})_{N_i-1,N_k}^{N_i,N_k-1} = f_{ik}^{(1)} \sqrt{N_i N_k} \qquad \text{for nondiagonal elements,} \tag{3.3}$$

where

$$f_{ik}^{(1)} = \int \varphi_i^*(\xi) f^{(1)} \varphi_k(\xi)\, d\xi.$$

As far as the numbers N_i are concerned, the effect of the operator $F^{(1)}$ can be expressed in terms of new operators a_i ($i = 1, \ldots, n$), where a_i decreases by one the number of particles in the ith state, and its only nonzero matrix element is

$$(a_i)_{N_i}^{N_i-1} = \sqrt{N_i}. \tag{3.4}$$

The only nonzero matrix element of a_i^+, the Hermitian conjugate of the operator a_i, is obviously

$$(a_i^+)_{N_i-1}^{N_i} = (a_i^*)_{N_i}^{N_i-1} = \sqrt{N_i}, \tag{3.5}$$

i.e., a_i^+ increases by one the number of particles in the ith state. It is easily verified that the operator $F^{(1)}$ can be written in the form

$$F^{(1)} = \sum_{i,k} f_{ik}^{(1)} a_i^+ a_k. \tag{3.6}$$

In fact, the matrix elements of this operator coincide with those given by (3.3). Formula (3.6) is precisely the expression for $F^{(1)}$ in the second-quantized representation.

According to (3.4) and (3.5), multiplication of the operators a_i and a_i^+ gives the diagonal operators

$$\begin{aligned} a_i^+ a_i &= N_i, \\ a_i a_i^+ &= N_i + 1. \end{aligned} \tag{3.7}$$

Formulas (3.4), (3.5) and (3.7) imply the following commutation relations for the operators a_i, a_i^+:

$$\begin{aligned} [a_i, a_k^+] &\equiv a_i a_k^+ - a_k^+ a_i = \delta_{ik}, \\ [a_i, a_k] &= [a_i^+, a_k^+] = 0. \end{aligned} \tag{3.8}$$

A similar representation is possible for symmetric operators of the form

$$F^{(2)} = \sum_{a,b} f_{ab}^{(2)}, \tag{3.9}$$

where $f_{ab}^{(2)}$ acts on functions of ξ_a and ξ_b. In the second-quantized representation, $F^{(2)}$ has the form

$$F^{(2)} = \sum_{i,k,l,m} f^{(2)}{}_{lm}^{ik}\, a_i^+ a_k^+ a_l a_m, \tag{3.10}$$

where

$$f^{(2)}{}_{lm}^{ik} = \int \varphi_i^*(\xi_1) \varphi_k^*(\xi_2) f^{(2)} \varphi_l(\xi_1) \varphi_m(\xi_2)\, d\xi_1\, d\xi_2.$$

The same kind of considerations apply to more complicated operators. If

$$H = \sum_a H_a^{(1)} + \sum_{a,b} U^{(2)}(\mathbf{r}_a, \mathbf{r}_b) + \sum_{a,b,c} U^{(3)}(\mathbf{r}_a, \mathbf{r}_b, \mathbf{r}_c) + \cdots, \tag{3.11}$$

where

$$H_a^{(1)} = -\frac{\nabla_a^2}{2m} + U(\mathbf{r}_a),$$

is the Hamiltonian of a system of interacting particles located in an external field, then, according to the foregoing, H has the following form in the second-quantized representation:

$$H = \sum_{i,k} H_{ik}^{(1)} a_i^+ a_k + \sum_{i,k,l,m} U^{(2)ik}_{lm}\, a_i^+ a_k^+ a_l a_m + \cdots \tag{3.12}$$

If for the φ_i we choose the eigenfunctions of the Hamiltonian $H_a^{(1)}$, the first term in (3.12) becomes

$$H^{(1)} = \sum_i \varepsilon_i a_i^+ a_i = \sum_i \varepsilon_i N_i. \tag{3.13}$$

In the case of Fermi statistics, the complete wave function of the system must be antisymmetric in all the variables. This implies that in the case of noninteracting particles, the occupation numbers can only take the values 0 or 1, and the wave function has the form

$$\Phi_{N_1 N_2} \cdots = \frac{1}{\sqrt{N!}} \sum_P (-1)^P \varphi_{p_1}(\xi_1)\varphi_{p_2}(\xi_2)\cdots\varphi_{p_N}(\xi_N), \tag{3.14}$$

where all the numbers $p_1, p_2, \ldots, p_N$ are different. The symbol $(-1)^P$ means that odd permutations appear in the sum (3.14) with a minus sign. For definiteness, the term in (3.14) for which

$$p_1 < p_2 < \cdots < p_N \tag{3.15}$$

will be chosen with a plus sign. Then the matrix elements of an operator $F^{(1)}$ of the type (3.2) have the form[20]

$$\overline{F^{(1)}} = \sum_i f_{ii} N_i \qquad \text{for diagonal elements,}$$

$$(F^{(1)})_{0_i 1_k}^{1_i 0_k} = \pm f_{ik}^{(1)} \qquad \text{for nondiagonal elements,} \tag{3.16}$$

where the plus or the minus sign is chosen, depending on whether the total number of particles in the states between the ith and kth states is even or odd. If we introduce the operators a_i and a_i^+ whose only nonzero matrix elements are

$$(a_i)_{1_i}^{0_i} = (a_i^+)_{0_i}^{1_i} = (-1)^{\sum_{l=1}^{i-1} N_l}, \tag{3.17}$$

[20] Again, the notation is essentially that of L7, Sec. 63; $1_i 0_k$ stands for $N_i = 1$, $N_k = 0$.

then, using these operators, we can write the operator $F^{(1)}$ in the form (3.6). Multiplication of the operators a_i and a_i^+ gives

$$\begin{aligned} a_i^+ a_i &= N_i, \\ a_i a_i^+ &= 1 - N_i. \end{aligned} \tag{3.18}$$

Therefore,

$$\{a_i, a_i^+\} \equiv a_i a_i^+ + a_i^+ a_i = 1,$$

and all the other anticommutators vanish, i.e.,

$$\begin{aligned} \{a_i, a_k^+\} &= \delta_{ik}, \\ \{a_i, a_k\} &= \{a_i^+, a_k^+\} = 0. \end{aligned} \tag{3.19}$$

More complicated operators, in particular the Hamiltonian, can be written in terms of the operators a_i and a_i^+ in the same way as in the case of bosons.

4. The Dilute Bose Gas

A simple example of a quantum liquid is a "weakly nonideal" gas, i.e., a gas in which the role of interactions between particles is relatively small. As we shall see, this requires that the scattering amplitude of the particles should be small compared to the average wavelength $\lambda = 1/\bar{p}$, which for a degenerate gas is of the same order of magnitude as the average distance between particles. Under these conditions, because of the smallness of the momenta of the colliding particles, we can consider only s-wave scattering to a first approximation. If we denote the amplitude of the s-wave scattering by a, the amplitude of the p-wave scattering[21] will be of order $a(a/\lambda)^2$. Thus, for example, if the s-wave scattering contributes terms of order a/λ and higher to the total energy, then the p-wave scattering contributes terms of order no lower than $(a/\lambda)^3$. It follows that, up to terms of first order, the scattering can be regarded as isotropic. To the same approximation, we can neglect triple collisions. For a Fermi gas it can be shown that the contribution from triple collisions is even less,[22] of order $(a/\lambda)^5$.

We shall assume that the interaction between particles is repulsive, i.e., that the scattering amplitude is positive. In the case of a Bose gas, this assumption is connected with the fact that even for infinitesimal attractive forces, a Bose gas cannot stay dilute at low temperatures. In a Fermi gas,

---

[21] If the range of the forces is characterized by r_0, then, according to quantum mechanics (see e.g., L7), for $\lambda \gg r_0$ the scattering amplitudes with different angular momenta l will be of order $r_0(r_0/\lambda)^{2l}$.

[22] The last statement is a consequence of the fact that the wave function of three colliding fermions must be antisymmetric. This requires that the third particle have odd orbital angular momentum with respect to the particle among the other two which has the same spin projection as itself. As a result, at least one extra factor of $(a/\lambda)^2$ appears.

attraction between particles leads to superfluidity, and this case will not be considered here.

We now calculate the energy of the ground state and the energy spectrum for a dilute Bose gas (at $T = 0$).[23] The case of a dilute Fermi gas will be considered in the next section. For simplicity, we assume that the particles of the Bose gas have spin zero. In this case, we can write the interaction energy in the form

$$H_{\text{int}} = \frac{U}{2V} \sum_{\mathbf{p}_1 + \mathbf{p}_2 = \mathbf{p}_3 + \mathbf{p}_4} a^+_{\mathbf{p}_4} a^+_{\mathbf{p}_3} a_{\mathbf{p}_2} a_{\mathbf{p}_1}, \tag{4.1}$$

where U has been brought out in front of the summation sign, since the interaction between every pair of particles is identical, and the scattering amplitude does not depend on angle (s-wave scattering). To a first approximation, the quantity U is connected with the scattering amplitude by the relation

$$U = \frac{4\pi a}{m}. \tag{4.2}$$

This formula is easily obtained by the following argument: By definition, the amplitude of s-wave scattering is connected with the effective scattering cross-section for two identical particles by the formula

$$d\sigma = (2a)^2 \, d\Omega$$

in the center-of-mass system (see L7, Secs. 108, 114). On the other hand, the quantity $d\sigma$ can be determined by using the Hamiltonian (4.1). Thus, in the Born approximation (see L7, Sec. 110), we obtain

$$d\sigma = \left(\frac{m}{4\pi}\right)^2 (2U)^2 \, d\Omega,$$

which implies the formula (4.2).

As shown by Bogoliubov (B5), in the case of the ground state or the weakly excited states of a dilute Bose gas, the interaction-energy operator (4.1) can be simplified considerably, and as a result, it is possible to diagonalize the Hamiltonian and obtain the energy spectrum. The basic idea involved in making this simplification is the following: In the ground state, the particles of an ideal Bose gas occupy the lowest level, with zero energy, and are often said to be in the *condensate* (synonymously, the *condensed state*). In a dilute gas, because of the weakness of the interactions, the ground state will differ only slightly from the ground state of an ideal gas, i.e., the number of particles (N_0) in the condensate will still greatly exceed the number of particles in other levels, so that $N - N_0 \ll N_0$. The same applies to weakly excited states. Since the matrix elements of the Bose operators a_i are equal to $\sqrt{N_i}$, it is clear that we need only take into account interaction of particles in the

[23] The energy spectrum of a dilute Bose gas was first found by Bogoliubov (B5), and the energy of the ground state was determined in the work of Huang, Yang and Luttinger (H3), and Brueckner and Sawada (B10). Here, we mainly follow the papers B5 and B10.

condensate with each other and interaction of "excited" particles with particles in the condensate, neglecting interaction of "excited" particles with each other. This means that from the whole sum (4.1), it is sufficient to retain only the following terms:

$$H_{\text{int}} = \frac{U}{2V}\Big[a_0^+a_0^+a_0a_0 + \sum_{\mathbf{p}\neq 0}(2a_{\mathbf{p}}^+a_0^+a_{\mathbf{p}}a_0 + 2a_{-\mathbf{p}}^+a_0^+a_{-\mathbf{p}}a_0 + a_{\mathbf{p}}^+a_{-\mathbf{p}}^+a_0a_0 + a_0^+a_0^+a_{\mathbf{p}}a_{-\mathbf{p}})\Big]. \tag{4.3}$$

Since N_0 is a very large number, we are justified in regarding the operators a_0, a_0^+ as c-numbers, replacing them by $\sqrt{N_0}$. In fact, the commutators of these operators with each other or with the other operators a_i, a_i^+ are either 0 or 1, i.e., in any event are small compared to the matrix elements of the operators a_0, a_0^+. Thus, we obtain

$$H_{\text{int}} = \frac{U}{2V}\Big[N_0^2 + 2N_0\sum_{\mathbf{p}\neq 0}(a_{-\mathbf{p}}^+a_{-\mathbf{p}} + a_{\mathbf{p}}^+a_{\mathbf{p}}) + N_0\sum_{\mathbf{p}\neq 0}(a_{\mathbf{p}}^+a_{-\mathbf{p}}^+ + a_{\mathbf{p}}a_{-\mathbf{p}})\Big]. \tag{4.4}$$

The total number of particles in the system can be written in the form

$$N = N_0 + \frac{1}{2}\sum_{\mathbf{p}\neq 0}(a_{\mathbf{p}}^+a_{\mathbf{p}} + a_{-\mathbf{p}}^+a_{-\mathbf{p}}). \tag{4.5}$$

This enables us to express the number N_0 appearing in (4.4) in terms of N. Retaining only terms in H_{int} which are at least of degree 1 in N, and adding the kinetic-energy operator, we obtain the following Hamiltonian:

$$H = \frac{UN^2}{2V} + \frac{1}{2}\sum_{\mathbf{p}\neq 0}\Big[\Big(\frac{p^2}{2m} + \frac{UN}{V}\Big)(a_{\mathbf{p}}^+a_{\mathbf{p}} + a_{-\mathbf{p}}^+a_{-\mathbf{p}}) + \frac{UN}{V}(a_{\mathbf{p}}^+a_{-\mathbf{p}}^+ + a_{\mathbf{p}}a_{-\mathbf{p}})\Big]. \tag{4.6}$$

The second term in the Hamiltonian is not diagonal. To diagonalize it, we carry out a linear transformation of the operators $a_{\mathbf{p}}$ and $a_{\mathbf{p}}^+$, writing

$$a_{\mathbf{p}} = \frac{1}{\sqrt{1 - A_{\mathbf{p}}^2}}(\alpha_{\mathbf{p}} + A_{\mathbf{p}}\alpha_{-\mathbf{p}}^+),$$
$$a_{\mathbf{p}}^+ = \frac{1}{\sqrt{1 - A_{\mathbf{p}}^2}}(\alpha_{\mathbf{p}}^+ + A_{\mathbf{p}}\alpha_{-\mathbf{p}}). \tag{4.7}$$

The new operators $\alpha_{\mathbf{p}}$ and $\alpha_{\mathbf{p}}^+$ obey the same commutation relations as the old operators. Expressing the operators $a_{\mathbf{p}}$ and $a_{\mathbf{p}}^+$ in formula (4.6) in terms of $\alpha_{\mathbf{p}}$ and $\alpha_{\mathbf{p}}^+$, we obtain

$$\begin{aligned} H = \frac{UN^2}{2V} &+ \sum_{\mathbf{p}\neq 0}\frac{1}{1 - A_{\mathbf{p}}^2}\Big[\Big(\frac{p^2}{2m} + \frac{UN}{V}\Big)A_{\mathbf{p}}^2 + \frac{UN}{V}A_{\mathbf{p}}\Big] \\ &+ \frac{1}{2}\sum_{\mathbf{p}\neq 0}\frac{1}{1 - A_{\mathbf{p}}^2}\Big[\Big(\frac{p^2}{2m} + \frac{UN}{V}\Big)(1 + A_{\mathbf{p}}^2) + 2\frac{UN}{V}A_{\mathbf{p}}\Big](\alpha_{\mathbf{p}}^+\alpha_{\mathbf{p}} + \alpha_{-\mathbf{p}}^+\alpha_{-\mathbf{p}}) \\ &+ \frac{1}{2}\sum_{\mathbf{p}\neq 0}\frac{1}{1 - A_{\mathbf{p}}^2}\Big[\Big(\frac{p^2}{2m} + \frac{UN}{V}\Big)2A_{\mathbf{p}} + \frac{UN}{V}(1 + A_{\mathbf{p}}^2)\Big](\alpha_{\mathbf{p}}^+\alpha_{-\mathbf{p}}^+ + \alpha_{\mathbf{p}}\alpha_{-\mathbf{p}}). \end{aligned} \tag{4.8}$$

In order for the nondiagonal terms to vanish, the coefficients $A_{\mathbf{p}}$ must satisfy the relation

$$\left(\frac{p^2}{2m} + \frac{UN}{V}\right)2A_{\mathbf{p}} + \frac{UN}{V}(1 + A_{\mathbf{p}}^2) = 0.$$

This gives

$$A_{\mathbf{p}} = \frac{V}{UN}\left[-\frac{p^2}{2m} - \frac{UN}{V} + \sqrt{\left(\frac{p^2}{2m} + \frac{UN}{V}\right)^2 - \left(\frac{UN}{V}\right)^2}\right].$$

The plus sign in front of the square root is necessary, if the excited states are to have positive energy. Substituting this expression for the coefficient $A_{\mathbf{p}}$ into (4.8), we finally obtain

$$\begin{aligned} H = \frac{UN^2}{2V} - \frac{1}{2}\sum_{\mathbf{p}\neq 0}\left[\left(\frac{p^2}{2m} + \frac{UN}{V}\right) - \sqrt{\left(\frac{p^2}{2m} + \frac{UN}{V}\right)^2 - \left(\frac{UN}{V}\right)^2}\right] \\ + \frac{1}{2}\sum_{\mathbf{p}\neq 0}\sqrt{\left(\frac{p^2}{2m} + \frac{UN}{V}\right)^2 - \left(\frac{UN}{V}\right)^2}(\alpha_{\mathbf{p}}^+\alpha_{\mathbf{p}} + \alpha_{-\mathbf{p}}^+\alpha_{-\mathbf{p}}). \end{aligned} \tag{4.9}$$

The expression (4.9) consists of three terms. The sum of the first two terms is a certain constant. The third term represents a diagonal operator, which can be written in the form

$$\sum_{\mathbf{p}\neq 0} n_{\mathbf{p}}\varepsilon(\mathbf{p}), \tag{4.10}$$

where $n_{\mathbf{p}}$ denotes the occupation number of the operator $\alpha_{\mathbf{p}}$. The smallest value of the energy is obtained when all the $n_{\mathbf{p}}$ equal zero, and therefore (4.10) is the energy of the excitation. This expression has the same form as the energy of a system of noninteracting particles, given by (3.13). It follows that the weakly excited states of a dilute Bose gas can be described by using the model of elementary excitations, with an energy spectrum[24]

$$\varepsilon(\mathbf{p}) = \sqrt{\left(\frac{p^2}{2m} + \frac{UN}{V}\right)^2 - \left(\frac{UN}{V}\right)^2}. \tag{4.11}$$

In the limit of small momenta, this expression becomes

$$\varepsilon(\mathbf{p}) \approx \frac{\sqrt{4\pi aN/V}}{m}p, \tag{4.12}$$

which corresponds to the phonon part of the spectrum of the Bose liquid. For large momenta, the energy $\varepsilon(\mathbf{p})$ goes into the energy

$$\varepsilon(\mathbf{p}) \approx \frac{p^2}{2m} \tag{4.13}$$

of a free particle, which is also in accord with the results of Sec. 2.

---

[24] We note that the Born approximation has been used to derive (4.11). However, formula (4.11), which can be expressed in terms of the scattering amplitude a by using (4.2), is actually valid not only in the Born approximation, but also whenever the condition $a/\lambda \ll 1$ holds. This will be proved in Chap. 5. The same applies to formulas (4.16), (5.20) and (5.21).

The sum of the first two terms in the right-hand side of (4.9) obviously represents the ground-state energy of the Bose liquid. It is not hard to see that for large momenta the sum over **p** in this expression diverges like $\sum_{\mathbf{p}} p^{-2}$. This is connected with the fact that the energy cannot actually be expanded in powers of U. In fact, the presence of the constant U causes the energy to be infinite, as can be seen directly from (4.9). In the case being considered, it is essential that the scattering amplitude a should be finite and indeed small, which makes an expansion of the energy with respect to a possible.

The relation (4.2) between U and the scattering amplitude is not exact, but is only valid up to first-order terms. Since we are also interested in higher-order terms in the energy, formula (4.2) has to be corrected. Using second-order perturbation theory to examine the scattering of two particles of the condensate, with a transition to the state $\mathbf{p}, -\mathbf{p}$, we find that

$$U - \frac{U^2}{V}\sum_{\mathbf{p}\neq 0}\frac{1}{p^2/m} = \frac{4\pi a}{m}. \tag{4.14}$$

Then, expressing U in terms of a and substituting the resulting expression into (4.9), we obtain the following expression for the energy of the ground state:

$$\begin{aligned} E = {} & \frac{2\pi a}{m}\frac{N^2}{V} + \frac{8\pi^2 a^2}{m^2}\left(\frac{N}{V}\right)^2 \sum_{\mathbf{p}\neq 0}\frac{1}{p^2/m} \\ & - \frac{1}{2}\sum_{\mathbf{p}\neq 0}\left(\frac{p^2}{2m} + \frac{4\pi a N}{mV}\right)\left(1 - \sqrt{1 - \left[\frac{4\pi a N/mV}{(p^2/2m) + (4\pi a N/mV)}\right]^2}\right). \end{aligned} \tag{4.15}$$

The expression (4.15) converges for large p. Integrating with respect to momentum, we obtain

$$\frac{E}{V} = \frac{2\pi a}{m}\left(\frac{N}{V}\right)^2\left[1 + \frac{128}{15\sqrt{\pi}}a^{3/2}\left(\frac{N}{V}\right)^{1/2}\right], \tag{4.16}$$

where it should be noted that the expansion is in terms of $[a(N/V)^{1/3}]^{3/2}$. Using (4.16), we find that the velocity of sound equals

$$u = \sqrt{\frac{V^2}{mN}\frac{\partial^2 E}{\partial V^2}} = \frac{\sqrt{4\pi a N/V}}{m}. \tag{4.17}$$

As must be the case, this expression agrees with the coefficient of p in the formula (4.12) for the phonon part of the spectrum.

It was already stipulated at the beginning of this section that the amplitude a must be positive in a Bose gas. This is also clear from formula (4.17), since if a were negative, the velocity of sound would be imaginary (corresponding to an unstable state).

The momentum distribution of the excitations is given by the usual Bose formula

$$\bar{n}_{\mathbf{p}} = \alpha_{\mathbf{p}}^{+}\alpha_{\mathbf{p}} = \frac{1}{e^{\varepsilon(\mathbf{p})/T} - 1}. \tag{4.18}$$

The momentum distribution of the particles of the Bose liquid themselves can be found by calculating

$$\bar{N}_{\mathbf{p}} = \overline{a_{\mathbf{p}}^{+} a_{\mathbf{p}}}.$$

Substituting from formula (4.7), we obtain

$$\bar{N}_{\mathbf{p}} = \frac{\bar{n}_{\mathbf{p}} + A_{\mathbf{p}}^2(\bar{n}_{\mathbf{p}} + 1)}{1 - A_{\mathbf{p}}^2}. \tag{4.19}$$

Of course, this expression applies only to the case $\mathbf{p} \neq 0$. The number of particles with zero energy is obtained from the formula

$$N_0 = N - \sum_{\mathbf{p} \neq 0} \bar{N}_{\mathbf{p}}.$$

At absolute zero $\bar{n}_{\mathbf{p}} = 0$, and hence we find from (4.19) that

$$\bar{N}_{\mathbf{p}} = \frac{(8\pi^2 a^2/m^2)(N/V)^2}{\varepsilon(\mathbf{p})[\varepsilon(\mathbf{p}) + (p^2/2m) + (4\pi a N/mV)]}, \tag{4.20}$$

$$\frac{N_0}{N} = 1 - \frac{8}{3\sqrt{\pi}} a^{3/2}\left(\frac{N}{V}\right)^{1/2}. \tag{4.21}$$

Thus, it is clear that the particles of a nonideal Bose gas do not all have zero momentum, even in the ground state.

5. The Dilute Fermi Gas

Next, we consider the case of a Fermi gas. We shall determine the ground-state energy, the effective mass of the excitations, and the function f, with an accuracy up to terms of order $(a/\lambda)^2$, where a is the amplitude of the s-wave scattering.[25]

For a Fermi gas, the operator corresponding to the interaction energy cannot be written in the form (4.1), as in the boson case. In fact, the sum in (4.1) vanishes if we regard the indices as specifying not only momenta but also spin projections, because of the anticommutativity of the Fermi operators. This is connected with the fact that the Hamiltonian (4.1) does not take account of the specific nature of fermion scattering. According to quantum mechanics (see L7, Sec. 114), for identical particles with spin $\frac{1}{2}$, s-wave scattering can occur only when the spins are antiparallel, and then the amplitude is twice what it is when the particles are different. Bearing this fact in mind, we can write the interaction energy in the form

$$H_{\text{int}} = \frac{U}{V} \sum_{\mathbf{p}_1 + \mathbf{p}_2 = \mathbf{p}_3 + \mathbf{p}_4} a_{\mathbf{p}_3, 1/2}^{+} a_{\mathbf{p}_4, -1/2}^{+} a_{\mathbf{p}_2, -1/2} a_{\mathbf{p}_1, 1/2}, \tag{5.1}$$

---

[25] The ground-state energy was calculated in the papers by Huang and Yang (H2), and by Lee and Yang (L10), while the effective mass of the excitations was found by Abrikosov and Khalatnikov (A8), and also by Galitski (G1). The function f was calculated in the paper A8.

or equivalently,

$$H_{\text{int}} = \frac{U}{2V} \sum_{\mathbf{p}_1+\mathbf{p}_2=\mathbf{p}_3+\mathbf{p}_4} a^+_{\mathbf{p}_3\alpha} a^+_{\mathbf{p}_4\beta} a_{\mathbf{p}_2\beta} a_{\mathbf{p}_1\alpha}. \tag{5.1'}$$

To a first approximation, the quantity U is connected with the amplitude of s-wave scattering by the same formula

$$U = \frac{4\pi a}{m} \tag{5.2}$$

as before.

We now apply perturbation theory to the operator H_{int}. The first-order contribution to the ground-state energy is given by the diagonal matrix element of H_{int}, i.e., by

$$E^{(1)} = \frac{U}{V} \sum_{i,k} n_i n_k Q_{ik}, \tag{5.3}$$

where the indices i and k correspond to given momenta and spin projections, and the n_i are the occupation numbers for $T = 0$,[26] equal to 1 for $p < p_0$ and to 0 for $p > p_0$ [$p_0 = (3\pi^2 N/V)^{1/3}$]. The factor Q_{ik} in (5.3) takes account of the fact that the particles in states i and k have antiparallel spins. It is convenient to choose this factor in the form

$$Q_{ik} = \frac{1}{4}(1 - \boldsymbol{\sigma}_i \cdot \boldsymbol{\sigma}_k), \tag{5.4}$$

where $\frac{1}{2}\boldsymbol{\sigma}_i$ is the spin operator of the particle in state i. Substituting (5.2) and (5.4) into (5.3), we find that

$$E^{(1)} = \frac{\pi a}{m}\frac{N^2}{V}. \tag{5.5}$$

To find the second-order correction, we use the perturbation-theory formula

$$E_s^{(2)} = \sum_{r \neq s} \frac{|(H_{\text{int}})_{rs}|^2}{E_s - E_r}. \tag{5.6}$$

Substituting (5.1) into (5.6), we obtain the following sum:

$$\frac{U^2}{V^2} \sum_{i,j,k,l} \frac{n_i n_j (1 - n_k)(1 - n_l) Q_{ij} Q_{kl}}{(\mathbf{p}_i^2 + \mathbf{p}_j^2 - \mathbf{p}_k^2 - \mathbf{p}_l^2)/2m}. \tag{5.7}$$

Since our aim is to find an expression for the energy in powers of a, just as in Sec. 4 we must take account of the fact that the relation (5.2) between U and the scattering amplitude is not exact, but only valid up to first-order terms in U. If we take account of second-order terms, we obtain

$$U + \frac{U^2}{V} \sum_{k,l} \frac{Q_{kl}}{(\mathbf{p}_i^2 + \mathbf{p}_j^2 - \mathbf{p}_k^2 - \mathbf{p}_l^2)/2m} = \frac{4\pi a}{m}, \tag{5.2'}$$

instead of (5.2). Using (5.2') to express U in terms of a, and substituting the

[26] Here the n_i denote occupation numbers for noninteracting particles. It is not hard to see that for $T = 0$ they coincide with the occupation numbers of the quasi-particles and differ from N_i, the occupation numbers for a system of interacting particles.

result into (5.3), we find that terms proportional to a^2 appear in $E^{(1)}$. Naturally, these terms should be included in the second-order correction. Bearing this in mind, we obtain the following expression for the second approximation to the energy:

$$E^{(2)} = \frac{16a^2\pi^2}{m^2V^2} \sum_{i,j,k,l} \left[\frac{n_i n_j(1 - n_k)(1 - n_l)Q_{ij}Q_{kl}}{(\mathbf{p}_i^2 + \mathbf{p}_j^2 - \mathbf{p}_k^2 - \mathbf{p}_l^2)/2m} - \frac{n_i n_j Q_{ij}Q_{kl}}{(\mathbf{p}_i^2 + \mathbf{p}_j^2 - \mathbf{p}_k^2 - \mathbf{p}_l^2)/2m}\right]. \tag{5.7'}$$

Unlike (5.7), this expression does not diverge for large $\mathbf{p}$. Hence, just as in the case of the Bose gas, renormalization of U eliminates divergence of the energy.

We note that at first glance the relation (5.2′) seems contradictory, since its left-hand side depends on the angle between $\mathbf{p}_i$ and $\mathbf{p}_j$, while its right-hand side does not. This might be construed as evidence that we have chosen the wrong form for the interaction-energy operator. However, actually this fact need cause no misgivings, since in (5.2′) we have not taken account of terms involving higher angular momenta, which, as already noted, make a small contribution to the energy.

In the expression (5.7′), the term with four n_i, which arises in the first summand, equals zero, since the numerator is symmetric and the denominator antisymmetric under the permutation $i, j \rightleftarrows k, l$, while all the regions of summation are the same. The remaining two terms involving products of three n_i are equal. Thus, we finally obtain

$$E^{(2)} = -\frac{32a^2\pi^2}{m^2V^2} \sum_{i,j,k,l} \frac{n_i n_j n_k Q_{ij} Q_{kl}}{(\mathbf{p}_i^2 + \mathbf{p}_j^2 - \mathbf{p}_k^2 - \mathbf{p}_l^2)/2m}. \tag{5.8}$$

Going over from summation to integration, we can write this expression in the form

$$E^{(2)} = -\frac{32a^2\pi^2}{m^2(2\pi)^9} \times \int_{|\mathbf{p}_1|<p_0} d\mathbf{p}_1 \int_{|\mathbf{p}_2|<p_0} d\mathbf{p}_2 \int_{|\mathbf{p}_3|<p_0} d\mathbf{p}_3 \int d\mathbf{p}_4 \frac{\delta(\mathbf{p}_1 + \mathbf{p}_2 - \mathbf{p}_3 - \mathbf{p}_4)}{(\mathbf{p}_1^2 + \mathbf{p}_2^2 - \mathbf{p}_3^2 - \mathbf{p}_4^2)/2m}. \tag{5.9}$$

According to Sec. 2, the energy of the excitations is given by the formula[27]

$$\varepsilon_i = \frac{\delta E}{\delta n_i}. \tag{5.10}$$

[27] It might appear at first glance that this formula is incorrect, since ε is the variational derivative of E with respect to the distribution function of the quasi-particles, rather than of the particles themselves. However, formula (5.10) really involves the derivative with respect to the distribution of noninteracting particles, and not with respect to the true distribution of the particles. As already noted (see footnote 26, p. 37), for $T = 0$ the distribution of noninteracting particles coincides with the distribution of the quasi-particles of an interacting system.

Variation of the expressions (5.3) and (5.8) with respect to n_i gives

$$\varepsilon(\mathbf{p}) = \frac{p^2}{2m} + \frac{2\pi a N}{mV} + \frac{16\pi^2 a^2}{m^2(2\pi)^9} \int_{|\mathbf{p}_1|<p_0} d\mathbf{p}_1 \int_{|\mathbf{p}_2|<p_0} d\mathbf{p}_2 \int d\mathbf{p}_3 \left[\frac{\delta(\mathbf{p}_1 + \mathbf{p}_2 - \mathbf{p} - \mathbf{p}_0)}{(\mathbf{p}^2 + \mathbf{p}_3^2 - \mathbf{p}_1^2 - \mathbf{p}_2^2)/2m} - 2\frac{\delta(\mathbf{p}_1 + \mathbf{p} - \mathbf{p}_2 - \mathbf{p}_3)}{(\mathbf{p}^2 + \mathbf{p}_1^2 - \mathbf{p}_2^2 - \mathbf{p}_3^2)/2m}\right]. \tag{5.11}$$

Thus, to calculate the energy of the ground state and the effective mass of the excitations, we have to evaluate the integrals (5.9) and (5.11). These calculations are quite formidable, because of the number of integrations involved and the inconvenience of the regions of integration. However, we resort to a simpler method, based on the use of the function f. After introducing this function

$$f_{ik} = \frac{\delta^2 E}{\delta n_i \delta n_k}, \tag{5.12}$$

we can find the effective mass and the velocity of low-frequency sound from formulas (2.12) and (2.19) of Sec. 2. Then, from the velocity of sound, we can find the energy of the ground state by carrying out a suitable integration.

In this way, the problem reduces to determining the quantity f. Varying the expressions (5.3) and (5.8) first with respect to n_i, and then with respect to n_k, we find the following expression for f:

$$f = \frac{8a\pi}{m} Q_{\sigma\sigma'} - \frac{64\pi^2 a^2}{m^2(2\pi)^3} \int_{|\mathbf{p}_1|<p_0} d\mathbf{p}_1 \int d\mathbf{p}_2 \left[Q_{\sigma\sigma'} \frac{\delta(\mathbf{p} + \mathbf{p}' - \mathbf{p}_1 - \mathbf{p}_2)}{(\mathbf{p}^2 + \mathbf{p}'^2 - \mathbf{p}_1^2 - \mathbf{p}_2^2)/2m} + \frac{1}{4}\frac{\delta(\mathbf{p} + \mathbf{p}_1 - \mathbf{p}' - \mathbf{p}_2)}{(\mathbf{p}^2 + \mathbf{p}_1^2 - \mathbf{p}'^2 - \mathbf{p}_2^2)/2m} + \frac{1}{4}\frac{\delta(\mathbf{p}' + \mathbf{p}_1 - \mathbf{p} - \mathbf{p}_2)}{(\mathbf{p}'^2 + \mathbf{p}_1^2 - \mathbf{p}^2 - \mathbf{p}_2^2)/2m}\right] \tag{5.13}$$

To carry out the calculation, we can immediately set $|\mathbf{p}| = |\mathbf{p}'| = p_0$, thereby making the integration much simpler than in (5.9) and (5.11). As a result, we obtain

$$f(\chi) = \frac{2\pi a}{m}\left[1 + 2a\left(\frac{3N}{\pi V}\right)^{1/3}\left(2 + \frac{\cos\chi}{2\sin\frac{1}{2}\chi}\ln\frac{1 + \sin\frac{1}{2}\chi}{1 - \sin\frac{1}{2}\chi}\right)\right] - \frac{2\pi a}{m}(\boldsymbol{\sigma}\cdot\boldsymbol{\sigma}')\left[1 + 2a\left(\frac{3N}{\pi V}\right)^{1/3}\left(1 - \frac{\sin\frac{1}{2}\chi}{2}\ln\frac{1 + \sin\frac{1}{2}\chi}{1 - \sin\frac{1}{2}\chi}\right)\right]. \tag{5.14}$$

The following special feature of formula (5.14) is worthy of attention: When the angle χ is near π, the function f has a logarithmic singularity for particles with opposite spins, i.e.,

$$f(\chi) \propto (1 - \boldsymbol{\sigma}\cdot\boldsymbol{\sigma}')\ln\frac{1}{\pi - \chi}. \tag{5.15}$$

It is clear that, strictly speaking, the approximation used here is not applicable in this case. The singularity of the function f for $\chi = \pi$ reflects a singularity in the scattering amplitude of excitations whose collision angles equal π (see Chap. 4). In this case, the correct expression can be obtained by summation of the principal terms of the perturbation series, i.e., the terms in which the logarithm appears to the highest power (one less than the power of a). If we set the angle χ exactly equal to π and write

$$\lambda^2 = p^2 + p'^2 - 2p_0^2 \neq 0,$$

the summation leads to the appearance of the factor

$$\frac{1}{1 + a\left(\frac{3N}{\pi V}\right)^{1/3}\left(\ln \frac{p_0^2}{\lambda} + \frac{i\pi}{2}\right)} \tag{5.16}$$

in the expression for f (the real part in the denominator is written with logarithmic accuracy). Since $a > 0$ by hypothesis, this expression converges to zero as $\lambda \to 0$.

The case $a < 0$ is also theoretically possible in a Fermi gas. In fact, because of the Pauli principle, in this case the gas remains dilute (unlike the Bose gas), and at first glance all the formulas remain applicable. But if we examine (5.16), it becomes clear that the scattering amplitude has a pole for some small imaginary value of λ. This is connected with the instability of the ground state with respect to the formation of bound pairs of quasi-particles with opposite momenta and spins (the *Cooper phenomenon*), which is the basic cause of superconductivity in metals (see Chap. 7). Here, however, we confine ourselves to the case $a > 0$.

Thus, the expression which we have found for f is not valid for angles near π. However, since the singularity is logarithmic, it has an effect only in the immediate neighborhood of the singular point. Since we are only interested in integrated expressions, involving the function f in combination with regular functions, this logarithmic singularity is not important.

Substituting (5.14) into (2.12), we find that the effective mass m^* is given by

$$\frac{m}{m^*} = 1 - \frac{8}{15}(7 \ln 2 - 1)a^2\left(\frac{3N}{\pi V}\right)^{2/3}. \tag{5.17}$$

Similarly, from the expression (2.19) for the velocity of sound we obtain

$$u^2 = \frac{\pi^{4/3}}{3^{1/3}}\left(\frac{N}{V}\right)^{2/3}\frac{1}{m^2} + 2\frac{\pi a}{m^2}\frac{N}{V}\left[1 + \frac{4}{15}a\left(\frac{3N}{\pi V}\right)^{1/3}(11 - 2\ln 2)\right]. \tag{5.18}$$

Using this formula, we can easily find the ground-state energy of the Fermi liquid. In fact, according to (2.16),

$$u^2 = \frac{N}{m}\frac{\partial\mu}{\partial N},$$

so that integrating (5.18) twice, we obtain

$$E = \int \mu\, dN = E^{(0)} + \frac{\pi a}{m}\frac{N^2}{V}\left[1 + \frac{6}{35}a\left(\frac{3N}{\pi V}\right)^{1/3}(11 - 2\ln 2)\right]. \tag{5.19}$$

The results (5.17) and (5.19) can also be obtained directly, by carrying out the integrations in (5.9) and (5.11). This proves the validity of the basic ideas underlying our model of a Fermi liquid. A detailed derivation of these ideas will be given in Chap. 4.

Finally, just as in the case of a Bose gas, we are interested in the momentum distribution of the particles, for which we have to calculate the matrix element

$$\bar{N}_{\mathbf{p},1/2} = \bar{N}_{\mathbf{p},-1/2} = \langle \Psi^* a^+_{\mathbf{p},1/2} a_{\mathbf{p},1/2} \Psi \rangle, \tag{5.20}$$

where Ψ is the true wave function of the interacting particles. The result of a perturbation-theory calculation of Ψ up to terms of the second order is

$$\Psi_s = \Psi_s^{(0)} + \sum_{r \neq s} \frac{(H_{\text{int}})_{rs} \Psi_r^{(0)}}{E_s - E_r} + \sum_{r \neq s} \sum_{k \neq s} \frac{(H_{\text{int}})_{rk} (H_{\text{int}})_{ks} \Psi_r^{(0)}}{(E_s - E_k)(E_s - E_r)}$$
$$- (H_{\text{int}})_{ss} \sum_{r \neq s} \frac{(H_{\text{int}})_{rs} \Psi_r^{(0)}}{(E_s - E_r)^2} - \frac{\Psi_s^{(0)}}{2} \sum_{r \neq s} \frac{|(H_{\text{int}})_{rs}|^2}{(E_s - E_r)^2} \tag{5.21}$$

(see L7, Sec. 38). Substituting (5.21) into (5.20), and bearing in mind that the operator

$$a^+_{\mathbf{p},1/2} a_{\mathbf{p},1/2}$$

is diagonal in the representation determined by the functions $\Psi_s^{(0)}$, we find that

$$\bar{N}_{\mathbf{p},1/2} - n_{\mathbf{p},1/2} = \sum_{k \neq 0} \frac{|(H_{\text{int}})_{k0}|^2 (n^{(k)}_{\mathbf{p},1/2} - n_{\mathbf{p},1/2})}{(E_0 - E_k)^2}, \tag{5.22}$$

where $n^{(k)}_{\mathbf{p},1/2}$ is the number of particles with momentum $\mathbf{p}$ and spin projection $\frac{1}{2}$ in the state $\Psi_k^{(0)}$ of a noninteracting system, and $n_{\mathbf{p},1/2}$ is the corresponding quantity for the ground state. As already noted, the distribution $n_{\mathbf{p},1/2}$ of noninteracting particles coincides with the distribution of excitations for $T = 0$.

Substituting H_{int} from (5.1) into (5.22), we obtain

$$\bar{N}_{\mathbf{p},1/2} - n_{\mathbf{p},1/2}$$
$$= \begin{cases} -\dfrac{16\pi^2 a^2}{(2\pi)^6 m^2} \displaystyle\int_{|\mathbf{p}_1| < p_0} d\mathbf{p}_1 \int_{|\mathbf{p}_2| > p_0} d\mathbf{p}_2 \int_{|\mathbf{p}_3| > p_0} d\mathbf{p}_3 \, \frac{\delta(\mathbf{p} + \mathbf{p}_1 - \mathbf{p}_2 - \mathbf{p}_3)}{[(\mathbf{p}^2 + \mathbf{p}_1^2 - \mathbf{p}_2^2 - \mathbf{p}_3^2)/2m]^2} \\ \qquad \text{for} \quad |\mathbf{p}| < p_0, \\ \dfrac{16\pi^2 a^2}{(2\pi)^6 m^2} \displaystyle\int_{|\mathbf{p}_1| > p_0} d\mathbf{p}_1 \int_{|\mathbf{p}_2| < p_0} d\mathbf{p}_2 \int_{|\mathbf{p}_3| < p_0} d\mathbf{p}_3 \, \frac{\delta(\mathbf{p} + \mathbf{p}_1 - \mathbf{p}_2 - \mathbf{p}_3)}{[(\mathbf{p}^2 + \mathbf{p}_1^2 - \mathbf{p}_2^2 - \mathbf{p}_3^2)/2m]^2} \\ \qquad \text{for} \quad |\mathbf{p}| > p_0. \end{cases} \tag{5.23}$$

Thus, it turns out that the momentum distribution of the particles differs from the distribution $n_{\mathbf{p},1/2}$ of the quasi-particles only by terms of the second order in a. The integral of (5.23) over all $\mathbf{p}$ is zero, corresponding to the fact that the number of particles of the liquid equals the number of quasi-particles. It is interesting to observe that the function $\bar{N}_{\mathbf{p},1/2}$ has a jump for $|\mathbf{p}| = p_0$. In Chap. 3, it will be shown that this is a general property of Fermi liquids.

Evaluation of the integrals (5.23) leads to rather formidable expressions, which we shall not give in full here.[28] Instead, we only give a few limiting values:

$$
\begin{aligned}
N_{0,1/2} &= 1 - 2a^2\left(\frac{3}{\pi}\frac{N}{V}\right)^{2/3}\left(1 - \frac{1}{2}\ln 2\right),\\
N_{p_0-0,1/2} &= 1 - 2a^2\left(\frac{3}{\pi}\frac{N}{V}\right)^{2/3}\left(\frac{1}{3} + \ln 2\right),\\
N_{p_0+0,1/2} &= 2a^2\left(\frac{3}{\pi}\frac{N}{V}\right)^{2/3}\left(\ln 2 - \frac{1}{3}\right),\\
N_{p_0-0,1/2} - N_{p_0+0,1/2} &= 1 - 4a^2\left(\frac{3}{\pi}\frac{N}{V}\right)^{2/3}\ln 2,\\
N_{p\gg p_0,1/2} &= \frac{16a^2}{9}\left(\frac{3}{\pi}\frac{N}{V}\right)^{2/3}\left(\frac{p_0}{p}\right)^4.
\end{aligned}
\tag{5.24}
$$

Thus, for $p < p_0$, the function N_p is close to 1, and falls off slowly as p is increased from 0 to p_0. Then N_p drops discontinuously to a value of order $a^2(N/V)^{2/3}$, and for $p \gg p_0$, N_p falls off according to the law

$$
a^2\left(\frac{N}{V}\right)^{2/3}\left(\frac{p_0}{p}\right)^4.
$$

---

[28] The calculations have been carried out by Belyakov (B4).

2

METHODS OF QUANTUM FIELD THEORY FOR $T=0$

6. The Interaction Representation

The method of second quantization, as presented in the preceding chapter, is not in a satisfactory form for solving a great many problems. In fact, the method can only be used when the interaction between particles is weak, in which case either perturbation theory is applicable or the Hamiltonian simplifies to such an extent that it can easily be diagonalized. However, we often have to deal with situations where it is not permissible to consider only the first few terms of the perturbation series. In these cases, a method is needed which gives comparatively simple and intuitive rules for writing down any term of the perturbation series.

By using appropriate physical conditions, we can quite often pick out from the set of all terms of the perturbation series a sequence (usually infinite) of so-called "principal terms," which exceed the other terms in order of magnitude. Then the problem reduces to summing this sequence. However, in the general case, where all the terms of the perturbation series are of the same order, the theoretical problem consists in finding various general relations, e.g., formula (2.1).[1] For these purposes, the most convenient method is the *diagram technique*, borrowed from quantum field theory (see, e.g., A10) and presented in this chapter.

[1] Formula (2.1), connecting the Fermi momentum p_0 and the number N of particles in the liquid, underlies Landau's theory of the Fermi liquid.

We begin our discussion of the methods of quantum field theory by representing the apparatus of second quantization in another form. To this end, we introduce the "particle-field" operators

$$\psi(\xi) = \sum_i \varphi_i(\xi) a_i, \qquad \psi^+(\xi) = \sum_i \varphi_i^*(\xi) a_i^+, \tag{6.1}$$

where a_i^+, a_i are the second-quantization operators with which we are already familiar from Chap. 1, and $\varphi_i(\xi)$ is the wave function of the particle in state i. The operators $\psi(\xi)$ and $\psi^+(\xi)$ can be interpreted as the operators for annihilation and creation, respectively, of a particle at a given point of ξ-space. The results of Sec. 3 imply that these operators obey the commutation relations

$$\begin{aligned} \psi(\xi)\psi^+(\xi') \mp \psi^+(\xi')\psi(\xi) &= \delta(\xi - \xi'), \\ \psi(\xi)\psi(\xi') \mp \psi(\xi')\psi(\xi) &= 0, \\ \psi^+(\xi)\psi^+(\xi') \mp \psi^+(\xi')\psi^+(\xi) &= 0, \end{aligned} \tag{6.2}$$

where the upper sign corresponds to Bose statistics and the lower sign to Fermi statistics. In the new representation, a one-particle operator $F^{(1)}$ takes the form

$$F^{(1)} = \int \psi^+(\xi) f^{(1)} \psi(\xi)\, d\xi, \tag{6.3}$$

while two-particle and more complicated operators can be written similarly.

It is not hard to write the Hamiltonian in terms of the operators ψ and ψ^+. For example, the Hamiltonian takes the form

$$\begin{aligned} H = \int \left[\frac{1}{2m} \nabla\psi_\alpha^+(\mathbf{r}) \cdot \nabla\psi_\alpha(\mathbf{r}) + U(\mathbf{r})\psi_\alpha^+(\mathbf{r})\psi_\alpha(\mathbf{r})\right] d\mathbf{r} \\ + \frac{1}{2} \int \psi_\alpha^+(\mathbf{r})\psi_\beta^+(\mathbf{r}') U^{(2)}(\mathbf{r}, \mathbf{r}')\psi_\beta(\mathbf{r}')\psi_\alpha(\mathbf{r})\, d\mathbf{r}\, d\mathbf{r}' + \cdots \end{aligned} \tag{6.4}$$

for a system of particles with spin $\frac{1}{2}$, in the absence of a magnetic field. Here, it is assumed that the interaction between the particles is spin-independent. The subscripts α and β indicate the spin projections, and the expression is assumed to be summed over pairs of identical indices. For a system of bosons with spin zero, we have the same Hamiltonian except that the operators ψ, ψ^+ have no subscripts. The generalization to more complicated cases presents no difficulties. In external appearance, (6.4) is the same as the expression for the average energy of a system of N particles in identical states $\psi_\alpha(\mathbf{r})$, normalized by the relation

$$\int |\psi_\alpha(\mathbf{r})|^2\, d\mathbf{r} = N.$$

Using this property, we can always easily find the Hamiltonian in the second-quantized representation.

Besides the Hamiltonian, another operator of basic importance is the operator for the density of particles at a given point. Since this operator has the form

$$n(\mathbf{r}) = \sum_a \delta(\mathbf{r} - \mathbf{r}_a)$$

in the usual representation, here we obtain

$$n(\mathbf{r}) = \int \psi_\alpha^+(\mathbf{r}_a)\delta(\mathbf{r} - \mathbf{r}_a)\psi_\alpha(\mathbf{r}_a)\, d\mathbf{r}_a = \psi_\alpha^+(\mathbf{r})\psi_\alpha(\mathbf{r}). \tag{6.5}$$

Correspondingly, the operator for the total number of particles has the form

$$N = \int n(\mathbf{r})\, d\mathbf{r} = \int \psi_\alpha^+(\mathbf{r})\psi_\alpha(\mathbf{r})\, d\mathbf{r}.$$

Now suppose we have a system of particles with a Hamiltonian H, and we want to determine how the state of the system changes in time. To do this, we must solve the Schrödinger equation

$$i\frac{\partial \Phi}{\partial t} = H\Phi, \tag{6.6}$$

where Φ is the wave function of the system. The solution of equation (6.6) can be written in the symbolic form[2]

$$\Phi(t) = e^{-iHt}\Phi_H, \tag{6.7}$$

where Φ_H is a function which is independent of time. Using (6.7), we can find how the matrix elements of any operator F vary in time:

$$F_{nm}(t) = \langle \Phi_n^*(t)F\Phi_m(t)\rangle = \langle \Phi_{Hn}^* e^{iHt}Fe^{-iHt}\Phi_{Hm}\rangle. \tag{6.8}$$

The last expression can be interpreted as the matrix element of the operator

$$\tilde{F}(t) = e^{iHt}Fe^{-iHt} \tag{6.9}$$

with respect to the functions Φ_H. This corresponds to going over to a new representation called the *Heisenberg representation.* The representation considered earlier, in which the operators F [e.g., $\psi(\mathbf{r})$ and $\psi^+(\mathbf{r})$] are time-independent, is called the *Schrödinger representation.* The most important property of the Heisenberg representation is that the wave functions Φ_H do not depend on time. Instead, the time dependence is transferred to the operators, and according to (6.9), we have

$$\frac{\partial \tilde{F}}{\partial t} = i(H\tilde{F} - \tilde{F}H) \equiv i[H, \tilde{F}]. \tag{6.10}$$

In the Schrödinger representation, the state of affairs is just the opposite: The operators are time-independent (if we are not concerned with variable external fields), while the wave function depends on time. For the Hamiltonian itself, both representations are the same, as can be seen from (6.9).

---

[2] The operator e^{-iHt} is a symbolic way of writing the series

$$1 - iHt + \cdots + \frac{1}{n!}(-iHt)^n + \cdots$$

If we consider a stationary state of the system, the wave function Φ_{Hn} satisfies the equation

$$H\Phi_{Hn} = E_n\Phi_{Hn}. \tag{6.11}$$

In this case, (6.8) gives

$$F_{nm}(t) = \langle \Phi^*_{Hn} F \Phi_{Hm} \rangle e^{i(E_n - E_m)t}. \tag{6.12}$$

Consider, for example, a system of noninteracting particles without spin. For $\varphi_i(\xi)$ we choose the free-particle functions

$$\frac{1}{\sqrt{V}} e^{i\mathbf{p}\cdot\mathbf{r}},$$

where V is the volume. Then the operator ψ has the form

$$\psi(\mathbf{r}) = \frac{1}{\sqrt{V}} \sum_{\mathbf{p}} a_{\mathbf{p}} e^{i\mathbf{p}\cdot\mathbf{r}} \tag{6.13}$$

in the Schrödinger representation. Using formula (6.4), we find that the Hamiltonian has the form (3.13), i.e.,

$$H = \sum_{\mathbf{p}} \varepsilon_0(\mathbf{p}) n_{\mathbf{p}},$$

where $\varepsilon_0(\mathbf{p})$ is the free-particle energy. Therefore, according to (6.9), the operator $\tilde{\psi}(\mathbf{r}, t)$ equals

$$\begin{aligned} \tilde{\psi}(\mathbf{r}, t) &= \frac{1}{\sqrt{V}} \sum_{\mathbf{p}} e^{i \sum_{\mathbf{p}'} \varepsilon_0(\mathbf{p}') n_{\mathbf{p}'} t} a_{\mathbf{p}} e^{-i \sum_{\mathbf{p}''} \varepsilon_0(\mathbf{p}'') n_{\mathbf{p}''} t} e^{i\mathbf{p}\cdot\mathbf{r}} \\ &= \frac{1}{\sqrt{V}} \sum_{\mathbf{p}} a_{\mathbf{p}} e^{i[\mathbf{p}\cdot\mathbf{r} - \varepsilon_0(\mathbf{p})t]} \end{aligned} \tag{6.14}$$

in the Heisenberg representation.

It should be noted that in general the Heisenberg operators $\tilde{\psi}(\mathbf{r}, t)$ do not satisfy the commutation rules (6.2) for the corresponding Schrödinger operators. However, in the case where the operators $\tilde{\psi}(\mathbf{r}, t)$ are taken at the same instant of time, it follows from (6.9) and (6.2) that their commutation rules are the same as the commutation rules for the Schrödinger operators $\psi(\mathbf{r})$.

In addition to the two representations just considered, there is still another, of great importance for our subsequent work. This is a representation of an intermediate type, called the *interaction representation*, whose properties underlie the methods of quantum field theory. First we isolate from the Hamiltonian the part H_{int} corresponding to interaction between particles:

$$H = H_0 + H_{\text{int}}. \tag{6.15}$$

Then we carry out the following transformation of the Schrödinger wave function of the system:

$$\Phi_i = e^{iH_0t}\Phi. \tag{6.16}$$

Differentiating the function Φ_i with respect to time, we obtain

$$i\frac{\partial\Phi_i}{\partial t} = -H_0\Phi_i + e^{iH_0t}(H_0 + H_{\text{int}})\Phi = e^{iH_0t}H_{\text{int}}e^{-iH_0t}\Phi_i, \qquad (6.17)$$

and hence

$$i\frac{\partial\Phi_i}{\partial t} = \mathbf{H}_{\text{int}}(t)\Phi_i,$$
$$\mathbf{H}_{\text{int}}(t) = e^{iH_0t}H_{\text{int}}e^{-iH_0t}. \qquad (6.18)$$

The functions Φ_i determine the interaction representation. Every operator $\mathbf{F}(t)$ in this representation is obtained from the corresponding Schrödinger operator by using the same formula (6.18) as in the case of $\mathbf{H}_{\text{int}}(t)$, and hence $\mathbf{F}(t)$ satisfies the equation

$$\frac{\partial\mathbf{F}(t)}{\partial t} = i[H_0, \mathbf{F}(t)], \qquad (6.19)$$

i.e., the same equation as a Heisenberg operator for a system of noninteracting particles. Thus, we arrive at the conclusion that in the interaction representation all operators have the same form as the Heisenberg operators for the corresponding noninteracting system, while the wave function satisfies a Schrödinger equation with the Hamiltonian $\mathbf{H}_{\text{int}}(t)$. The possibility of using "free" operators constitutes the great advantage of this representation.

We now determine how the function $\Phi_i(t)$ depends on time in the interaction representation. In doing so, since the operators $\mathbf{H}_{\text{int}}(t)$ do not commute at different times, we cannot simply write down the solution of equation (6.17) in the form

$$\Phi_i(t) = \text{const}\cdot\exp\left\{-i\int^t \mathbf{H}_{\text{int}}(t')\,dt'\right\}.$$

Therefore, we argue as follows: Suppose we know the value of Φ_i at some time t_0. Then we go from the differential equation (6.17) to an integral equation by integrating both sides of (6.17) with respect to time between the limits t_0 and t ($t > t_0$). This gives

$$\Phi_i(t) = \Phi_i(t_0) - i\int_{t_0}^t \mathbf{H}_{\text{int}}(t')\Phi_i(t')\,dt'.$$

We write the solution of this equation as a series

$$\Phi_i(t) = \Phi_i^{(0)}(t) + \Phi_i^{(1)}(t) + \cdots$$

in the quantity $\mathbf{H}_{\text{int}}$, where

$$\Phi_i^{(0)}(t) = \Phi_i(t_0)$$

is the zeroth approximation. The first approximation is

$$\Phi_i^{(1)}(t) = -i\int_{t_0}^t \mathbf{H}_{\text{int}}(t_1)\,dt_1\Phi_i(t_0),$$

the second approximation is

$$\Phi_i^{(2)}(t) = -\int_{t_0}^t \mathbf{H}_{\text{int}}(t_1)\,dt_1\int_{t_0}^{t_1}\mathbf{H}_{\text{int}}(t_2)\,dt_2\Phi_i(t_0),$$

etc., and the nth approximation is

$$\Phi_i^{(n)}(t) = (-i)^n \int_{t_0}^{t} \mathbf{H}_{\text{int}}(t_1)\, dt_1 \int_{t_0}^{t_1} \mathbf{H}_{\text{int}}(t_2)\, dt_2 \cdots \int_{t_0}^{t_{n-1}} \mathbf{H}_{\text{int}}(t_n)\, dt_n \Phi_i(t_0).$$

It follows from the structure of the series for $\Phi_i(t)$ that the whole result can be represented in the form

$$\Phi_i(t) = S(t, t_0)\Phi_i(t_0), \tag{6.20}$$

where the matrix $S(t, t_0)$ is given by the series

$$S(t, t_0) = 1 - i\int_{t_0}^{t} \mathbf{H}_{\text{int}}(t_1)\, dt_1 + \cdots$$
$$+ (-i)^n \int_{t_0}^{t} \mathbf{H}_{\text{int}}(t_1)\, dt_1 \cdots \int_{t_0}^{t_{n-1}} \mathbf{H}_{\text{int}}(t_n)\, dt_n + \cdots \tag{6.21}$$

It is characteristic of the series (6.21) that the operators $\mathbf{H}_{\text{int}}$ taken at later times always appear to the left of the operators taken at earlier times, since the inequality

$$t > t_1 > t_2 > \cdots > t_n > t_0$$

always holds.

The expression (6.21) can be made more symmetric. Consider the nth term of the series

$$(-i)^n \int_{t > t_1 > \cdots > t_0} \mathbf{H}_{\text{int}}(t_1)\mathbf{H}_{\text{int}}(t_2) \cdots \mathbf{H}_{\text{int}}(t_n)\, dt_1\, dt_2 \cdots dt_n.$$

Obviously, this expression does not change if we subject the variables of integration to any permutation

$$t_1, t_2, \ldots, t_n \to t_{p_1}, t_{p_2}, \ldots, t_{p_n}.$$

Carrying out all such permutations of the variables $t_1, \ldots, t_n$, adding the resulting expressions and dividing by $n!$, the total number of permutations, we can extend the range of integration of each variable from t_0 to t_1. However, in doing this, it is important that the operators $\mathbf{H}_{\text{int}}$ in the integral should always be arranged from left to right in order of decreasing times. Letting T denote the operation which accomplishes this ordering, we can write the expression for the nth term of the series (6.21) in the form

$$S^{(n)}(t, t_0) = \frac{(-i)^n}{n!} \int_{t_0}^{t} \cdots \int_{t_0}^{t} T(\mathbf{H}_{\text{int}}(t_1) \cdots \mathbf{H}_{\text{int}}(t_n))\, dt_1 \cdots dt_n.$$

It is now easy to see that (6.21) can be written as

$$S(t, t_0) = T \exp\left\{-i \int_{t_0}^{t} \mathbf{H}_{\text{int}}(t')\, dt'\right\}. \tag{6.22}$$

In fact, we need just expand the exponential in (6.23) in series and use the definition of the "ordering operation" T. Obviously, the operator $S(t, t_0)$ has the property

$$S(t_2, t_1)S(t_1, t_0) = S(t_2, t_0), \qquad t_2 > t_1 > t_0. \tag{6.23}$$

Formulas (6.16) and (6.18) establish the connection between the Schrödinger representation and the interaction representation. Moreover, formula (6.20) allows us to find the connection between the interaction representation and the Heisenberg representation. Suppose the wave function transforms in the way described by the formula

$$\Phi_i(t) = Q(t)\Phi_H, \tag{6.24}$$

where $Q(t)$ is a unitary operator. Then (6.20) implies that

$$Q(t) = S(t, t_0)Q(t_0),$$

from which, according to (6.24), it follows that

$$Q(t) = S(t, \alpha)P,$$

where α is some instant of time, and P is a time-independent operator. To determine P, we substitute into (6.24) the formulas (6.16) and (6.7), which express Φ_i and Φ_H in terms of the Schrödinger function Φ. This gives

$$e^{iH_0t} = S(t, \alpha)Pe^{iHt}.$$

Bearing in mind that $S(\alpha, \alpha) = 1$, we obtain

$$P = e^{iH_0\alpha}e^{-iH\alpha}.$$

At this stage it is convenient to introduce the assumption of the so-called "adiabatically turned-on interaction."[3] Suppose that at the time $t = -\infty$ there is no interaction between particles, and that afterwards the interaction is "turned on" infinitely slowly. Then $P \to 1$ as $\alpha \to -\infty$, and hence

$$\Phi_i(t) = S(t)\Phi_H, \tag{6.25}$$

where

$$S(t) = S(t, -\infty). \tag{6.26}$$

Using the property (6.23), we find that

$$S(t_2, t_1) = S(t_2)S^{-1}(t_1). \tag{6.27}$$

According to (6.25), the relation between operators in the interaction representation and Heisenberg operators has the form

$$\tilde{F}(t) = S^{-1}(t)\mathbf{F}(t)S(t). \tag{6.28}$$

Later on, we shall often encounter chronological products of several Heisenberg operators, averaged over the ground state of the system Φ_H^0:

$$\langle \Phi_H^{0*}T(\tilde{A}(t)\tilde{B}(t')\tilde{C}(t'')\cdots)\Phi_H^0\rangle. \tag{6.29}$$

In this connection, we shall generalize somewhat the definition of T-ordering for Fermi operators, compared to the definition given in deriving formula (6.23), but for Bose operators we shall retain the old definition. Thus, by a *T-product* of the Fermi operators $A(t_1)$, $B(t_2)$, $C(t_3), \ldots$, we now mean the product of the operators written from left to right, multiplied by the factor

---

[3] It should be pointed out at once that the "adiabatically turned-on interaction" used here is a purely formal device, which allows us to obtain the correct result in the quickest way, but is hardly necessary (see, e.g., G4).

$(-1)^P$, where P is the number of permutations needed to obtain the chronological product from $A(t_1)B(t_2)C(t_3)\cdots$. For example, if $F_1(t_1)$ and $F_2(t_2)$ are Fermi operators, while $B_1(t_3)$ and $B_2(t_4)$ are Bose operators, we have

$$T(F_1(t_1)F_2(t_2)) = \begin{cases} F_1(t_1)F_2(t_2) & \text{for} \quad t_1 > t_2, \\ -F_2(t_2)F_1(t_1) & \text{for} \quad t_1 < t_2, \end{cases}$$
$$T(B_1(t_3)F_1(t_1)) = \begin{cases} B_1(t_3)F_1(t_1) & \text{for} \quad t_3 > t_1, \\ F_1(t_1)B_1(t_3) & \text{for} \quad t_3 < t_1, \end{cases}$$
$$T(B_1(t_3)B_2(t_4)) = \begin{cases} B_1(t_3)B_2(t_4) & \text{for} \quad t_3 > t_4, \\ B_2(t_4)B_1(t_3) & \text{for} \quad t_3 < t_4. \end{cases}$$

Applied to the operators $\mathbf{H}_{\text{int}}(t)$, the new definition of T-ordering coincides with the old definition, since Fermi operators always occur in pairs in $\mathbf{H}_{\text{int}}$. Of course, all prescriptions involving the operation of T-ordering are the same both for operators in the Heisenberg representation and operators in the interaction representation.

Suppose the arrangement of times in (6.29) is such that

$$t > t' > t'' > \cdots$$

Using (6.28), we go over to operators in the interaction representation, obtaining

$$\begin{aligned} &\langle \Phi_H^{0*} S^{-1}(t)\mathbf{A}(t)S(t)S^{-1}(t')\mathbf{B}(t')S(t')\cdots\Phi_H^0\rangle \\ &\qquad = \langle \Phi_H^{0*} S^{-1}(\infty)S(\infty, t)\mathbf{A}(t)S(t, t')\mathbf{B}(t')\cdots\Phi_H^0\rangle \\ &\qquad = \langle \Phi_H^{0*} S^{-1}(\infty)T(\mathbf{A}(t)\mathbf{B}(t')\mathbf{C}(t'')\cdots S(\infty))\Phi_H^0\rangle. \end{aligned} \tag{6.30}$$

It is clear that the transformation of (6.29) into the form (6.30) is independent of the order of the times $t, t', t'', \ldots$, i.e., that it is valid in any case.

We still have to determine the quantity

$$\Phi_H^{0*} S^{-1}(\infty) = [S(\infty)\Phi_H^0]^*,$$

which involves subjecting the ground-state function to the operator $S(\infty)$. It follows from (6.20) and (6.25) that

$$\Phi_H^0 = \Phi_i(-\infty), \qquad S(\infty)\Phi_H^0 = \Phi\,(\infty).$$

Thus, $S(\infty)\Phi_H^0$ is the function $\Phi_i(\infty)$ obtained from the ground-state function $\Phi_i(-\infty)$ by adiabatically turning on the interaction between the particles. As is well known, the ground state of a system, i.e., the state in which the energy is a minimum, is necessarily nondegenerate. But, according to the general principles of quantum mechanics (see L7, p. 146), a system in a nondegenerate stationary state cannot make a transition into another state under the action of an infinitely slow perturbation. Therefore, we conclude that the function $\Phi_i(\infty) = S(\infty)\Phi_H^0$ can differ from Φ_H^0 only by a phase factor

$$S(\infty)\Phi_H^0 = e^{iL}\Phi_H^0. \tag{6.31}$$

This implies the final result

$$\langle \Phi_H^{0*} T(\tilde{A}(t)\tilde{B}(t')\tilde{C}(t'')\cdots)\Phi_H^0\rangle = \frac{\langle \Phi_H^{0*} T(\mathbf{A}(t)\mathbf{B}(t')\mathbf{C}(t'')\cdots S(\infty))\Phi_H^0\rangle}{\langle \Phi_H^{0*} S(\infty)\Phi_H^0\rangle}. \tag{6.32}$$

We emphasize that this conclusion is valid only for averages over the ground state of the system, since any other energy level of the system is multiply degenerate and in general can make a transition to another state as a result of collisions between particles. Thus, in averaging over an excited state, formula (6.30) is valid, but not (6.32).

In this chapter, we shall consider systems at absolute zero, i.e., systems in the ground state. For simplicity, we shall denote the corresponding averages by just $\langle\cdots\rangle$, and we shall no longer use boldface letters to denote operators in the interaction representation. Cases where Schrödinger operators are needed will be pointed out explicitly, and the fact that such operators [e.g., $\psi(\mathbf{r})$] depend only on the coordinates will be emphasized.

7. The Green's Function[4]

7.1. Definition. Free-particle Green's functions. In the method of quantum field theory, the *one-particle Green's function* is one of the most important quantities characterizing the microscopic properties of a system.[5] This function is defined by

$$G_{\alpha\beta}(x, x') = -i\langle T(\tilde{\psi}_\alpha(x)\tilde{\psi}^+_\beta(x'))\rangle, \tag{7.1}$$

where x or x' denotes a set of four variables (the coordinates $\mathbf{r}$ and the time t), while α and β denote spin indices.

From a knowledge of the Green's function, we can calculate the average value of one-particle operators of the type (3.2) over the ground state. In fact, according to formula (6.3), we have

$$\overline{F^{(1)}} = \pm i \int \Big[\lim_{\substack{\mathbf{r}'\to\mathbf{r} \\ t'\to t+0}} f^{(1)}_{\alpha\beta}(x) G_{\alpha\beta}(x, x') \Big] \, d\mathbf{r}$$

where the plus sign applies to Bose statistics, and the minus sign to Fermi statistics. For example, the particle (number) density and the particle current density are equal to

$$n(x) = \pm i \lim_{\substack{\mathbf{r}'\to\mathbf{r} \\ t'\to t+0}} G_{\alpha\alpha}(x, x'),$$

$$j(x) = \pm \frac{1}{m} \lim_{\substack{\mathbf{r}'\to\mathbf{r} \\ t'\to t+0}} (\nabla_{\mathbf{r}} - \nabla_{\mathbf{r}'}) G_{\alpha\alpha}(x, x'),$$

respectively.

---

[4] This section is to a large extent based on the work of Galitski and Migdal (G2).

[5] In field theory, the term "Green's function" has a different meaning from its meaning in the theory of linear differential equations. Although the Green's function satisfies an equation whose right-hand side contains a δ-function, the equation is in general nonlinear (see Sec. 10). An exception is the case of free-particle Green's functions, which are actually Green's functions of linear equations for $\tilde{\psi}(\mathbf{r}, t)$, the Heisenberg field operators. The term "Green's function," at first used only in this case, has since been extended to apply to the expression (7.1) for any interacting system.

It will be shown later that by using the Green's function we can find the energy as a function of the volume, and hence we can also find the equation of state of the system (i.e., the dependence of pressure on density) at absolute zero. Moreover, it will be shown that the poles of the Fourier transform of the Green's function (7.1) determine the spectrum of the excitations. This enables us to find the thermodynamic functions of the system for temperatures which are nonzero (but sufficiently low, of course).

Another very important fact is that the Green's function can be calculated by using the so-called *diagram technique* (see Secs. 8 and 9), which has great advantages over the ordinary form of perturbation theory.

This section is devoted to an analysis of the general properties of the Green's function. For simplicity, we shall drop the indices α and β. This entails no error, since in the absence of ferromagnetism and of an external magnetic field, $G_{\alpha\beta}$ must be of the form

$$G_{\alpha\beta} = G\delta_{\alpha\beta},$$

and here we consider only this case.

In the present chapter, we consider properties of systems of fermions, since, as is well known, Bose systems at absolute zero have many special features, connected with the presence of the condensate (Bose systems will be studied in Chap. 5). An exception is the case of phonons, i.e., the vibrational quanta of a solid body. Since the number of phonons is not specified, there is no condensate in a phonon gas and its properties can be studied by the usual methods.

In the absence of external fields, the Green's functions of homogeneous and spatially infinite systems depend only on the coordinate differences $\mathbf{r} - \mathbf{r}'$ and the time difference $t - t'$. Suppose we represent the Green's function as a Fourier integral, writing

$$G(x - x') = \int G(\mathbf{p}, \omega) e^{i[\mathbf{p}\cdot(\mathbf{r}-\mathbf{r}') - \omega(t-t')]} \frac{d^4p}{(2\pi)^4} \tag{7.2}$$

$$(d^4p = d\mathbf{p}\, d\omega).$$

Then the function $G(\mathbf{p}, \omega)$ can be found very easily for a system of non-interacting particles. In the case of a system of fermions, substituting the expression (6.14) for the free-field Heisenberg operators into (7.1), and bearing in mind that all levels with $|\mathbf{p}| < p_0$ are occupied, while those with $|\mathbf{p}| > p_0$ are empty, we obtain

$$G^{(0)}(x) = -\frac{i}{V}\sum_{\mathbf{p}} e^{i[\mathbf{p}\cdot\mathbf{r} - \varepsilon_0(\mathbf{p})t]} \begin{cases} 1 - n_{\mathbf{p}} & \text{for} \quad t > 0, \\ \quad - n_{\mathbf{p}} & \text{for} \quad t < 0, \end{cases} \tag{7.3}$$

where

$$n_{\mathbf{p}} = a_{\mathbf{p}}^{+} a_{\mathbf{p}} = \begin{cases} 1 & \text{for} \quad |\mathbf{p}| < p_0, \\ 0 & \text{for} \quad |\mathbf{p}| > p_0. \end{cases}$$

Going over to the momentum representation, we have, according to (7.3),

$$G^{(0)}(\mathbf{p}, \omega) = -i\Big\{\theta(|\mathbf{p}| - p_0)\int_0^\infty e^{i[\omega - \varepsilon_0(\mathbf{p})]t}\,dt - \theta(p_0 - |\mathbf{p}|)\int_0^\infty e^{-i[\omega - \varepsilon_0(\mathbf{p})]t}\,dt\Big\}, \tag{7.4}$$

where

$$\theta(z) = \begin{cases} 1 & \text{for} \quad z > 0, \\ 0 & \text{for} \quad z < 0. \end{cases}$$

The expression for $G^{(0)}(\mathbf{p}, \omega)$ contains two integrals of the type

$$\int_0^\infty e^{ist}\,dt.$$

We define an integral of this type as the limit

$$\lim_{\delta\to+0}\int_0^\infty e^{ist-\delta t}\,dt = i\lim_{\delta\to+0}\frac{1}{s+i\delta}. \tag{7.5}$$

The quantity $i\delta$ in the denominator characterizes the way we bypass the pole $s = 0$ in evaluating integrals containing the function $1/(s + i\delta)$, and in fact

$$\int F(s)\frac{ds}{s+i\delta} = \mathrm{P}\int\frac{F(s)}{s}\,ds - i\pi F(0),$$

where P denotes the operation of taking the principal value of an integral. Thus, we can formally write

$$\frac{1}{s+i\delta} = \frac{\mathrm{P}}{s} - i\pi\delta(s).$$

Sometimes the notation $\delta_+(s)$ is used to denote the quantity

$$\frac{1}{\pi}\frac{i}{s+i\delta},$$

i.e.,

$$\delta_+(s) = \delta(s) - \frac{\mathrm{P}}{i\pi s}. \tag{7.6}$$

Using (7.4) and (7.5), we obtain

$$G^{(0)}(\mathbf{p}, \omega) = \frac{\theta(|\mathbf{p}| - p_0)}{\omega - \varepsilon_0(\mathbf{p}) + i\delta} + \frac{\theta(p_0 - |\mathbf{p}|)}{\omega - \varepsilon_0(\mathbf{p}) - i\delta}.$$

Noting that the only difference between the formula for $G^{(0)}$ when $|\mathbf{p}| < p_0$ and the formula when $|\mathbf{p}| > p_0$ consists in a change of the sign of δ, we can finally write

$$G^{(0)}(\mathbf{p}, \omega) = \frac{1}{\omega - \varepsilon_0(\mathbf{p}) + i\delta\,\mathrm{sgn}\,(|\mathbf{p}| - p_0)}. \tag{7.7}$$

Next, we consider a system of phonons, confining ourselves to the simplest case, i.e., longitudinal vibrations in an isotropic continuous medium. First of all, we define what we mean by the operators of the phonon field.

Let the displacement of the points of the medium be denoted by $\mathbf{q}(\mathbf{r}, t)$. Then the momentum of a unit volume equals $\rho\dot{\mathbf{q}}(\mathbf{r}, t)$, where ρ is the density. According to quantum mechanics, the quantities $\mathbf{q}$ and $\dot{\mathbf{q}}$ are operators with the commutation rules

$$\rho[\dot{q}_i(\mathbf{r}, t), q_j(\mathbf{r}', t)] = -i\delta(\mathbf{r} - \mathbf{r}')\delta_{ij}. \tag{7.8}$$

The integral of (7.8) over a small volume $d\mathbf{r}$ gives the usual commutation rule involving the coordinate and momentum.

We now expand $\mathbf{q}$ in plane waves. In the case under consideration, the value of the wave vector $\mathbf{k}$ uniquely determines the frequency, which we denote by $\omega_0(\mathbf{k})$. Thus, we have

$$\mathbf{q}(\mathbf{r}, t) = \frac{1}{\sqrt{V}} \sum_{\mathbf{k}} \frac{\mathbf{k}}{|\mathbf{k}|} \{q_{\mathbf{k}} e^{i[\mathbf{k}\cdot\mathbf{r} - \omega_0(\mathbf{k})t]} + q_{\mathbf{k}}^+ e^{-i[\mathbf{k}\cdot\mathbf{r} - \omega_0(\mathbf{k})t]}\}. \tag{7.9}$$

Since we are considering longitudinal waves, the Fourier components of the vector $\mathbf{q}$ are directed along the wave vector $\mathbf{k}$. Therefore, in (7.9) we have written $q_{\mathbf{k}}\mathbf{k}/|\mathbf{k}|$ instead of $\mathbf{q}_{\mathbf{k}}$.

Our next step is to introduce the operators $b_{\mathbf{k}}$, related to $q_{\mathbf{k}}$ by the formula

$$q_{\mathbf{k}} = \frac{b_{\mathbf{k}}}{\sqrt{2\rho\omega_0(\mathbf{k})}}. \tag{7.10}$$

Then it follows from (7.8) that the operators $b_{\mathbf{k}}$ satisfy the usual commutation relations for Bose annihilation and creation operators. The kinetic-energy operator equals

$$K = \frac{\rho}{2} \int \dot{\mathbf{q}}[(\mathbf{r}, t)]^2 \, d\mathbf{r}. \tag{7.11}$$

Using the fact that the average kinetic energy of the vibrations equals the average potential energy, we obtain the formula

$$\bar{H} = 2\bar{K} = \sum_{\mathbf{k}} \omega_0(\mathbf{k}) \left(n_{\mathbf{k}} + \frac{1}{2}\right), \tag{7.12}$$

where $n_{\mathbf{k}} = b_{\mathbf{k}}^+ b_{\mathbf{k}}$.

We might choose the displacement operators $\mathbf{q}$ as the operators of the free field. However, it is more convenient to define these operators somewhat differently, anticipating our study of the interaction between phonons and electrons in a metal (see Sec. 9.1). Thus, we write

$$\tilde{\varphi}(x) = \frac{i}{\sqrt{V}} \sum_{\mathbf{k}} \sqrt{\frac{\omega_0(\mathbf{k})}{2}} \{b_{\mathbf{k}} e^{i[\mathbf{k}\cdot\mathbf{r} - \omega_0(\mathbf{k})t]} - b_{\mathbf{k}}^+ e^{-i[\mathbf{k}\cdot\mathbf{r} - \omega_0(\mathbf{k})t]}\}. \tag{7.13}$$

This formula applies to longitudinal phonons in the Debye model (see Sec. 1), if the summation over $\mathbf{k}$ is restricted by the condition $|\mathbf{k}| < k_D$. We again emphasize that the operators of the phonon field are real, since they correspond to real displacements of the atoms of the lattice. This property is obviously preserved if we take into account the interaction of phonons with themselves and with other particles.

The Green's function of the phonons is usually denoted by the letter D. The definition of this function is analogous to formula (7.1), i.e.,

$$D(x, x') = -i\langle T(\tilde{\varphi}(x)\tilde{\varphi}(x'))\rangle. \tag{7.14}$$

Substituting the free-field operators (7.13) into (7.14), and bearing in mind that there are no phonons in the ground state, we obtain

$$D^{(0)}(x) = -\frac{i}{V}\sum_{\mathbf{k}}\frac{\omega_0(\mathbf{k})}{2}\begin{cases} e^{i[\mathbf{k}\cdot\mathbf{r}-\omega_0(\mathbf{k})t]} & \text{for} \quad t > 0, \\ e^{-i[\mathbf{k}\cdot\mathbf{r}-\omega_0(\mathbf{k})t]} & \text{for} \quad t < 0. \end{cases} \tag{7.15}$$

The Fourier components corresponding to (7.15) are

$$\begin{aligned} D^{(0)}(\mathbf{k}, \omega) &= \frac{\omega_0(\mathbf{k})}{2}\left[\frac{1}{\omega - \omega_0(\mathbf{k}) + i\delta} - \frac{1}{\omega + \omega_0(\mathbf{k}) - i\delta}\right] \\ &= \frac{\omega_0^2(\mathbf{k})}{\omega^2 - \omega_0^2(\mathbf{k}) + i\delta}. \end{aligned} \tag{7.16}$$

7.2. Analytic properties. We now consider the basic properties of the Green's function of a system of interacting particles. We begin with the case of Fermi systems. Going over to Schrödinger operators, we obtain

$$\begin{aligned} G(\mathbf{r} - \mathbf{r}', t - t') &= -i\langle e^{iHt}\psi(\mathbf{r})e^{-iH(t-t')}\psi^+(\mathbf{r}')e^{-iHt'}\rangle \\ &= -i\sum_s \langle \Phi_H^{0*} e^{iHt}\psi(\mathbf{r})e^{-iHt}\Phi_s\rangle\langle\Phi_s^* e^{iHt'}\psi(\mathbf{r}')e^{-iHt'}\Phi_H^0\rangle \\ &= -i\sum_s \psi_{0s}(\mathbf{r})\psi_{s0}^+(\mathbf{r}')e^{-i(E_s - E_0)(t-t')} \quad \text{for} \quad t > t', \end{aligned}$$

$$G(\mathbf{r} - \mathbf{r}', t - t') = i\sum_{s'} \psi_{0s'}^+(\mathbf{r}')\psi_{s'0}(\mathbf{r})e^{i(E_{s'} - E_0)(t-t')} \quad \text{for} \quad t < t'.$$

For a homogeneous system, the coordinate dependence of the matrix elements $\psi_{nm}(\mathbf{r})$ and $\psi_{nm}^+(\mathbf{r})$ is of the form

$$\psi_{nm}(\mathbf{r}) = \psi_{nm}(0)e^{-i\mathbf{p}_{nm}\cdot\mathbf{r}}, \qquad \psi_{nm}^+(\mathbf{r}) = \psi_{nm}^+(0)e^{-i\mathbf{p}_{nm}\cdot\mathbf{r}},$$

where $\mathbf{p}_{nm} = \mathbf{p}_n - \mathbf{p}_m$, $\mathbf{p}_n$ is the momentum of the system in the state n, and $\mathbf{p}_m$ is its momentum in the state m.[6] Setting $\mathbf{p}_0 = 0$, we have

$$\begin{aligned} G(\mathbf{r} - \mathbf{r}', t - t') &= -i\sum_s |\psi_{0s}(0)|^2 e^{i\mathbf{p}_s\cdot(\mathbf{r}-\mathbf{r}')}e^{-i(E_s - E_0)(t-t')} \quad \text{for} \quad t > t' \\ G(\mathbf{r} - \mathbf{r}', t - t') &= i\sum_{s'} |\psi_{s'0}(0)|^2 e^{-i\mathbf{p}_{s'}\cdot(\mathbf{r}-\mathbf{r}')}e^{i(E_{s'} - E_0)(t-t')} \quad \text{for} \quad t < t'. \end{aligned} \tag{7.17}$$

---

[6] This follows from the fact that according to quantum mechanics (see L7, Sec. 13), the spatial-displacement operator equals $e^{i\hat{\mathbf{p}}\cdot\mathbf{r}}$, where $\hat{\mathbf{p}}$ is the momentum operator and therefore

$$\psi(\mathbf{r}) = e^{-i\hat{\mathbf{p}}\cdot\mathbf{r}}\psi(0)e^{i\hat{\mathbf{p}}\cdot\mathbf{r}}.$$

We note in passing that if $\psi(\mathbf{r})$ is written in the form

$$\psi(\mathbf{r}) = \frac{1}{\sqrt{V}}\sum_{\mathbf{p}} a_{\mathbf{p}}e^{i\hat{\mathbf{p}}\cdot\mathbf{r}},$$

then obviously

$$\psi_{nm}(0) = \frac{1}{\sqrt{V}}(a_{-\mathbf{p}_{nm}})_{nm}.$$

The operator $\psi^+(\mathbf{r})$ increases the number of particles by unity, and hence for $t > t'$ the summation with respect to s is over states with particle numbers equal to $N + 1$. On the other hand, for $t < t'$ the summation with respect to s' is over states with particle numbers equal to $N - 1$.

We now introduce the notation

$$E_s - E_0(N) = \varepsilon_s + \mu, \tag{7.18}$$

where

$$\varepsilon_s = E_s - E_0(N + 1) \tag{7.19}$$

is the excitation energy of the system (which by definition is positive), and where

$$\mu = E_0(N + 1) - E_0(N)$$

is the chemical potential for $T = 0$. Similarly, we write

$$\begin{aligned} E_{s'} - E_0(N) &= E_{s'} - E_0(N - 1) - [E_0(N) - E_0(N - 1)] \\ &= \varepsilon_{s'} - \mu', \end{aligned} \tag{7.18'}$$

where in the last formula the quantities $\varepsilon_{s'}$ and μ' pertain to a system of $N - 1$ particles. However, we can assume that

$$\varepsilon_s = \varepsilon_{s'}, \qquad \mu = \mu',$$

which only introduces an error of order $1/N$. We also introduce the functions

$$\begin{aligned} A(\mathbf{p}, E)\, dE &= (2\pi)^3 \sum_s |\psi_{0s}(0)|^2 \delta(\mathbf{p} - \mathbf{p}_s) \quad \text{for} \quad E < \varepsilon_s < E + dE, \\ B(\mathbf{p}, E)\, dE &= (2\pi)^3 \sum_{s'} |\psi_{s'0}(0)|^2 \delta(\mathbf{p} + \mathbf{p}_s) \quad \text{for} \quad E < \varepsilon_{s'} < E + dE. \end{aligned} \tag{7.20}$$

Next, we represent the function G as a Fourier integral:[7]

$$G(\mathbf{p}, \omega) = \int_0^\infty \left[\frac{A(\mathbf{p}, E)}{\omega - E - \mu + i\delta} + \frac{B(\mathbf{p}, E)}{\omega + E - \mu - i\delta}\right] dE. \tag{7.21}$$

The coefficients A and B in this formula are real and positive. By using the representation (7.21), we can investigate the analytic properties of the function $G(\mathbf{p}, \omega)$. Separating the real and imaginary parts of the function G, we obtain

$$\operatorname{Re} G(\mathbf{p}, \omega) = \mathrm{P}\int_0^\infty \left[\frac{A(\mathbf{p}, E)}{\omega - E - \mu} + \frac{B(\mathbf{p}, E)}{\omega + E - \mu}\right] dE, \tag{7.22}$$

$$\operatorname{Im} G(\mathbf{p}, \omega) = \begin{cases} -\pi A(\mathbf{p}, \omega - \mu) & \text{for} \quad \omega > \mu, \\ \pi B(\mathbf{p}, \mu - \omega) & \text{for} \quad \omega < \mu. \end{cases} \tag{7.23}$$

Thus, the imaginary part of the Green's function changes sign for $\omega = \mu$. Comparing (7.23) and (7.22), we find the following relation between the

[7] Formulas of this type were first obtained by Lehmann (L11) in a paper on quantum field theory.

real and imaginary parts of $G(\mathbf{p}, \omega)$:

$$\operatorname{Re} G(\mathbf{p}, \omega) = \frac{P}{\pi} \int_{-\infty}^{\infty} \frac{\operatorname{Im} G(\mathbf{p}, \omega') \operatorname{sgn}(\omega' - \mu)}{\omega' - \omega} d\omega'. \tag{7.24}$$

From formulas (7.21') and (7.20) we can obtain an asymptotic formula for G as $\omega \to \infty$:

$$G(\mathbf{p}, \omega) \to \frac{1}{\omega} \int_0^{\infty} [A(\mathbf{p}, E) + B(\mathbf{p}, E)]dE$$

$$= \frac{1}{\omega}\left[(2\pi)^3 \sum_s |\psi_{0s}(0)|^2 \delta(\mathbf{p} - \mathbf{p}_s) + (2\pi)^3 \sum_{s'} |\psi_{s'0}(0)|^2 \delta(\mathbf{p} + \mathbf{p}_{s'})\right].$$

It is not hard to see that the coefficient of $1/\omega$ equals the Fourier component of the anticommutator

$$\{\psi(\mathbf{r}), \psi^+(\mathbf{r}')\} = \psi(\mathbf{r})\psi^+(\mathbf{r}') + \psi^+(\mathbf{r}')\psi(\mathbf{r}) = \delta(\mathbf{r} - \mathbf{r}'),$$

i.e., equals 1. To verify this, it is sufficient to average the anticommutator over the ground state (which does not affect its value), transform the resulting average as in (7.17), and then take the Fourier transform with respect to $\mathbf{r} - \mathbf{r}'$. In this way, we obtain

$$G(\mathbf{p}, \omega) \to \frac{1}{\omega} \quad \text{as} \quad \omega \to \infty. \tag{7.21'}$$

We now examine the properties of G as a function of the complex variable ω. It follows from (7.24) that the function $G(\mathbf{p}, \omega)$ is not analytic. In fact, the relation between the real and imaginary parts which differs from (7.24) by having 1 instead of $\operatorname{sgn}(\omega' - \mu)$ corresponds to a function which is analytic in the upper half-plane, and the relation which has -1 instead of $\operatorname{sgn}(\omega' - \mu)$ corresponds to a function which is analytic in the lower half-plane. Together with G, we consider two functions G_R and G_A, which are analytic in the upper and lower half-planes, respectively, and are defined by the relations

$$\begin{aligned} \operatorname{Re} G &= \operatorname{Re} G_R = \operatorname{Re} G_A, \\ \operatorname{Im} G_R &= \operatorname{Im} G \operatorname{sgn}(\omega - \mu), \\ \operatorname{Im} G_A &= -\operatorname{Im} G \operatorname{sgn}(\omega - \mu). \end{aligned} \tag{7.25}$$

(for real ω).[8] The formulas (7.25) show that G_R coincides with G^* on the real half-line $\omega - \mu < 0$, while G_A coincides with G^* for $\omega - \mu > 0$. Thus, we can write

$$\begin{aligned} G_R(\mathbf{p}, \omega) &= \begin{cases} G(\mathbf{p}, \omega) & \text{for} \quad \omega > \mu, \\ G^*(\mathbf{p}, \omega) & \text{for} \quad \omega < \mu, \end{cases} \\ G_A(\mathbf{p}, \omega) &= \begin{cases} G^*(\mathbf{p}, \omega) & \text{for} \quad \omega > \mu, \\ G(\mathbf{p}, \omega) & \text{for} \quad \omega < \mu. \end{cases} \end{aligned} \tag{7.25'}$$

[8] Whether the indices R and A, indicating retarded and advanced quantities, respectively, are chosen to be subscripts, as here, or superscripts, as in Secs. 17 and 28, is purely a matter of typographic convenience.

It follows from (7.25′) that G_R is the analytic continuation of G from the half-line $\omega > \mu$, and G_A is the analytic continuation of G from the half-line $\omega < \mu$.

In the coordinate representation, the functions G_R and G_A are defined as follows:

$$G_R(x - x') = \begin{cases} -i\langle\tilde{\psi}(x)\tilde{\psi}^+(x') + \tilde{\psi}^+(x')\tilde{\psi}(x)\rangle & \text{for} \quad t > t', \\ 0 & \text{for} \quad t < t', \end{cases}$$
$$G_A(x - x') = \begin{cases} 0 & \text{for} \quad t > t', \\ i\langle\tilde{\psi}^+(x')\tilde{\psi}(x) + \tilde{\psi}(x)\tilde{\psi}^+(x')\rangle & \text{for} \quad t < t'. \end{cases} \tag{7.26}$$

In fact, performing the same operations as used to derive (7.21), we obtain

$$G_R(\mathbf{p}, \omega) = \int_0^\infty \left[\frac{A(\mathbf{p}, E)}{\omega - E - \mu + i\delta} + \frac{B(\mathbf{p}, E)}{\omega + E - \mu + i\delta}\right] dE,$$
$$G_A(\mathbf{p}, \omega) = G_R^*(\mathbf{p}, \omega). \tag{7.27}$$

By comparing the real and imaginary parts of G_R and G_A with formulas (7.22) and (7.23), it is easily seen that these functions satisfy the definitions (7.25). The functions G_R and G_A are called the *retarded* and *advanced* Green's functions, respectively.

Finally, we consider the case of phonons. The operator of the phonon field is real, i.e.,

$$\tilde{\varphi}(x) = \tilde{\chi}(x) + \tilde{\chi}^+(x).$$

Moreover, it should be kept in mind that the chemical potential μ vanishes (see Sec. 1) and that there are no particles in the ground state. An argument like that given above leads to the formula

$$D(\mathbf{r} - \mathbf{r}', t - t') = \begin{cases} -i\sum_s |\chi_{0s}(0)|^2 e^{-i(E_s - E_0)(t-t')} e^{i\mathbf{k}_s\cdot(\mathbf{r}-\mathbf{r}')} & \text{for} \quad t > t', \\ -i\sum_s |\chi_{0s}(0)|^2 e^{i(E_s - E_0)(t-t')} e^{-i\mathbf{k}_s\cdot(\mathbf{r}-\mathbf{r}')} & \text{for} \quad t < t'. \end{cases} \tag{7.28}$$

Introducing the function

$$P(\mathbf{k}, E)\, dE = (2\pi)^3 \sum_s |\chi_{0s}(0)|^2 \delta(\mathbf{k} - \mathbf{k}_s)$$
$$= (2\pi)^3 \sum_s |\chi_{0s}(0)|^2 \delta(\mathbf{k} + \mathbf{k}_s),$$

where the summation with respect to s is over states whose energies E_s satisfy the inequality $E < E_s - E_0 < E + dE$, and expanding (7.28) as a Fourier integral, we obtain

$$D(\mathbf{k}, \omega) = \int_0^\infty P(\mathbf{k}, E) \left[\frac{1}{\omega - E + i\delta} - \frac{1}{\omega + E - i\delta}\right] dE. \tag{7.29}$$

The imaginary part of this function is always negative, i.e.,

$$\text{Im } D(\mathbf{k}, \omega) = -\pi P(\mathbf{k}, |\omega|), \tag{7.30}$$

and the relation between the real and imaginary parts of (7.29) is given by the same formula as for the function $G(\mathbf{p}, \varepsilon)$. It follows that the analytic properties of the phonon Green's function are the same as those of the Green's function for a system of fermions with $\mu = 0$. Therefore, we can construct two analytic functions D_R and D_A satisfying the conditions (7.25) with $\mu = 0$. In the coordinate representation, these functions have the form

$$D_R(x - x') = \begin{cases} -i\langle\tilde{\varphi}(x)\tilde{\varphi}(x') - \tilde{\varphi}(x')\tilde{\varphi}(x)\rangle & \text{for} \quad t > t', \\ 0 & \text{for} \quad t < t', \end{cases}$$
$$D_A(x - x') = \begin{cases} 0 & \text{for} \quad t > t', \\ -i\langle\tilde{\varphi}(x')\tilde{\varphi}(x) - \tilde{\varphi}(x)\tilde{\varphi}(x')\rangle & \text{for} \quad t < t'. \end{cases} \tag{7.31}$$

7.3. The physical meaning of the poles. As already noted, knowledge of the Green's function allows us to find a great many physical characteristics of the system, in particular the spectrum of the elementary excitations. Consider a Fermi system which at the initial time t' is described by the wave function

$$\Psi_0(t') = \psi_{\mathbf{p}}^+(t')\Phi_i(t'), \tag{7.32}$$

where $\psi_{\mathbf{p}}^+(t')$ is the creation operator for a particle with momentum $\mathbf{p}$ in the interaction representation, i.e., $a_{\mathbf{p}}^+ e^{i\varepsilon_0(\mathbf{p})t'}$, while $\Phi_i(t')$ is the wave function of the ground state of the system of particles in the interaction representation. At the time $t > t'$, the wave function of the system has the form

$$\Psi(t) = S(t, t')\psi_{\mathbf{p}}^+(t')\Phi_i(t').$$

The probability amplitude of the state $\Psi_0(t)$ equals

$$\begin{aligned} \langle\Psi_0^*(t)\Psi(t)\rangle &= \langle\Phi_i^*(t)\psi_{\mathbf{p}}(t)S(t, t')\psi_{\mathbf{p}}^+(t')\Phi_i(t')\rangle \\ &= \langle\Phi_H^{0*}S^{-1}(t)\psi_{\mathbf{p}}(t)S(t, t')\psi_{\mathbf{p}}^+(t')S(t')\Phi_H^0\rangle \\ &= \langle\tilde{\psi}_{\mathbf{p}}(t)\tilde{\psi}_{\mathbf{p}}^+(t')\rangle = iG(\mathbf{p}, t - t') \quad \text{for} \quad t - t' > 0, \end{aligned} \tag{7.33}$$

where we have gone over from the interaction representation to the Heisenberg representation.

To obtain the function $G(\mathbf{p}, t)$, we have to evaluate the integral

$$G(\mathbf{p}, t) = \int_{-\infty}^{\infty} G(\mathbf{p}, \omega)e^{-i\omega t}\frac{d\omega}{2\pi}. \tag{7.34}$$

Since the function $G(\mathbf{p}, \omega)$ is not analytic, we divide this integral into two parts, one from $-\infty$ to μ and the other from μ to ∞. In the first integral, G coincides with the analytic function G_A, while in the second integral G coincides with G_R. As already noted above, the function G_A has no singularities in the lower half-plane. Therefore, we can deform the contour of integration in the first integral (see Fig. 2). If the horizontal part of the contour is shifted sufficiently far into the lower half-plane, then, because

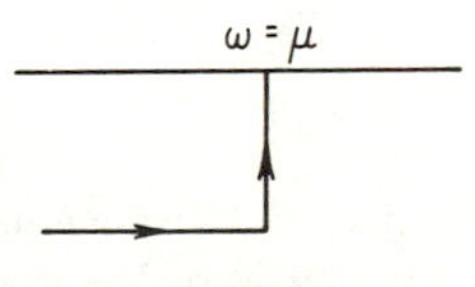

FIGURE 2

of the factor $e^{-i\omega t}$ in (7.34), the integral over this part of the contour will be very small, so that only the integral

$$\int_{\mu - i\infty}^{\mu} G_A e^{-i\omega t} \frac{d\omega}{2\pi}$$

remains.

Next, we consider the second integral. In general, the function G_R has singularities in the lower half-plane. Suppose that in the fourth quadrant of the complex variable $\omega - \mu$, the singularity closest to the real axis is a simple pole at the point $\omega = \varepsilon(\mathbf{p}) - i\gamma$, where $\gamma \ll \varepsilon(\mathbf{p}) - \mu$. We deform the contour of integration in the way shown in Fig. 3. Of course, the horizontal part of this contour cannot be shifted to $-i\infty$. However, by choosing a sufficiently large time t, we can make this integral small. Thus, all that remains is the integral along the vertical part of the contour and along the part encircling the pole, i.e.,

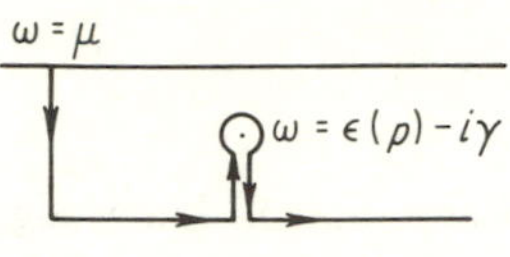

FIGURE 3

$$\int_{\mu}^{\mu - i\infty} G_R e^{-i\omega t} \frac{d\omega}{2\pi} - iae^{-i\varepsilon(\mathbf{p})t - \gamma t},$$

where a is the residue of G_R at the pole. We shall show below that when

$$t \gg [\varepsilon(\mathbf{p}) - \mu]^{-1},$$

the contributions from both integrals along the vertical parts of the contours shown in Figs. 2 and 3 are small. Thus, in the limit of large times t, we have

$$iG(\mathbf{p}, t) \approx ae^{-i\varepsilon(\mathbf{p})t - \gamma t}. \tag{7.35}$$

If in the initial state there were a single free particle with momentum $\mathbf{p}$ and energy $\varepsilon(\mathbf{p})$, then the quantity analogous to (7.33) would be

$$e^{-i\varepsilon_0(\mathbf{p})(t - t')}.$$

It follows that the state (7.32) corresponds to a wave packet which behaves like a quasi-particle with enrgy $\varepsilon(\mathbf{p})$ and decays in time according to the law $e^{-\gamma(t - t')}$. Thus, the energy and attenuation of the quasi-particles is determined by the real and imaginary parts of the pole of the function G_R in the lower half-plane. Moreover, the amplitude of the wave packet is related to the residue of G_R at this pole.

We now show that the omitted parts of the integrals can be regarded as small in the range of momenta such that $\varepsilon(\mathbf{p}) \approx \mu$. Recalling that $G_A = G_R^*$, we find that the sum of the integrals along the vertical parts of the contours shown in Figs. 2 and 3 equals

$$\int_{\mu}^{\mu - i\infty} (G_R - G_R^*) e^{-i\omega t} \frac{d\omega}{2\pi} = 2i \int_{\mu}^{\mu - i\infty} \operatorname{Im} G_R \cdot e^{-i\omega t} \frac{d\omega}{2\pi}.$$

According to the phenomenological considerations presented in Sec. 2, the condition $\gamma \ll \varepsilon(\mathbf{p}) - \mu$ holds only when $\varepsilon(\mathbf{p}) \approx \mu$.[9] Therefore, assuming

[9] The validity of the assumption that $\gamma \ll \varepsilon(\mathbf{p}) - \mu$ as $\varepsilon(\mathbf{p}) - \mu \to 0$ can be proved rigorously in many specific cases (see e.g., Secs. 21 and 22).

that

$$t \gg \frac{1}{\varepsilon(\mathbf{p}) - \mu},$$

we can replace G_R by

$$\frac{a}{\omega - \varepsilon(\mathbf{p}) + i\gamma}.$$

Introducing a new variable $u = i(\omega - \mu)$, we obtain

$$-\frac{2\gamma a e^{-i\mu t}}{2\pi} \int_0^\infty \frac{e^{-ut}\, du}{\gamma^2 + [\varepsilon(\mathbf{p}) - \mu + iu]^2}.$$

Since $t \gg [\varepsilon(\mathbf{p}) - \mu]^{-1}$, this integral turns out to be

$$-\frac{\gamma a e^{-i\mu t}}{\pi t[\varepsilon(\mathbf{p}) - \mu]^2}.$$

If we assume that t is not too large compared to $1/\gamma$, this quantity is much less than the result of going around the pole in Fig. 3.

We can also develop a similar argument for the state with wave function

$$\Psi_0''(t) = \psi_{\mathbf{p}}(t)\Phi_i(t). \tag{7.36}$$

Considering this state at a later time t', we obtain

$$\langle \Psi_0''^*(t')\Psi''(t')\rangle = -iG(\mathbf{p}, t - t') \quad \text{for} \quad t - t' < 0.$$

In calculating $G(\mathbf{p}, t)$ by formula (7.34), it is the poles of the function $G_A(\mathbf{p}, \omega)$ in the upper half-plane that are important, since $t < 0$. When $|t| \gg |\varepsilon(\mathbf{p}) - \mu|^{-1}$, we find in the same way as before that

$$-iG(\mathbf{p}, t) \approx a e^{-i\varepsilon(\mathbf{p})t - \gamma t},$$

where $\varepsilon(\mathbf{p}) < \mu, \gamma < 0$, i.e., we obtain a wave packet corresponding to a hole with $\varepsilon(\mathbf{p}) < \mu$. Therefore, the energy and attenuation of the holes are given by the poles of the function $G_A(\mathbf{p}, \omega)$ in the upper half-plane. We note that the sign of γ for "particles" is the opposite of its sign for "holes."

The same results also apply to phonons. In this case, it is not hard to see from formula (7.29) that each pole of the function $D_R(\mathbf{k}, \omega)$ in the lower half-plane corresponds to a pole of the function $D_A(\mathbf{k}, \omega)$ in the upper half-plane, where the two poles in question are symmetrically located with respect to the point $\omega = 0$. Thus, both ways of determining the spectrum of the excitations lead to the same result.

The Green's function can be used not only to obtain the energy spectrum, but also to find the relation between the chemical potential and the number of particles per unit volume, as well as the ground-state energy and the momentum distribution of the particles. (Of course, with our restrictions, all this applies only to fermions.) It follows from the very definition of the Green's function (7.1) that

$$\begin{aligned}\frac{N}{V} &= \langle \tilde{\psi}_\alpha^+(x)\tilde{\psi}_\alpha(x)\rangle \\ &= -i \lim_{\substack{\mathbf{r}\to\mathbf{r}' \\ t'\to t+0}} G_{\alpha\alpha}(x - x') = -2i \lim_{\substack{\mathbf{r}\to\mathbf{r}' \\ t'\to t+0}} G(x - x'),\end{aligned}$$

where $G_{\alpha\beta} = \delta_{\alpha\beta}G$. Going over to the momentum representation for G, we obtain

$$\frac{N}{V} = -2i \lim_{t\to+0} \int G(\mathbf{p}, \omega)e^{i\omega t} \frac{d\mathbf{p}\,d\omega}{(2\pi)^4}. \tag{7.37}$$

Since the integral in (7.37) depends only on μ, this gives the dependence of N on μ. Then, solving for the function $\mu(N)$ and using the formula

$$\mu = \left(\frac{\partial E_0}{\partial N}\right)_V,$$

we can find the ground-state energy. Admittedly, this way of doing things is not very convenient in practice. We shall return to the problem of determining the ground-state energy in Sec. 9.

To find the momentum distribution of the particles, it is sufficient to calculate the expression

$$N_{1/2}(\mathbf{p}) = N_{-1/2}(\mathbf{p}) = \langle\Phi_0^* a^+_{\mathbf{p},1/2} a_{\mathbf{p},1/2}\Phi_0\rangle = \langle a^+_{\mathbf{p},1/2} a_{\mathbf{p},1/2}\rangle,$$

where $\Phi_0 = e^{-iE_0t}\Phi_H^0$ is the Schrödinger wave function of the ground state of the system. Comparing this expression with (7.17) [see also footnote 6, p. 55], we find that

$$N_{1/2}(\mathbf{p}) = N_{-1/2}(\mathbf{p}) = -2i \lim_{t\to+0} \int_{-\infty}^{\infty} G(\mathbf{p}, \omega)e^{i\omega t} \frac{d\omega}{2\pi}. \tag{7.38}$$

From formula (7.38) we can obtain a very interesting property of the momentum distribution [see Migdal (M5)]. Defining the *Fermi momentum* p_0 by the equation $\varepsilon(p_0) = \mu$, we examine $N_{1/2}(\mathbf{p})$ in the neighborhood of $|\mathbf{p}| = p_0$. According to the postulates of Sec. 2, the excitations of a Fermi liquid are "particles" and "holes" with momenta near p_0, and the attenuation of these quasi-particles is small compared to $|\varepsilon(\mathbf{p}) - \mu|$. This information can be used to determine the poles of the functions G_A and G_R. For $|\mathbf{p}| < p_0$, the function G_A has a pole in the upper half-plane near the real axis, while for $|\mathbf{p}| > p_0$, this pole disappears and G_R acquires a pole in the lower half-plane.

We now represent the integral (7.38) as the sum of two contour integrals involving G_A and G_R, just as was done for the integral (7.34). We shift the horizontal parts of the contours shown in Figs. 2 and 3 into the lower half-plane in such a way that their distances from the real axis greatly exceed $\varepsilon(\mathbf{p}) - \mu$. Then the integrals along these parts of the contours will be insensitive to small changes of the momentum $\mathbf{p}$. As for the integrals along the vertical parts of the contours, they can be reduced to the integral

$$2\int_{\mu}^{\mu - iL} \operatorname{Im} G_R(\mathbf{p}, \omega)\, \frac{d\omega}{2\pi}.$$

This integral can be divided into a part coming from a region whose distance from the point $\varepsilon = \mu$ exceeds $\varepsilon(\mathbf{p}) - \mu$, and a part coming from a region which lies near the point $\varepsilon = \mu$. The integral over the "distant region" depends only slightly on $|\mathbf{p}|$, whereas we can set

$$G_R \approx \frac{a}{\omega - \varepsilon(\mathbf{p}) + i\gamma}$$

in the integral over the "nearby region" and then convince ourselves that this integral is negligibly small (of order $\gamma/[\varepsilon(\mathbf{p}) - \mu]$). Therefore, the entire difference between the expressions for $N_{1/2}(\mathbf{p})$ for $|\mathbf{p}| < p_0$ and for $|\mathbf{p}| > p_0$ is due to the fact that in the first case the part of the contour going around the pole is absent, while in the second case it is present. It follows that

$$N_{1/2}(p_0 - 0) - N_{1/2}(p_0 + 0) = a. \tag{7.39}$$

According to (7.21), the constant a must be positive. Thus, we arrive at the conclusion that the momentum distribution of the particles and the momentum distribution of the quasi-particles both have a jump at the point $|\mathbf{p}| = p_0$. According to a general assumption of the theory of the Fermi liquid, the Fermi momentum p_0 of the quasi-particles is related to the number of particles by formula (2.1). (The validity of this assumption will be proved in Chap. 4.) Thus, the jump in the momentum distribution of the interacting particles takes place at the same point as for noninteracting particles. From the fact that

$$0 \leqslant N_{1/2}(\mathbf{p}) \leqslant 1,$$

we can deduce a bound on the size of the jump:

$$0 \leqslant a \leqslant 1. \tag{7.40}$$

An example is afforded by the momentum distribution of the particles in a dilute Fermi gas, found in Sec. 5.

7.4. The Green's function of a system in an external field. We now consider systems in a time-independent external field. In this case, the Green's function depends on the variables $t - t'$, $\mathbf{r}$ and $\mathbf{r}'$, and instead of (7.17), we have

$$\begin{aligned} G(\mathbf{r}, \mathbf{r}', t - t') &= -i\sum_s \psi(\mathbf{r})_{0s}\psi^*_{0s}(\mathbf{r}')e^{-i(E_s - E_0)(t-t')} \quad \text{for} \quad t > t', \\ G(\mathbf{r}, \mathbf{r}', t - t') &= i\sum_s \psi^*(\mathbf{r}')_{s0}\psi_{s0}(\mathbf{r})e^{i(E_s - E_0)(t-t')} \quad \text{for} \quad t < t', \end{aligned} \tag{7.41}$$

where $\psi(\mathbf{r})$ and $\psi(\mathbf{r}')$ are Schrödinger operators. If we proceed as in the case where there is no external field, we obtain a formula of the type (7.21), involving certain complex functions A and B. This inconvenience can be avoided by using the symmetric combination

$$\frac{1}{2}\left[G(\mathbf{r}, \mathbf{r}', t - t') + G(\mathbf{r}', \mathbf{r}, t - t')\right]. \tag{7.42}$$

As far as the ω-dependence of its Fourier components is concerned, this function has all the properties of the function G in the absence of an external field. All the formulas (7.21)–(7.27) hold for the function (7.42), except that here all quantities will depend on the parameters $\mathbf{r}$ and $\mathbf{r}'$ instead of the parameter $\mathbf{p}$.

If we consider noninteracting fermions in an external field, it is convenient to choose the operators $\psi(\mathbf{r})$ in the form

$$\psi(\mathbf{r}) = \sum_s a_s \varphi_s(\mathbf{r}),$$

where the $\varphi_s(\mathbf{r})$ are the eigenfunctions of the particle in the field. In this case, instead of (7.17), we find

$$G(\mathbf{r}, \mathbf{r}', t - t') = -i \sum_s \varphi_s^*(\mathbf{r}')\varphi_s(\mathbf{r}) e^{-i\varepsilon_s(t-t')} \begin{cases} 1 - n_s & \text{for} \quad t > 0, \\ \quad - n_s & \text{for} \quad t < 0, \end{cases}$$

where

$$n_s = \begin{cases} 1 & \text{for} \quad \varepsilon_s < \mu, \\ 0 & \text{for} \quad \varepsilon_s > \mu, \end{cases} \tag{7.43}$$

and ε_s denotes the energy of a particle in the state φ_s. Taking the Fourier component with respect to time, we obtain

$$G(\mathbf{r}, \mathbf{r}', \omega) = \sum_{\varepsilon_s > \mu} \frac{\varphi_s^*(\mathbf{r}')\varphi_s(\mathbf{r})}{\omega - \varepsilon_s + i\delta} + \sum_{\varepsilon_s < \mu} \frac{\varphi_s^*(\mathbf{r}')\varphi_s(\mathbf{r})}{\omega - \varepsilon_s - i\delta}. \tag{7.44}$$

We now introduce a quantity analogous to the quantities A and B in (7.20):

$$A(\mathbf{r}, \mathbf{r}', E)\, dE = \sum_s \varphi_s^*(\mathbf{r}')\varphi_s(\mathbf{r}) \quad \text{for} \quad E < \varepsilon_s < E + dE.$$

Setting $\mathbf{r} = \mathbf{r}'$, we integrate this relation over dV. Then, because of the normalization condition for the functions $\varphi_s(\mathbf{r})$ in the right-hand side, we simply obtain the number of levels dN in the interval dE, i.e.,

$$\int A(\mathbf{r}, \mathbf{r}, E)\, d\mathbf{r} = \frac{dN(E)}{dE}.$$

In terms of the function A, (7.44) can be written as

$$G(\mathbf{r}, \mathbf{r}', \omega) = \int_{-\infty}^{\infty} \frac{A(\mathbf{r}, \mathbf{r}', E)}{\omega - E + i\delta \operatorname{sgn}(E - \mu)}\, dE.$$

It follows that the imaginary part of $G(\mathbf{r}, \mathbf{r}, \omega)$ equals

$$\operatorname{Im} G(\mathbf{r}, \mathbf{r}, \omega) = \begin{cases} -\pi A(\mathbf{r}, \mathbf{r}, \omega) & \text{for} \quad \omega > \mu, \\ \pi A(\mathbf{r}, \mathbf{r}, \omega) & \text{for} \quad \omega < \mu. \end{cases}$$

(A is real and positive when $\mathbf{r} = \mathbf{r}'$). Thus, we finally obtain

$$\frac{dN(E)}{dE} = -\frac{1}{\pi} \operatorname{sgn}(E - \mu) \int \operatorname{Im} G(\mathbf{r}, \mathbf{r}, E)\, d\mathbf{r}. \tag{7.45}$$

8. Basic Principles of the Diagram Technique

8.1. Transformation from the variable N to the variable μ. Before undertaking the calculation of the Green's function, we transform to new variables. Until now, we have considered a system with a given number of particles. In what follows, it will be convenient to consider this number to be variable and to regard the chemical potential as given. As a matter of fact, we have already used this change of variables for the case of phonons, where we had $\mu = 0$ and the number of particles in the system was not specified. However, in the case of Fermi systems, we actually did specify the number of particles, and the chemical potential μ appearing in any formula was regarded as a function of this number. In practical calculations, we shall find it more convenient to regard μ as an independent variable, going over to a given number of particles only in the final result.

A transformation from one independent variable to another can be carried out in the following way: As is well known, the wave functions and energy levels of the system can be obtained from the variational principle

$$\langle \Psi^* \hat{H} \Psi \rangle = \min, \tag{8.1}$$

subject to the condition

$$\langle \Psi^* \hat{N} \Psi \rangle = \text{const}, \tag{8.2}$$

where $\hat{H}$ and $\hat{N}$ are the Hamiltonian and the particle-number operator, respectively. Instead of (8.1) and (8.2), we can use the method of Lagrange multipliers, finding the absolute minimum of the expression

$$\langle \Psi^* (\hat{H} - \mu \hat{N}) \Psi \rangle,$$

where μ is a constant, determined by using the condition (8.2). Thus, transforming from a given N to a given μ reduces to replacing the Hamiltonian by the operator $\hat{H} - \mu \hat{N}$. Since the operator $\hat{N}$ commutes with the Hamiltonian, we easily see that the formula

$$\tilde{\psi}_\mu(x) = e^{-i\mu \hat{N} t} \tilde{\psi}_N(x) e^{i\mu \hat{N} t} = e^{i\mu t} \tilde{\psi}_N(x) \tag{8.3}$$

describes the transformation of the operators $\tilde{\psi}(x)$, since the operator $\tilde{\psi}_N$ decreases the number of particles by one. Similarly, for the operators $\tilde{\psi}_\mu^+(x)$ we have

$$\tilde{\psi}_\mu^+(x) = e^{-i\mu t} \tilde{\psi}_N^+(x). \tag{8.4}$$

The Green's function is defined as

$$G_\mu(x, x') = G_N(x, x') e^{i\mu(t - t')}, \tag{8.5}$$

from which it follows that all the conclusions of the preceding section also apply to G_μ, provided we make the substitution

$$\omega_{(N)} \to \omega_{(\mu)} + \mu. \tag{8.6}$$

Since to calculate the number of particles and their momentum distribution, we only need the values of G for $t = t'$, the corresponding formulas (7.37) and (7.38) obviously do not change. The poles of the new Green's function gives the energy of the excitations, referred to the level of the chemical potential. As already remarked, for practical calculations it is more convenient to use the functions G_μ. Therefore, as a rule, this will be the definition of the Green's function which we shall have in mind from now on, and the Green's function will be denoted simply by G. Whenever the number of particles is regarded as fixed in analyzing the general properties of the function G (as in the preceding section), this will be explicitly stated.

8.2. Wick's theorem. We now turn to the calculation of the Green's function. Using formula (6.32) for the transformation to the interaction representation, we can represent the perturbation series in a simple and concise form. As applied to the Green's function, formula (6.32) has the form

$$G(x, x') = \frac{-i\langle T(\psi(x)\psi^+(x')S(\infty))\rangle}{\langle S(\infty)\rangle}, \tag{8.7}$$

where

$$S(\infty) = T\exp\left\{-i\int_{-\infty}^{\infty} H_{\text{int}}\, dt\right\}. \tag{8.8}$$

We emphasize once again that the ψ-operators appearing in (8.7) (including those appearing in H_{int}) obey the equations for noninteracting particles.

Expanding the quantity $S(\infty)$ in the numerator of (8.7) in powers of H_{int}, we obtain

$$S(\infty) = 1 - i\int_{-\infty}^{\infty} H_{\text{int}}\, dt + \frac{(-i)^2}{2}\int_{-\infty}^{\infty}\int_{-\infty}^{\infty} T(H_{\text{int}}(t_1)H_{\text{int}}(t_2))\, dt_1\, dt_2 + \cdots,$$

$$G(x, x') = -\frac{i}{\langle S(\infty)\rangle}\sum_{n=0}^{\infty}\frac{(-i)^n}{n!}\int_{-\infty}^{\infty}\cdots\int_{-\infty}^{\infty} dt_1\cdots dt_n \tag{8.9}$$
$$\times \langle T(\psi(x)\psi^+(x')H_{\text{int}}(t_1)\cdots H_{\text{int}}(t_n))\rangle.$$

For the time being, we shall not expand the quantity $\langle S(\infty)\rangle$ appearing in the denominator of (8.7). As a rule, the interaction Hamiltonian H_{int} is an integral over the space variables (and sometimes over the time variables as well) of products of a certain number of ψ-operators (specific examples will be given below). Thus, every term of the series (8.9) contains an average of a chronological product of certain particle-field operators in the interaction representation, and hence we ought to begin by considering expressions of the type

$$\langle T(ABCD\cdots XYZ)\rangle,$$

where $A, B, C, D, \ldots, X, Y, Z$ are field operators in the interaction representation.

Each of the field operators can be decomposed into a sum of two terms. One of these terms, which might be called the "annihilation operator," gives zero when it acts upon the ground state. In the phonon operator (7.13) this operator is the sum involving $b_{\mathbf{k}}$, and in the Fermi operator (6.14) it is the part of the sum with $|\mathbf{p}| > p_0$. The other term, which might be called the "creation operator," has the property that its Hermitian conjugate gives zero when it acts upon the ground state. A product

$$N(ABCD\cdots XYZ)$$

of several operators will be called *normal* if all the "creation operators" appear to the left of the "annihilation operators," and if the sign of the product corresponds to the parity of the permutation of the Fermi operators. Moreover, the difference

$$A^c B^c = T(AB) - N(AB)$$

will be called the "pairing" of the two operators A and B.

We now show that the T-product can always be expressed in terms of all possible N-products with all possible "pairings," i.e.,

$$\begin{aligned} T(ABCD\cdots XYZ) = N(ABCD\cdots XYZ) + N(A^cB^cCD\cdots XYZ) \\ + N(A^cBC^cD\cdots XYZ) + \cdots + N(A^cB^aC^a\cdots X^cY^bZ^b). \end{aligned} \tag{8.10}$$

This relation is called *Wick's theorem* (see e.g., A9). First of all, we note that a simultaneous permutation of the operators in both sides of (8.10) does not affect the validity of (8.10). Therefore, without loss of generality, we can assume that the arguments of the operators in (8.10) are already arranged in decreasing time order. To obtain an N-product from a T-product, we have to successively interchange all the creation operators in the T-product with the annihilation operators appearing to their left. In so doing, we obtain a sum of N-products of the type written in (8.10). However, this sum contains only pairings of operators whose order in the T-product differs from their order in the N-product. But since pairings of operators vanish for which both these orders are equivalent, we can assume that the right-hand side of (8.10) contains normal products with all possible pairings. This proves the relation (8.10).

Using (6.14) and (7.13), we can easily verify that the pairings of two Fermi operators $\psi^+(x')$ and $\psi(x)$, as well as the pairings of two phonon operators $\varphi(x)$ and $\varphi(x')$, are c-numbers, while all other pairings are zero. For example, we have

$$\begin{aligned} \psi^{+c}(x')\psi^c(x) &= \frac{1}{V}\sum_{\mathbf{p},\mathbf{p}'}\Big\{ a^+_{\mathbf{p}'}a_{\mathbf{p}} - \underset{|\mathbf{p}|>p_0}{a^+_{\mathbf{p}'}a_{\mathbf{p}}} + \underset{|\mathbf{p}|<p_0}{a_{\mathbf{p}}a^+_{\mathbf{p}'}} \Big\} e^{i[\mathbf{p}\cdot\mathbf{r}-\mathbf{p}'\cdot\mathbf{r}'-\varepsilon_0(\mathbf{p})t+\varepsilon_0(\mathbf{p}')t']} \\ &= \frac{1}{V}\sum_{|\mathbf{p}|<p_0} e^{i\mathbf{p}\cdot(\mathbf{r}-\mathbf{r}')-i\varepsilon_0(\mathbf{p})(t-t')} \qquad \text{for} \quad t'>t, \end{aligned}$$

$$\begin{aligned} \psi^{+c}(x')\psi^c(x) &= \frac{1}{V}\sum_{\mathbf{p},\mathbf{p}'}\Big\{ -a_{\mathbf{p}}a^+_{\mathbf{p}'} - \underset{|\mathbf{p}|>p_0}{a^+_{\mathbf{p}'}a_{\mathbf{p}}} + \underset{|\mathbf{p}|<p_0}{a_{\mathbf{p}}a^+_{\mathbf{p}'}} \Big\} e^{i[\mathbf{p}\cdot\mathbf{r}-\mathbf{p}'\cdot\mathbf{r}'-\varepsilon_0(\mathbf{p})t+\varepsilon_0(\mathbf{p}')t']} \\ &= -\frac{1}{V}\sum_{|\mathbf{p}|>p_0} e^{i\mathbf{p}\cdot(\mathbf{r}-\mathbf{r}')-i\varepsilon_0(\mathbf{p})(t-t')} \quad \text{for} \quad t'<t. \end{aligned}$$

By the definition of the normal product, its average over the ground state vanishes, and hence

$$A^c B^c = \langle T(AB) \rangle.$$

Then, taking the average of (8.10) over the ground state, we obtain

$$\begin{aligned} \langle T(ABCD\cdots XYZ)\rangle = \langle T(AB)\rangle\langle T(CD)\rangle\cdots\langle T(YZ)\rangle \\ \pm \langle T(AC)\rangle\langle T(BD)\rangle\cdots\langle T(YZ)\rangle \pm\cdots \end{aligned} \tag{8.11}$$

Thus, our average decomposes into a sum of all possible products of averages over the ground state of separate pairs of operators. As always, the sign in front of each term corresponds to the parity of the permutation of Fermi operators. In particular, it follows from formula (8.11) that among the operators $A, B, C, D, \ldots$, there must be an even number of operators of each field. Taking account of the definition of the Green's function (7.1), we arrive at the conclusion that the average of a T-product of any number of field operators can be expressed as a sum of products of free Green's functions.

8.3. Feynman diagrams. We now return to the original expression (8.9). Since H_{int} is an integral of pairs of operators ψ, every term of the sum in (8.9) can be transformed according to formula (8.11). The result can be represented graphically by using so-called *Feynman diagrams.* It is best to illustrate this technique by a specific example: Suppose the system under consideration consists of identical fermions with spin-independent interaction forces acting between pairs of particles. Then, according to Sec. 6, H_{int} has the form

$$H_{\text{int}} = \frac{1}{2}\int \psi_\alpha^+(\mathbf{r}_1)\psi_\beta^+(\mathbf{r}_2)U(\mathbf{r}_1 - \mathbf{r}_2)\psi_\beta(\mathbf{r}_2)\psi_\alpha(\mathbf{r}_1)\, d\mathbf{r}_1\, d\mathbf{r}_2. \tag{8.12}$$

If we introduce the function

$$V(x_1 - x_2) = U(\mathbf{r}_1 - \mathbf{r}_2)\delta(t_1 - t_2),$$

the operator $\int H_{\text{int}}\, dt$ will contain two four-dimensional integrals.

We now consider the terms of the sum (8.9). The first term is the Green's function for noninteracting particles. The next term is of the form

$$\delta G^{(1)} = -\frac{1}{2\langle S(\infty)\rangle}\int d^4x_1\, d^4x_2$$
$$\times \langle T(\psi_\alpha(x)\psi_\beta^+(x')\psi_\gamma^+(x_1)\psi_\delta^+(x_2)\psi_\delta(x_2)\psi_\gamma(x_1))\rangle V(x_1 - x_2).$$

According to (8.11), the matrix element in the integrand equals

$$\begin{aligned}
&\langle T(\psi_\alpha(x)\psi_\gamma^+(x_1))\rangle \langle\psi_\delta^+(x_2)\psi_\delta(x_2)\rangle \langle T(\psi_\gamma(x_1)\psi_\beta^+(x'))\rangle \\
&- \langle T(\psi_\alpha(x)\psi_\gamma^+(x_1))\rangle \langle\psi_\delta^+(x_2)\psi_\gamma(x_1)\rangle \langle T(\psi_\delta(x_2)\psi_\beta^+(x'))\rangle \\
&+ \langle T(\psi_\alpha(x)\psi_\delta^+(x_2))\rangle \langle\psi_\gamma^+(x_1)\psi_\gamma(x_1)\rangle \langle T(\psi_\delta(x_2)\psi_\beta^+(x'))\rangle \\
&- \langle T(\psi_\alpha(x)\psi_\delta^+(x_2))\rangle \langle\psi_\gamma^+(x_1)\psi_\delta(x_2)\rangle \langle T(\psi_\gamma(x_1)\psi_\beta^+(x'))\rangle \\
&+ \langle T(\psi_\alpha(x)\psi_\beta^+(x'))\rangle \langle\psi_\gamma^+(x_1)\psi_\gamma(x_1)\rangle \langle\psi_\delta^+(x_2)\psi_\delta(x_2)\rangle \\
&- \langle T(\psi_\alpha(x)\psi_\beta^+(x'))\rangle \langle\psi_\gamma^+(x_1)\psi_\delta(x_2)\rangle \langle\psi_\delta^+(x_2)\psi_\gamma(x_1)\rangle,
\end{aligned}$$

which, by the definition of the Green's function (7.1), can be written as

$$\begin{aligned}
&iG^{(0)}_{\alpha\gamma}(x, x_1)G^{(0)}_{\delta\delta}(x_2, x_2)G^{(0)}_{\gamma\beta}(x_1, x') \\
&-iG^{(0)}_{\alpha\gamma}(x, x_1)G^{(0)}_{\gamma\delta}(x_1, x_2)G^{(0)}_{\delta\beta}(x_2, x') \\
&+iG^{(0)}_{\alpha\delta}(x, x_2)G^{(0)}_{\gamma\gamma}(x_1, x_1)G^{(0)}_{\delta\beta}(x_2, x') \\
&-iG^{(0)}_{\alpha\delta}(x, x_2)G^{(0)}_{\delta\gamma}(x_2, x_1)G^{(0)}_{\gamma\beta}(x_1, x') \\
&-iG^{(0)}_{\alpha\beta}(x, x')G^{(0)}_{\gamma\gamma}(x_1, x_1)G^{(0)}_{\delta\delta}(x_2, x_2) \\
&+iG^{(0)}_{\alpha\beta}(x, x')G^{(0)}_{\delta\gamma}(x_2, x_1)G^{(0)}_{\gamma\delta}(x_1, x_2).
\end{aligned} \tag{8.13}$$

Thus, the expression under consideration separates into a sum of terms, each of which contains three Green's functions for noninteracting particles.

Following Feynman, we can associate with each such term a certain diagram, which is constructed according to the following principle: We use points in the plane to represent the set of space-time coordinates and spin projections on which the ψ-functions (appearing in a given expression) depend. Then the points appearing as arguments in a single function $G^{(0)}$ are connected by a solid line, and the points x_1, x_2 appearing in $V(x_1 - x_2)$ are connected by a wavy line. When this is done, $\delta G^{(1)}$ corresponds to the six diagrams shown in Fig. 4,[10] where each diagram has two external points x, x' and two internal points x_1, x_2. We integrate over the coordinates of the internal points, and we also sum over their spin variables. A similar correspondence between formulas and diagrams can also be set up for perturbation-theory terms of higher order, as well as for other forms of the interaction Hamiltonian. Diagrams like these are known as *Feynman diagrams.*

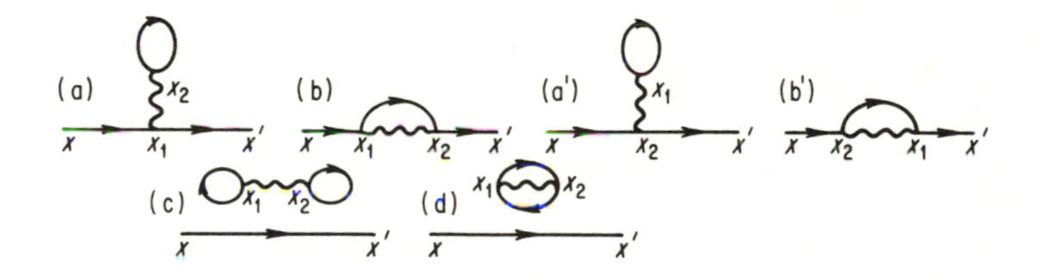

FIGURE 4

In this way, a certain analytical expression is associated with each Feynman diagram, and calculation of the perturbation series reduces to forming all possible Feynman diagrams and calculating the corresponding integrals. The rules by which the diagrams and the corresponding formulas are constructed depend on the specific form of the interaction. Nevertheless, in every case a certain regularity of behavior is observed, which greatly simplifies the calculations.

All the Feynman diagrams for the function G can be divided into two groups, *connected* diagrams and *disconnected* diagrams, where by a connected diagram, we mean one in which all points are connected by lines to the external points x and x'. For example, in Fig. 4 the diagrams labeled (a), (b), (a') and (b') are connected, while those labeled (c) and (d) are disconnected. In the general case where we are dealing with some term or other of the perturbation series (8.9), the connected diagrams are those in which $\psi(x)$ is paired with a ψ^+ in $H_{\text{int}}(t_{p_1})$, a ψ in $H_{\text{int}}(t_{p_1})$ is paired with a ψ^+ in $H_{\text{int}}(t_{p_2})$, and so on, until finally we arrive in this way at $\psi^+(x')$ without

[10] For simplicity, the spin variables are not indicated in Fig. 4.

omitting a single H_{int} [see Fig. 5(a)]. The remaining diagrams in which one or several operators H_{int} are not connected by any pairings to $\psi(x)$ and $\psi^+(x')$ are said to be *disconnected* [see Fig. 5(b)].

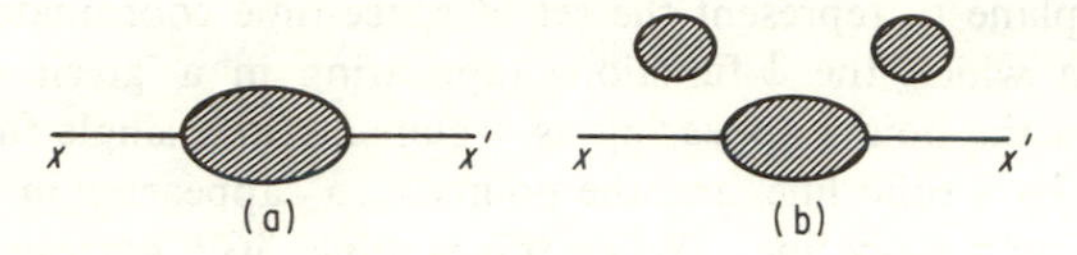

FIGURE 5

We now consider the correction to the Green's function corresponding to a disconnected diagram. This correction obviously consists of two factors. The first factor contains all the H_{int} connected to $\psi(x)$ and $\psi^+(x')$, i.e., it contains the expression corresponding to the connected block in Fig. 5(b) attached to the external points x and x'. The second factor describes the remaining part of the diagram. Thus, the expression for the correction in question equals

$$-i\frac{(-i)^n}{n!}\int dt_1\cdots dt_m\langle T(\psi(x)\psi^+(x')H_{\text{int}}(t_1)\cdots H_{\text{int}}(t_m))\rangle_{\text{con}}$$
$$\times\int dt_{m+1}\cdots dt_n\langle T(H_{\text{int}}(t_{m+1})\cdots H_{\text{int}}(t_n))\rangle.$$

Here, $\langle\cdots\rangle_{\text{con}}$ and $\langle\cdots\rangle$ denote a definite decomposition into pairs of operators ψ, ψ^+ in accordance with Wick's theorem, where $\langle\cdots\rangle_{\text{con}}$ emphasizes the fact that in the corresponding expression, the pairings lead to a connected diagram.

It is not hard to see that among all the diagrams, there are some whose contribution is exactly the same. In fact, changing the pairing in such a way as to simply redistribute the different H_{int} among the angular brackets $\langle\cdots\rangle_{\text{con}}$ and $\langle\cdots\rangle$ merely corresponds to relabeling the variables of integration and does not change the size of the contribution to G. Obviously, the number of such diagrams equals

$$\frac{n!}{m!(n-m)!},$$

i.e., the number of decompositions of the operators H_{int} into groups of m and $n - m$ operators, respectively. The total contribution of all these diagrams equals

$$(-i)\frac{(-i)^m}{m!}\int dt_1\cdots dt_m\langle T(\psi(x)\psi^+(x')H_{\text{int}}(t_1)\cdots H_{\text{int}}(t_m))\rangle_{\text{con}}$$
$$\times\frac{(-i)^{n-m}}{(n-m)!}\int dt_{m+1}\cdots dt_n\langle T(H_{\text{int}}(t_{m+1})\cdots H(t_n))\rangle.$$

We now sum the contributions from all diagrams of any order containing a certain connected part and an arbitrary disconnected part. It is clear that the result is

$$(-i)\frac{(-i)^m}{m!}\int dt_1\cdots dt_m\langle T(\psi(x)\psi^+(x')H_{\text{int}}(t_1)\cdots H_{\text{int}}(t_m))\rangle_{\text{con}}$$
$$\times\left[1 - i\int dt_{m+1}H_{\text{int}}(t_{m+1}) - \frac{1}{2}\int dt_{m+1}\,dt_{m+2}T(H_{\text{int}}(t_{m+1})H_{\text{int}}(t_{m+2}))\right.$$
$$\left.+\cdots+\frac{(-1)^k}{k!}\int dt_{m+1}\cdots dt_{m+k}\langle T(H_{\text{int}}(t_{m+1})\cdots H_{\text{int}}(t_{m+k}))\rangle+\cdots\right].$$

Returning to the original formula (8.7), we see that if the quantity $\langle S(\infty)\rangle$ in the denominator is expanded in powers of H_{int}, we obtain precisely the same expression as that between brackets in the last equation. Thus, we have

$$\langle T(\psi(x)\psi^+(x')S(\infty))\rangle = \langle T(\psi(x)\psi^+(x')S(\infty))\rangle_{\text{con}}\langle S(\infty)\rangle,$$

and according to (8.7),

$$G(x, x') = -i\langle T(\psi(x)\psi^+(x')S(\infty))\rangle_{\text{con}}. \tag{8.14}$$

This rule is valid not only for the Green's function, but also for calculating any expression of the type (6.32), containing an arbitrary number of field operators. This conclusion will be important in what follows. In practice, the rule just proved allows us to omit the factor $\langle S(\infty)\rangle$ in the denominator of formula (8.9) and at the same time ignore the contributions of disconnected diagrams.

A further simplification is made possible by the fact that identical contributions are made by all types of pairings in the expression

$$-i\frac{(-i)^m}{m!}\int dt_1\cdots dt_m\langle T(\psi(x)\psi^+(x')H_{\text{int}}(t_1)\cdots H_{\text{int}}(t_m))\rangle_{\text{con}}$$

which differ only by permutations of H_{int}. This allows us to omit the factor $1/m!$ and consider only pairings that lead to topologically nonequivalent diagrams, i.e., diagrams that cannot be obtained from each other by permutation of the operators H_{int}. The contribution from each such diagram no longer contains a factor which depends in an essential way on the order m of the diagram. As a result, each diagram can be decomposed into elements which can be regarded as separate corrections to the appropriate Green's function. Obviously, a factor of λ^m, where λ is a constant, does not depend on m in an essential way, since this factor does not prevent us from decomposing the diagram into elements. On the other hand, the appearance of a factor $1/m$ would prevent us from making this decomposition and summing over the parts of the diagram separately.

9. Rules for Constructing Diagrams for Interactions of Various Types

9.1. The diagram technique in coordinate space. Examples. We now consider in detail the rules for constructing Feynman diagrams in various special cases. Of basic importance in every Feynman diagram are the lines representing Green's functions of fermions or phonons. We shall represent a fermion Green's function by a solid line and a phonon Green's function by a dashed line. Moreover, we shall often equip a line with an arrow indicating its direction. For example, the line in Fig. 6(a) goes from the point with

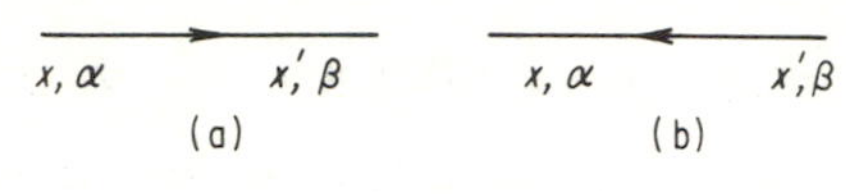

FIGURE 6

coordinates x and spin projection α to the point with coordinates x' and spin projection β, and this line denotes the Green's function

$$G^{(0)}_{\alpha\beta}(x, x') \equiv G^{(0)}_{\alpha\beta}(x - x').$$

On the other hand, the line in Fig. 6(b) denotes the Green's function

$$G^{(0)}_{\beta\alpha}(x', x) \equiv G^{(0)}_{\beta\alpha}(x' - x).$$

x x'

FIGURE 7

We can omit the arrow on a phonon line (see Fig. 7), since, as we have seen in Sec. 7, $D^{(0)}$ is an even function of $x - x'$. We integrate over the coordinates of points where lines meet (the integration is over all space and from $t = -\infty$ to $t = \infty$), and we also sum over all spin variables of such vertices.

We now study some specific examples:

A. Two-particle interactions. The simplest Feynman diagrams for two-particle interactions have already been considered (see Fig. 4) in our discussion of the correspondence between formulas and diagrams. As already explained, disconnected diagrams should be discarded together with the factor $\langle S(\infty)\rangle^{-1}$. Thus, to the first order, the only diagrams left in Fig. 4 are those labeled (a), (b), (a′) and (b′). However, since we integrate over the coordinates x_1 and x_2 (while summing over the corresponding spin variables), it turns out that diagrams (a) and (a′) are equivalent, and so are diagrams (b) and (b′). This compensates the factor of $\frac{1}{2}$ in H_{int}. (A similar situation occurs in the higher-order approximations.) Thus, our prescription consists in omitting the factor $\frac{1}{2}$ and considering only topologically nonequivalent diagrams [e.g., diagrams (a) and (b)].

Next, we note the following fact: As already mentioned, the sign with which a diagram appears depends on the parity of the permutation of the Fermi operators ψ. It is not hard to see that a change in sign is connected with the formation of a closed loop in the diagram. Therefore, the sign of a diagram is determined by the factor $(-1)^F$, where F is the number of closed loops.

Another situation worthy of mention is the case where the times in both arguments of a single function $G^{(0)}$ are the same. This happens only when two operators from a single Hamiltonian H_{int} are paired. Since the order of the operators in H_{int} is specified (all the ψ^+ appear to the left of all the ψ), such a $G^{(0)}$ must be interpreted as the limit

$$\lim_{\delta \to +0} G(t, t + \delta) \equiv \lim_{\delta \to +0} G(-\delta) = i\langle \psi^+(\mathbf{r}_1)\psi(\mathbf{r}_2)\rangle.$$

We now state the general rules which are used to calculate the correction of order n:

1. Form all connected, topologically nonequivalent diagrams with $2n$ vertices and two external points, where two solid lines and one wavy line meet at each vertex;
2. With each solid line associate a Green's function $G^{(0)}_{\alpha\beta}(x, x')$, where x, α are the coordinates of the initial point of the line, and x', β are the coordinates of its end point;
3. With each wavy line associate a potential
$$V(x - x') = U(\mathbf{r} - \mathbf{r}')\delta(t - t');$$
4. Integrate over all the vertex coordinates $x(d^4x = d\mathbf{r}\, dt)$, and sum over all internal spin variables α;
5. Multiply the resulting expression by $i^n(-1)^F$, where F is the number of closed loops;
6. If there are any Green's functions $G^{(0)}$ whose time arguments are the same, interpret them as
$$\lim_{t \to +0} G^{(0)}(\mathbf{r}_1 - \mathbf{r}_2, -t).$$

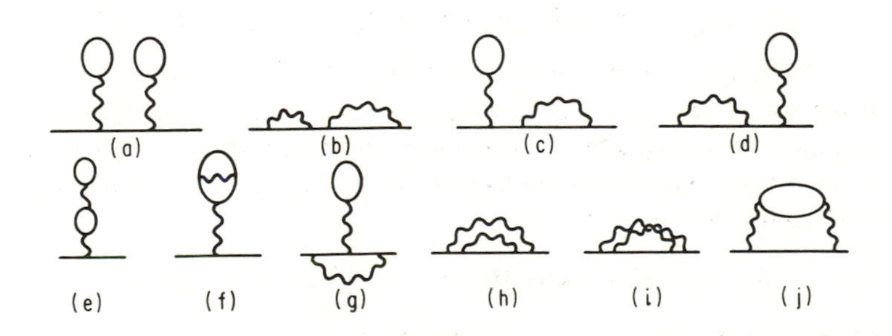

FIGURE 8

We now illustrate this procedure by calculating the second-order correction. The appropriate connected, topologically nonequivalent diagrams are shown in Fig. 8. According to the above rules, the corresponding analytical expressions are the following:

$$-\int d^4x_1\, d^4x_2\, d^4x_3\, d^4x_4\, G^{(0)}_{\alpha\gamma_1}(x - x_1)G^{(0)}_{\gamma_1\gamma_2}(x_1 - x_2) \times G^{(0)}_{\gamma_2\beta}(x_2 - x')G^{(0)}_{\gamma_3\gamma_3}(0)G^{(0)}_{\gamma_4\gamma_4}(0)V(x_1 - x_3)V(x_2 - x_4), \tag{a}$$

$$-\int d^4x_1\, d^4x_2\, d^4x_3\, d^4x_4\, G^{(0)}_{\alpha\gamma_1}(x - x_1)G^{(0)}_{\gamma_1\gamma_2}(x_1 - x_2)G^{(0)}_{\gamma_2\gamma_3}(x_2 - x_3) \times G^{(0)}_{\gamma_3\gamma_4}(x_3 - x_4)G^{(0)}_{\gamma_4\beta}(x_4 - x')V(x_1 - x_2)V(x_3 - x_4), \tag{b}$$

$$+\int d^4x_1\, d^4x_2\, d^4x_3\, d^4x_4\, G^{(0)}_{\alpha\gamma_1}(x - x_1)G^{(0)}_{\gamma_1\gamma_2}(x_1 - x_2)G^{(0)}_{\gamma_2\gamma_3}(x_2 - x_3) \times G^{(0)}_{\gamma_3\beta}(x_3 - x')G^{(0)}_{\gamma_4\gamma_4}(0)V(x_1 - x_4)V(x_2 - x_3), \tag{c}$$

$$+\int d^4x_1\, d^4x_2\, d^4x_3\, d^4x_4\, G^{(0)}_{\alpha\gamma_1}(x - x_1)G^{(0)}_{\gamma_1\gamma_2}(x_1 - x_2)G^{(0)}_{\gamma_2\gamma_3}(x_2 - x_3) \times G^{(0)}_{\gamma_3\beta}(x_3 - x')G^{(0)}_{\gamma_4\gamma_4}(0)V(x_1 - x_2)V(x_3 - x_4), \tag{d}$$

$$-\int d^4x_1\, d^4x_2\, d^4x_3\, d^4x_4\, G^{(0)}_{\alpha\gamma_1}(x - x_1)G^{(0)}_{\gamma_1\beta}(x_1 - x')G^{(0)}_{\gamma_2\gamma_3}(x_2 - x_3) \times G^{(0)}_{\gamma_3\gamma_2}(x_3 - x_2)G^{(0)}_{\gamma_4\gamma_4}(0)V(x_1 - x_2)V(x_3 - x_4), \tag{e}$$

$$+\int d^4x_1\, d^4x_2\, d^4x_3\, d^4x_4\, G^{(0)}_{\alpha\gamma_1}(x - x_1)G^{(0)}_{\gamma_1\beta}(x_1 - x')G^{(0)}_{\gamma_2\gamma_3}(x_2 - x_3) \times G^{(0)}_{\gamma_3\gamma_4}(x_3 - x_4)G^{(0)}_{\gamma_4\gamma_2}(x_4 - x_2)V(x_1 - x_2)V(x_3 - x_4), \tag{f}$$

$$+\int d^4x_1\, d^4x_2\, d^4x_3\, d^4x_4\, G^{(0)}_{\alpha\gamma_1}(x - x_1)G^{(0)}_{\gamma_1\gamma_2}(x_1 - x_2)G^{(0)}_{\gamma_2\gamma_3}(x_2 - x_3) \times G^{(0)}_{\gamma_3\beta}(x_3 - x')G^{(0)}_{\gamma_4\gamma_4}(0)V(x_1 - x_3)V(x_2 - x_4), \tag{g}$$

$$-\int d^4x_1\, d^4x_2\, d^4x_3\, d^4x_4\, G^{(0)}_{\alpha\gamma_1}(x - x_1)G^{(0)}_{\gamma_1\gamma_2}(x_1 - x_2)G^{(0)}_{\gamma_2\gamma_3}(x_2 - x_3) \times G^{(0)}_{\gamma_3\gamma_4}(x_3 - x_4)G^{(0)}_{\gamma_4\beta}(x_4 - x')V(x_1 - x_4)V(x_2 - x_3), \tag{h}$$

$$-\int d^4x_1\, d^4x_2\, d^4x_3\, d^4x_4\, G^{(0)}_{\alpha\gamma_1}(x - x_1)G^{(0)}_{\gamma_1\gamma_2}(x_1 - x_2)G^{(0)}_{\gamma_2\gamma_3}(x_2 - x_3) \times G^{(0)}_{\gamma_3\gamma_4}(x_3 - x_4)G^{(0)}_{\gamma_4\beta}(x_4 - x')V(x_1 - x_3)V(x_2 - x_4), \tag{i}$$

$$+\int d^4x_1\, d^4x_2\, d^4x_3\, d^4x_4\, G^{(0)}_{\alpha\gamma_1}(x - x_1)G^{(0)}_{\gamma_1\gamma_2}(x_1 - x_2)G^{(0)}_{\gamma_2\beta}(x_2 - \beta) \times G^{(0)}_{\gamma_3\gamma_4}(x_3 - x_4)G^{(0)}_{\gamma_4\gamma_3}(x_4 - x_3)V(x_1 - x_3)V(x_2 - x_4). \tag{j}$$

In the case of two-particle interactions, we can set up our perturbation theory in a somewhat different and more symmetric form, which is more convenient when the interaction depends on the spins. Such an interaction has a Hamiltonian of the form

$$H_{\text{int}} = \frac{1}{2}\int \psi^+_\alpha(\mathbf{r}_1)\psi^+_\beta(\mathbf{r}_2)U_{\alpha\beta\gamma\delta}(\mathbf{r}_1 - \mathbf{r}_2)\psi_\delta(\mathbf{r}_2)\psi_\gamma(\mathbf{r}_1)\, d\mathbf{r}_1\, d\mathbf{r}_2. \tag{9.1}$$

We represent the integral $\int H_{\text{int}}\, dt$ appearing in the operator S in a form which is symmetric in all the variables:

$$\int H_{\text{int}}\, dt = \frac{1}{4}\int d^4x_1\, d^4x_2\, d^4x_3\, d^4x_4\, \psi^+_{\gamma_1}(x_1)\psi^+_{\gamma_2}(x_2) \times \Gamma^{(0)}_{\gamma_1\gamma_2,\gamma_3\gamma_4}(x_1, x_2; x_3, x_4)\psi_{\gamma_4}(x_4)\psi_{\gamma_3}(x_3). \tag{9.2}$$

Since the ψ-operators anticommute, we can regard $\Gamma^{(0)}$ as an antisymmetric function with respect to the permutations $x_1, \gamma_1 \rightleftarrows x_2, \gamma_2$ and $x_3, \gamma_3 \rightleftarrows x_4, \gamma_4$. The function $\Gamma^{(0)}$ can be obtained from the expression

$$U_{\gamma_1\gamma_2,\gamma_3\gamma_4}(\mathbf{r}_1 - \mathbf{r}_2)\delta(t_1 - t_2)\delta(x_1 - x_3)\delta(x_2 - x_4)$$

by subtracting a similar expression in which the indices 3 and 4 have been permuted.[11]

Using $\Gamma^{(0)}$, we now calculate the first-order correction to the Green's function, i.e.,

$$\delta G^{(1)} = -\frac{1}{4}\int d^4x_1\, d^4x_2\, d^4x_3\, d^4x_4\, \Gamma^{(0)}_{\gamma_1\gamma_2,\gamma_3\gamma_4}(x_1, x_2; x_3, x_4) \times \langle T(\psi_\alpha(x)\psi^+_\beta(x')\psi^+_{\gamma_1}(x_1)\psi^+_{\gamma_2}(x_2)\psi_{\gamma_4}(x_4)\psi_{\gamma_3}(x_3))\rangle,$$

where in the averaging symbol $\langle\cdots\rangle$ we drop everywhere the label "con" (referring to connected diagrams). Since $\Gamma^{(0)}$ is antisymmetric in all its arguments, the expression for $\delta G^{(1)}$ reduces to a single term

$$i\int d^4x_1\, d^4x_2\, d^4x_3\, d^4x_4\, G^{(0)}_{\alpha\gamma_1}(x - x_1)G^{(0)}_{\gamma_3\gamma_2}(x_3 - x_2) \times G^{(0)}_{\gamma_4\beta}(x_4 - x')\Gamma^{(0)}_{\gamma_1\gamma_2,\gamma_3\gamma_4}(x_1, x_2; x_3, x_4).$$

We shall use an open square to represent $\Gamma^{(0)}$ in our diagrams. Thus, the first-order diagram has the form shown in Fig. 9.

FIGURE 9

In the second approximation, there are three kinds of connected and topologically distinct diagrams (see Fig. 10), and the corresponding expressions are

$$-\int d^4x_1\cdots d^4x_8\, G^{(0)}_{\alpha\gamma_1}(x - x_1)G^{(0)}_{\gamma_3\beta}(x_3 - x') \times G^{(0)}_{\gamma_4\gamma_5}(x_4 - x_5)G^{(0)}_{\gamma_7\gamma_2}(x_7 - x_2)G^{(0)}_{\gamma_8\gamma_6}(x_8 - x_6) \times \Gamma^{(0)}_{\gamma_1\gamma_2,\gamma_3\gamma_4}(x_1, x_2; x_3, x_4)\Gamma^{(0)}_{\gamma_5\gamma_6,\gamma_7\gamma_8}(x_5, x_6; x_7, x_8), \tag{a}$$

$$-\int d^4x_1\cdots d^4x_8\, G^{(0)}_{\alpha\gamma_1}(x - x_1)G^{(0)}_{\gamma_3\gamma_5}(x_3 - x_5) \times G^{(0)}_{\gamma_7\beta}(x_7 - x')G^{(0)}_{\gamma_4\gamma_2}(x_4 - x_2)G^{(0)}_{\gamma_8\gamma_6}(x_8 - x_6) \times \Gamma^{(0)}_{\gamma_5\gamma_6,\gamma_7\gamma_8}(x_5, x_6; x_7, x_8)\Gamma^{(0)}_{\gamma_1\gamma_2,\gamma_3\gamma_4}(x_1, x_2; x_3, x_4) \tag{b}$$

$$-\frac{1}{2}\int d^4x_1\cdots d^4x_8\, G^{(0)}_{\alpha\gamma_1}(x - x_1)G^{(0)}_{\gamma_3\gamma_5}(x_3 - x_5) \times G^{(0)}_{\gamma_7\gamma_2}(x_7 - x_2)G^{(0)}_{\gamma_8\beta}(x_8 - x')G^{(0)}_{\gamma_4\gamma_6}(x_4 - x_6) \times \Gamma^{(0)}_{\gamma_1\gamma_2,\gamma_3\gamma_4}(x_1, x_2; x_3, x_4)\Gamma^{(0)}_{\gamma_5\gamma_6,\gamma_7\gamma_8}(x_5, x_6; x_7, x_8). \tag{c}$$

Note that the last term contains the factor $\frac{1}{2}$.

The calculation of the nth-order correction goes as follows:

---

[11] In this notation, the term corresponding to the "transition" $x_1, \gamma_1 \to x_3, \gamma_3$; $x_2, \gamma_2 \to x_4, \gamma_4$ is taken with a plus sign [cf. (9.1)]. This must be kept in mind when determining the signs of the diagrams (see below).

1. Form all connected, topologically nonequivalent diagrams containing n squares (in the present case, all diagrams obtained by permuting vertex coordinates of the squares are topologically equivalent);
2. With each line associate a Green's function $G^{(0)}_{\alpha\beta}(x - x')$;
3. With each square associate a function $\Gamma^{(0)}_{\gamma_1\gamma_2,\gamma_3\gamma_4}(x_1, x_2; x_3, x_4)$;
4. Integrate over the coordinates of the vertices of all the squares, and sum over the corresponding spin variables;
5. Multiply the resulting expression by

$$\frac{m}{2^n}(i)^n,$$

where m is the number of different diagrams in the unsymmetrized technique corresponding to the given expression. The sign of each expression is also determined by comparison with the unsymmetrized technique.

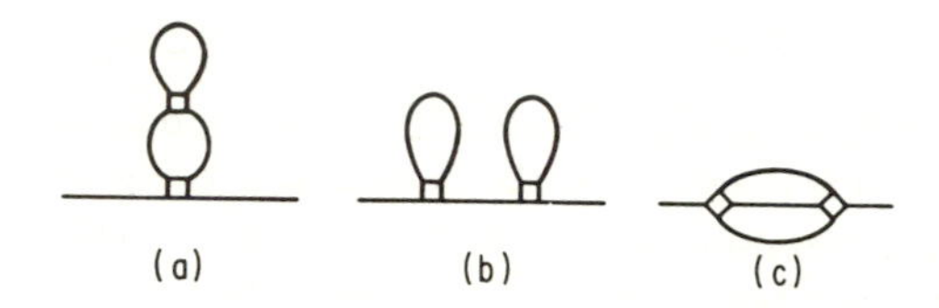

FIGURE 10

The following example illustrates the last rule: Consider diagram (a) of Fig. 10. In the unsymmetrized technique, this diagram corresponds to diagrams (e), (f), (g) and (h) of Fig. 8, and hence $m = 4$. Moreover, diagram (c) of Fig. 10 corresponds to just two diagrams of Fig. 8, i.e., diagrams (i) and (j), and hence in this case $m = 2$, and the corresponding expression appears with the coefficient $\frac{1}{2}$.[12]

To illustrate how the sign of an expression is chosen, we again use diagram (a) of Fig. 10. The quantity $\Gamma^{(0)}$ was obtained by antisymmetrizing an expression in which the points 3 and 1, as well as the points 2 and 4, coincide. If we now regard these coordinates as the same in the expression corresponding to diagram (a) of Fig. 10, we immediately obtain diagram (e) of Fig. 8, which contains two loops and appears with the coefficient $(i)^2$. In practical calculations, it is simplest to first write the arguments in all the $\Gamma^{(0)}$ and only afterwards write the arguments in the $G^{(0)}$, keeping in mind the correspondence with one of the diagrams shown in Fig. 8.

[12] In complicated diagrams, this program may be hard to carry out. It is then simpler to obtain an analytical expression directly from formula (8.14), using diagrams only to suggest various possible kinds of pairings.

According to the rules just given, the following expression corresponds to the third-order diagram shown in Fig. 11:

$$
\begin{aligned}
-\frac{(i)^3}{2}\int d^4x_1\cdots d^4x_{12}\, & G^{(0)}_{\alpha\gamma_1}(x-x_1)G^{(0)}_{\gamma_3\gamma_5}(x_3-x_5)\\
\times\, & G^{(0)}_{\gamma_7\gamma_9}(x_7-x_9)G^{(0)}_{\gamma_{11}\beta}(x_{11}-x')G^{(0)}_{\gamma_4\gamma_{10}}(x_4-x_{10})\\
\times\, & G^{(0)}_{\gamma_{12}\gamma_6}(x_{12}-x_6)G^{(0)}_{\gamma_8\gamma_2}(x_8-x_2)\Gamma^{(0)}_{\gamma_1\gamma_2,\gamma_3\gamma_4}(x_1,x_2;x_3,x_4)\\
\times\, & \Gamma^{(0)}_{\gamma_5\gamma_6,\gamma_7\gamma_8}(x_5,x_6;x_7,x_8)\Gamma^{(0)}_{\gamma_9\gamma_{10},\gamma_{11}\gamma_{12}}(x_9,x_{10};x_{11},x_{12}).
\end{aligned}
$$

Obviously, if we substitute for $\Gamma^{(0)}$ its expression in terms of the potential $U_{\alpha\beta,\gamma\delta}(\mathbf{r}_1-\mathbf{r}_2)$, all the above expressions go over into the corresponding expressions of the unsymmetrized theory.

Expressions involving a spin-independent point interaction, i.e., an interaction with potential

$$U_{\alpha\beta,\gamma\delta}(\mathbf{r}_1-\mathbf{r}_2)=\lambda\delta_{\alpha\gamma}\delta_{\beta\delta}\delta(\mathbf{r}_1-\mathbf{r}_2),$$

take a particularly simple form. In this case, $\Gamma^{(0)}$ becomes

$$
\begin{aligned}
\Gamma^{(0)}_{\gamma_1\gamma_2,\gamma_3\gamma_4} &= \lambda(\delta_{\gamma_1\gamma_3}\delta_{\gamma_2\gamma_4}-\delta_{\gamma_1\gamma_4}\delta_{\gamma_2\gamma_3})\delta(x_1-x_2)\delta(x_1-x_3)\delta(x_1-x_4)\\
&= \lambda L_{\gamma_1\gamma_2,\gamma_3\gamma_4}\delta(x_1-x_2)\delta(x_1-x_3)\delta(x_1-x_4),
\end{aligned}\tag{9.3}
$$

and only one integration survives among the four integrations over the vertices of the open squares in Figs. 9 and 10. Therefore, these squares can be replaced by points. For example, then the diagrams in Figs. 9, 10(c) and 11 take the forms shown in Fig. 12, and the corresponding expressions are given by

FIGURE 11

$$i\lambda L_{\gamma_1\gamma_2,\gamma_3\gamma_4}G^{(0)}_{\gamma_3\gamma_2}(0)\int d^4x_1\, G^{(0)}_{\alpha\gamma_1}(x-x_1)G^{(0)}_{\gamma_4\beta}(x_1-x'), \tag{a}$$

$$
\begin{aligned}
-\frac{\lambda^2}{2}L_{\gamma_1\gamma_2,\gamma_3\gamma_4}L_{\gamma_5\gamma_6,\gamma_7\gamma_8}\int d^4x_1\,d^4x_2\, & G^{(0)}_{\alpha\gamma_1}(x-x_1)G^{(0)}_{\gamma_3\gamma}(x_1-x_2)\\
\times\, & G^{(0)}_{\gamma_7\gamma_2}(x_2-x_1)G^{(0)}_{\gamma_8\beta}(x_2-x')G^{(0)}_{\gamma_4\gamma_6}(x_1-x_2),
\end{aligned}\tag{b}
$$

$$
\begin{aligned}
-\frac{i\lambda^3}{2}L_{\gamma_1\gamma_2,\gamma_3\gamma_4}L_{\gamma_5\gamma_6,\gamma_7\gamma_8}L_{\gamma_9\gamma_{10},\gamma_{11}\gamma_{12}}\int d^4x_1\,d^4x_2\,d^4x_3\, & G^{(0)}_{\alpha\gamma_1}(x-x_1)\\
\times\, & G^{(0)}_{\gamma_3\gamma_5}(x_1-x_2)G^{(0)}_{\gamma_7\gamma_9}(x_2-x_3)G^{(0)}_{\gamma_{11}\beta}(x_3-x')\\
\times\, & G^{(0)}_{\gamma_4\gamma_{10}}(x_1-x_3)G^{(0)}_{\gamma_{12}\gamma_6}(x_3-x_2)G^{(0)}_{\gamma_8\gamma_2}(x_2-x_1).
\end{aligned}\tag{c}
$$

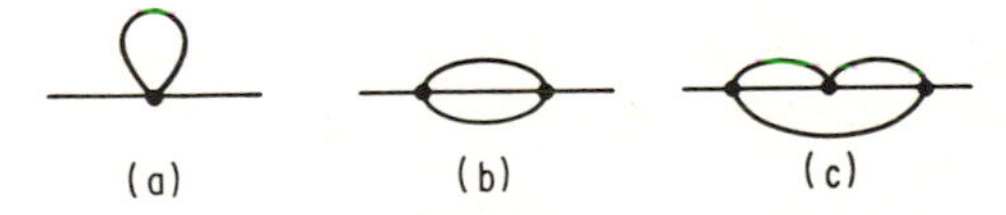

FIGURE 12

B. Electron-phonon interactions. With a view to later applications, it will be assumed that we are dealing with the isotropic model of a metal, in which electrons interact with phonons, and the interaction mechanism is due to the polarization of the medium produced by the lattice vibrations. As a result of the polarization, the energy of the electrons is changed by an amount

$$-e \int n(\mathbf{r})K(\mathbf{r} - \mathbf{r}') \operatorname{div} \mathbf{P}(\mathbf{r}') \, d\mathbf{r} \, d\mathbf{r}'. \tag{9.4}$$

Here, $n(\mathbf{r})$ is the electron density at the point $\mathbf{r}$, $\mathbf{P}$ is the polarization vector, and $K(\mathbf{r} - \mathbf{r}')$ is an interaction function. For values of $|\mathbf{r} - \mathbf{r}'|$ which are small compared to the lattice spacing, we have

$$K(\mathbf{r} - \mathbf{r}') \approx \frac{1}{|\mathbf{r} - \mathbf{r}'|}.$$

For larger distances, $K(\mathbf{r} - \mathbf{r}')$ falls off rapidly to zero, because of the screening of the polarization charge by electrons. This allows us to replace $K(\mathbf{r} - \mathbf{r}')$ by $a^2\delta(\mathbf{r} - \mathbf{r}')$, where a is a constant of the order of the lattice spacing. The polarization vector $\mathbf{P}$ is proportional to the displacement of the medium, i.e.,

$$\mathbf{P}(\mathbf{r}) = C\mathbf{q}(\mathbf{r}),$$

where C is a constant of order ZeN/V. (A unit volume contains N/V ions, each with charge Ze.)

Since the energy of the interaction between the electrons and the lattice vibrations contains $\operatorname{div} \mathbf{P} = C \operatorname{div} \mathbf{q}$, it follows that the electrons interact only with the *longitudinal* vibrations. According to (9.4), the interaction-energy operator can be written in the form

$$ea^2C \int \psi^+(\mathbf{r})\psi(\mathbf{r}) \operatorname{div} \mathbf{q}(\mathbf{r}) \, d\mathbf{r}.$$

Since the operators $q_{\mathbf{k}}$ are linear combinations of the annihilation and creation operators, we might as well define the field operators $\varphi(\mathbf{r})$ in such a way as to give H_{int} a more convenient form. It is easily seen that if we choose (7.13) as our definition of the field operators $\varphi(x)$, the Hamiltonian for electron-phonon interaction takes the form

$$H_{\text{int}} = g \int \psi_\alpha^+(\mathbf{r})\psi_\alpha(\mathbf{r})\varphi(\mathbf{r}) \, d\mathbf{r}, \tag{9.5}$$

where the (*interaction*) *coupling constant* is

$$g = \frac{ea^2C}{u_0\sqrt{\rho}},$$

and $u_0 = \omega_0(\mathbf{k})/|\mathbf{k}|$ is the velocity of sound. Replacing all the constants on the right by their orders of magnitude, expressed in terms of electron parameters, we obtain

$$g^2 = \frac{2\pi^2\eta}{mp_0}, \tag{9.6}$$

where m is the mass of the electron. With this definition, the constant η is dimensionless, and comparison with experimental data for metals shows that $\eta \sim 1$.

To find the Green's functions, we need only take into account even terms in the expansion of $S(\infty)$ in powers of H_{int}. Since we can average the electron and phonon operators separately, the diagrams for the electron Green's functions turn out to be the same as for two-particle interactions between fermions. The only difference is that we must now change all wavy lines to dashed lines, corresponding to phonon Green's functions, while simultaneously making the change

$$V(x_1 - x_2) \to g^2 D^{(0)}(x_1 - x_2).$$

As for the phonon Green's functions, the first nonvanishing correction appears in the second order (with respect to H_{int}), as shown by the diagrams in Fig. 13. The corresponding expressions are

$$-g^2 i \int d^4x_1\, d^4x_2\, D^{(0)}(x - x_1) D^{(0)}(x_2 - x') G^{(0)}_{\alpha\beta}(x_1 - x_2) G^{(0)}_{\beta\alpha}(x_2 - x_1), \qquad \text{(a)}$$

$$+ g^2 i \int d^4x_1\, d^4x_2\, D^{(0)}(x - x_1) D^{(0)}(x_2 - x') G^{(0)}_{\alpha\alpha}(0) G^{(0)}_{\beta\beta}(0). \qquad \text{(b)}$$

The second of these expressions must equal zero. To see this, we note that by definition the function $D^{(0)}$ contains the quantities φ which are proportional to div $\mathbf{q}$, where $\mathbf{q}$ is the displacement vector. It follows that the function $D^{(0)}(x - x_1)$ is proportional to

$$T(\varphi(x) \operatorname{div} \mathbf{q}(x_1)) = \operatorname{div}_{\mathbf{r}_1} \langle T(\varphi(x)\mathbf{q}(x_1))\rangle.$$

Since in the expression corresponding to diagram (b) of Fig. 13, the coordinate $\mathbf{r}_1$ appears only in $D^{(0)}(x - x_1)$, and since, as just shown, this function has the form of a divergence, the integral with respect to $\mathbf{r}_1$ transforms into a surface integral, and hence vanishes, regardless of whether the displacement on the boundary is assumed to be zero or to obey periodic boundary conditions. For the same reason, all diagrams for the D-function in which the external points are separated [in the sense of Fig. 13(b)] lead to vanishing analytical expressions.

(a) (b)

FIGURE 13

We now state the general rules which are used to calculate the corrections of order $2n$ to the electron and phonon Green's functions:

1. Form all connected,[13] topologically nonequivalent diagrams with $2n$ vertices;
2. With each solid line associate a free-particle Green's function $G^{(0)}_{\alpha\beta}(x - x')$, and with each dashed line associate a function $D^{(0)}(x - x')$;
3. Integrate over the coordinates of all vertices, and sum over the corresponding spin variables;
4. Multiply the resulting expression by $g^{2n}(i)^n(-1)^F$, where F is the number of closed loops formed by fermion $G^{(0)}$-lines.

[13] Here diagrams of the type shown in Fig. 13(b) are considered to be disconnected.

As an example, the expression corresponding to the diagram shown in Fig. 14 is

$$g^4 \int d^4x_1 d^4x_2 d^4x_3 d^4x_4 \, D^{(0)}(x - x_1) D^{(0)}(x_2 - x_3) D^{(0)}(x_4 - x')$$
$$\times \, G^{(0)}_{\gamma_1\gamma_2}(x_1 - x_2) G^{(0)}_{\gamma_2\gamma_4}(x_2 - x_4) G^{(0)}_{\gamma_4\gamma_3}(x_4 - x_3) G^{(0)}_{\gamma_3\gamma_1}(x_3 - x_1).$$

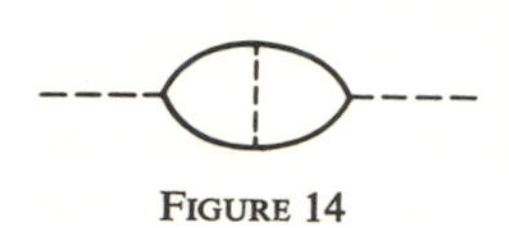

FIGURE 14

C. External fields. As our last example, we consider the interaction of particles with an external field. Then, according to Sec. 6, the interaction Hamiltonian takes the form

$$H_{\text{int}} = \int \psi^+_\alpha(\mathbf{r}) V_{\alpha\beta}(\mathbf{r}, t) \psi_\beta(\mathbf{r}) \, d\mathbf{r}. \tag{9.7}$$

The indices α, β on the potential V allow us to take into account the influence of an external magnetic field on the spins of the particles. In this case,

$$V_{\alpha\beta}(\mathbf{r}, t) = - \mu_0 \boldsymbol{\sigma}_{\alpha\beta} \cdot \mathbf{H}(\mathbf{r}, t),$$

where μ_0 is the magnetic moment of the particles, and $\boldsymbol{\sigma}$ denotes the Pauli matrices.

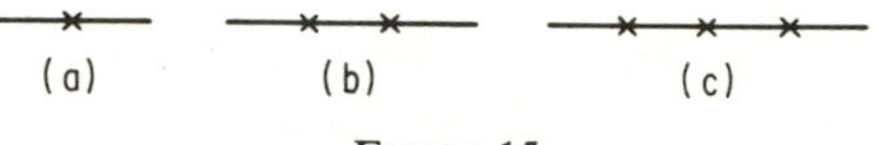

FIGURE 15

It is not hard to see that in the present case, all the diagrams have the simple form shown in Fig. 15, where each cross denotes a potential $V_{\alpha\beta}(x)$. For example, the expression associated with diagram (b) is

$$\int d^4x_1 \, d^4x_2 \, G^{(0)}_{\alpha\gamma_1}(x - x_1) G^{(0)}_{\gamma_2\gamma_3}(x_1 - x_2) G^{(0)}_{\gamma_4\beta}(x_2 - x') V_{\gamma_1\gamma_2}(x_1) V_{\gamma_3\gamma_4}(x_2).$$

The rules for forming diagrams and the corresponding analytical expressions are trivial, and the diagrams of all orders have the same multiplicative factor 1. The only thing worthy of mention is that since space and time are no longer homogeneous, the Green's functions now depend on x and x' separately, and not just on the difference $x - x'$.

9.2. The diagram technique in momentum space. Examples. Using the technique just presented, we can easily write down any term of the perturbation series in integral form. However, evaluation of the resulting integrals is very difficult, since $G^{(0)}$ and $D^{(0)}$ are discontinuous functions of the time argument. Thus, to calculate the corrections to the Green's functions, we would have to carry out integration with respect to time over a set of regions whose number would increase "catastrophically fast" as the order of the approximation is increased. The way out of this situation is to expand all quantities in Fourier integrals. We now illustrate this procedure by discussing the same examples as in Sec. 9.1.

A. Two-particle interactions. Consider the expression corresponding to diagram (b) of Fig. 4:

$$i\int d^4x_1\, d^4x_2 G^{(0)}_{\alpha\gamma_1}(x - x_1)G^{(0)}_{\gamma_1\gamma_2}(x_1 - x_2)G^{(0)}_{\gamma_2\beta}(x_2 - x')V(x_1 - x_2). \tag{9.8}$$

We expand all quantities in Fourier integrals, using the formulas

$$\begin{aligned} G^{(0)}_{\alpha\gamma}(x_1 - x_2) &= \int \frac{d^4p}{(2\pi)^4}\, G^{(0)}_{\alpha\gamma}(p)e^{ip(x_1 - x_2)}, \\ V(x_1 - x_2) &= \int \frac{d^4q}{(2\pi)^4}\, V(q)e^{iq(x_1 - x_2)}, \end{aligned} \tag{9.9}$$

where p, q are the four-dimensional vectors $p = (\mathbf{p}, \omega)$, $q = (\mathbf{q}, \omega)$, and the product $p(x_1 - x_2)$ equals

$$\mathbf{p}\cdot(\mathbf{r}_1 - \mathbf{r}_2) - \omega(t_1 - t_2).$$

The expression for the free fermion Green's function $G^{(0)}_{\alpha\gamma}(p)$ has already been found in Sec. 7 [i.e., formula (7.7) with ω replaced by $\omega + \mu$]. Substituting (9.9) into (9.8), we obtain

$$i(2\pi)^{-16}\int d^4p_1\, d^4p_2\, d^4p_3\, d^4q\, d^4x_1\, d^4x_2 G^{(0)}_{\alpha\gamma_1}(p_1)G^{(0)}_{\gamma_1\gamma_2}(p_2)G^{(0)}_{\gamma_2\beta}(p_3)V(q)$$
$$\times\, e^{ip_1(x - x_1) + ip_2(x_1 - x_2)}e^{ip_3(x_2 - x') + iq(x_1 - x_2)}.$$

Then integration with respect to x_1 and x_2 gives

$$\begin{aligned} &i(2\pi)^{-8}\int d^4p_1\, d^4p_2\, d^4p_3\, d^4q G^{(0)}_{\alpha\gamma_1}(p_1)G^{(0)}_{\gamma_1\gamma_2}(p_2)G^{(0)}_{\gamma_2\beta}(p_3)V(q) \\ &\qquad \times\, \delta(p_1 - p_2 - q)\delta(p_2 + q - p_3)e^{ip_1x - ip_3x'}. \end{aligned} \tag{9.10}$$

We now take the Fourier components of (9.10) with respect to x and x', obtaining

$$\delta G^{(1)}_{\alpha\beta}(p, p') = i\int d^4p_2\, d^4q G^{(0)}_{\alpha\gamma_1}(p)G^{(0)}_{\gamma_1\gamma_2}(p_2)G^{(0)}_{\gamma_2\beta}(p')V(q)$$
$$\times\, \delta(p - p_2 - q)\delta(p_2 + q - p').$$

Comparing this expression with diagram (b) of Fig. 4, we see that each solid line now corresponds to a function $G^{(0)}(p)$, each wavy line to a function $V(q)$, and each vertex to a δ-function $\delta(\Sigma p) = \delta(\Sigma\mathbf{p})\,\delta(\Sigma\omega)$ expressing conservation of energy and momentum, where the integration is over the momenta of the internal lines. Carrying out the integration with respect to p_2, and bearing in mind that $G^{(0)}_{\alpha\beta}(p) = G^{(0)}(p)\delta_{\alpha\beta}$, we find that

$$\delta G^{(1)}_{\alpha\beta}(p, p') = \delta G^{(1)}(p)\delta(p - p')(2\pi)^4\delta_{\alpha\beta},$$
$$\delta G^{(1)}(p) = iG^{(0)}(p)\int \frac{d^4q}{(2\pi)^4}\, G^{(0)}(p - q)V(q)G^{(0)}(p).$$

This expression for $\delta G^{(1)}(p)$, which represents a correction to the Fourier component of the function $G(x - x')$ with respect to the variable $x - x'$, can be given a very lucid interpretation in terms of Feynman diagrams. In fact, imagine a particle with momentum p, which in the course of its motion emits an "interaction quantum" with momentum q, thereby acquiring momentum $p - q$ itself. After a certain time, the particle absorbs this quantum, getting back momentum p.

A similar interpretation can be made for the other diagrams. For example, the correction $\delta G^{(1')}(p)$ corresponding to diagram (a) of Fig. 4 has the form

$$\delta G^{(1')} = -2iG^{(0)}(p)V(0)\int\frac{d^4p_1}{(2\pi)^4}\,G^{(0)}(p_1)e^{i\omega t}G^{(0)}(p),$$

where $t \to +0$. The factor $e^{i\omega t}$ in the integrand appears at the place where a Green's function with identical arguments appears in the coordinate representation. [As already noted, such a function has to be defined as the limit $G^{(0)}(-0)$.] The coefficient 2 is the result of taking the trace with respect to the spins. The diagrams for $\delta G^{(1)}$ and $\delta G^{(1')}$ in the momentum representation are shown in Fig. 16.

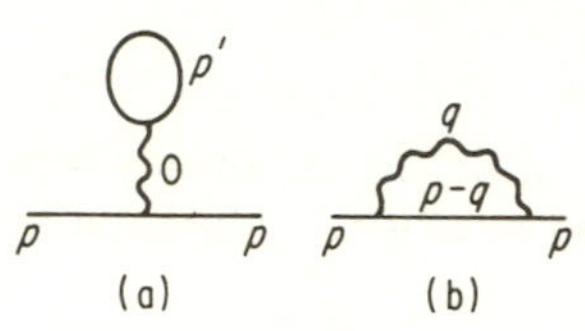

FIGURE 16

Next, we consider diagrams of any order n, containing $2n$ vertices, $2n + 1$ solid lines and n wavy lines. We substitute Fourier expansions like (9.8) for $G^{(0)}$ and V in the appropriate coordinate-representation formulas, and then integrate over the $2n$ coordinates of the vertices. This leads to $2n$ factors of the type $\delta(\Sigma p)$ expressing conservation of energy and momentum. One of these conservation laws reduces to equality of the external momenta, and as a result, all the terms of the expansion of $G(x, x')$ in the perturbation series depend only on the difference $x - x'$ (this is an obvious consequence of the homogeneity of space). Because of the presence of the remaining $2n - 1$ δ-functions, only n integrations survive among the $3n - 1$ integrations over the four-momenta of the internal lines (both solid and wavy).

The general rules for obtaining the expression corresponding to a given diagram of order n are as follows:

1. With each line associate a given four-momentum in such a way that the two external lines have the external momentum, and the momenta of the internal lines satisfy the conservation laws at each vertex;
2. With each solid line associate a quantity

$$G^{(0)}_{\alpha\beta}(p) = \frac{\delta_{\alpha\beta}}{\omega - \xi(\mathbf{p}) + i\delta\,\mathrm{sgn}\,\xi(\mathbf{p})},$$

   where

$$\xi(\mathbf{p}) = \varepsilon_0(\mathbf{p}) - \mu = \frac{\mathbf{p}^2}{2m} - \mu, \quad \delta \to +0;$$

3. With each wavy line associate a quantity

$$V(q) \equiv U(\mathbf{q});$$

4. Integrate over the n independent four-momenta;
5. Multiply the resulting expression by $(2\pi)^{-4n}(i)^n(-1)^F$, where F is the number of closed loops.

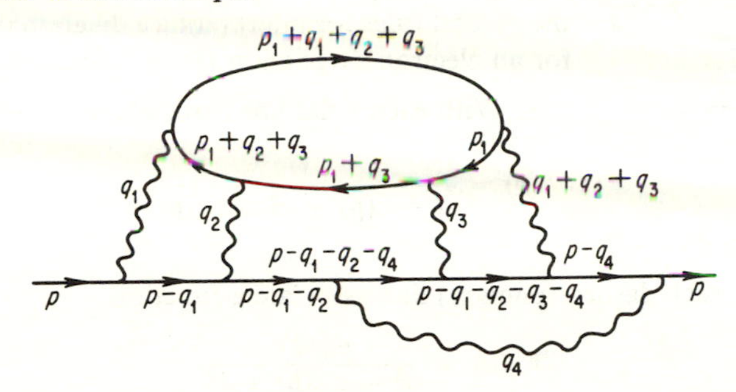

FIGURE 17

Using these rules, we can easily write down any correction to the Green's function. For example, the correction corresponding to the diagram shown in Fig. 17 equals

$$-i\delta_{\alpha\beta}(G^{(0)}(p)^2)(2\pi)^{-20}\int d^4q_1\, d^4q_2\, d^4q_3\, d^4q_4\; U(q_1)U(q_2)U(q_3)$$
$$\times\; U(q_1+q_2+q_3)U(q_4)G^{(0)}(p-q_1)G^{(0)}(p-q_1-q_2)$$
$$\times\; G^{(0)}(p-q_1-q_2-q_4)G^{(0)}(p-q_1-q_2-q_3-q_4)$$
$$\times\; G^{(0)}(p-q_4)\int d^4p_1\; G^{(0)}(p_1)G^{(0)}(p_1+q_3)G^{(0)}(p_1+q_2+q_3)$$
$$\times\; G^{(0)}(p_1+q_1+q_2+q_3).$$

Next, we consider the other, symmetrized version of the diagram technique for two-particle interactions. On p. 74, we introduced the symmetrized quantity $\Gamma^{(0)}_{\gamma_1\gamma_2,\gamma_3\gamma_4}(x_1, x_2; x_3, x_4)$, which by its very definition depends only on the coordinate differences. Therefore, the Fourier components of $\Gamma^{(0)}$ contain $\delta(p_1+p_2-p_3-p_4)$, and hence from the outset it is convenient to define the Fourier components of $\Gamma^{(0)}$ as

$$(2\pi)^4\delta(p_1+p_2-p_3-p_4)\Gamma^{(0)}(p_1, p_2; p_3, p_1+p_2-p_3)$$
$$= \int d^4x_1\, d^4x_2\, d^4x_3\, d^4x_4\; \Gamma^{(0)}(x_1, x_2; x_3, x_4)e^{-ip_1x_1-ip_2x_2+ip_3x_3+ip_4x_4}.$$

The Fourier transform of the first-order correction corresponding to the diagram shown in Fig. 9 has the form

$$-i(G^{(0)}(p))^2\int\frac{d^4p_1}{(2\pi)^4}\,\Gamma^{(0)}_{\alpha\gamma,\beta\gamma}(p, p_1; p, p_1)G^{(0)}(p_1).$$

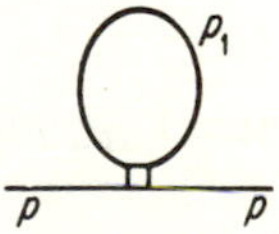

FIGURE 18

In momentum space, this diagram has the form shown in Fig. 18. The general rules for constructing diagrams are exactly the same as before. In particular, the coefficient associated with the

nth-order diagram differs from the corresponding coefficient in the coordinate representation only by the factor $(2\pi)^{-4n}$.

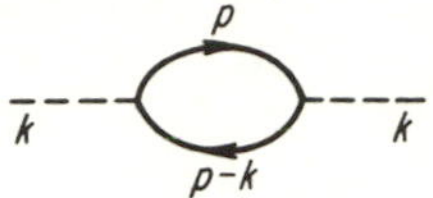

FIGURE 19

B. Electron-phonon interactions. The following are the general rules for interpreting a diagram of order $2n$ for an electron or phonon Green's function:

1. With each solid line associate a function

$$G^{(0)}(p) = \frac{1}{\omega - \xi(\mathbf{p}) + i\delta \operatorname{sgn} \xi(\mathbf{p})},$$

where $\delta \to +0$;

2. With each dashed (phonon) line associate a function

$$D^{(0)}(k) = \frac{\omega_0^2(\mathbf{k})}{\omega^2 - \omega_0^2(\mathbf{k}) + i\delta},$$

where $\delta \to +0$ [see (7.16)];

3. Integrate over the n independent momenta;

4. Multiply the resulting expression by

$$g^{2n}(2\pi)^{-4n}(i)^n(-1)^F,$$

where F is the number of closed loops.

For example, the expression corresponding to the second-order diagram shown in Fig. 19 is

$$-2[D^{(0)}(k)]^2 g^2 i \int \frac{d^4p}{(2\pi)^4} G^{(0)}(p)G^{(0)}(p - k).$$

C. External fields. As already mentioned, space becomes inhomogeneous in the presence of an external field, and $G(x, x')$ ceases to be a function of only the difference variable $x - x'$. Therefore, we introduce the function $G_{\alpha\beta}(p, p')$, which is the Fourier transform of $G_{\alpha\beta}(x, x')$ with respect to both variables:

$$G_{\alpha\beta}(x, x') = \int \frac{d^4p}{(2\pi)^4} \frac{d^4p'}{(2\pi)^4} G_{\alpha\beta}(p, p')e^{ipx - ip'x'}.$$

Taking the Fourier transform of the expression

$$\int dx_1\, G^{(0)}_{\alpha\gamma_1}(x - x_1)G^{(0)}_{\gamma_2\beta}(x_1 - x')V_{\gamma_1\gamma_2}(x_1),$$

corresponding to diagram (a) of Fig. 15, we obtain

$$G^{(0)}(p)V_{\alpha\beta}(p - p')G^{(0)}(p'),$$

where $V_{\alpha\beta}(p)$ is the Fourier component of $V_{\alpha\beta}(x)$:

$$V_{\alpha\beta}(x) = \int \frac{d^4p}{(2\pi)^4} V_{\alpha\beta}(p)e^{ipx}.$$

The corresponding diagram in momentum space is shown as diagram (a) of

Fig. 20. The expression corresponding to the second-order diagram [diagram (b) of Fig. 20] is

$$G^{(0)}(p) \int \frac{d^4 p_1}{(2\pi)^4} V_{\alpha\gamma}(p-p_1) G^{(0)}(p_1) V_{\gamma\beta}(p_1-p') G^{(0)}(p').$$

Thus, we have the following general rules for interpreting the nth-order diagram for $G(p, p')$:

1. Associate $G^{(0)}(p)$ with the left-hand external line and $G^{(0)}(p')$ with the right-hand external line;

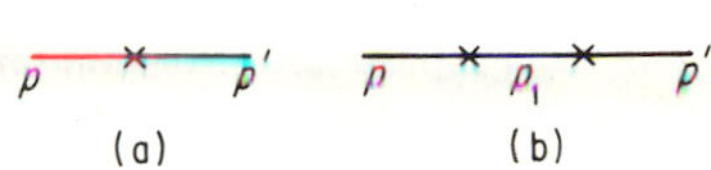

FIGURE 20

2. Let each cross denote the Fourier component of the external field with respect to the momentum difference of the $G^{(0)}$-lines appearing immediately to the left and to the right of the cross;
3. Integrate over the momenta of all the $G^{(0)}$-lines except the two external lines, and sum over all the spin variables on which V depends (except the external variables);
4. After performing the integration and summation, multiply the resulting expression by $(2\pi)^{-4(n-1)}$.

10. Dyson's Equation. The Vertex Part. Many-Particle Green's Functions

10.1. Sums of diagrams. Dyson's equation. In most problems of quantum statistics, it is usually not permissible to take into account only the first few terms of the perturbation series. Instead, we have to sum various infinite series of terms, corresponding to so-called "principal diagrams," which, because of the conditions of the problem, make contributions of the same order of magnitude. A remarkable property of the diagram technique for calculating Green's functions (presented earlier in this chapter) is that calculation of the sum of an infinite (or finite) set of terms of the perturbation series can characteristically be described in terms of "graphical summation" of diagrams. This means that the diagram representing a certain sum may be regarded as made up of elements, each of which is in turn the result of a summation. For example, the lines of a diagram may represent sums of certain infinite sequences of terms in the perturbation series for the Green's function, and hence may themselves represent "sums of diagrams." Analytical expressions are associated with diagrams according to the same rules used to calculate the expressions by perturbation theory, e.g., each line representing a sum of diagrams is assigned the corresponding analytical expression.

The possibility of carrying out graphical summation is a consequence of the rules given above for calculating corrections to the Green's function by using appropriate diagrams. A moment's reflection shows that every such correction is constructed, so to speak, from individual "bricks" (Green's functions and vertex operators), with integration over coordinates (or momenta) serving as a kind of "mortar" joining the various bricks together. Thus, we can construct diagrams not only from "primitive" elements (zeroth-order Green's functions $G^{(0)}$ and elementary vertices), but also from entire "blocks" at a time, where the blocks themselves consist of a large number of primitive elements.

For example, consider diagram (a) of Fig. 21. One way of writing the analytical expression corresponding to this diagram is to use the ordinary form of the diagram technique. However, the following approach is also possible: We first calculate the contribution of the subdiagram surrounded by the dashed line in Fig. 21(a). Then we write the expression corresponding to diagram (b), but with the upper line (marked in the figure) we associate instead of $G^{(0)}$ a more complicated expression. It is easily verified by direct calculation that both approaches give the identical result.

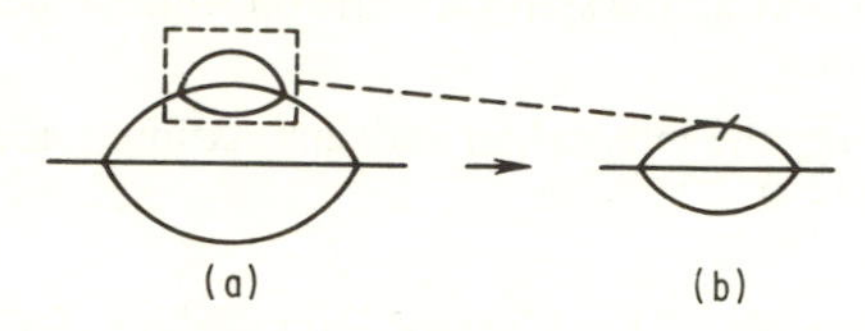

FIGURE 21

This procedure is completely general. Thus, from a diagram for G we can always delete any part which contains two external lines and is joined to the rest of the diagram by two $G^{(0)}$-lines. Then we can calculate the contribution from the deleted part and afterwards write the expression for the whole diagram by using an "abbreviated diagram," in which some appropriate line is assigned the whole contribution from the deleted part.

Any part of a diagram which can be joined to the rest of the diagram by two $G^{(0)}$-lines (or two $D^{(0)}$-lines) is called a *self-energy part*. Moreover, by an *irreducible* self-energy part, we mean a self-energy part which cannot be divided into two parts joined by only one $G^{(0)}$-line. For example, the self-energy parts in Figs. 9, 10(a) and 10(c) are irreducible, but the one in Fig. 10(b) is reducible. Every diagram for a function G or D consists of a basic line with irreducible self-energy parts strung along it. Moreover, there can be infinitely many such self-energy parts, and they can appear in any order.

In general, it is impossible to sum all the diagrams for the Green's functions. However, we can carry out a partial summation, and as a result, all that is left is a sum involving various irreducible self-energy parts. In fact, consider any diagram for a G-function. The diagram begins with a $G^{(0)}$-line, and some irreducible self-energy part comes next. If we cut these first two elements off the diagram, the rest of the diagram again begins with a $G^{(0)}$-line and contains an arbitrary number of arbitrary self-energy parts. Thus, the rest of the diagram again represents the exact G-function. This leads to the basic equation

$$G = G^{(0)} + G^{(0)} \Sigma G$$

for G, or equivalently,

$$G^{-1} = (G^{(0)})^{-1} - \Sigma, \tag{10.1}$$

where

$$\Sigma = \Sigma_1 + \Sigma_2 + \Sigma_3 + \cdots \tag{10.2}$$

is the sum of all the different irreducible self-energy parts. We call Σ the *exact irreducible self-energy part* or the *mass operator*.

The quantity Σ can be calculated by using diagrams which differ from the diagrams for G only by the absence of the two external $G^{(0)}$-lines (and which cannot be divided into two parts joined by only one $G^{(0)}$-line). However, in cases where it is necessary to sum an infinite series, instead of just calculating the first few diagrams, it is usually more convenient to express Σ in terms of another set of diagrams, called the *vertex part*. This procedure depends on the specific nature of the interaction, and will be illustrated by using the examples considered in Sec. 9.

A. Two-particle interactions. In this case, it is most convenient to use the symmetrized form of the theory. The first-order term in Σ corresponds to the diagram in Fig. 9, without the external $G^{(0)}$-lines. We begin by separating from the higher-order terms all diagrams in which the self-energy part is connected to the basic G-line by a single open square $\Gamma^{(0)}$, as in Fig. 10(a). It is entirely clear that the set of all such diagrams for Σ can be obtained from the first-order diagram by inserting all possible self-energy parts into the internal $G^{(0)}$-line, which as a result gives a line corresponding to the exact Green's function. Thus, the expression corresponding to the set of all diagrams for Σ which are connected to the basic G-line by a single square $\Gamma^{(0)}$ equals

$$\Sigma^{(1)}_{\alpha\beta}(p) = i \int \frac{d^4 p_1}{(2\pi)^4} \Gamma^{(0)}_{\alpha\gamma,\delta\beta}(p, p_1; p_1, p) G_{\delta\gamma}(p_1). \tag{10.3}$$

From now on, we shall represent the exact Green's function by a thick line. Then the quantity $\Sigma^{(1)}$ can be represented by the diagram shown in Fig. 22.

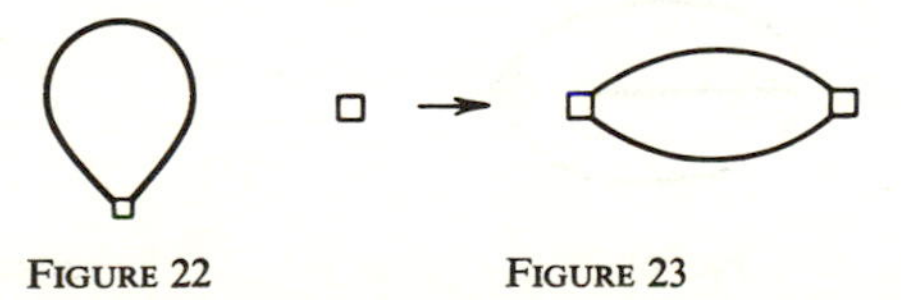

FIGURE 22 FIGURE 23

The simplest diagram which does not appear in the set of diagrams just described is the self-energy part in Fig. 10(c). Moreover, although some of the more complicated diagrams can be obtained by insertion of self-energy parts into internal $G^{(0)}$-lines, this method does not lead to the diagram shown in Fig. 11. Nevertheless, we can derive Fig. 11 from Fig. 10(c) if we first separate the $\Gamma^{(0)}$-square on the left from the rest of the diagram by cutting the three internal $G^{(0)}$-lines leaving the square, and then replace the square on the right by another diagram, as shown in Fig. 23. It is not hard to see that all the diagrams for Σ which are not included in (10.3) can be obtained from Fig. 10(c) by insertion of self-energy parts into internal $G^{(0)}$-lines and replacement of the square on the right by the set of all diagrams which have four external points and cannot be decomposed into disconnected parts.[14] We call this set the *vertex part*, denoted by $\Gamma_{\alpha\beta,\gamma\delta}(p_1, p_2; p_3, p_4)$, and we represent it in diagrams by a shaded square. It should be noted that just as in the case of $\Gamma^{(0)}$, the four momenta appearing in Γ must satisfy the conservation laws $p_1 + p_2 = p_3 + p_4$.

Thus, the second part of Σ can be represented by the diagram shown in Fig. 24, and equals

$$\begin{aligned}\Sigma^{(2)}_{\alpha\beta} = &-\frac{1}{2}\int\frac{d^4p_1\,d^4p_2}{(2\pi)^8}\,\Gamma^{(0)}_{\alpha\xi,\eta\delta}(p, p_1; p_2, p + p_1 - p_2)\\ &\times G_{\eta\mu}(p_2)G_{\nu\xi}(p_1)G_{\delta\gamma}(p + p_1 - p_2)\Gamma_{\mu\gamma,\nu\beta}(p_2, p + p_1 - p_2; p_1, p).\end{aligned} \tag{10.4}$$

Substituting $\Sigma = \Sigma^{(1)} + \Sigma^{(2)}$ into equation (10,1), we obtain

$$\begin{aligned}&[\omega - \xi(\mathbf{p})]G_{\alpha\beta}(p) - i\int\frac{d^4p_1}{(2\pi)^4}\,\Gamma^{(0)}_{\alpha\xi,\eta\gamma}(p, p_1; p_1, p)G_{\eta\xi}(p_1)G_{\gamma\beta}(p)\\ &\quad + \frac{1}{2}\int\frac{d^4p_1\,d^4p_2}{(2\pi)^8}\,\Gamma^{(0)}_{\alpha\xi,\eta\delta}(p, p_1; p_2, p_1 + p - p_2)G_{\eta\mu}(p_2)G_{\nu\xi}(p_1)\\ &\quad \times G_{\delta\gamma}(p + p_1 - p_2)\Gamma_{\mu\gamma,\nu\rho}(p_2, p + p_1 - p_2; p_1, p)G_{\rho\beta}(p) = \delta_{\alpha\beta}.\end{aligned} \tag{10.5}$$

This equation, connecting the G-function with the vertex part, is called *Dyson's equation*, and has been derived here by summation of diagrams. Below, we shall give an analytical derivation of Dyson's equation, and we shall examine the vertex part in more detail.

B. Electron-phonon interactions. The simplest diagram for Σ in the electron Green's function is shown in Fig. 25(a). In just the same way as before, we can easily see that this diagram is the only skeleton diagram, i.e.,

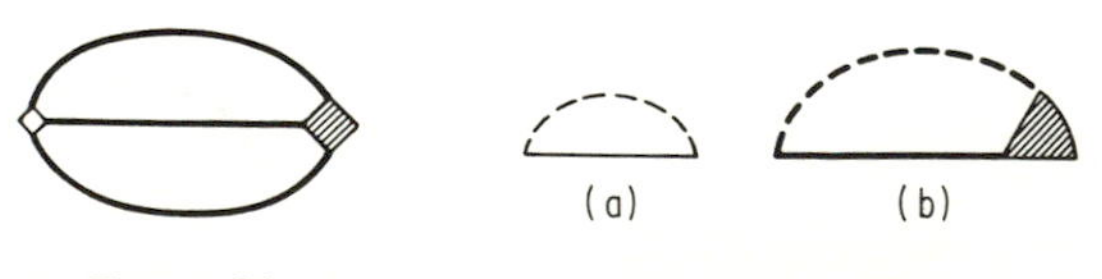

FIGURE 24 FIGURE 25

all the more complicated diagrams can be obtained by insertion of self-energy parts into internal $G^{(0)}$ or $D^{(0)}$-lines, and by replacement of the vertex on the right by the set of all diagrams with three external points, one associated with a

[14] The diagrams shown in Figs. 9 and 10(c), which are basic for constructing more complicated diagrams, are sometimes called *skeleton diagrams*.

phonon and two with electrons. We again call this quantity the *vertex part*, denoted by $\Gamma(p, p - k; k)$, and we represent it in diagrams by a shaded triangle.

Thus, in the case of electron-phonon interactions, the exact irreducible self-energy part for electrons, denoted by Σ, is represented by diagram (b) of Fig. 25, and equals

$$\Sigma = ig \int \frac{d^4k}{(2\pi)^4} G(p - k)D(k)\Gamma(p - k, p; k), \tag{10.6}$$

where we have set $G_{\alpha\beta} = G\delta_{\alpha\beta}$. Substituting (10.6) into (10.1), we obtain Dyson's equation for the electron Green's function:

$$[\omega - \xi(p)]G(p) - ig \int \frac{d^4k}{(2\pi)^4} G(p - k)D(k)\Gamma(p - k, p; k)G(p) = 1. \tag{10.7}$$

Similarly, the self-energy part for phonons, denoted by Π, can be obtained from the skeleton diagram shown in Fig. 26(a) by replacing the electron $G^{(0)}$-lines by exact G-lines and one of the constants g by the vertex part. Then Fig. 26(a) goes into Fig. 26(b), which corresponds to

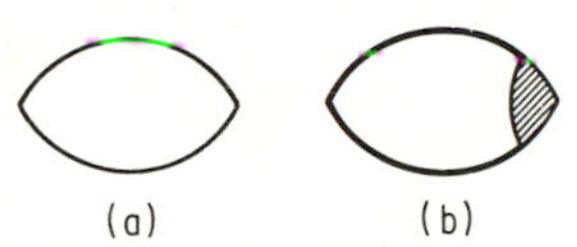

FIGURE 26

$$\Pi(k) = -2ig \int \frac{d^4p}{(2\pi)^4} G(p)G(p - k)\Gamma(p, p - k; k). \tag{10.8}$$

In this case, Dyson's equation for the phonon Green's function has the form

$$\begin{aligned} &[\omega_0^2(\mathbf{k})]^{-1}[\omega^2 - \omega_0^2(\mathbf{k})]D(k) \\ &\qquad + 2ig \int \frac{d^4p}{(2\pi)^4} G(p)G(p - k)\Gamma(p, p - k; k)D(k) = 1. \end{aligned} \tag{10.9}$$

C. External fields. We can also write an equation like Dyson's equation for a system of fermions in an external field. Bearing in mind that all the diagrams for G are chain-shaped, as in Fig. 15, we conclude that the Fourier component of the potential $V_{\alpha\beta}$ plays the role of Σ. In this case, Dyson's equation has the form

$$[\omega - \xi(\mathbf{p})]G_{\alpha\beta}(p, p') - \int \frac{d^4p_1}{(2\pi)^4} V_{\alpha\gamma}(p - p_1)G_{\gamma\beta}(p_1, p') = \delta_{\alpha\beta}. \tag{10.10}$$

10.2. Vertex parts. Many-particle Green's functions. Dyson's equation can also be obtained directly from the equations of motion for the Heisenberg operators:

$$i\frac{\partial\tilde{\psi}_\alpha}{\partial t} = [\tilde{\psi}_\alpha(x), \hat{H} - \hat{N}\mu],$$

$$\hat{H} = -\int \psi_\alpha^+(\mathbf{r}) \frac{\nabla^2}{2m} \psi_\alpha(\mathbf{r})\, d\mathbf{r} + H_{\text{int}}.$$

The operators $\hat{H}$ and $\hat{N}$ can be regarded as expressions involving either the

Schrödinger operators $\psi_\alpha(\mathbf{r})$ or the Heisenberg operators $\tilde{\psi}_\alpha(\mathbf{r}, t)$, since $\hat{H}$ and $\hat{N}$ are the same in both representations. Separating H_{int} from $\hat{H} - \mu\hat{N}$, and using the commutation rules for the operators $\tilde{\psi}$ and $\tilde{\psi}^+$ taken at the same time, we obtain

$$i\frac{\partial\tilde{\psi}_\alpha}{\partial t} = \left(-\frac{\nabla^2}{2m} - \mu\right)\tilde{\psi}_\alpha(x) + [\tilde{\psi}_\alpha(x), H_{\text{int}}].$$

We now differentiate the G-function with respect to the first time argument:

$$i\frac{\partial}{\partial t}G_{\alpha\beta}(x, x') = \frac{\partial}{\partial t}\langle T(\tilde{\psi}_\alpha(x)\tilde{\psi}_\beta^+(x'))\rangle.$$

Representing $T(\tilde{\psi}_\alpha(x)\tilde{\psi}_\beta^+(x'))$ in the form

$$\theta(t - t')\tilde{\psi}_\alpha(x)\tilde{\psi}_\beta^+(x') - \theta(t' - t)\tilde{\psi}_\beta^+(x')\tilde{\psi}_\alpha(x),$$

where

$$\theta(t) = \begin{cases} 1 & \text{for} \quad t > 0, \\ 0 & \text{for} \quad t < 0, \end{cases}$$

we find that

$$\begin{aligned} i\frac{\partial}{\partial t}G_{\alpha\beta}(x, x') &= \theta(t - t')\frac{\partial\tilde{\psi}_\alpha(x)}{\partial t}\tilde{\psi}_\beta^+(x') - \theta(t' - t)\tilde{\psi}_\beta^+(x')\frac{\partial\tilde{\psi}_\alpha(x)}{\partial t} \\ &\quad + \delta(t - t')(\tilde{\psi}_\alpha(\mathbf{r}, t)\tilde{\psi}_\beta(\mathbf{r}', t) + \tilde{\psi}_\beta^+(\mathbf{r}', t)\tilde{\psi}_\alpha(\mathbf{r}, t)) \\ &= \left\langle T\left(\frac{\partial\tilde{\psi}_\alpha(x)}{\partial t}\tilde{\psi}_\beta^+(x')\right)\right\rangle + \delta(x - x')\delta_{\alpha\beta} \end{aligned}$$

where we have used the commutation rules. The final result is

$$\left(i\frac{\partial}{\partial t} + \frac{\nabla^2}{2m} + \mu\right)G_{\alpha\beta}(x, x') = \delta(x - x')\delta_{\alpha\beta} - i\langle T([\tilde{\psi}_\alpha(x), H_{\text{int}}]\tilde{\psi}_\beta^+(x'))\rangle. \tag{10.11}$$

Since the form of the right-hand side depends on the specific nature of the interaction, we now consider various special cases.

A. Two-particle interactions. In this case, the operator H_{int} is given by formula (9.2). Carrying out the calculations and writing the result in symmetrized form [just as was done in deriving (9.3)], we obtain the following expression for the last term in (10.11):

$$-\frac{i}{2}\int d^4x_2\, d^4x_3\, d^4x_4\, \Gamma^{(0)}_{\alpha\gamma_2,\gamma_3\gamma_4}(x, x_2; x_3, x_4)\langle T(\tilde{\psi}_{\gamma_2}^+(x_2)\tilde{\psi}_{\gamma_4}(x_4)\tilde{\psi}_{\gamma_3}(x_3)\tilde{\psi}_\beta^+(x'))\rangle.$$

Thus, the problem reduces to finding the following average of a chronological product of four $\tilde{\psi}$-operators, a quantity we call a *two-particle Green's function*:

$$G^{\text{II}}_{\alpha\beta,\gamma\delta}(x_1, x_2; x_3, x_4) = \langle T(\tilde{\psi}_\alpha(x_1)\tilde{\psi}_\beta(x_2)\tilde{\psi}_\gamma^+(x_3)\tilde{\psi}_\delta^+(x_4))\rangle. \tag{10.12}$$

According to (6.32), G^{II} can be expressed in terms of ψ-operators in the interaction representation:

$$G^{\text{II}}_{\alpha\beta,\gamma\delta}(x_1, x_2; x_3, x_4) = \frac{\langle T(\psi_\alpha(x_1)\psi_\beta(x_2)\psi_\gamma^+(x_3)\psi_\delta^+(x_4)S(\infty))\rangle}{\langle S(\infty)\rangle}. \tag{10.13}$$

The calculation of (10.13) is completely analogous to that of an ordinary Green's function, and goes as follows: The operator $S(\infty)$ in the numerator is expanded as a power series in H_{int}. Using Wick's theorem, we represent each term of this series as a sum of terms containing products of $G^{(0)}$-functions, and then we associate a Feynman diagram with each of the latter terms. Unlike the diagrams for the ordinary Green's function, all these diagrams have four external lines. It is not hard to see that just as before, we need only take into account connected diagrams, i.e., diagrams which contain no lines that are not connected to one of the external lines, provided that at the same time we discard the factor $\langle S(\infty)\rangle$ appearing in the denominator of (10.13). Moreover, it is still true that all expressions depend on the order of the diagrams only through factors of the form λ^n. This makes it possible to deal with parts of diagrams separately and carry out partial summations.

All the connected diagrams for G^{II} can be divided into two groups. One of these groups contains the diagrams in which the points x_1 and x_3, as well as the points x_2 and x_4, are connected by a sequence of pairings, while the points x_1 and x_4, say, are isolated from each other. Such diagrams decompose into two separate parts which are not connected to each other by any lines at all. Moreover, we assign to the same group all diagrams in which x_1 is connected to x_4 and x_2 to x_3, while x_1 and x_3 are not connected. The simplest diagrams of this kind are of the zeroth order in H_{int} and are shown in Fig. 27. They correspond to the expressions

$$G^{(0)}_{\alpha\gamma}(x_1 - x_3)G^{(0)}_{\beta\delta}(x_2 - x_4), \quad \text{(a)}$$

$$-G^{(0)}_{\alpha\delta}(x_1 - x_4)G^{(0)}_{\beta\gamma}(x_2 - x_3). \quad \text{(b)}$$

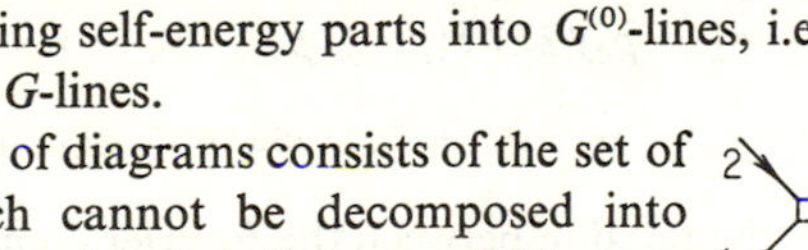

FIGURE 27

It is not hard to see that all the more complicated diagrams of this group are obtained by inserting self-energy parts into $G^{(0)}$-lines, i.e., by replacing thin $G^{(0)}$-lines by thick G-lines.

The other group of diagrams consists of the set of all diagrams which cannot be decomposed into separate parts. The simplest diagram of this type is of the first order in H_{int} and has the form shown in Fig. 28(a). It corresponds to the expression

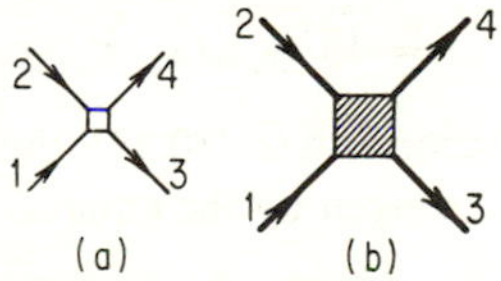

FIGURE 28

$$i\int G^{(0)}_{\alpha\gamma_1}(x_1 - x_1')G^{(0)}_{\beta\gamma_2}(x_2 - x_2')G^{(0)}_{\gamma_3\gamma}(x_3' - x_3)G^{(0)}_{\gamma_4\delta}(x_4' - x_4)$$
$$\times\ \Gamma^{(0)}_{\gamma_1\gamma_2,\,\gamma_3\gamma_4}(x_1', x_2'; x_3', x_4')\, d^4x_1'\, d^4x_2'\, d^4x_3'\, d^4x_4'.$$

The more complicated diagrams are obtained from Fig. 28(a) by insertion of diagrams into the external $G^{(0)}$-lines and replacement of the square by more complicated structures with four external points, e.g., like that shown in Fig. 23. Then $G^{(0)}$ becomes G in the expression just written, and $\Gamma^{(0)}$ becomes Γ, where Γ corresponds to the set of all possible diagrams with four external points. In other words, Fig. 28(a) goes into Fig. 28(b). It follows from these considerations that the quantity $G^{\text{II}}_{\alpha\beta,\gamma\delta}(x_1, x_2; x_3, x_4)$ can be conveniently written in the form

$$\begin{aligned} G^{\text{II}}_{\alpha\beta,\gamma\delta}(x_1, x_2; x_3, x_4) &= G_{\alpha\gamma}(x_1 - x_3)G_{\beta\delta}(x_2 - x_4) - G_{\alpha\delta}(x_1 - x_4)G_{\beta\gamma}(x_2 - x_3) \\ &\quad + i\int d^4x_1'\, d^4x_2'\, d^4x_3'\, d^4x_4'\, G_{\alpha\gamma_1}(x_1 - x_1')G_{\beta\gamma_2}(x_2 - x_2') \\ &\quad \times G_{\gamma_3\gamma}(x_3' - x_3)G_{\gamma_4\delta}(x_4' - x_4)\Gamma_{\gamma_1\gamma_2,\gamma_3\gamma_4}(x_1', x_2'; x_3', x_4'), \end{aligned} \tag{10.14}$$

where the quantity Γ corresponds to the vertex part just introduced.

The last term in equation (10.11) equals

$$\frac{i}{2}\int d^4x_2\, d^4x_3\, d^4x_4\, \Gamma^{(0)}_{\alpha\gamma_2,\gamma_3\gamma_4}(x, x_2; x_3, x_4)G^{\text{II}}_{\gamma_3\gamma_4,\gamma_2\beta}(x_3, x_4; x_2, x').$$

Using (10.14) to expand $G^{\text{II}}_{\gamma_3\gamma_4,\gamma_2\beta}$ and bearing in mind that $\Gamma^{(0)}$ is antisymmetric in the arguments with indices 3 and 4, we find that equation (10.11) becomes

$$\begin{aligned} &\left(i\frac{\partial}{\partial t} - H_0 + \mu\right)G_{\alpha\beta}(x - x') - i\int d^4x_2\, d^4x_3\, d^4x_4 \\ &\quad \times \Gamma^{(0)}_{\alpha\gamma_2,\gamma_3\gamma_4}(x, x_2; x_3, x_4)G_{\gamma_3\gamma_2}(x_3 - x_2)G_{\gamma_4\beta}(x_4 - x') \\ &\quad + \frac{1}{2}\int d^4x_2 \cdots d^4x_8\, \Gamma^{(0)}_{\alpha\gamma_2,\gamma_3\gamma_4}(x, x_2; x_3, x_4) \\ &\quad \times G_{\gamma_4\gamma_6}(x_4 - x_6)G_{\gamma_3\gamma_5}(x_3 - x_5)G_{\gamma_7\gamma_2}(x_7 - x_2)G_{\gamma_8\beta}(x_8 - x') \\ &\quad \times \Gamma_{\gamma_5\gamma_6,\gamma_7\gamma_8}(x_5, x_6; x_7, x_8) = \delta(x - x')\delta_{\alpha\beta}. \end{aligned} \tag{10.15}$$

Because of the homogeneity of space, the quantities Γ and G^{II} depend only on three coordinate differences. Therefore, the Fourier components of these quantities should be defined in the same way as in the case of $\Gamma^{(0)}$, e.g.,

$$\begin{aligned} &\Gamma_{\alpha\beta,\gamma\delta}(p_1, p_2; p_3, p_1 + p_2 - p_3)(2\pi)^4\delta(p_1 + p_2 - p_3 - p_4) \\ &\quad = \int \Gamma_{\alpha\beta,\gamma\delta}(x_1, x_2; x_3, x_4)e^{-ip_1x_1 - ip_2x_2 + ip_3x_3 + ip_4x_4}\, d^4x_1\, d^4x_2\, d^4x_3\, d^4x_4. \end{aligned} \tag{10.16}$$

According to (10.14), the relation between the Fourier components of G^{II} and Γ is given by the equation

$$\begin{aligned} G^{\text{II}}_{\alpha\beta,\gamma\delta}(p_1, p_2; p_3, p_1 + p_2 - p_3) &= G_{\alpha\gamma}(p_1)G_{\beta\delta}(p_2)\delta(p_1 - p_3)(2\pi)^4 \\ &\quad - G_{\alpha\delta}(p_1)G_{\beta\gamma}(p_2)\delta(p_2 - p_3)(2\pi)^4 \\ &\quad + iG_{\alpha\gamma_1}(p_1)G_{\beta\gamma_2}(p_2)G_{\gamma_3\gamma}(p_3)G_{\gamma_4\delta}(p_1 + p_2 - p_3) \\ &\quad \times \Gamma_{\gamma_1\gamma_2,\gamma_3\gamma_4}(p_1, p_2; p_3, p_1 + p_2 - p_3). \end{aligned} \tag{10.17}$$

Taking the Fourier transform of equation (10.15), we obtain equation (10.5). In this way, we have derived Dyson's equation by an analytical method, where the quantity Γ is determined by formulas (10.12), (10.17) and (10.16).

The quantity Γ can be calculated by summation of certain diagrams. Examples of such diagrams are shown in Fig. 23, and also in Fig. 29. The fact that the rules for associating analytical expressions with the diagrams for Γ are the same as in the case of the function G is an immediate consequence of the fact that the diagrams for Γ can be regarded as a subset of the diagrams for G. This can also be easily verified directly by using the analytical definition of Γ and then arguing in complete analogy with the methods of the preceding section.

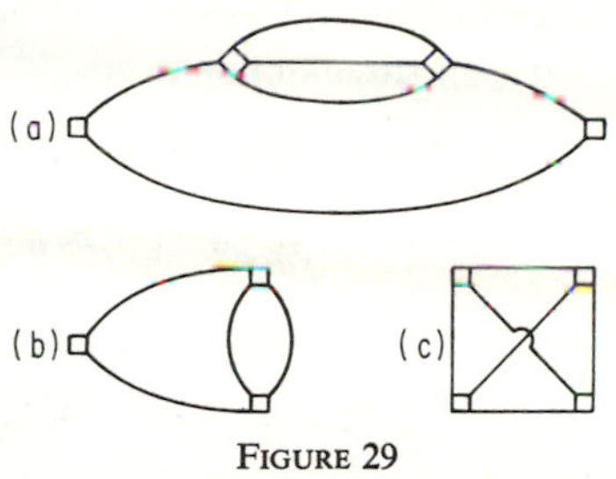

FIGURE 29

In calculating Γ, it is usually convenient to first carry out partial summation of separate parts. To this end, we introduce the concept of a *compact diagram*, by which we mean a diagram containing no self-energy parts. For example, the diagrams in Fig. 23, and diagrams (b) and (c) of Fig. 29, are compact, while diagram (a) of Fig. 29 is not compact. All the diagrams for Γ can be obtained from compact diagrams by inserting self-energy parts into internal $G^{(0)}$-lines, i.e., by replacing $G^{(0)}$-lines by exact G-lines. Thus, to calculate Γ, we need only consider compact diagrams, associating exact G-functions with the solid lines in each such diagram.

B. Electron-phonon interactions. Choosing (9.5) for H_{int}, we find that the last term in equation (10.11) is

$$-ig\langle T(\tilde{\psi}_\alpha(x)\tilde{\psi}_\beta^+(x')\tilde{\varphi}(x))\rangle.$$

The quantity

$$G_{\alpha\beta}(x_1, x_2; x_3) = \langle T(\tilde{\psi}_\alpha(x_1)\tilde{\psi}_\beta^+(x_2)\tilde{\varphi}(x_3))\rangle \tag{10.18}$$

can be associated with a set of Feynman diagrams with one external phonon line and two external electron lines. The simplest of these diagrams is obtained in the first-order perturbation-theory approximation [see diagram (a) of Fig. 30], and corresponds to the quantity

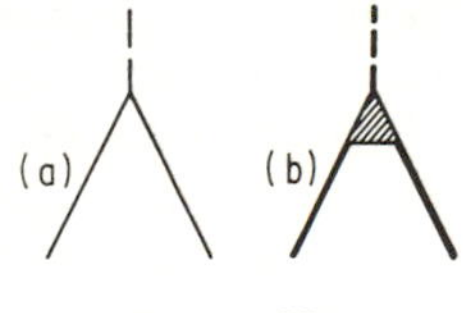

FIGURE 30

$$-g\delta_{\alpha\beta}\int d^4y\, G^{(0)}(x_1 - y)G^{(0)}(y - x_2)D^{(0)}(y - x).$$

Arguing in the same way as before, we can associate diagram (b) of Fig. 30 with the quantity $G_{\alpha\beta}$ equal to

$$G_{\alpha\beta}(x_1, x_2; x_3) = \delta_{\alpha\beta}G(x_1, x_2; x_3) \tag{10.19}$$

$$= -\delta_{\alpha\beta}\int d^4x_1'\, d^4x_2'\, d^4x_3'\, G(x_1 - x_1')G(x_2' - x_2)D(x_3' - x_3)\Gamma(x_1', x_2'; x_3').$$

The function Γ corresponds to the set of all diagrams with three external points (associated with one phonon and two electrons). Thus, Γ represents the vertex part for electron-phonon interactions [see diagram (b) of Fig. 30]. Because of the homogeneity of space, the quantities Γ and G depend only on two coordinate differences. Therefore, for example, the Fourier transform of Γ can be written in the form

$$\Gamma(p, p-k; k)(2\pi)^4\delta(p-p'-k) = \int d^4x_1\, d^4x_2\, d^4x_3\, \Gamma(x_1, x_2; x_3)e^{-ipx_1+ip'x_2+ikx_3}. \tag{10.20}$$

The relation between the Fourier components of Γ and G is

$$G(p, p-k; k) = -G(p)G(p-k)D(k)\Gamma(p, p-k; k). \tag{10.21}$$

Using formulas (10.18) and (10.19) to write an expression for the last term of equation (10.11), in the case of electron-phonon interactions, we find an equation for G in coordinate space. Using (10.20) to take the Fourier transform of this last equation, we obtain Dyson's equation (10.7).

All that has been said about the calculation of the vertex part for two-particle interactions remains valid in the present case. To calculate Γ, we have to form all compact diagrams and associate analytical expressions with them, according to the same rules used to calculate G. Then each solid line will denote an exact G-function and each dashed line an exact D-function. Examples are shown in Fig 31.

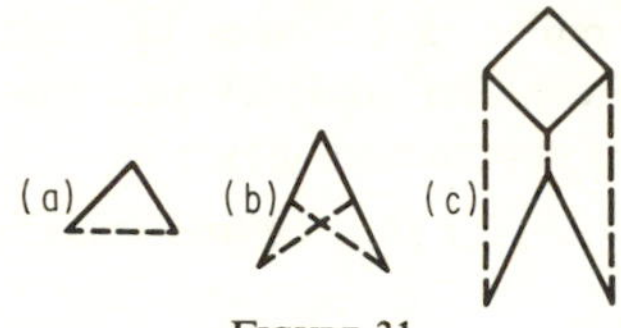

FIGURE 31

A few remarks are in order concerning the meaning of the functions G^{II} and $G(x_1, x_2; x_3)$, which we introduced in deriving Dyson's equation. These functions, and also other averages of chronological products of a large number of field operators, are called *many-particle Green's functions*, while the functions G and D themselves are called *one-particle Green's functions*. Many-particle Green's functions, just like one-particle Green's functions, determine the macroscopic properties of systems. In particular, the two-particle Green's function G^{II} determines the behavior of a system of electrons in an external electromagnetic field (see Chap. 7). Since these functions depend on a large number of arguments, it is very difficult to study their analytic properties. The situation becomes simpler if we can assume that some of these arguments are equal. For example, if $x_1 = x_3$, $x_2 = x_4$ in the function G^{II}, the analytic properties of the Fourier transform of G^{II} with respect to the variables $x_1 - x_2$ are the same as those of the phonon Green's function $D(\mathbf{k}, \omega)$. Since it is usually just these special cases that are of interest, it is simplest to determine the analytic properties of the corresponding particular Green's functions without analyzing the general case.

The poles of the Fourier components of many-particle Green's functions determine the excitation spectrum of the system, just as do the poles of $G(p)$ and $D(k)$, all of which occur among the poles of the many-particle

Green's functions. However, in addition to the poles of $G(p)$ and $D(k)$, new poles can occur, corresponding to other branches of the excitation spectrum. We shall not undertake a general analysis of this problem. A specific example will be considered in Chap. 4, Sec. 19, where we shall find the equation for the poles of a two-particle Green's function for a Fermi system and show that these poles determine excitations of the Bose type.

In principle, one might attempt to calculate many-particle Green's functions by writing equations analogous to Dyson's equation, relating these functions to higher-order functions. However, this procedure gives no useful results in practice, and it is simplest to sum the diagrams directly. Then it frequently turns out that a definite sequence of diagrams is most important. In such cases, summing the diagrams usually does not involve much work.

10.3. The ground-state energy. We conclude this section by deriving some formulas which enable us to obtain the correction to the ground-state energy due to interaction between particles. Subtracting the corresponding equation for the function $G^{(0)}$ from equation (10.11), we obtain

$$\left(i\frac{\partial}{\partial t} + \frac{\nabla^2}{2m} + \mu\right)[G_{\alpha\beta}(x - x') - G^{(0)}_{\alpha\beta}(x - x')] = -i\langle T([\tilde{\psi}_\alpha(x), H_{\text{int}}]\tilde{\psi}^+_\beta(x'))\rangle.$$

Letting $\mathbf{r} \to \mathbf{r}'$, $t' \to t + 0$, and then integrating both sides with respect to $\mathbf{r}$, we find that

$$\nu\langle H_{\text{int}}\rangle = -i\int d\mathbf{r} \lim_{\substack{\mathbf{r}'\to\mathbf{r}\\ t'\to t+0}} \left(i\frac{\partial}{\partial t} + \frac{\nabla^2}{2m} + \mu\right)[G_{\alpha\alpha}(x - x') - G^{(0)}_{\alpha\alpha}(x - x')],$$

where ν is the number of ψ^+-operators appearing in H_{int}.

Suppose the interaction Hamiltonian is proportional to some constant g (such a constant can always be introduced). As a function of μ, the ground-state energy (more exactly, the thermodynamic potential $\Omega = E - \mu N$) equals

$$\Omega = \langle \hat{H} - \mu\hat{N}\rangle.$$

According to a familiar statistical formula (see L8), we have

$$\frac{\partial\Omega}{\partial g} = \left\langle \frac{\partial}{\partial g}(\hat{H} - \mu\hat{N})\right\rangle = \frac{1}{g}\langle H_{\text{int}}\rangle.$$

Integrating this relation between the limits 0 and g, we obtain

$$\Omega - \Omega_0 = \int_0^g \frac{dg_1}{g_1}\langle H_{\text{int}}\rangle, \tag{10.22}$$

where Ω_0 is the potential for noninteracting particles. Substituting into (10.22) the expression found above for $\langle H_{\text{int}}\rangle$ in terms of the Green's function, we find that

$$\Omega - \Omega_0 = -\frac{i}{\nu}\int_0^g \frac{dg_1}{g_1}\int d\mathbf{r} \lim_{\substack{\mathbf{r}'\to\mathbf{r}\\ t'\to t+0}} \left(i\frac{\partial}{\partial t} + \frac{\nabla^2}{2m} + \mu\right)[G_{\alpha\alpha}(x - x') - G^{(0)}_{\alpha\alpha}(x - x')]. \tag{10.23}$$

Setting $G_{\alpha\beta}(x - x') = \delta_{\alpha\beta}G(x - x')$, going over to the momentum representation and using the equation for $G^{(0)}$, we finally obtain

$$\Omega - \Omega_0 = -\frac{2i}{\nu} V \int_0^g \frac{dg_1}{g_1} \int \frac{d^4p}{(2\pi)^4} G^{(0)-1}(\mathbf{p})[G(p) - G^{(0)}(p)]e^{i\omega t},$$

where $t \to +0$ and V is the volume of the system.

Another useful formula can be obtained from the relation (see L8)

$$\left(\frac{\partial\Omega}{\partial m}\right)_{T,V,\mu} = \left\langle\frac{\partial\hat{H}}{\partial m}\right\rangle.$$

Since

$$\frac{\partial\hat{H}}{\partial m} = \frac{1}{2m^2}\int \psi_\alpha^+(\mathbf{r})\,\nabla^2\psi_\alpha(\mathbf{r})\,d\mathbf{r},$$

it follows that

$$\frac{\partial\Omega}{\partial m} = -\frac{i}{2m^2}\int \Big[\nabla_x^2 \underset{\substack{x' \to x \\ t' \to t+0}}{G_{\alpha\alpha}} (x - x')\Big] d\mathbf{r}.$$

Transforming to Fourier components, we obtain

$$\frac{\partial\Omega}{\partial m} = \frac{iV}{m^2}\int \frac{d^4p}{(2\pi)^4} p^2 G(p)e^{i\omega t}, \tag{10.24}$$

where $t \to +0$.

Finally, this is an appropriate place to recall the formula

$$\frac{\partial\Omega}{\partial\mu} = -N = 2iV\int \frac{d^4p}{(2\pi)^4} G(p)e^{i\omega t}, \tag{10.25}$$

derived in Sec. 7.3.

3

THE DIAGRAM TECHNIQUE FOR $T \neq 0$

To a great extent, the discussion and calculations presented in this chapter duplicate the corresponding material in the preceding chapter. However, we deem it advisable to preserve this parallelism because of the importance of these two chapters for our subsequent considerations. At the same time, this will enable readers already familiar with the methods of quantum field theory, who are interested in studying the diagram technique only for $T \neq 0$, to begin reading the book at this point.

11. Temperature Green's Functions

11.1 General properties. So far, we have studied the properties of many-particle systems at the absolute zero of temperature. For "finite" temperatures (i.e., for $T \neq 0$), the problem becomes much more complicated. The usual "classical" method of statistical physics consists in calculating directly the thermodynamic quantities characterizing a system, as functions of the system's temperature and density. In doing this, the answer is expressed in terms of the powers of some small parameter, since no problem of this type can actually be solved exactly. Applying the usual thermodynamic perturbation theory (see L8, Sec. 32), we can easily write down the first two terms of the perturbation series for the free energy F:

$$F = F_0 + \sum_n V_{nn} e^{(F_0 - E_n^{(0)})/T} + \frac{1}{2} \sum_{n,m} \frac{|V_{nm}|^2}{E_n^{(0)} - E_m^{(0)}} \left[e^{(F_0 - E_n^{(0)})/T} - e^{(F_0 - E_m^{(0)})/T} \right] + \frac{1}{2T} \left[\sum_n V_{nn} e^{(F_0 - E_n^{(0)})/T} \right]^2 + \cdots$$

However, even writing down subsequent terms (not to mention making direct calculations) is quite a difficult problem. Moreover, summing an infinite sequence of terms of this series is completely hopeless. For this reason, use of the diagram technique is particularly attractive when dealing with statistical problems at finite temperatures, since this technique works with Green's functions and allows us to form a very intuitive picture of the structure and character of any approximation.

The diagram technique presented in the preceding chapter cannot be directly generalized to the case of finite temperatures. However, a diagram technique can be constructed for special quantities, the so-called "temperature Green's functions," which, unlike ordinary Green's functions, do not depend on the time t, but rather on a fictitious "imaginary time" $\tau = it$, varying in the interval from 0 to i/T [see Matsubara (M1)]. As in the technique for $T = 0$, in Matsubara's method it is not the thermodynamic quantities themselves that are calculated, but rather the temperature Green's functions $\mathscr{G}(\mathbf{r}, \tau)$ just referred to. Any term of the perturbation series for these functions is described by a corresponding Feynman diagram, and can be calculated by the rules of the Feynman technique, i.e., a free-particle temperature Green's function $\mathscr{G}^{(0)}(\mathbf{r}, \tau)$ is associated with every line of the diagram, an interaction operator is associated with every vertex of the diagram, and so on. The only difference compared to the case $T = 0$ is that instead of integrating with respect to t from $-\infty$ to ∞ at each vertex of the diagram, we integrate with respect to τ from 0 to $1/T$.

The temperature Green's function figuring in the diagram technique is defined as

$$\mathscr{G}_{\alpha\beta}(\mathbf{r}_1, \tau_1; \mathbf{r}_2, \tau_2) = \begin{cases} -\operatorname{Sp}\{e^{(\Omega+\mu\hat{N}-\hat{H})/T} e^{(\hat{H}-\mu\hat{N})(\tau_1-\tau_2)}\psi_\alpha(\mathbf{r}_1) e^{-(\hat{H}-\mu\hat{N})(\tau_1-\tau_2)}\psi_\beta^+(\mathbf{r}_2)\} & \text{for } \tau_1 > \tau_2, \\ \pm\operatorname{Sp}\{e^{(\Omega+\mu\hat{N}-\hat{H})/T} e^{-(\hat{H}-\mu\hat{N})(\tau_1-\tau_2)}\psi_\beta^+(\mathbf{r}_2) e^{(\hat{H}-\mu\hat{N})(\tau_1-\tau_2)}\psi_\alpha(\mathbf{r}_1)\} & \text{for } \tau_1 < \tau_2. \end{cases} \tag{11.1}$$

Here $\psi_\alpha(\mathbf{r})$, $\psi_\alpha^+(\mathbf{r})$ are the Schrödinger operators of the system, the plus sign corresponds to the case of fermions, and the minus sign corresponds to the case of bosons. The symbol Sp (for *Spur* = *trace*) denotes the operation of taking the sum of all diagonal elements of the matrix written after it, where the sum extends over both the number of particles in the system and all possible states of the system with a given number of particles. Thus, $\mathscr{G}$ is by definition a function of the temperature T and of the chemical potential μ. The quantity Ω appearing in the exponential in (11.1) is the thermodynamic potential written in terms of the variables T, V and μ, i.e.,

$$d\Omega = -S\,dT - P\,dV - N\,d\mu.$$

We recall that the operator

$$\operatorname{Sp}\{e^{(\Omega+\mu\hat{N}-\hat{H})/T}\ldots\}$$

is the ordinary Gibbs statistical average, which we shall often denote by just the symbol $\langle \cdots \rangle$. Similarly, the temperature Green's function $\mathscr{D}$ for phonons is defined as

$$\mathscr{D}(\mathbf{r}_1, \tau_1; \mathbf{r}_2, \tau_2) = \begin{cases} -\mathrm{Sp}\,\{e^{(\Omega-\hat{H})/T}e^{\hat{H}(\tau_1-\tau_2)}\varphi(\mathbf{r}_1)e^{-\hat{H}(\tau_1-\tau_2)}\varphi(\mathbf{r}_2)\} & \text{for} \quad \tau_1 > \tau_2, \\ -\mathrm{Sp}\,\{e^{(\Omega-\hat{H})/T}e^{-\hat{H}(\tau_1-\tau_2)}\varphi(\mathbf{r}_2)e^{\hat{H}(\tau_1-\tau_2)}\varphi(\mathbf{r}_1)\} & \text{for} \quad \tau_1 < \tau_2, \end{cases} \tag{11.2}$$

where $\varphi(\mathbf{r})$ is the Schrödinger operator of the phonon field.

It is an immediate consequence of the definitions (11.1) and (11.2) that the temperature Green's functions involve τ_1 and τ_2 only through the "time" difference $\tau_1 - \tau_2$. Of course, if the system is also isolated and homogeneous, $\mathscr{G}$ and $\mathscr{D}$ involve $\mathbf{r}_1$ and $\mathbf{r}_2$ only through the coordinate differences $\mathbf{r}_1 - \mathbf{r}_2$, e.g., $\mathscr{G} = \mathscr{G}(\mathbf{r}_1 - \mathbf{r}_2, \tau_1 - \tau_2)$. Moreover, $\mathscr{G}(\tau)$ is a discontinuous function of the variable τ, and undergoes a jump at $\tau = 0$, where the size of the jump can be calculated directly from the definition of $\mathscr{G}$. In the case of fermions, we have

$$\Delta\mathscr{G} = \mathscr{G}(\tau) - \mathscr{G}(-\tau)|_{\tau\to+0} = -\mathrm{Sp}\,\{e^{(\Omega+\mu\hat{N}-\hat{H})/T}[\psi_\alpha(\mathbf{r}_1)\psi_\beta^+(\mathbf{r}_2) + \psi_\beta^+(\mathbf{r}_2)\psi_\alpha(\mathbf{r}_1)]\},$$

and according to the commutation rules for ψ and ψ^+,

$$\Delta\mathscr{G} = -\delta_{\alpha\beta}\delta(\mathbf{r}_1 - \mathbf{r}_2).$$

The size of the jump of the $\mathscr{G}$-function is the same for bosons as for fermions.

The expressions (11.1) and (11.2) can be written in a form analogous to the definition of the Green's function for $T = 0$. To do this, we use the formulas[1]

$$\begin{aligned} \tilde{\psi}_\alpha(\mathbf{r}, \tau) &= e^{(\hat{H}-\mu\hat{N})\tau}\,\psi_\alpha(\mathbf{r})e^{-(\hat{H}-\mu\hat{N})\tau}, \\ \bar{\tilde{\psi}}_\alpha(\mathbf{r}, \tau) &= e^{(\hat{H}-\mu\hat{N})\tau}\,\psi_\alpha^+(\mathbf{r})e^{-(\hat{H}-\mu\hat{N})\tau}, \\ \tilde{\varphi}(\mathbf{r}, \tau) &= e^{\hat{H}\tau}\varphi(\mathbf{r})e^{-\hat{H}\tau} \end{aligned} \tag{11.3}$$

to introduce "Heisenberg" field operators depending on the "time" τ. In terms of these operators, complicated expressions of the type (11.1) take the form

$$\begin{aligned} \mathscr{G}_{\alpha\beta}(\mathbf{r}_1, \tau_1; \mathbf{r}_2, \tau_2) &= -\mathrm{Sp}\,\{e^{(\Omega+\mu\hat{N}-\hat{H})/T}T_\tau(\tilde{\psi}_\alpha(\mathbf{r}_1, \tau_1)\bar{\tilde{\psi}}_\beta(\mathbf{r}_2, \tau_2))\} \\ &\equiv -\langle T_\tau(\tilde{\psi}_\alpha(\mathbf{r}_1, \tau_1)\bar{\tilde{\psi}}_\beta(_2,\mathbf{r}\ \tau_2))\rangle \end{aligned} \tag{11.4}$$

[cf. (7.1) and (7.14)].

The symbol T_τ in (11.4) denotes the operation of T-ordering, with which we are already familiar from the preceding chapter (Sec. 6). The operators appearing in a T_τ-product are arranged from left to right in order of decreasing "time" τ. (The symbol denoting the T_τ-product is equipped with an index τ to distinguish it from the temperature T.) We recall that in the case of fermions

$$T_\tau(\psi_1\psi_2\cdots) = \delta_P\psi_{i_1}\psi_{i_2}\cdots,$$

[1] We note at once that ψ and $\bar{\psi}$ are no longer Hermitian conjugates of each other.

where the ψ-operators on the right are arranged in chronological order, and δ_P equals $+1$ or -1, depending on whether or not the permutation

$$1, 2, \cdots \to i_1, i_2, \cdots$$

is even or odd. In particular, we have

$$T_\tau(\tilde{\psi}(1)\bar{\tilde{\psi}}(2)) = \quad \tilde{\psi}(1)\bar{\tilde{\psi}}(2) \quad \text{for} \quad \tau_1 > \tau_2,$$

$$T_\tau(\tilde{\psi}(1)\bar{\tilde{\psi}}(2)) = -\bar{\tilde{\psi}}(2)\tilde{\psi}(1) \quad \text{for} \quad \tau_1 < \tau_2.$$

Similar relations are used to define many-particle Green's functions in Matsubara's technique. Thus, the two-particle temperature Green's function has the form

$$\mathscr{G}^{\text{II}}_{\alpha\beta,\,\gamma\delta}(1, 2; 3, 4) = \langle T_\tau(\tilde{\psi}_\alpha(1)\tilde{\psi}_\beta(2)\bar{\tilde{\psi}}_\gamma(3)\bar{\tilde{\psi}}_\delta(4))\rangle. \tag{11.5}$$

The generalization to the case of Green's functions depending on a larger number of variables is obvious.

In principle, the function $\mathscr{G}$ defines all the thermodynamic properties of the system. For example, using the formula

$$N = \pm \int \mathscr{G}_{\alpha\alpha}(\mathbf{r}, \tau; \mathbf{r}, \tau + 0)\, d\mathbf{r} \tag{11.6}$$

(which is an immediate consequence of the definition of $\mathscr{G}$) and the relation

$$\hat{N} = \int \psi^+_\alpha(\mathbf{r})\psi_\alpha(\mathbf{r})\, d\mathbf{r},$$

we can calculate the number of particles in the system as a function of its chemical potential μ, or, solving (11.6) for μ, we can calculate the chemical potential as a function of the temperature and density $n = N/V$. Then integrating the familiar thermodynamic relation

$$\frac{\partial f}{\partial n} = \mu(n, T),$$

we can find $f(n, T)$, the free energy per unit volume.

If only two-particle interactions occur in the system, as described by the Hamiltonian

$$\hat{H} = -\frac{1}{2m}\int \psi^+_\alpha(\mathbf{r})\, \nabla^2\psi_\alpha(\mathbf{r})\, d\mathbf{r}$$
$$+ \frac{1}{2}\int \psi^+_\alpha(\mathbf{r}_1)\psi^+_\beta(\mathbf{r}_2)U(\mathbf{r}_1 - \mathbf{r}_2)\psi_\beta(\mathbf{r}_2)\psi_\alpha(\mathbf{r}_1)\, d\mathbf{r}_1\, d\mathbf{r}_2,$$

then the energy of the system can be expanded in terms of the two-particle temperature Green's function:

$$E(\mu, T) = \langle \hat{H} \rangle = \mp \frac{1}{2m}\int \nabla^2_{\mathbf{r}_1}\mathscr{G}_{\alpha\alpha}(1, 2)\Big|_{\substack{\mathbf{r}_1 = \mathbf{r}_2 \\ \tau_2 = \tau_1 + 0}} d\mathbf{r}_1$$
$$- \frac{1}{2}\int U(\mathbf{r}_1 - \mathbf{r}_2)\mathscr{G}^{\text{II}}_{\alpha\beta,\,\beta\alpha}(1, 2; 3, 4)\Big|_{\substack{\mathbf{r}_3 = \mathbf{r}_2,\, \mathbf{r}_4 = \mathbf{r}_1 \\ \tau_3 = \tau_4 + 0,\, \tau_4 = \tau_1 + 0 \\ \tau_1 = \tau_2 + 0}} d\mathbf{r}_1\, d\mathbf{r}_2.$$

Later on, we shall give many more formulas relating the temperature Green's functions to thermodynamic quantities.

The class of problems that can be solved by using temperature Green's functions is not confined just to thermodynamics. The Green's function determines various correlation properties of the system, e.g., those appearing in the interaction of condensed matter with neutrons, X-rays, etc. For example, the two-particle Green's function is connected in an obvious way with the density correlation function

$$F(\mathbf{r}_1, \mathbf{r}_2) = \overline{n(\mathbf{r}_1)n(\mathbf{r}_2)} \equiv \langle \psi_\alpha^+(\mathbf{r}_1)\psi_\alpha(\mathbf{r}_1)\psi_\beta^+(\mathbf{r}_2)\psi_\beta(\mathbf{r}_2)\rangle,$$

which determines the elastic scattering of X-rays and neutrons. Moreover, below we shall establish the relation between temperature Green's functions and the corresponding time-dependent quantities, thereby making possible the study of various transport phenomena.

Next, we note an important property of the temperature Green's function $\mathscr{G}$. As already remarked, $\mathscr{G}$ is a function of the "time" difference $\tau_1 - \tau_2 = \tau$, and as such, is defined in the interval from $-1/T$ to $1/T$. In the expression (11.1) for $\mathscr{G}(\tau < 0)$, we carry out a cyclic permutation of the operators appearing behind the trace sign:[2]

$$\begin{aligned}\mathscr{G}(\tau < 0) &= \pm \operatorname{Sp}\{e^{\Omega/T}e^{(\hat{H}-\mu\hat{N})\tau}\psi(\mathbf{r}_1)e^{-(\hat{H}-\mu\hat{N})[\tau+(1/T)]}\psi^+(\mathbf{r}_2)\} \qquad (11.7)\\ &= \pm \operatorname{Sp}\{e^{(\Omega+\mu\hat{N}-\hat{H})/T}e^{(\hat{H}-\mu\hat{N})[\tau+(1/T)]}\psi(\mathbf{r}_1)e^{-(\hat{H}-\mu\hat{N})[\tau+(1/T)]}\psi^+(\mathbf{r}_2)\}.\end{aligned}$$

Comparing (11.7) with the formula for $\mathscr{G}(\tau > 0)$, and noting that

$$0 < \tau + \frac{1}{T} < \frac{1}{T}$$

for $\tau < 0$, we obtain the relation

$$\mathscr{G}(\tau < 0) = \mp\mathscr{G}\left(\tau + \frac{1}{T}\right). \qquad (11.8)$$

connecting $\mathscr{G}$ for negative "times" with its values for $\tau > 0$. Similarly, of course, we have

$$\mathscr{D}(\tau < 0) = \mathscr{D}\left(\tau + \frac{1}{T}\right). \qquad (11.8')$$

Another useful relation follows from the obvious fact that the phonon $\mathscr{D}$-function is real [the operators $\varphi(\mathbf{r})$ are real!]. We calculate the quantity $\mathscr{D}^*(\tau < 0)$:

$$\mathscr{D}(\tau < 0) = \mathscr{D}^*(\tau < 0) = -\operatorname{Sp}\{e^{\Omega/T}\varphi(\mathbf{r}_1)e^{\hat{H}\tau}\varphi(\mathbf{r}_2)e^{-\hat{H}\tau}e^{-\hat{H}/T}\}.$$

[2] The possibility of making such a permutation follows immediately from the definition of the trace of the matrix of a product of several operators:

$$\begin{aligned}\operatorname{Sp}(ABC\cdots DF) &= \sum_{i,k,\ldots} A_{ik}B_{kl}C_{lm}\cdots D_{np}F_{pi}\\ &= \sum_{i,k,\ldots} B_{kl}C_{lm}\cdots D_{np}F_{pi}A_{ik} = \operatorname{Sp}(BC\cdots DFA).\end{aligned}$$

Then, comparing this expression with $\mathscr{D}(\tau > 0)$, we arrive at the conclusion that the temperature Green's function for phonons is an even function of τ:

$$\mathscr{D}(\tau) = \mathscr{D}(-\tau). \tag{11.9}$$

This result is valid for the Green's function of any real field.

11.2 Temperature Green's functions for free particles. In a perturbation theory based on the diagram technique, free-particle Green's functions play an important role. In the absence of interactions, the statistical average in (11.1) can be carried out over the states of the individual particles separately. The energy levels E_n of the system can be written as a sum over the energies of the individual particles (in states with a given momentum $\mathbf{p}$ and spin projection α), and the same is true of the thermodynamic potential Ω:

$$E_n^{(0)} = \sum_{\mathbf{p},\alpha} n_{\mathbf{p}\alpha}\varepsilon_0(\mathbf{p}), \qquad \Omega_0 = \sum_{\mathbf{p},\alpha} \Omega_{\mathbf{p}\alpha}^{(0)}.$$

In the case of Fermi statistics, the occupation numbers of the states can only take the values 0 and 1, because of the Pauli principle.

To calculate the free-particle Green's functions, it is most convenient to use the definition (11.1). Substituting into (11.1) the Fourier expansions of the operators ψ and ψ^+

$$\psi_\alpha(\mathbf{r}_1) = \frac{1}{\sqrt{V}} \sum_{\mathbf{p}_1} a_{\mathbf{p}_1\alpha} e^{i\mathbf{p}_1\cdot\mathbf{r}_1}, \qquad \psi_\beta^+(\mathbf{r}_2) = \frac{1}{\sqrt{V}} \sum_{\mathbf{p}_2} a_{\mathbf{p}_2\beta}^+ e^{-i\mathbf{p}_2\cdot\mathbf{r}_2},$$

we obtain

$$\mathscr{G}_{\alpha\beta}^{(0)}(\tau > 0) = -\frac{1}{V} \sum_{\mathbf{p}_1,\mathbf{p}_2} e^{i(\mathbf{p}_1\cdot\mathbf{r}_1 - \mathbf{p}_2\cdot\mathbf{r}_2)} \times \operatorname{Sp}\{e^{(\Omega_0 + \mu\hat{N} - \hat{H}_0)/T} e^{(\hat{H}_0 - \mu\hat{N})\tau} a_{\mathbf{p}_1\alpha} e^{-(\hat{H}_0 - \mu\hat{N})\tau} a_{\mathbf{p}_2\beta}^+\}.$$

Moreover, bearing in mind that in the interaction representation the Hamiltonian $\hat{H}$ has the form

$$\hat{H}_0 = \sum_{\mathbf{p},\alpha} \hat{n}_{\mathbf{p}\alpha}\varepsilon_0(\mathbf{p}), \qquad \hat{N} = \sum_{\mathbf{p},\alpha} \hat{n}_{\mathbf{p}\alpha},$$

we can easily verify the identities

$$\begin{aligned} e^{(\hat{H}_0 - \mu\hat{N})\tau} a_{\mathbf{p}\alpha} e^{-(\hat{H}_0 - \mu\hat{N})\tau} &= a_{\mathbf{p}\alpha} e^{-[\varepsilon_0(\mathbf{p}) - \mu]\tau}, \\ e^{(\hat{N}_0 - \mu\hat{N})\tau} a_{\mathbf{p}\alpha}^+ e^{-(\hat{H}_0 - \mu\hat{N})\tau} &= a_{\mathbf{p}\alpha}^+ e^{[\varepsilon_0(\mathbf{p}) - \mu]\tau} \end{aligned} \tag{11.10}$$

(it is sufficient to calculate the only nonzero matrix element in the left and right-hand sides). Thus we have

$$\mathscr{G}^{(0)}(\tau > 0) = -\frac{1}{V} \sum_{\mathbf{p}_1,\mathbf{p}_2} e^{i(\mathbf{p}_1\cdot\mathbf{r}_1 - \mathbf{p}_2\cdot\mathbf{r}_2) - [\varepsilon_0(\mathbf{p}_1) - \mu]\tau} \operatorname{Sp}\{e^{(\Omega_0 + \mu\hat{N} - \hat{H}_0)/T} a_{\mathbf{p}_1\alpha} a_{\mathbf{p}_2\beta}^+\}.$$

Since the product $a_{\mathbf{p}_1\alpha}a^+_{\mathbf{p}_2\beta}$ has nonzero diagonal matrix elements only for $\mathbf{p}_1 = \mathbf{p}_2$, $\alpha = \beta$, it follows that

$$\mathscr{G}^{(0)}_{\alpha\beta}(\mathbf{r}_1 - \mathbf{r}_2, \tau > 0) = -\,\delta_{\alpha\beta}\frac{1}{V}\sum_{\mathbf{p}} e^{i\mathbf{p}\cdot(\mathbf{r}_1-\mathbf{r}_2)-[\varepsilon_0(\mathbf{p})-\mu]\tau}\langle a_{\mathbf{p}\alpha}a^+_{\mathbf{p}\alpha}\rangle.$$

The quantity $\langle a_{\mathbf{p}\alpha}a^+_{\mathbf{p}\alpha}\rangle$ can be expressed in terms of the equilibrium occupation numbers $n(\mathbf{p})$, which depend on the temperature and the chemical potential. For fermions we have

$$\langle a_{\mathbf{p}\alpha}a^+_{\mathbf{p}\alpha}\rangle = 1 - n(\mathbf{p}), \qquad n(\mathbf{p}) = \{e^{[\varepsilon_0(\mathbf{p})-\mu]/T} + 1\}^{-1}, \tag{11.11}$$

and for bosons

$$\langle a_{\mathbf{p}\alpha}a^+_{\mathbf{p}\alpha}\rangle = 1 + n(\mathbf{p}), \qquad n(\mathbf{p}) = \{e^{[\varepsilon_0(\mathbf{p})-\mu]/T} - 1\}^{-1}. \tag{11.12}$$

We now let the volume V of the system approach infinity, going over from summation over the momenta to integration, in the usual way. The final result is

$$\mathscr{G}^{(0)}_{\alpha\beta}(\mathbf{r}, \tau > 0) = -\delta_{\alpha\beta}\frac{1}{(2\pi)^3}\int d\mathbf{p}\, e^{i\mathbf{p}\cdot\mathbf{r}-[\varepsilon_0(\mathbf{p})-\mu]\tau}[1 \mp n(\mathbf{p})], \tag{11.13a}$$

where the upper sign corresponds to fermions, and the lower sign to bosons. To calculate $\mathscr{G}^{(0)}$ for $\tau < 0$, it is simplest to use the relation (11.8):

$$\begin{aligned}\mathscr{G}^{(0)}_{\alpha\beta}(\mathbf{r}, \tau < 0) &= \mp\, \mathscr{G}^{(0)}_{\alpha\beta}\left(\mathbf{r}, \tau + \frac{1}{T}\right)\\ &= \pm\,\delta_{\alpha\beta}\frac{1}{(2\pi)^3}\int d\mathbf{p}\, e^{i\mathbf{p}\cdot\mathbf{r}-[\varepsilon_0(\mathbf{p})-\mu]\tau}n(\mathbf{p}).\end{aligned} \tag{11.13b}$$

The Green's function for free phonons is calculated similarly. Substituting into (11.2) the Fourier expansion of the operator $\varphi(\mathbf{r})$

$$\varphi(\mathbf{r}) = \frac{i}{\sqrt{V}}\sum_{\mathbf{k}}\sqrt{\frac{\omega_0(\mathbf{k})}{2}}\,(b_{\mathbf{k}}e^{i\mathbf{k}\cdot\mathbf{r}} - b^+_{\mathbf{k}}\,e^{-i\mathbf{k}\cdot\mathbf{r}}),$$

where $\omega_0(\mathbf{k})$ is the energy of the phonon, we find after making the appropriate calculations that

$$\begin{aligned}\mathscr{D}^{(0)}(\mathbf{r}, \tau) = -\,\frac{1}{2(2\pi)^3}\int d\mathbf{k}\,\omega_0(\mathbf{k})\{&[N(\mathbf{k}) + 1]e^{i\mathbf{k}\cdot\mathbf{r}-\omega_0(\mathbf{k})|\tau|}\\ &+ N(\mathbf{k})e^{i\mathbf{k}\cdot\mathbf{r}+\omega_0(\mathbf{k})|\tau|}\},\end{aligned} \tag{11.14}$$

where

$$N(\mathbf{k}) = [e^{\omega_0(\mathbf{k})/T} - 1]^{-1}.$$

According to (11.9), $\mathscr{D}^{(0)}$ is an even function of τ.

12. Perturbation Theory

12.1. The interaction representation. If the particles making up the system are not free, then in the expression for the temperature Green's function (11.1), we can go over to a special interaction representation, like the interaction representation of quantum field theory [Matsubara (M1)]. To this end, we introduce a matrix $\mathscr{S}(\tau)$, where $0 < \tau < 1/T$, analogous to the S-matrix of field theory and defined by the relations

$$\begin{aligned} e^{-(\hat{H}-\mu\hat{N})\tau} &= e^{-(\hat{H}_0-\mu\hat{N})\tau}\mathscr{S}(\tau), \\ e^{(\hat{H}-\mu\hat{N})\tau} &= \mathscr{S}^{-1}(\tau)e^{(\hat{H}_0-\mu\hat{N})\tau}. \end{aligned} \tag{12.1}$$

Moreover, we introduce the particle-field operators

$$\begin{aligned} \psi(\mathbf{r}, \tau) &= e^{(\hat{H}_0-\mu\hat{N})\tau}\,\psi(\mathbf{r})e^{-(\hat{H}_0-\mu\hat{N})\tau}, \\ \bar{\psi}(\mathbf{r}, \tau) &= e^{(\hat{H}_0-\mu\hat{N})\tau}\,\psi^{+}(\mathbf{r})e^{-(\hat{H}_0-\mu\hat{N})\tau} \end{aligned} \tag{12.2}$$

in the interaction representation, which for $\hat{H} = \hat{H}_0$ coincide with the Heisenberg operators mentioned in Sec. 11. Other operators in the interaction representation are introduced by analogy with (12.2). In particular, we have

$$\begin{aligned} \hat{H}(\tau) &= e^{(\hat{H}_0-\mu\hat{N})\tau}\hat{H}e^{-(\hat{H}_0-\mu\hat{N})\tau}, \\ \hat{H}_{\text{int}}(\tau) &= e^{(\hat{H}_0-\mu\hat{N})\tau}\hat{H}_{\text{int}}e^{-(\hat{H}_0-\mu\hat{N})\tau}. \end{aligned}$$

This definition implies that the operators $\hat{H}(\tau)$, $\hat{H}_{\text{int}}(\tau)$ are obtained from $\hat{H}$, $\hat{H}_{\text{int}}$ by replacing $\psi(\mathbf{r})$, $\psi^{+}(\mathbf{r})$ by $\psi(\mathbf{r}, \tau)$, $\bar{\psi}(\mathbf{r}, \tau)$, respectively. Moreover, we note that $\hat{H}_0(\tau)$ and $\hat{N}(\tau)$ actually do not depend on τ (the free-particle Hamiltonian commutes with the operator $\hat{N}$):

$$\begin{aligned} \hat{H}_0(\tau) &= e^{(\hat{H}_0-\mu\hat{N})\tau}\hat{H}_0e^{-(\hat{H}_0-\mu\hat{N})\tau} = \hat{H}_0, \\ \hat{N}(\tau) &= e^{(\hat{H}_0-\mu\hat{N})\tau}\hat{N}e^{-(\hat{H}_0-\mu\hat{N})\tau} = \hat{N}. \end{aligned}$$

The matrix $\mathscr{S}(\tau)$ satisfies a simple equation, which differs from the corresponding equation (6.27) for the S-matrix by the substitution $t \to -i\tau$. However, we now give a fresh derivation of this equation. Differentiating the first of the equations (12.1) with respect to τ, we obtain

$$-(\hat{H} - \mu\hat{N})e^{-(\hat{H}-\mu\hat{N})\tau} = e^{-(\hat{H}_0-\mu\hat{N})\tau}\frac{\partial\mathscr{S}(\tau)}{\partial\tau} - (\hat{H}_0 - \mu\hat{N})e^{-(\hat{H}_0-\mu\hat{N})\tau}\mathscr{S}(\tau).$$

Multiplying both sides of this equation by $e^{(\hat{H}_0-\mu\hat{N})\tau}$, we have

$$\frac{\partial\mathscr{S}(\tau)}{\partial\tau} = -\hat{H}_{\text{int}}(\tau)\mathscr{S}(\tau). \tag{12.3}$$

The solution of equation (12.3) satisfying $\mathscr{S}(0) = 1$, which follows from the definition of $\mathscr{S}$, has the form

$$\mathscr{S}(\tau) = T_\tau \exp\left\{-\int_0^\tau \hat{H}_{\text{int}}(\tau')\,d\tau'\right\}. \tag{12.4}$$

As already noted, the symbol T_τ means that all operators must be arranged from left to right in order of decreasing τ. We can easily verify the validity of (12.4) by direct differentiation, taking account of the operation T_τ just mentioned.

Besides $\mathscr{S}(\tau)$, we consider another matrix $\mathscr{S}(\tau_1, \tau_2)$, where $\tau_1 > \tau_2$ is defined by

$$\mathscr{S}(\tau_1, \tau_2) = T_\tau \exp\left\{-\int_{\tau_2}^{\tau_1} \hat{H}_{\text{int}}(\tau')\, d\tau'\right\},$$
$$\mathscr{S}(\tau) = \mathscr{S}(\tau, 0).$$

Obviously, $\mathscr{S}(\tau_1, \tau_2)$ has the properties

$$\begin{aligned} \mathscr{S}(\tau_1, \tau_3) &= \mathscr{S}(\tau_1, \tau_2)\mathscr{S}(\tau_2, \tau_3) \quad \text{for} \quad \tau_1 > \tau_2 > \tau_3, \\ \mathscr{S}(\tau_1, \tau_2) &= \mathscr{S}(\tau_1)\mathscr{S}^{-1}(\tau_2) \qquad\quad \text{for} \quad \tau_1 > \tau_2. \end{aligned} \tag{12.5}$$

We now go over to the interaction representation in formula (11.1) for the Green's function. Expressing all exponentials containing $\hat{H}$ in terms of $\hat{H}_0$ and $\mathscr{S}$, we have

$$\begin{aligned} \mathscr{G}(\tau > 0) = -e^{\Omega/T} \operatorname{Sp}\{&e^{-(\hat{H}_0 - \mu\hat{N})/T}\mathscr{S}(1/T)\mathscr{S}^{-1}(\tau_1)e^{(\hat{H}_0 - \mu\hat{N})\tau_1}\psi(\mathbf{r}_1) \\ &\times e^{-(\hat{H}_0 - \mu\hat{N})\tau_1}\mathscr{S}(\tau_1)\mathscr{S}^{-1}(\tau_2)e^{(\hat{H}_0 - \mu\hat{N})\tau_2}\psi^+(\mathbf{r}_2)e^{-(\hat{H}_0 - \mu\hat{N})\tau_2}\mathscr{S}(\tau_2)\}, \end{aligned}$$

or, because of (12.1) and (12.5),

$$\mathscr{G}(\tau > 0) = -e^{\Omega/T} \operatorname{Sp}\{e^{-(\hat{H}_0 - \mu\hat{N})/T}\mathscr{S}(1/T, \tau_1)\psi(\mathbf{r}_1, \tau_1)\mathscr{S}(\tau_1, \tau_2)\bar{\psi}(\mathbf{r}_2, \tau_2)\mathscr{S}(\tau_2)\}. \tag{12.6a}$$

Similarly, for $\tau < 0$ we can write $\mathscr{G}$ in the form

$$\begin{aligned} \mathscr{G}(\tau < 0) = \pm\, e^{\Omega/T} \operatorname{Sp}\{&e^{-(\hat{H}_0 - \mu\hat{N})/T}\mathscr{S}(1/T, \tau_2) \\ &\times \bar{\psi}(\mathbf{r}_2, \tau_2)\mathscr{S}(\tau_2, \tau_1)\psi(\mathbf{r}_1, \tau_1)\mathscr{S}(\tau_1)\}. \end{aligned} \tag{12.6b}$$

The expressions (12.6a) and (12.6b) can be combined into a single formula

$$\mathscr{G}(\tau) = -e^{\Omega/T} \operatorname{Sp}\{e^{-(\hat{H}_0 - \mu\hat{N})/T}T_\tau(\psi(\mathbf{r}_1, \tau_1)\bar{\psi}(\mathbf{r}_2, \tau_2)\mathscr{S}(1/T))\}, \tag{12.6c}$$

which follows at once from (12.5) and the definition of the operation of T_τ-ordering.

We must still transform the quantity $e^{\Omega/T}$. To do so, we note that

$$e^{-\Omega/T} = \operatorname{Sp}\{e^{-(\hat{H} - \mu\hat{N})/T}\},$$

by definition, which immediately implies

$$e^{-\Omega/T} = \operatorname{Sp}\{e^{-(\hat{H}_0 - \mu\hat{N})/T}\mathscr{S}(1/T)\}.$$

Thus, we can finally write the expression for $\mathscr{G}$ in the interaction representation in the form

$$\mathscr{G}(\mathbf{r}_1, \tau_1; \mathbf{r}_2, \tau_2) = \frac{-\operatorname{Sp}\{e^{-(\hat{H}_0 - \mu\hat{N})/T}T_\tau(\psi(\mathbf{r}_1, \tau_1)\bar{\psi}(\mathbf{r}_2, \tau_2)\mathscr{S}(1/T))\}}{\operatorname{Sp}\{e^{-(\hat{H}_0 - \mu\hat{N})/T}\mathscr{S}(1/T)\}},$$

or, introducing the symbol for the Gibbs average over the states of a system of noninteracting particles,

$$\mathscr{G}(\mathbf{r}_1, \tau_1; \mathbf{r}_2, \tau_2) = -\frac{\langle T_\tau(\psi(\mathbf{r}_1, \tau_1)\bar{\psi}(\mathbf{r}_2, \tau_2)\mathscr{S})\rangle_0}{\langle\mathscr{S}\rangle_0}, \tag{12.7}$$

where

$$\langle\cdots\rangle_0 = \operatorname{Sp}\{e^{(\Omega_0 + \mu\hat{N} - \hat{H}_0)/T}\cdots\}, \qquad \mathscr{S} \equiv \mathscr{S}(1/T). \tag{12.8}$$

Repeating step by step all the above calculations, we can obtain expressions for the phonon Green's function and for many-particle Green's functions in the interaction representation. For the phonon Green's function, we find

$$\mathscr{D}(1, 2) = -\frac{\langle T_\tau(\varphi(1)\varphi(2)\mathscr{S})\rangle_0}{\langle\mathscr{S}\rangle_0}, \tag{12.9}$$

and for the two-particle Green's function

$$\mathscr{G}^{\text{II}}(1, 2; 3, 4) = -\frac{\langle T_\tau(\psi(1)\psi(2)\bar{\psi}(3)\bar{\psi}(4)\mathscr{S})\rangle_0}{\langle\mathscr{S}\rangle_0}. \tag{12.10}$$

The formulas for Green's functions depending on a larger number of variables differ from (12.7), (12.9) and (12.10) only by the number of ψ-operators inside the T_τ-product.

Finally, we give the formula relating the thermodynamic potential Ω and the matrix $\mathscr{S}$:

$$\Omega = \Omega_0 - T \ln \langle\mathscr{S}\rangle_0. \tag{12.11}$$

Here, Ω_0 denotes the potential Ω in the absence of any interaction:

$$\Omega_0 = -T \ln \text{Sp}\,\{e^{-(\hat{H}_0 - \mu\hat{N})/T}\}.$$

12.2. Wick's theorem. We now turn to our basic problem, i.e., the calculation of the Green's function for a system of interacting particles. If the interaction between the particles can be regarded as weak, then, using the expression for the temperature Green's function in the interaction representation, we can write the perturbation series with respect to $\hat{H}_{\text{int}}$ in an exceptionally concise form. The interaction Hamiltonian enters the Green's function only by way of the matrix $\mathscr{S}$. Expanding the exponential in the right-hand side of (12.4) in a power series in $\hat{H}_{\text{int}}(\tau)$, we obtain

$$\begin{aligned}\mathscr{S} &= 1 - \int_0^{1/T} \hat{H}_{\text{int}}(\tau')\,d\tau' \\ &\quad + \frac{1}{2}\int_0^{1/T}\int_0^{1/T} d\tau'\,d\tau''\,T_\tau(\hat{H}_{\text{int}}(\tau')\hat{H}_{\text{int}}(\tau'')) - \cdots \\ &= \sum_{n=0}^{\infty}\frac{(-1)^n}{n!}\int_0^{1/T}\cdots\int_0^{1/T} d\tau_1\cdots d\tau_n\,T_\tau(\hat{H}_{\text{int}}(\tau_1)\cdots\hat{H}_{\text{int}}(\tau_n)).\end{aligned} \tag{12.12}$$

Then, substituting this expansion into the numerator of formula (12.7), we find the perturbation series for the Green's function

$$\begin{aligned}&\mathscr{G}_{\alpha\beta}(\mathbf{r}_1, \tau_1; \mathbf{r}_2, \tau_2) \\ &\quad = -\frac{1}{\langle\mathscr{S}\rangle_0}\sum_{n=0}^{\infty}\frac{(-1)^n}{n!}\int_0^{1/T}\cdots\int_0^{1/T} d\tau_1'\cdots d\tau_n' \\ &\qquad \times \langle T_\tau(\psi_\alpha(\mathbf{r}_1, \tau_1)\bar{\psi}_\beta(\mathbf{r}_2, \tau_2)\hat{H}_{\text{int}}(\tau_1')\cdots\hat{H}_{\text{int}}(\tau_n'))\rangle_0.\end{aligned} \tag{12.13}$$

Of course, the first term of (12.13) is just the free Green's function

$$\mathscr{G}^{(0)} = -\langle T_\tau(\psi(1)\bar{\psi}(2))\rangle_0,$$

calculated in Sec. 11. We shall not expand the matrix $\mathscr{S}$ in the expression $\langle\mathscr{S}\rangle_0$ appearing in the denominator of (12.13), since it will turn out that

$\langle \mathscr{S} \rangle_0$ cancels an identical factor in the numerator. Moreover, $\langle \mathscr{S} \rangle_0$ is a constant, independent of $\mathbf{r}$ and τ, and can have no effect on our subsequent considerations.

In all real problems, H_{int} is a product of a certain number (usually not large) of operators $\psi(\mathbf{r}, t)$, $\bar{\psi}(\mathbf{r}, t)$, and perhaps $\varphi(\mathbf{r}, t)$, integrated over the space variables. Therefore, the problem of calculating the Green's function by using perturbation theory reduces to calculating the average value of the T_τ-product of a certain number of ψ-operators, evaluated at various points of space and "time":

$$\langle T_\tau(\psi_\alpha(\mathbf{r}, \tau) \cdots \bar{\psi}_{\alpha'}(\mathbf{r}', \tau') \cdots)\rangle_0. \tag{12.14}$$

We have already encountered a problem of this kind in the preceding chapter, while calculating ordinary Green's functions at the absolute zero of temperature. It was shown there that the average of any number of operators reduces to a sum of products of all possible averages of pairs of operators, where the latter, by definition, equal the free-particle Green's functions (Wick's theorem). As we shall see in a moment, the same situation prevails in the present case.

To convince ourselves of this, we replace the ψ-operators in (12.14) by their Fourier expansions[3] (with respect to $\mathbf{r}$):

$$\begin{aligned} \psi(\mathbf{r}, \tau) &= \frac{1}{\sqrt{V}} \sum_{\mathbf{p}} a_{\mathbf{p}}(\tau) e^{i\mathbf{p}\cdot\mathbf{r} - [\varepsilon_0(\mathbf{p}) - \mu]\tau}, \\ \bar{\psi}(\mathbf{r}, \tau) &= \frac{1}{\sqrt{V}} \sum_{\mathbf{p}} a_{\mathbf{p}}^+(\tau) e^{-i\mathbf{p}\cdot\mathbf{r} + [\varepsilon_0(\mathbf{p}) - \mu]\tau}. \end{aligned} \tag{12.15}$$

The operators $a_{\mathbf{p}}(\tau)$ and $a_{\mathbf{p}}^+(\tau)$ in (12.15) represent ordinary annihilation and creation operators, and actually do not depend on τ. However, we preserve the argument τ to indicate the places the various operators should occupy in T_τ-products. Substituting the expansions (12.15) into (12.14), we obtain the following expression [except for the exponentials in (12.15)]:

$$\frac{1}{\sqrt{V}} \sum_{\mathbf{p}_1} \frac{1}{\sqrt{V}} \sum_{\mathbf{p}_2} \cdots \frac{1}{\sqrt{V}} \sum_{\mathbf{p}_1'} \frac{1}{\sqrt{V}} \sum_{\mathbf{p}_2'} \cdots \times \langle T_\tau(a_{\mathbf{p}_1}(\tau_1) a_{\mathbf{p}_2}(\tau_2) \cdots a_{\mathbf{p}_1'}^+(\tau_1') a_{\mathbf{p}_2'}^+(\tau_2') \cdots)\rangle_0. \tag{12.16}$$

In the sum over $\mathbf{p}_1, \mathbf{p}_2, \cdots, \mathbf{p}_1', \mathbf{p}_2', \cdots$, the only nonzero terms are those containing equal numbers of annihilation and creation operators referring to the same momenta. In particular, the terms containing only one annihilation operator and one creation operator with the same momenta are nonzero, as are the terms

$$\frac{1}{V} \sum_{\substack{\mathbf{p}_1 \\ \mathbf{p}_1 \neq \mathbf{p}_2 \neq \cdots}} \frac{1}{V} \sum_{\mathbf{p}_2} \cdots \langle T_\tau(a_{\mathbf{p}_1}(\tau_1) a_{\mathbf{p}_2}(\tau_2) \cdots a_{\mathbf{p}_1}^+(\tau_1') a_{\mathbf{p}_2}^+(\tau_2') \cdots)\rangle_0, \tag{12.17a}$$

[3] The validity of (12.15) can be verified most easily by using the definition of the operators in the interaction representation and the identities (11.10).

and other terms differing from (12.17a) by a permutation of the momenta $\mathbf{p}_1, \mathbf{p}_2, \cdots$ indexing the a^+-operators. When there are several annihilation and creation operators with the same momenta (say two), the corresponding nonzero terms in the sum have the form

$$\frac{1}{V}\cdot\frac{1}{V}\sum_{\mathbf{p}_1}\frac{1}{V}\sum_{\substack{\mathbf{p}_2 \\ \mathbf{p}_1 \neq \mathbf{p}_2 \neq \cdots}} \cdots \langle T_\tau(a_{\mathbf{p}_1}(\tau_1)a_{\mathbf{p}_1}(\tau_2)a_{\mathbf{p}_2}(\tau_3)\cdots a^+_{\mathbf{p}_1}(\tau_1')a^+_{\mathbf{p}_1}(\tau_2')a^+_{\mathbf{p}_2}(\tau_3')\cdots)\rangle_0. \tag{12.17b}$$

The expressions (12.17a) have a special feature which distinguishes them from all the others, i.e., the number of factors $1/V$ in (12.17a) is the same as the number of summations, whereas in all the other expressions this number of factors is larger. Suppose, after having carried out the average $\langle\cdots\rangle_0$, we let the volume V of our system approach infinity, while keeping the particle density N/V constant.[4] In the limit as $V \to \infty$, the sum (12.17a) remains finite, and in fact becomes an expression involving integrals (with respect to the momenta) of various combinations of Fermi or Bose functions.[5] On the other hand, in expressions of the form (12.17b), besides these integrals with respect to the momenta there remains a certain number of extra factors $1/V$, and hence these expressions vanish as $V \to \infty$. Thus, the only terms in the sum (12.16) which remain in the limit as $V \to \infty$ are the terms of the form (12.17a), where all the annihilation and creation operators are indexed by different momenta. This means that in calculating the expression

$$\langle T_\tau(a_{\mathbf{p}_1}(\tau_1)a_{\mathbf{p}_2}(\tau_2)\cdots a^+_{\mathbf{p}_1'}(\tau_1')a^+_{\mathbf{p}_2'}(\tau_2')\cdots)\rangle_0,$$

we can actually average each pair of operators $a_\mathbf{p}$ and $a^+_\mathbf{p}$ separately. As a result, the average value of a T_τ-product of a large number of operators can be expressed as a sum of all possible averages of pairs of operators. For example,

$$\begin{aligned}\langle T_\tau(a_{\mathbf{p}_1}(\tau_1)a_{\mathbf{p}_2}(\tau_2)a^+_{\mathbf{p}_1'}(\tau_1')a^+_{\mathbf{p}_2'}(\tau_2'))\rangle_0 &\\ = \langle T_\tau(a_{\mathbf{p}_1}(\tau_1)a^+_{\mathbf{p}_2'}(\tau_2'))\rangle_0\langle T_\tau(a_{\mathbf{p}_2}(\tau_2)a^+_{\mathbf{p}_1'}(\tau_1'))\rangle_0 &\\ \mp \langle T_\tau(a_{\mathbf{p}_1}(\tau_1)a^+_{\mathbf{p}_1'}(\tau_1'))\rangle_0\langle T_\tau(a_{\mathbf{p}_2}(\tau_2)a^+_{\mathbf{p}_2'}(\tau_2'))\rangle_0,&\end{aligned} \tag{12.18a}$$

where the minus sign corresponds to Fermi statistics, and the plus sign to Bose statistics.[6]

---

[4] When this is done, the sums are replaced by integrals, according to the rule

$$\frac{1}{V}\sum\cdots \to \frac{1}{(2\pi)^3}\int\cdots$$

[5] We have already encountered the simplest example of this kind in Sec. 11, in calculating the free Green's function $\mathcal{G}^{(0)} = -\langle T_\tau(\psi(1)\bar{\psi}(2))\rangle_0$.

[6] It can be shown that this rule also holds for the average of a T_τ-product of several $a_\mathbf{p}$ and $a^+_\mathbf{p}$ with the same momenta [as in (12.17b)], excluding the case of the operators a_0 and a_0^+ for a Bose system below the condensation temperature (see p. 110). For simplicity, the proof will not be given here.

In the coordinate representation, these results mean that the average of a T_τ-product of a certain number of ψ-operators decomposes into a sum of products of all possible averages involving pairs of operators $\psi, \bar{\psi}$. For example, instead of (12.18a), we have

$$\begin{aligned}
&\langle T_\tau(\psi(\mathbf{r}_1, \tau_1)\psi(\mathbf{r}_2, \tau_2)\bar{\psi}(\mathbf{r}_1', \tau_1')\bar{\psi}(\mathbf{r}_2', \tau_2'))\rangle_0 \\
&\quad = \frac{1}{V^2} \sum_{\mathbf{p}_1', \mathbf{p}_2', \mathbf{p}_1, \mathbf{p}_2} e^{i(\mathbf{p}_1\cdot\mathbf{r}_1 + \mathbf{p}_2\cdot\mathbf{r}_2 - \mathbf{p}_1'\cdot\mathbf{r}_1' - \mathbf{p}_2'\cdot\mathbf{r}_2')} e^{-[\varepsilon_0(\mathbf{p}_1)-\mu]\tau_1 - [\varepsilon_0(\mathbf{p}_2)-\mu]\tau_2} \\
&\qquad \times e^{[\varepsilon_0(\mathbf{p}_1')-\mu]\tau_1' + [\varepsilon_0(\mathbf{p}_2')-\mu]\tau_2'} \langle T_\tau(a_{\mathbf{p}_1}(\tau_1)a_{\mathbf{p}_2}(\tau_2)a^+_{\mathbf{p}_1'}(\tau_1')a^+_{\mathbf{p}_2'}(\tau_2'))\rangle_0 \\
&\quad = \frac{1}{V} \sum_{\mathbf{p}_1, \mathbf{p}_2'} e^{i(\mathbf{p}_1\cdot\mathbf{r}_1 - \mathbf{p}_2'\cdot\mathbf{r}_2') - [\varepsilon_0(\mathbf{p}_1)-\mu]\tau_1 + [\varepsilon_0(\mathbf{p}_2')-\mu]\tau_2'} \langle T_\tau(a_{\mathbf{p}_1}(\tau_1)a^+_{\mathbf{p}_2'}(\tau_2'))\rangle_0 \\
&\qquad \times \frac{1}{V} \sum_{\mathbf{p}_2, \mathbf{p}_1'} e^{i(\mathbf{p}_2\cdot\mathbf{r}_2 - \mathbf{p}_1'\cdot\mathbf{r}_1') - [\varepsilon_0(\mathbf{p}_2)-\mu]\tau_2 + [\varepsilon_0(\mathbf{p}_1')-\mu]\tau_1'} \langle T_\tau(a_{\mathbf{p}_2}(\tau_2)a^+_{\mathbf{p}_1'}(\tau_1'))\rangle_0 \\
&\qquad \mp \frac{1}{V} \sum_{\mathbf{p}_1, \mathbf{p}_2'} e^{i(\mathbf{p}_1\cdot\mathbf{r}_1 - \mathbf{p}_1'\cdot\mathbf{r}_1') - [\varepsilon_0(\mathbf{p}_1)-\mu]\tau_1 + [\varepsilon_0(\mathbf{p}_1')-\mu]\tau_1'} \langle T_\tau(a_{\mathbf{p}_1}(\tau_1)a^+_{\mathbf{p}_1'}(\tau_1'))\rangle_0 \\
&\qquad \times \frac{1}{V} \sum_{\mathbf{p}_2, \mathbf{p}_2'} e^{i(\mathbf{p}_2\cdot\mathbf{r}_2 - \mathbf{p}_2'\cdot\mathbf{r}_2' - [\varepsilon_0(\mathbf{p}_2)-\mu]\tau_2 + [\varepsilon_0(\mathbf{p}_2')-\mu]\tau_2'} \langle T_\tau(a_{\mathbf{p}_2}(\tau_2)a^+_{\mathbf{p}_2'}(\tau_2'))\rangle_0 \\
&\quad = \langle T_\tau(\psi(\mathbf{r}_1, \tau_1)\bar{\psi}(\mathbf{r}_2', \tau_2'))\rangle_0 \langle T_\tau(\psi(\mathbf{r}_2, \tau_2)\bar{\psi}(\mathbf{r}_1', \tau_1'))\rangle_0 \\
&\qquad \mp \langle T_\tau(\psi(\mathbf{r}_1, \tau_1)\bar{\psi}(\mathbf{r}_1', \tau_1'))\rangle_0 \langle T_\tau(\psi(\mathbf{r}_2, \tau_2)\bar{\psi}(\mathbf{r}_2', \tau_2'))\rangle_0,
\end{aligned} \tag{12.18b}$$

and similar relations hold for a larger number of operators.

The averages appearing in the right-hand side of (12.18b) are just free-particle temperature Green's functions (except possibly for sign). Thus, in calculating temperature Green's functions, we encounter the same situation as found in the case $T = 0$. In fact, the Green's function $\mathcal{G}$ satisfies the expansion (12.13), which, except for the factors of i^n and the limits of τ-integration, is identical with the expansion (8.9) for the function G. Moreover, to calculate the averages $\langle T_\tau(\cdots)\rangle_0$ appearing in (12.13), we can use Wick's theorem, just as before, which allows us to express these averages in terms of averages of pairs of annihilation and creation operators. It should be noted that in the present technique, there is no concept of a normal product, and Wick's theorem does not hold for T_τ-products themselves, but only for average values of these products.

Using Wick's theorem to write down any term of the series (12.13), and replacing $\langle T_\tau(\psi_1\bar{\psi})\rangle_0$ by the free Green's function

$$\mathcal{G}^{(0)}_{\alpha\beta}(\mathbf{r}_1 - \mathbf{r}_2, \tau_1 - \tau_2) = -\langle T_\tau(\psi_\alpha(\mathbf{r}_1, \tau_1)\bar{\psi}_\beta(\mathbf{r}_2, \tau_2))\rangle_0,$$

we arrive at expressions which have exactly the same structure as the corresponding series for $T = 0$. This allows us to describe the various approximations of the perturbation series by using the same Feynman diagrams as used in the preceding chapter, and the only change is in the rules by which analytical expressions are associated with elements of the diagrams. In the present case, with each line of a diagram we have to associate a free-particle temperature Green's function $\mathscr{G}^{(0)}$, instead of a function $G^{(0)}$. Moreover, integration with respect to the time t (from $-\infty$ to ∞) at every vertex of a diagram has to be replaced by integration with respect to the "time" τ (from 0 to $1/T$).

So far, we have tacitly assumed that as the volume of the system goes to infinity (with the density kept constant), all the free-particle Green's functions and the integrals involving them remain finite. In particular, this justifies our neglect of terms of the form (12.17b) in the limit as $V \to \infty$. The situation changes drastically in the case of Bose systems below the condensation temperature T_c or Fermi systems exhibiting superconductivity. In the case of a Bose gas with $T < T_c$, the annihilation and creation operators for particles in the state of zero momentum are proportional to the square root of the volume:

$$a_0 \sim a_0^+ \propto \sqrt{N} \propto \sqrt{V}.$$

Then the terms of the type (12.17b), with several a_0 and a_0^+, remain finite as $V \to \infty$, and moreover do not obey Wick's theorem (cf. footnote 6, p. 108). A similar situation occurs for superconductors. In both cases, we have to use a special technique, which will be described in separate chapters (Chaps. 5 and 7).

We now return to the case where the usual diagram technique is applicable. Just as in the preceding chapter, the diagrams for the Green's functions have two external lines. One of these external lines begins at the point $\mathbf{r}_1, \tau_1$ corresponding to the coordinates of the operator $\psi_\alpha(\mathbf{r}_1, \tau_1)$, while the other external line terminates at the point $\mathbf{r}_2, \tau_2$ corresponding to the operator $\bar{\psi}_\beta(\mathbf{r}_2, \tau_2)$. As before, the diagrams for the function $\mathscr{G}$ can be divided into two groups, i.e., connected diagrams and disconnected diagrams. Using an argument exactly analogous to that given for the case $T = 0$, we can verify that the contribution from the disconnected diagrams cancels the denominator in formula (12.7). As a result, we find that

$$\mathscr{G}_{\alpha\beta}(\mathbf{r}_1, \tau_1; \mathbf{r}_2, \tau_2) = -\langle T_\tau(\psi_\alpha(\mathbf{r}_1, \tau_1)\bar{\psi}_\beta(\mathbf{r}_2, \tau_2)\mathscr{S})\rangle_{\text{con}}, \qquad (12.19)$$

where $\langle \cdots \rangle_{\text{con}}$ denotes the contribution from connected diagrams only. A similar result holds for many-particle Green's functions, since the derivation nowhere uses the fact that the diagrams have two external lines. In other words, $\langle \mathscr{S} \rangle_0$ can be omitted in the denominators of the corresponding formulas [of the type (12.10)], and we need only take account of connected diagrams when calculating averages.

Just as in the preceding chapter, each diagram appears in the series for $\mathscr{G}$ with some coefficient of the form λ^m, which does not depend in an essential way on the order m of the diagram (see p. 71). This fact is very important when summing infinite sequences of diagrams.

13. The Diagram Technique in Coordinate Space. Examples

The basic result of the preceding section is our proof of the fact that in calculating temperature Green's functions, we can apply the usual technique of Feynman diagrams. The basic elements of every such diagram are the lines representing the Green's functions of free particles or of phonons. Just as in Chap. 2, we represent the Green's function of a particle by a solid line (see Fig. 32). The direction of the line is indicated by an arrow, i.e., the line

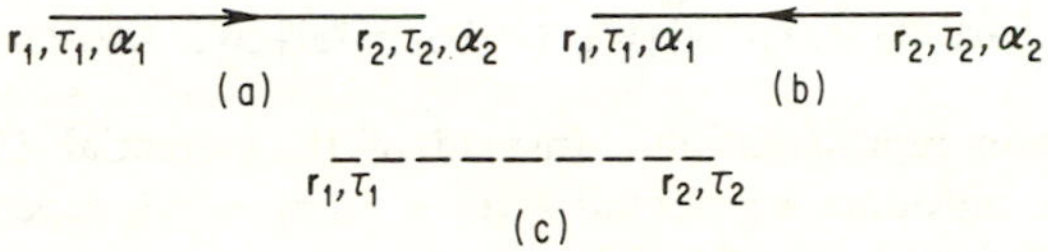

$\mathbf{r}_1, \tau_1, \alpha_1$ — $\mathbf{r}_2, \tau_2, \alpha_2$ (a) $\mathbf{r}_1, \tau_1, \alpha_1$ — $\mathbf{r}_2, \tau_2, \alpha_2$ (b)

$\mathbf{r}_1, \tau_1$ — $\mathbf{r}_2, \tau_2$ (c)

FIGURE 32

begins at the point with coordinates $\mathbf{r}_1$, τ_1 and spin projection α_1 (the point corresponding to the operator ψ in the definition of the function $\mathscr{G}$), and terminates at the point $\mathbf{r}_2$, τ_2, α_2 (corresponding to the operator $\bar{\psi}$). In arguments of Green's functions, the coordinates of initial points are written on the left and coordinates of end points are written on the right. Thus, the line in Fig. 32(a) represents the Green's function

$$\mathscr{G}^{(0)}_{\alpha_1\alpha_2}(\mathbf{r}_1, \tau_1; \mathbf{r}_2, \tau_2) \equiv \mathscr{G}^{(0)}_{\alpha_1\alpha_2}(\mathbf{r}_1 - \mathbf{r}_2, \tau_1 - \tau_2),$$

while the line in Fig. 32(b) represents the Green's function

$$\mathscr{G}^{(0)}_{\alpha_2\alpha_1}(\mathbf{r}_2, \tau_2; \mathbf{r}_1, \tau_1) \equiv \mathscr{G}^{(0)}_{\alpha_2\alpha_1}(\mathbf{r}_2 - \mathbf{r}_1, \tau_2 - \tau_1).$$

We use a dashed line to represent the Green's function of a phonon, as in Fig. 32(c). The direction of a phonon line need not be indicated, since as we have seen in Sec. 11, $\mathscr{D}^{(0)}$ is an even function of $\mathbf{r}_1 - \mathbf{r}_2$ and $\tau_1 - \tau_2$.

As before, we integrate over the coordinates of the "vertices," i.e., the points where the lines intersect. The integration with respect to $\mathbf{r}$ ranges over all space, and the integration with respect to τ is between the limits 0 and $1/T$. At the vertices, we also sum over the spin variables.

The specific form of the diagrams depends on the nature of the interaction between the particles. To construct diagrams, we have to use Wick's theorem, according to which averages of T_τ-products of operators appearing in the perturbation series (12.13) for the Green's function can be represented as a sum of products of averages involving pairs of operators. These latter averages are related to the free-particle Green's functions by the formulas

$$\begin{aligned} \langle T_\tau(\psi_\alpha(\mathbf{r}_1, \tau_1)\bar{\psi}_\beta(\mathbf{r}_2, \tau_2))\rangle &= -\mathscr{G}^{(0)}_{\alpha\beta}(\mathbf{r}_1 - \mathbf{r}_2, \tau_1 - \tau_2), \\ \langle T_\tau(\bar{\psi}_\beta(\mathbf{r}_2, \tau_2)\psi_\alpha(\mathbf{r}_1, \tau_1))\rangle &= \pm\mathscr{G}^{(0)}_{\alpha\beta}(\mathbf{r}_1 - \mathbf{r}_2, \tau_1 - \tau_2), \end{aligned} \tag{13.1}$$

where the plus sign corresponds to fermions, and the minus sign to bosons.

Similarly, the average of a product of two phonon operators can be expressed in terms of the function $\mathscr{D}^{(0)}$:

$$\langle T_\tau(\varphi(\mathbf{r}_1, \tau_1)\varphi(\mathbf{r}_2, \tau_2))\rangle = -\mathscr{D}^{(0)}(\mathbf{r}_1 - \mathbf{r}_2, \tau_1 - \tau_2). \tag{13.2}$$

We now consider different kinds of interactions.

A. Two-particle interactions. Suppose the particles of the system are acted upon by two-particle forces described by a potential $U(\mathbf{r}_1 - \mathbf{r}_2)$. Then the Hamiltonian has the form

$$\hat{H}_{\text{int}}(\tau) = \frac{1}{2}\int d\mathbf{r}_1\, d\mathbf{r}_2\, \bar{\psi}_\alpha(\mathbf{r}_1, \tau)\bar{\psi}_\beta(\mathbf{r}_2, \tau)U(\mathbf{r}_1 - \mathbf{r}_2)\psi_\beta(\mathbf{r}_2, \tau)\psi_\alpha(\mathbf{r}_1, \tau) \tag{13.3}$$

in the interaction representation. Instead of the potential $U(\mathbf{r}_1 - \mathbf{r}_2)$, it is convenient to introduce a potential $\mathscr{V}(\mathbf{r}_1 - \mathbf{r}_2, \tau_1 - \tau_2)$, depending on the "time" τ and defined by the formula

$$\mathscr{V}(\mathbf{r}_1 - \mathbf{r}_2, \tau_1 - \tau_2) = U(\mathbf{r}_1 - \mathbf{r}_2)\delta(\tau_1 - \tau_2). \tag{13.4}$$

Using (13.4), we can write the expression (12.4) for the matrix $\mathscr{S}$ in the symmetric form

$$\mathscr{S} = T_\tau \exp\Big\{-\frac{1}{2}\int d\mathbf{r}_1\, d\mathbf{r}_2\, d\tau_1\, d\tau_2\, \bar{\psi}_\alpha(\mathbf{r}_1, \tau_1)\bar{\psi}_\beta(\mathbf{r}_2, \tau_2)$$
$$\times \mathscr{V}(\mathbf{r}_1 - \mathbf{r}_2, \tau_1 - \tau_2)\psi_\beta(\mathbf{r}_2, \tau_2)\psi_\alpha(\mathbf{r}_1, \tau_1)\Big\}.$$

Next, we calculate the correction to the Green's function which is of the first order in U, obtaining[7]

$$\mathscr{G}^{(1)}_{\alpha\beta}(x - y) = \frac{1}{2}\int d^4z_1\, d^4z_2$$
$$\times \langle T_\tau(\psi_\alpha(x)\bar{\psi}_\beta(y)\mathscr{V}(z_1 - z_2)\bar{\psi}_{\gamma_1}(z_1)\bar{\psi}_{\gamma_2}(z_2)\psi_{\gamma_2}(z_2)\psi_{\gamma_1}(z_1))\rangle. \tag{13.5}$$

According to Wick's theorem, the average $\langle\cdots\rangle$ in the right-hand side of (13.5) can be written as a sum of four terms

$$\langle T_\tau(\psi_\alpha(x)\bar{\psi}_\beta(y))\rangle\langle\bar{\psi}_{\gamma_1}(z_1)\psi_{\gamma_1}(z_1)\rangle\langle\bar{\psi}_{\gamma_2}(z_2)\psi_{\gamma_2}(z_2)\rangle, \tag{a}$$

$$\mp \langle T_\tau(\psi_\alpha(x)\bar{\psi}_\beta(y))\rangle\langle\bar{\psi}_{\gamma_2}(z_2)\psi_{\gamma_1}(z_1)\rangle\langle\bar{\psi}_{\gamma_1}(z_1)\psi_{\gamma_2}(z_2)\rangle, \tag{b}$$

$$\langle T_\tau(\psi_\alpha(x)\bar{\psi}_{\gamma_1}(z_1))\rangle\langle T_\tau(\psi_{\gamma_1}(z_1)\bar{\psi}_\beta(y))\rangle\langle\bar{\psi}_{\gamma_2}(z_2)\psi_{\gamma_2}(z_2)\rangle, \tag{c}$$

$$\mp \langle T_\tau(\psi_\alpha(x)\bar{\psi}_{\gamma_1}(z_1))\rangle\langle\bar{\psi}_{\gamma_2}(z_2)\psi_{\gamma_1}(z_1)\rangle\langle T_\tau(\psi_{\gamma_2}(z_2)\bar{\psi}_\beta(y))\rangle, \tag{d}$$

and of four more terms, obtained from (a)–(d) by making the substitution $z_1 \to z_2, \gamma_1 \to \gamma_2$. The contribution from these last four terms to the integral in (13.5) is obviously the same as the contribution from (a)–(d), and this simply leads to the disappearance of the factor $\frac{1}{2}$ in front of the integral.

---

[7] In the rest of this section, we use lightface Latin letters to denote the set of four variables, $x = (\mathbf{r}, \tau)$. Thus, $\mathscr{G}(x - y) = \mathscr{G}(\mathbf{x} - \mathbf{y}, \tau_1 - \tau_2)$ and $d^4x = d\mathbf{r}\, d\tau$.

Using (13.1) to replace the averages $\langle T_\tau(\cdots)\rangle$ by suitable Green's functions $\mathscr{G}^{(0)}$, we find that the first-order correction is a sum of the following four terms:

$$-\mathscr{G}^{(0)}_{\alpha\beta}(x-y)\int d^4z_1\,d^4z_2\,\mathscr{G}^{(0)}_{\gamma_1\gamma_1}(0)\mathscr{G}^{(0)}_{\gamma_2\gamma_2}(0)\mathscr{V}(z_1-z_2), \tag{a}$$

$$\pm\,\mathscr{G}^{(0)}_{\alpha\beta}(x-y)\int d^4z_1\,d^4z_2\,\mathscr{G}^{(0)}_{\gamma_1\gamma_2}(z_1-z_2)\mathscr{G}^{(0)}_{\gamma_2\gamma_1}(z_2-z_1)\mathscr{V}(z_1-z_2), \tag{b}$$

$$\pm\int d^4z_1\,d^4z_2\,\mathscr{G}^{(0)}_{\alpha\gamma_1}(x-z_1)\mathscr{G}^{(0)}_{\gamma_1\beta}(z_1-y)\mathscr{G}^{(0)}_{\gamma_2\gamma_2}(0)\mathscr{V}(z_1-z_2), \tag{c}$$

$$-\int d^4z_1\,d^4z_2\,\mathscr{G}^{(0)}_{\alpha\gamma_1}(x-z_1)\mathscr{G}^{(0)}_{\gamma_1\gamma_2}(z_1-z_2)\mathscr{G}^{(0)}_{\gamma_2\beta}(z_2-y)\mathscr{V}(z_1-z_2). \tag{d}$$

It should be noted that the quantity $\mathscr{G}^{(0)}(\mathbf{r}_1-\mathbf{r}_2, 0)$ is always evaluated as $\lim_{\tau\to+0}\mathscr{G}^{(0)}(\mathbf{r}_1-\mathbf{r}_2, -\tau)$.

In constructing diagrams, we represent $\mathscr{V}(z_1-z_2)$ by a wavy line. Then the expressions (a)–(d) correspond to the diagrams shown in Fig. 33. Diagrams (a) and (b) are disconnected, and as shown in the preceding section, they should not be taken into account when calculating Green's functions. Thus, the only contribution to the first-order correction is made by diagrams (c) and (d), and by the "topologically equivalent" diagrams differing from them by permutation of the vertex coordinates z_1 and z_2 (recall that all topologically equivalent diagrams make the same contribution).

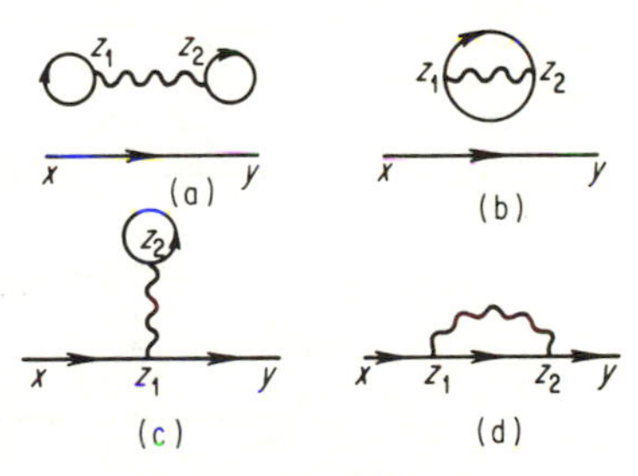

FIGURE 33

It should be pointed out that in the case of Fermi statistics, the expressions corresponding to diagrams (c) and (d) have opposite signs. This fact is associated with the presence of a closed loop in diagram (c). For a diagram of arbitrary order, it can be shown that any closed fermion loop (not necessarily formed of a single line, as in the present case) leads to a factor of -1 in the corresponding analytical expression.

We now state the general rules which are used to calculate the correction of order n:

1. Form all connected, topologically nonequivalent diagrams with $2n$ vertices and two external lines, where two solid lines and one wavy line meet at each vertex.
2. With each solid line associate a Green's function $\mathscr{G}^{(0)}_{\alpha\beta}(x-y)$, where x, α are the coordinates of the initial point of the line, and y, β are the coordinates of its end point.
3. With each wavy line associate a generalized potential $\mathscr{V}(x-y)$.

4. Integrate over all the vertex coordinates z ($d^4z = d\mathbf{r}\,d\tau$), and sum over all internal spin variables α.
5. Multiply the resulting expression by $(-1)^{n+F}$, where F is the number of closed fermion loops.
6. If there are any Green's functions $\mathscr{G}^{(0)}(0)$ whose "time" arguments are the same, interpret them as $\lim\limits_{\tau \to +0} \mathscr{G}^{(0)}(\mathbf{r}_1 - \mathbf{r}_2, -\tau)$.

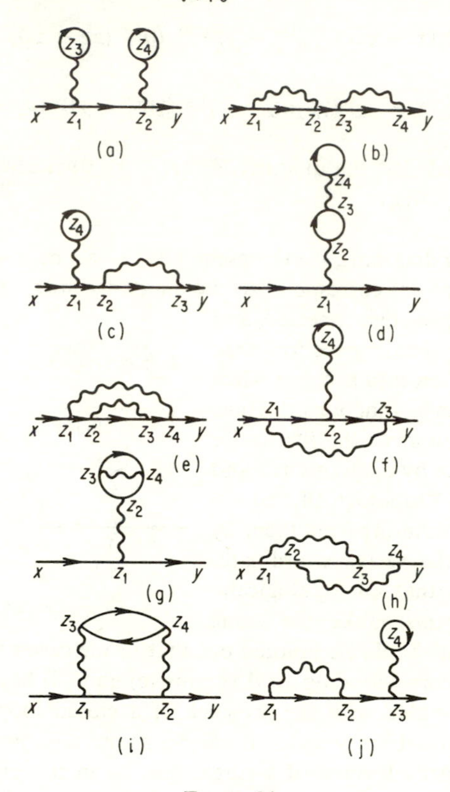

FIGURE 34

For example, consider the second-order correction. All possible connected, topologically nonequivalent diagrams with four vertices are shown in Fig. 34. Using the above rules, we can easily write down the following analytical expressions corresponding to these diagrams:

$$\begin{aligned}\int d^4z_1\, d^4z_2\, d^4z_3\, d^4z_4\, &\mathscr{G}^{(0)}_{\alpha\gamma_1}(x - z_1)\mathscr{G}^{(0)}_{\gamma_1\gamma_2}(z_1 - z_2)\mathscr{G}^{(0)}_{\gamma_2\beta}(z_2 - y)\\ &\times \mathscr{G}^{(0)}_{\gamma_3\gamma_3}(0)\mathscr{G}^{(0)}_{\gamma_4\gamma_4}(0)\mathscr{V}(z_1 - z_3)\mathscr{V}(z_2 - z_4),\end{aligned} \tag{a}$$

$$\begin{aligned}\int d^4z_1\, d^4z_2\, d^4z_3\, d^4z_4\, &\mathscr{G}^{(0)}_{\alpha\gamma_1}(x - z_1)\mathscr{G}^{(0)}_{\gamma_1\gamma_2}(z_1 - z_2)\mathscr{G}^{(0)}_{\gamma_2\gamma_3}(z_2 - z_3)\\ &\times \mathscr{G}^{(0)}_{\gamma_3\gamma_4}(z_3 - z_4)\mathscr{G}^{(0)}_{\gamma_4\beta}(z_4 - y)\mathscr{V}(z_1 - z_2)\mathscr{V}(z_3 - z_4),\end{aligned} \tag{b}$$

$$\mp \int d^4z_1\, d^4z_2\, d^4z_3\, d^4z_4\, \mathscr{G}^{(0)}_{\alpha\gamma_1}(x - z_1)\mathscr{G}^{(0)}_{\gamma_1\gamma_2}(z_1 - z_2)\mathscr{G}^{(0)}_{\gamma_2\gamma_3}(z_2 - z_3) \times \mathscr{G}^{(0)}_{\gamma_3\beta}(z_3 - y)\mathscr{G}^{(0)}_{\gamma_4\gamma_4}(0)\mathscr{V}(z_1 - z_4)\mathscr{V}(z_2 - z_3), \tag{c}$$

$$\int d^4z_1\, d^4z_2\, d^4z_3\, d^4z_4\, \mathscr{G}^{(0)}_{\alpha\gamma_1}(x - z_1)\mathscr{G}^{(0)}_{\gamma_1\beta}(z_1 - y)\mathscr{G}^{(0)}_{\gamma_2\gamma_3}(z_2 - z_3) \times \mathscr{G}^{(0)}_{\gamma_3\gamma_2}(z_3 - z_2)\mathscr{G}^{(0)}_{\gamma_4\gamma_4}(0)\mathscr{V}(z_1 - z_2)\mathscr{V}(z_3 - z_4), \tag{d}$$

$$\int d^4z_1\, d^4z_2\, d^4z_3\, d^4z_4\, \mathscr{G}^{(0)}_{\alpha\gamma_1}(x - z_1)\mathscr{G}^{(0)}_{\gamma_1\gamma_2}(z_1 - z_2)\mathscr{G}^{(0)}_{\gamma_2\gamma_3}(z_2 - z_3) \times \mathscr{G}^{(0)}_{\gamma_3\gamma_4}(z_3 - z_4)\mathscr{G}^{(0)}_{\gamma_4\beta}(z_4 - y)\mathscr{V}(z_1 - z_4)\mathscr{V}(z_2 - z_3), \tag{e}$$

$$\mp \int d^4z_1\, d^4z_2\, d^4z_3\, d^4z_4\, \mathscr{G}^{(0)}_{\alpha\gamma_1}(x - z_1)\mathscr{G}^{(0)}_{\gamma_1\gamma_2}(z_1 - z_2)\mathscr{G}^{(0)}_{\gamma_2\gamma_3}(z_2 - z_3) \times \mathscr{G}^{(0)}_{\gamma_3\beta}(z_3 - y)\mathscr{G}^{(0)}_{\gamma_4\gamma_4}(0)\mathscr{V}(z_1 - z_3)\mathscr{V}(z_2 - z_4), \tag{f}$$

$$\mp \int d^4z_1\, d^4z_2\, d^4z_3\, d^4z_4\, \mathscr{G}^{(0)}_{\alpha\gamma_1}(x - z_1)\mathscr{G}^{(0)}_{\gamma_1\beta}(z_1 - y)\mathscr{G}^{(0)}_{\gamma_2\gamma_3}(z_2 - z_3) \times \mathscr{G}^{(0)}_{\gamma_3\gamma_4}(z_3 - z_4)\mathscr{G}^{(0)}_{\gamma_4\gamma_2}(z_4 - z_2)\mathscr{V}(z_1 - z_2)\mathscr{V}(z_3 - z_4), \tag{g}$$

$$\int d^4z_1\, d^4z_2\, d^4z_3\, d^4z_4\, \mathscr{G}^{(0)}_{\alpha\gamma_1}(x - z_1)\mathscr{G}^{(0)}_{\gamma_1\gamma_2}(z_1 - z_2)\mathscr{G}^{(0)}_{\gamma_2\gamma_3}(z_2 - z_3) \times \mathscr{G}^{(0)}_{\gamma_3\gamma_4}(z_3 - z_4)\mathscr{G}^{(0)}_{\gamma_4\beta}(z_4 - y)\mathscr{V}(z_1 - z_3)\mathscr{V}(z_2 - z_4), \tag{h}$$

$$\mp \int d^4z_1\, d^4z_2\, d^4z_3\, d^4z_4\, \mathscr{G}^{(0)}_{\alpha\gamma_1}(x - z_1)\mathscr{G}^{(0)}_{\gamma_1\gamma_2}(z_1 - z_2)\mathscr{G}^{(0)}_{\gamma_2\beta}(z_2 - y) \times \mathscr{G}^{(0)}_{\gamma_3\gamma_4}(z_3 - z_4)\mathscr{G}^{(0)}_{\gamma_4\gamma_3}(z_4 - z_3)\mathscr{V}(z_1 - z_3)\mathscr{V}(z_2 - z_4), \tag{i}$$

$$\mp \int d^4z_1\, d^4z_2\, d^4z_3\, d^4z_4\, \mathscr{G}^{(0)}_{\alpha\gamma_1}(x - z_1)\mathscr{G}^{(0)}_{\gamma_1\gamma_2}(z_1 - z_2)\mathscr{G}^{(0)}_{\gamma_2\gamma_3}(z_2 - z_3) \times \mathscr{G}_{\gamma_3\beta}(z_3 - y)\mathscr{G}_{\gamma_4\gamma_4}(0)\mathscr{V}(z_1 - z_2)\mathscr{V}(z_3 - z_4). \tag{j}$$

In the case of two-particle interactions, perturbation theory can be cast in another, more symmetric form, which is particularly convenient when the interaction forces depend not only on the distances between the particles but also on their spins. The Hamiltonian for such an interaction has the form

$$\hat{H}_{\text{int}}(\tau) = \frac{1}{2}\int d\mathbf{r}_1\, d\mathbf{r}_2\, \bar{\psi}_\alpha(\mathbf{r}_1, \tau)\bar{\psi}_\beta(\mathbf{r}_2, \tau)U_{\alpha\beta,\delta\gamma}(\mathbf{r}_1 - \mathbf{r}_2)\psi_\gamma(\mathbf{r}_2, \tau)\psi_\delta(\mathbf{r}_1, \tau). \tag{13.6}$$

We write the integral

$$\int_0^{1/T} H_{\text{int}}(\tau)\, d\tau$$

appearing in the expression for $\mathscr{S}$ as

$$\frac{1}{4}\int_0^{1/T} d\tau_1 \cdots \int_0^{1/T} d\tau_4 \int d\mathbf{r}_1 d\mathbf{r}_2 d\mathbf{r}_3 d\mathbf{r}_4$$

$$\times\ \bar{\psi}_{\gamma_1}(\mathbf{r}_1, \tau_1)\bar{\psi}_{\gamma_2}(\mathbf{r}_2, \tau_2)\mathscr{T}^{(0)}_{\gamma_1\gamma_2,\gamma_3\gamma_4}(\mathbf{r}_1, \tau_1, \mathbf{r}_2, \tau_2; \mathbf{r}_3, \tau_3, \mathbf{r}_4, \tau_4)\psi_{\gamma_4}(\mathbf{r}_4, \tau_4)\psi_{\gamma_3}(\mathbf{r}_3, \tau_3),$$

which is symmetric in all four variables, or as

$$\frac{1}{4}\int d^4z_1\, d^4z_2\, d^4z_3\, d^4z_4\, \bar{\psi}_{\gamma_1}(z_1)\bar{\psi}_{\gamma_2}(z_2)\mathscr{T}^{(0)}_{\gamma_1\gamma_2,\gamma_3\gamma_4}(z_1, z_2; z_3, z_4)\psi_{\gamma_4}(z_4)\psi_{\gamma_3}(z_3), \tag{13.7}$$

where we introduce "four-dimensional" notation. The operators $\bar{\psi}_{\gamma_1}(z_1)$ and $\bar{\psi}_{\gamma_2}(z_2)$ [or $\psi_{\gamma_3}(z_3)$ and $\psi_{\gamma_4}(z_4)$] either anticommute or commute, depending on the statistics. Therefore, the quantity $\mathscr{T}^{(0)}$ can be regarded as antisymmetric under the permutations $z_1\gamma_1 \rightleftarrows z_2\gamma_2$ and $z_3\gamma_3 \rightleftarrows z_4\gamma_4$ in the case of fermions, or symmetric under these permutations in the case of bosons. Thus, in the case of Fermi statistics, $\mathscr{T}^{(0)}$ can be obtained from

$$U_{\gamma_1\gamma_2,\gamma_3\gamma_4}(\mathbf{r}_1 - \mathbf{r}_2)\delta(\tau_1 - \tau_2)\delta(\mathbf{r}_1 - \mathbf{r}_3)\delta(\tau_1 - \tau_3)\delta(\mathbf{r}_2 - \mathbf{r}_4)\delta(\tau_2 - \tau_4) \quad (13.8)$$

by antisymmetrization in the variables $z_1\gamma_1, z_2\gamma_2$ and $z_3\gamma_3, z_4\gamma_4$ while in the case of Bose statistics, $\mathscr{T}^{(0)}$ can be obtained from (13.8) by symmetrization in the same variables.

We now calculate the first-order correction to the Green's function, obtaining

$$\frac{1}{4}\int d^4z_1\, d^4z_2\, d^4z_3\, d^4z_4\, \mathscr{T}^{(0)}_{\gamma_1\gamma_2,\gamma_3\gamma_4}(z_1, z_2; z_3, z_4) \times \langle T_\tau(\psi_\alpha(x)\bar{\psi}_\beta(y)\bar{\psi}_{\gamma_1}(z_1)\bar{\psi}_{\gamma_2}(z_2)\psi_{\gamma_4}(z_4)\psi_{\gamma_3}(z_3))\rangle. \quad (13.9)$$

Applying Wick's theorem and using the symmetry properties of $\mathscr{T}^{(0)}$, we easily verify that (13.9) is the sum of the following two terms:

$$-\frac{1}{2}\mathscr{G}^{(0)}_{\alpha\beta}(x - y)\int d^4z_1\, d^4z_2\, d^4z_3\, d^4z_4\, \mathscr{G}^{(0)}_{\gamma_3\gamma_1}(z_3 - z_1) \times \mathscr{G}^{(0)}_{\gamma_4\gamma_2}(z_4 - z_2)\mathscr{T}^{(0)}_{\gamma_1\gamma_2,\gamma_3\gamma_4}(z_1, z_2; z_3, z_4), \quad \text{(a)}$$

$$-\int d^4z_1\, d^4z_2\, d^4z_3\, d^4z_4\, \mathscr{G}^{(0)}_{\alpha\gamma_1}(x - z_1)\mathscr{T}^{(0)}_{\gamma_1\gamma_2,\gamma_3\gamma_4}(z_1, z_2; z_3, z_4) \times \mathscr{G}^{(0)}_{\gamma_4\beta}(z_4 - y)\mathscr{G}^{(0)}_{\gamma_3\gamma_2}(z_3 - z_2). \quad \text{(b)}$$

If we represent $\mathscr{T}^{(0)}$ by an open square, the expressions (a) and (b) correspond to the diagrams shown in Fig. 35. Diagram (a) is disconnected, and its

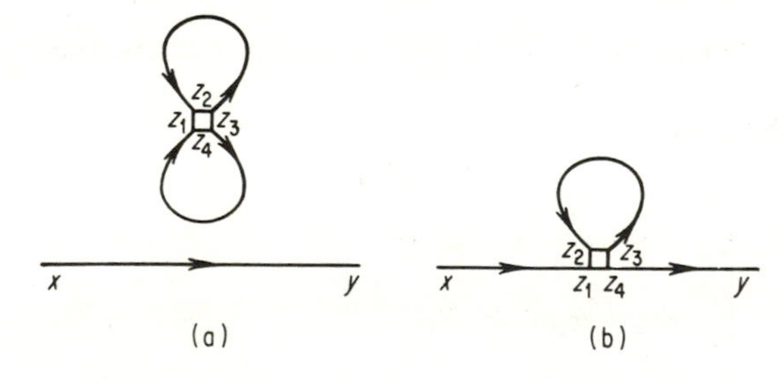

FIGURE 35

contribution does not have to be taken into account. Thus, in the first-order approximation of perturbation theory, we have a single diagram, whose contribution is given by the expression (b).

In the second-order approximation of perturbation theory, there are just three connected, topologically nonequivalent diagrams (see Fig. 36). It is easy to see that the expressions corresponding to these diagrams are

$$\int dz_1 \cdots dz_8\, \mathscr{G}^{(0)}_{\alpha\gamma_1}(x - z_1)\mathscr{T}^{(0)}_{\gamma_1\gamma_2,\gamma_3\gamma_4}(z_1, z_2; z_3, z_4)\, \mathscr{G}^{(0)}_{\gamma_4\gamma_5}(z_4 - z_5) \times \mathscr{G}^{(0)}_{\gamma_8\beta}(z_8 - y)\mathscr{G}^{(0)}_{\gamma_7\gamma_6}(z_7 - z_6) \times \mathscr{T}^{(0)}_{\gamma_5\gamma_6,\gamma_7\gamma_8}(z_5, z_6; z_7, z_8)\, \mathscr{G}^{(0)}_{\gamma_3\gamma_2}(z_3 - z_2), \tag{a}$$

$$\int dz_1 \cdots dz_8\, \mathscr{G}^{(0)}_{\alpha\gamma_1}(x - z_1)\mathscr{T}^{(0)}_{\gamma_1\gamma_2,\gamma_3\gamma_4}(z_1, z_2; z_3, z_4)\mathscr{G}^{(0)}_{\gamma_4\beta}(z_4 - y) \times \mathscr{G}^{(0)}_{\gamma_3\gamma_6}(z_3 - z_6)\mathscr{G}^{(0)}_{\gamma_7\gamma_2}(z_7 - z_2) \times \mathscr{T}^{(0)}_{\gamma_5\gamma_6,\gamma_7\gamma_8}(z_5, z_6; z_7, z_8)\mathscr{G}^{(0)}_{\gamma_8\gamma_5}(z_8 - z_5), \tag{b}$$

$$\frac{1}{2}\int dz_1 \cdots dz_8\, \mathscr{G}^{(0)}_{\alpha\gamma_1}(x - z_1)\mathscr{T}^{(0)}_{\gamma_1\gamma_2,\gamma_3\gamma_4}(z_1, z_2; z_3, z_4)\, \mathscr{G}^{(0)}_{\gamma_3\gamma_5}(z_3 - z_5) \times \mathscr{G}^{(0)}_{\gamma_4\gamma_6}(z_4 - z_6)\, \mathscr{G}^{(0)}_{\gamma_7\gamma_2}(z_7 - z_2) \times \mathscr{T}^{(0)}_{\gamma_5\gamma_6,\gamma_7\gamma_8}(z_5, z_6; z_7, z_8)\mathscr{G}^{(0)}_{\gamma_8\beta}(z_8 - y). \tag{c}$$

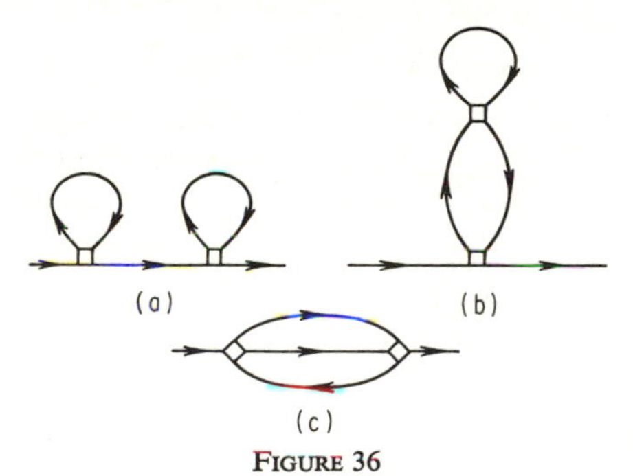

FIGURE 36

The following general rules are used to calculate the nth-order correction to the Green's function:

1. Form all connected, topologically nonequivalent diagrams containing n squares (in the present case, all diagrams obtained by permuting vertex coordinates of the squares are topologically equivalent).
2. With each line associate a Green's function $\mathscr{G}^{(0)}_{\alpha\beta}(x - y)$.
3. With each square associate a function $\mathscr{T}^{(0)}_{\gamma_1\gamma_2,\gamma_3\gamma_4}(z_1, z_2; z_3, z_4)$.
4. Integrate over the coordinates of the vertices of all the squares, and sum over the corresponding spin variables.
5. Multiply the resulting expression by precisely the same coefficient (A_n, say) as in the case of the diagram technique for $T = 0$ (see p. 76). However, it should be noted that the simplest way of finding this coefficient is to use Wick's theorem directly, expressing $\langle T_\tau(\psi_\alpha(x)\bar{\psi}_\beta(y)\cdots)\rangle$ in terms of averages of pairs of operators.

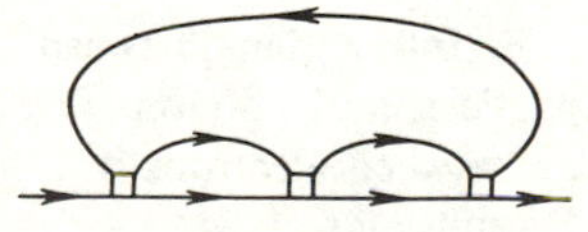
FIGURE 37

Using these rules, we can easily write down the expression corresponding to the diagram shown in Fig. 37:

$$\mp \frac{1}{4} \int d^4z_1 \cdots d^4z_{12}\, \mathscr{G}^{(0)}_{\alpha\gamma_1}(x - z_1)\mathscr{T}^{(0)}_{\gamma_1\gamma_2,\gamma_3\gamma_4}(z_1, z_2; z_3, z_4)$$
$$\times \mathscr{G}^{(0)}_{\gamma_3\gamma_5}(z_3 - z_5)\mathscr{G}^{(0)}_{\gamma_4\gamma_6}(z_4 - z_6)\mathscr{T}^{(0)}_{\gamma_5\gamma_6,\gamma_7\gamma_8}(z_5, z_6; z_7, z_8)$$
$$\times \mathscr{G}^{(0)}_{\gamma_7\gamma_9}(z_7 - z_9)\mathscr{G}^{(0)}_{\gamma_8\gamma_{10}}(z_8 - z_{10})\, \mathscr{T}^{(0)}_{\gamma_9\gamma_{10},\gamma_{11}\gamma_{12}}(z_9, z_{10}; z_{11}, z_{12})$$
$$\times \mathscr{G}^{(0)}_{\gamma_{12}\beta}(z_{12} - y)\mathscr{G}^{(0)}_{\gamma_{11}\gamma_2}(z_{11} - z_2).$$

The rather complicated expressions that come about when using this technique[8] become much simpler in the case of point interactions described by the potential

$$U_{\alpha\beta,\gamma\delta}(\mathbf{r}_1 - \mathbf{r}_2) = \lambda\delta_{\alpha\delta}\delta_{\beta\gamma}\delta(\mathbf{r}_1 - \mathbf{r}_2).$$

Then the function $\mathscr{T}^{(0)}$ takes the simple form

$$\mathscr{T}^{(0)}_{\gamma_1\gamma_2,\gamma_3\gamma_4} = \lambda(\delta_{\gamma_1\gamma_3}\delta_{\gamma_2\gamma_4} - \delta_{\gamma_1\gamma_4}\delta_{\gamma_2\gamma_3})\delta(z_1 - z_2)\delta(z_1 - z_3)\delta(z_1 - z_4)$$
$$= \lambda L_{\gamma_1\gamma_2,\gamma_3\gamma_4}\delta(z_1 - z_2)\delta(z_1 - z_3)\delta(z_1 - z_4),$$

and only one integral survives from the four integrals over the vertex coordinates of the squares, thanks to the presence of three δ-functions in $\mathscr{T}^{(0)}$. This allows us to replace the squares simply by points (vertices) in our

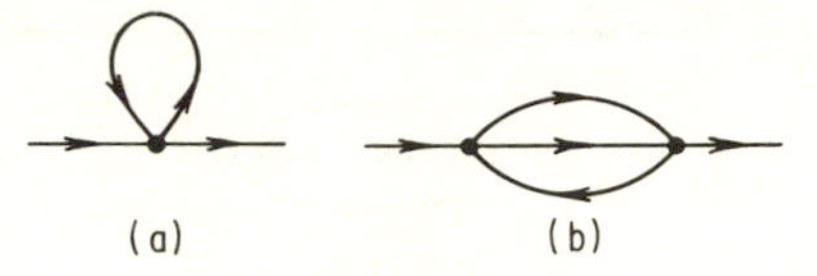

FIGURE 38

diagrams. For example, diagram (b) of Fig. 35 and diagram (c) of Fig. 36 take the simple form shown in Fig. 38, and the corresponding corrections can be written in the form

$$-\lambda L_{\gamma_1\gamma_2,\gamma_3\gamma_4} \int d^4z\, \mathscr{G}^{(0)}_{\alpha\gamma_1}(x - z)\mathscr{G}^{(0)}_{\gamma_4\beta}(z - y)\mathscr{G}^{(0)}_{\gamma_3\gamma_2}(0), \qquad \text{(a)}$$

$$\frac{\lambda^2}{2} L_{\gamma_1\gamma_2,\gamma_3\gamma_4}L_{\gamma_5\gamma_6,\gamma_7\gamma_8} \int d^4z_1\, d^4z_2\, \mathscr{G}^{(0)}_{\alpha\gamma_1}(x - z_1)\mathscr{G}^{(0)}_{\gamma_3\gamma_5}(z_1 - z_2)$$
$$\times \mathscr{G}^{(0)}_{\gamma_4\gamma_6}(z_1 - z_2)\mathscr{G}^{(0)}_{\gamma_7\gamma_2}(z_2 - z_1)\mathscr{G}^{(0)}_{\gamma_8\beta}(z_2 - y). \qquad \text{(b)}$$

The general rules for making calculations based on this diagram technique are obvious consequences of our previous considerations.

B. Interaction between particles and phonons. Interaction between particles and phonons (e.g., between particles of a liquid and sound waves, or between electrons in a metal and lattice vibrations) is described by the Hamiltonian

$$\hat{H}_{\text{int}}(\tau) = g \int \bar{\psi}_\alpha(\mathbf{r}, \tau)\psi_\alpha(\mathbf{r}, \tau)\varphi(\mathbf{r}, \tau)\, d\mathbf{r},$$

[8] Of course, this complexity is offset by the symmetry of the expressions.

where g is an interaction coupling constant. It is easy to see that the only nonzero corrections to the particle Green's function $\mathscr{G}$ and to the phonon Green's function $\mathscr{D}$ come from perturbation-theory terms of even order. (The expressions for the odd-order corrections contain an odd number of phonon operators φ.) Calculating the expressions for corrections to the particle Green's function $\mathscr{G}$, we find that they are exactly the same as the expressions for the corrections to $\mathscr{G}$ in our previous formulation of perturbation theory for two-particle interactions, except that the potential $\mathscr{V}(z_1 - z_2)$ is now replaced by $g^2\mathscr{D}^{(0)}(z_1 - z_2)$. It is natural that the corresponding corrections should be described by exactly the same diagrams as in Figs. 33 and 34, except that now we denote a phonon Green's function $\mathscr{D}^{(0)}$ by a dashed line instead of by a wavy line.

The second-order corrections to the phonon Green's function are described

FIGURE 39

by the two diagrams in Fig. 39. Calculations show that the correction corresponding to diagram (a) of Fig. 39 is

$$\pm g^2 \int d^4z_1\, d^4z_2\, \mathscr{D}^{(0)}(x - z_1)\mathscr{G}^{(0)}_{\alpha\beta}(z_1 - z_2)\mathscr{G}^{(0)}_{\beta\alpha}(z_2 - z_1)\mathscr{D}^{(0)}(z_2 - y),$$

while the correction corresponding to diagram (b) of Fig. 39 is

$$g^2 \int d^4z_1\, \mathscr{D}^{(0)}(x - z_1)\mathscr{G}_{\alpha\alpha}(0) \int d^4z_2\, \mathscr{D}^{(0)}(z_2 - y)\mathscr{G}_{\beta\beta}(0).$$

Just as in Sec. 9, we can convince ourselves that the expression corresponding to diagram (b) is zero. The same argument allows us to ignore all diagrams whose analytical expressions contain the integral

$$\int \mathscr{D}^{(0)}(z)\, d^4z.$$

This category includes all diagrams for $\mathscr{D}$ which decompose into two disconnected parts, where each part contains one external line. We can also ignore all diagrams for $\mathscr{G}$ of the type shown in Fig. 40, i.e., which have a part containing no external lines, joined to the rest of the part diagram by one phonon line.

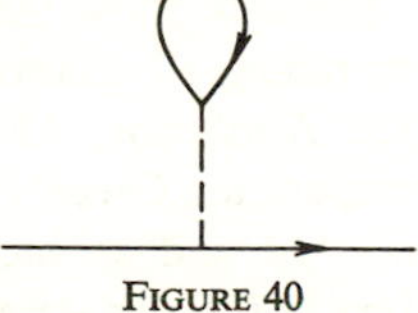

FIGURE 40

Examining the higher-order corrections to $\mathscr{D}$, and bearing in mind what has just been said about $\mathscr{G}$, we can state the following general rules for calculating the corrections of order $2n$ to the particle and phonon Green's functions, by using the diagram technique:

1. Form all connected,[9] topologically nonequivalent diagrams with $2n$ vertices.

[9] Here, diagrams of the types shown in Figs. 39(b) and 40 are considered to be disconnected.

2. With each solid line associate a free-particle Green's function $\mathscr{G}^{(0)}_{\alpha\beta}(x - y)$, and with each dashed line associate a function $\mathscr{D}^{(0)}(x - y)$.
3. Integrate over the coordinates of all vertices (with respect to both $\mathbf{r}$ and τ), and sum over the corresponding spin variables.

FIGURE 41

4. Multiply the resulting expression by $g^{2n}(-1)^{n+F}$, where F is the number of closed fermion loops.

For example, the fourth-order correction to the phonon Green's function corresponding to the diagram in Fig. 41 is

$$\mp g^4 \int d^4z_1\, d^4z_2\, d^4z_3\, d^4z_4\, \mathscr{D}^{(0)}(x - z_1)\mathscr{G}^{(0)}_{\gamma_1\gamma_2}(z_1 - z_2)\mathscr{G}^{(0)}_{\gamma_3\gamma_1}(z_3 - z_1)$$
$$\times\, \mathscr{D}^{(0)}(z_2 - z_3)\mathscr{G}^{(0)}_{\gamma_2\gamma_4}(z_2 - z_4)\mathscr{G}^{(0)}_{\gamma_4\gamma_3}(z_4 - z_3)\mathscr{D}^{(0)}(z_4 - y).$$

14. The Diagram Technique in Momentum Space

14.1. Transformation to momentum space. The diagram technique in coordinate space developed in the preceding section turns out to be quite unsuitable for making explicit calculations. It will be recalled that the success of field-theory methods at the absolute zero of temperature is chiefly due to the highly automatic way in which calculations can be performed. For $T = 0$, this was achieved by expanding all quantities appearing in the theory in four-dimensional Fourier integrals. However, in Matsubara's technique (described above), this automatic way of doing things is no longer possible, since the variable τ varies over a finite interval from 0 to $1/T$, so that a transformation to a Fourier-integral representation is impossible. The application of Matsubara's technique in the coordinate representation is very difficult because of the fact that $\mathscr{G}^{(0)}$ and $\mathscr{D}^{(0)}$ are discontinuous functions of the variable τ, and hence all integrals with respect to τ involve a large number of regions of integration, whose number increases very rapidly with n, the order of the approximation.

We now show that Matsubara's technique can be greatly simplified by expanding all quantities depending on τ in *Fourier series* with respect to τ [see Abrikosov, Gorkov and Dzyaloshinski (A5), Fradkin (F2)]. The temperature Green's function $\mathscr{G}$ (or $\mathscr{D}$) is a function of the difference variable $\tau_1 - \tau_2$, and as such is defined in the interval $[-1/T, 1/T]$. Expanding $\mathscr{G}(\tau)$ in Fourier series, we obtain

$$\mathscr{G}(\tau) = T \sum_n e^{-i\omega_n\tau}\mathscr{G}(\omega_n),$$
$$\mathscr{G}(\omega_n) = \frac{1}{2}\int_{-1/T}^{1/T} e^{i\omega_n\tau}\mathscr{G}(\tau)\, d\tau, \qquad \omega_n = n\pi T. \tag{14.1}$$

Our problem involves going over to a Fourier representation in the expressions (given in Sec. 13) for the corrections to the Green's functions. It is most desirable that the methods used to do this should not introduce any extra complications into the formulas, e.g., that no extra factors depending on the "frequencies" ω_n should appear.

As we now show, this way of dealing with the case $T \neq 0$ leads to only a slight modification of the situation for the case $T = 0$. As proved in Sec. 11, the Green's functions have the general property that their values for $\tau < 0$ are related by simple formulas to their values for $\tau > 0$ [see (11.8) and (11.8′)]. These formulas imply that the Fourier components $\mathscr{G}(\omega_n)$ [or $\mathscr{D}(\omega_n)$] of the Green's functions for bosons and phonons are nonzero only for "even" frequencies $\omega_n = 2n\pi T$, whereas for fermions $\mathscr{G}(\omega_n)$ is nonzero only for "odd" frequencies $\omega_n = (2n + 1)\pi T$. To see this, we first note that

$$\mathscr{G}(\omega_n) = \frac{1}{2}\int_{-1/T}^{1/T} e^{i\omega_n\tau}\mathscr{G}(\tau)\,d\tau$$
$$= \frac{1}{2}\int_{0}^{1/T} e^{i\omega_n\tau}\mathscr{G}(\tau)\,d\tau + \frac{1}{2}\int_{-1/T}^{0} e^{i\omega_n\tau}\mathscr{G}(\tau)\,d\tau.$$

Then, substituting (11.8) for $\mathscr{G}(\tau < 0)$ in the second integral, and making the change of variables $\tau' = \tau + (1/T)$, we obtain

$$\mathscr{G}(\omega_n) = \frac{1}{2}\int_{0}^{1/T} e^{i\omega_n\tau}\mathscr{G}(\tau)\,d\tau \mp \frac{1}{2}\int_{1/T}^{0} e^{i\omega_n\tau}\mathscr{G}\left(\tau + \frac{1}{T}\right)$$
$$= \frac{1}{2}(1 \mp e^{i\omega_n/T})\int_{0}^{1/T} e^{i\omega_n\tau}\mathscr{G}(\tau)\,d\tau,$$

which implies the assertion just made. In fact, we always have

$$\mathscr{G}(\omega_n) = \int_{0}^{1/T} e^{i\omega_n\tau}\mathscr{G}(\tau)\,d\tau, \tag{14.2}$$

where

$$\omega_n = \begin{cases}(2n + 1)\pi T & \text{for fermions,}\\ 2n\pi T & \text{for bosons.}\end{cases}$$

Next, we substitute the Fourier-series expansion (14.1) into all appropriate terms of the perturbation series, simultaneously taking Fourier transforms with respect to the space variables. In other words, we write

$$\mathscr{G}(\mathbf{r}) = \frac{1}{(2\pi)^3}\int e^{i\mathbf{p}\cdot\mathbf{r}}\mathscr{G}(\mathbf{p})\,d\mathbf{p},$$
$$\mathscr{G}(\mathbf{p}) = \int e^{-i\mathbf{p}\cdot\mathbf{r}}\mathscr{G}(\mathbf{r})\,d\mathbf{r}, \tag{14.3}$$

so that as far as the space variables are concerned, the transformation is carried out in exactly the same way as for $T = 0$. We note that an even number of fermion lines meet at every point whose coordinates are involved in an integration. It follows that in evaluating the integral

$$\int_{0}^{1/T} d\tau\, e^{i\tau\Sigma\omega_n}, \tag{14.4}$$

at each vertex, the sum of "frequencies" $\Sigma\, \omega_n$ in the exponent is always "even," i.e., $\Sigma\, \omega_n = 2N\pi T$, where N is an integer. In this case, the integral (14.4) equals

$$\frac{1}{T}\delta_{\Sigma\omega_n}, \qquad \delta_{\omega_n} = \begin{cases} 1 & \text{for } \omega_n = 0, \\ 0 & \text{for } \omega_n \neq 0. \end{cases} \tag{14.5}$$

Thus, the situation here is essentially the same as for $T = 0$. It will be recalled that for $T = 0$, integration over the space-time coordinates of the vertices gives rise to δ-functions of frequency and momentum, expressing conservation of the energy and momentum of certain virtual processes. For $T \neq 0$, each δ-function of the frequency is replaced by a Kronecker delta δ_{ω_n}, expressing conservation of the discrete "frequency" ω_n. As a result of all this, in describing the perturbation series in momentum space, we can continue to use the same Feynman diagrams as for $T = 0$. The only essential difference (apart from differences in coefficients) is the appearance of sums over the discrete frequencies ω_n instead of integrals over the continuous frequency ω in the expressions for the matrix elements.

Before considering specific examples, we derive formulas for the Fourier components of the zeroth-order Green's functions. In Sec. 11, we calculated zeroth-order Green's functions in coordinate space. According to (11.13a) after taking the Fourier transform (14.3) with respect to $\mathbf{r}$, we find that the Green's function of a free fermion has the form

$$\mathscr{G}^{(0)}_{\alpha\beta}(\mathbf{p}, \tau) = -\delta_{\alpha\beta}[1 - n(\mathbf{p})]e^{-[\varepsilon_0(\mathbf{p}) - \mu)\tau},$$
$$n(\mathbf{p}) = \{e^{[\varepsilon_0(\mathbf{p}) - \mu]/T} + 1\}^{-1}$$

for $\tau > 0$. Substituting this expression into (14.2), we obtain

$$\mathscr{G}^{(0)}_{\alpha\beta}(\mathbf{p}, \omega_n) = -\delta_{\alpha\beta}[1 - n(\mathbf{p})] \int_0^{1/T} e^{i\omega_n\tau - [\varepsilon_0(\mathbf{p}) - \mu]\tau}\, d\tau$$

$$= -\frac{\delta_{\alpha\beta}}{i\omega_n - \varepsilon_0(\mathbf{p}) + \mu}[1 - n(\mathbf{p})]\{e^{(2n+1)\pi i - [\varepsilon_0(\mathbf{p}) - \mu]/T} - 1\},$$

where $\omega_n = (2n + 1)\pi T$, i.e.,

$$\mathscr{G}^{(0)}_{\alpha\beta}(\mathbf{p}, \omega_n) = \delta_{\alpha\beta}\frac{1}{i\omega_n - \varepsilon_0(\mathbf{p}) + \mu}, \qquad \omega_n = (2n + 1)\pi T. \tag{14.6}$$

Similar calculations give

$$\mathscr{G}^{(0)}(\mathbf{p}, \omega_n) = \frac{1}{i\omega_n - \varepsilon_0(\mathbf{p}) + \mu}, \qquad \omega_n = 2n\pi T, \tag{14.7}$$

for bosons, and

$$\mathscr{D}^{(0)}(\mathbf{k}, \omega_n) = -\frac{\omega_0^2(\mathbf{k})}{\omega_n^2 + \omega_0^2(\mathbf{k})}, \qquad \omega_n = 2n\pi T \tag{14.8}$$

for phonons. Thus, the zeroth-order Green's functions for fermions and bosons differ only by the "parity" of the frequencies ω_n. The functions

(14.6)–(14.8) are obtained from the Green's functions (7.7) and (7.16) by making the substitution $\omega \to i\omega_n$. Later on, we shall show that there is a similar relation between the exact Green's functions for $T = 0$ and $T \neq 0$ (with certain stipulations, of course).

14.2. Examples. As we have seen, the calculation of temperature Green's functions can be carried out by using the technique of Feynman diagrams in momentum space. In doing so, with each line of a given diagram we associate a zeroth-order particle Green's function $\mathscr{G}^{(0)}(\mathbf{p}, \omega_n)$ or a zeroth-order phonon Green's function $\mathscr{D}^{(0)}(\mathbf{k}, \omega_n)$, and with each vertex we associate the quantity $\delta(\Sigma\, \mathbf{p})\delta_{\Sigma\, \omega_n}$ expressing conservation of momentum and conservation of the discrete "frequency" ω_n. Moreover, we integrate over the momenta and sum over the "frequencies" ω_n of all the internal lines.

The actual form of the diagrams and of the associated analytical expressions depends on the form of the interaction. We begin with the case of two-particle interactions.

A. Two-particle interactions. Consider the correction to the Green's function corresponding to diagram (d) of Fig. 33, which was found in Sec. 13 to be

$$-\int d^4z_1\, d^4z_2\, \mathscr{G}^{(0)}_{\alpha\gamma_1}(x - z_1)\mathscr{G}^{(0)}_{\gamma_1\gamma_2}(z_1 - z_2)\mathscr{G}^{(0)}_{\gamma_2\beta}(z_2 - y)\mathscr{V}(z_1 - z_2).$$

Calculating the Fourier components of this expression with respect to the coordinates and "time," we obtain

$$\delta\mathscr{G}^{(1)}(\mathbf{p}, \omega_n) = \frac{1}{2}\int d(\mathbf{x} - \mathbf{y}) \int_{-1/T}^{1/T} d(\tau_x - \tau_y)$$
$$\times\ \delta\mathscr{G}^{(1)}(\mathbf{x} - \mathbf{y}, \tau_x - \tau_y)e^{-i\mathbf{p}\cdot(\mathbf{x}-\mathbf{y}) + i\omega_n(\tau_x - \tau_y)}.$$

Next, we introduce the Fourier components of the potential $\mathscr{V}(z_1 - z_2)$:

$$\mathscr{V}(\mathbf{r}, \tau) = \frac{T}{(2\pi)^3}\sum_{\omega_n}\int d\mathbf{q}\, e^{i\mathbf{q}\cdot\mathbf{r} - i\omega_n\tau}\mathscr{V}(\mathbf{q}, \omega_n).$$

Since

$$T\sum_{n=-\infty}^{\infty} e^{2n\pi i T\tau} = \delta(\tau),$$

$$\mathscr{V}(\mathbf{q}, \omega_n) = U(\mathbf{q}),$$

we have

$$\delta\mathscr{G}^{(1)}_{\alpha\beta}(\mathbf{p}, \omega_n) = -\frac{1}{2}\left[\frac{T}{(2\pi)^3}\right]^4 \sum_{\substack{\omega_{n1},\, \omega_{n2} \\ \omega_{n3},\, \omega_{n4}}} \int d\mathbf{p}_1\, d\mathbf{p}_2\, d\mathbf{p}_3\, d\mathbf{p}_4\, \mathscr{G}^{(0)}_{\alpha\gamma_1}(\mathbf{p}_1, \omega_{n1})$$
$$\times\ \mathscr{G}^{(0)}_{\gamma_1\gamma_2}(\mathbf{p}_2, \omega_{n2})\, \mathscr{G}^{(0)}_{\gamma_2\beta}(\mathbf{p}_3, \omega_{n3})\mathscr{V}(\mathbf{q}, \omega_{n4})$$
$$\times \int d(\mathbf{x} - \mathbf{y})\, dz_1\, dz_2\, d(\tau_x - \tau_y)\, d\tau_1\, d\tau_2\, e^{-i\mathbf{p}\cdot(\mathbf{x}-\mathbf{y})}e^{i\omega_n(\tau_x - \tau_y)}$$
$$\times\ e^{i\mathbf{p}_1\cdot(\mathbf{x}-\mathbf{z}_1) + i\mathbf{p}_2\cdot(\mathbf{z}_1 - \mathbf{z}_2) + i\mathbf{p}_3\cdot(\mathbf{z}_2 - \mathbf{y})}e^{-i\omega_{n1}(\tau_x - \tau_1) - i\omega_{n2}(\tau_1 - \tau_2)}$$
$$\times\ e^{-i\omega_{n3}(\tau_2 - \tau_y)}e^{i\mathbf{q}\cdot(\mathbf{z}_1 - \mathbf{z}_2) - i\omega_{n4}(\tau_1 - \tau_2)}.$$

Making the change of variables $\mathbf{x} - \mathbf{y} \to \mathbf{x}$, $\tau_x - \tau_y \to \tau$ in the integrals over space and time, we find that

$$\frac{1}{2}\int d\mathbf{x}\, d\mathbf{z}_1\, d\mathbf{z}_2 \int_{-1/T}^{1/T} d\tau \int_0^{1/T}\int_0^{1/T} d\tau_1\, d\tau_2\, e^{i(-\mathbf{p}+\mathbf{p}_1)\cdot\mathbf{x}+i(-\mathbf{p}_1+\mathbf{p}_2+\mathbf{q})\cdot\mathbf{z}_1+i(-\mathbf{p}_2+\mathbf{p}_3-\mathbf{q})\cdot\mathbf{z}_2}$$

$$\times\, e^{i(\omega_n-\omega_{n1})\tau+i(\omega_{n1}-\omega_{n2}-\omega_{n4})\tau_1+i(\omega_{n2}-\omega_{n3}+\omega_{n4})\tau_2} e^{i(\mathbf{p}_1-\mathbf{p}_3)\cdot\mathbf{y}+i(-\omega_{n1}+\omega_{n3})\tau_y}$$

$$= \left[\frac{(2\pi)^3}{T}\right]^3 \delta(\mathbf{p}-\mathbf{p}_1)\delta(\mathbf{p}_1-\mathbf{p}_2-\mathbf{q})\delta(\mathbf{p}_2-\mathbf{p}_3+\mathbf{q})$$

$$\times\, \delta_{\omega_n-\omega_{n1}}\delta_{\omega_{n1}-\omega_{n2}-\omega_{n4}}\delta_{\omega_{n2}-\omega_{n3}+\omega_{n4}},$$

which implies that

$$\delta\mathscr{G}^{(1)}_{\alpha\beta}(\mathbf{p},\omega_n) = -\frac{T}{(2\pi)^3}\sum_{\omega_{n1}}\int d\mathbf{p}_1\, \mathscr{G}^{(0)}_{\alpha\gamma_1}(\mathbf{p},\omega_n)$$

$$\times\, \mathscr{G}^{(0)}_{\gamma_1\gamma_2}(\mathbf{p}_1,\omega_{n1})\mathscr{G}^{(0)}_{\gamma_2\beta}(\mathbf{p},\omega_n)\mathscr{V}(\mathbf{p}-\mathbf{p}_1,\omega_n-\omega_{n1}).$$

Substituting the expressions (14.6) and (14.7) for the zeroth-order Green's functions, we finally obtain

$$\delta\mathscr{G}^{(1)}_{\alpha\beta} = -\frac{\delta_{\alpha\beta}}{[i\omega_n-\varepsilon_0(\mathbf{p})+\mu]^2}\frac{T}{(2\pi)^3}\sum_{\omega_{n1}}\int d\mathbf{p}_1\,\frac{\mathscr{V}(\mathbf{p}-\mathbf{p}_1,\omega_n-\omega_{n1})}{i\omega_{n1}-\varepsilon_0(\mathbf{p}_1)+\mu}.$$

Similar calculations for the contribution of diagram (c) of Fig. 33 lead to the result

$$\pm\frac{\delta_{\alpha\beta}}{[i\omega_n-\varepsilon_0(\mathbf{p})+\mu]^2}\mathscr{V}(0,0)(2s+1)\frac{T}{(2\pi)^3}$$

$$\times\sum_{\omega_{n1}}\int d\mathbf{p}_1\,\frac{e^{i\omega_{n1}\tau}}{i\omega_{n1}-\varepsilon_0(\mathbf{p}_1)+\mu}, \tag{14.10}$$

where $\tau \to +0$, and s is the spin of the particle, equal to $\frac{1}{2}$ for fermions and 0 for bosons. Here we have introduced $e^{i\omega_n\tau}$ ($\tau \to +0$) inside the sum in accordance with the stipulation (made in Sec. 13) that a Green's function in coordinate space with identical time arguments is defined as

$$\mathscr{G}^{(0)}(0,0) = \lim_{\tau\to+0}\mathscr{G}^{(0)}(0,-\tau).$$

The Feynman diagrams corresponding to the expressions (14.9) and (14.10) are shown in Figs. 42(a) and 42(b), respectively. The external lines of these diagrams are labeled with $\mathbf{p}$, ω_n, the external momentum and frequency. The momenta and frequencies at each vertex satisfy conservation laws, i.e., the sum of all momenta and frequencies "entering" a given vertex equals the sum of all momenta and frequencies "leaving" the vertex.

We now consider the diagram for $\mathscr{G}(\mathbf{p},\omega_n)$ in the kth-order approximation of perturbation theory. Such a diagram has $2k$ vertices, $2k+1$ solid lines and k wavy lines. To calculate the Fourier components, we perform $2k$ integrations over the space and "time" coordinates of the vertices, and one integration over the difference coordinates of the external points. This leads to $2k + 1$ quantities of the type $\delta(\Sigma\, \mathbf{p})\delta_{\Sigma\omega_n}$, expressing $2k + 1$ conservation laws. It is easy to see that two of these conservation laws express the fact that the external lines have momentum $\mathbf{p}$ and frequency ω_n. The remaining $2k - 1$ conservation laws imply that only k integrations and summations actually survive among the $3k - 1$ integrations over the momenta and summations over the frequencies of the internal lines (both solid and wavy).

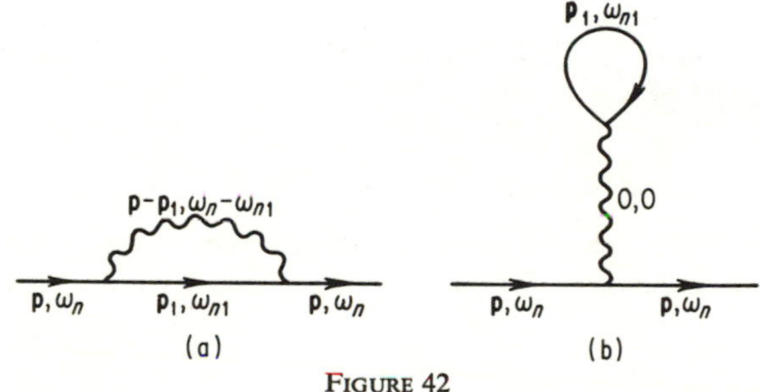

FIGURE 42

The general rules for obtaining the expression corresponding to a given diagram for the Green's function are as follows:

1. First, associate momenta and frequencies with the lines of the diagram in such a way that the external lines have the external momentum and frequency, while the momenta and frequencies of the internal lines satisfy the conservation laws $\Sigma\, \mathbf{p}' = 0, \Sigma\, \omega_n' = 0$, where the frequencies of Bose lines are always even $[\omega_n = 2n\pi T]$ and those of Fermi lines are odd $[\omega_n = (2n + 1)\pi T]$.
2. Integrate and sum over the independent internal momenta and frequencies.
3. With each solid internal line (of momentum $\mathbf{p}'$ and frequency ω_n') associate a quantity
$$\frac{1}{i\omega_n' - \varepsilon_0(\mathbf{p}') + \mu},$$
and with each wavy line (of momentum $\mathbf{q}'$ and frequency ω_n') associate a quantity
$$\mathscr{V}(\mathbf{q}, \omega_n') \equiv U(\mathbf{q}).$$
4. With both external lines (of momentum $\mathbf{p}$ and frequency ω_n) associate a quantity
$$\frac{\delta_{\alpha\beta}}{[i\omega_n - \varepsilon_0(\mathbf{p}) + \mu]^2}.$$
5. Multiply the resulting expression by
$$(-1)^k \frac{T^k}{(2\pi)^{3k}} (2s + 1)^F(\mp 1)^F,$$
where F is the number of closed loops formed by particle lines.

Using these rules, it is not hard to write down the correction corresponding to an arbitrarily complicated diagram. For example, the correction corresponding to the diagram shown in Fig. 43 equals

$$\pm \frac{\delta_{\alpha\beta}}{[i\omega - \varepsilon_0(\mathbf{p}) + \mu]^2} \frac{T^3}{(2\pi)^9} (2s + 1) \sum_{\omega_1, \omega_2, \omega_3} \int d\mathbf{p}_1 \, d\mathbf{p}_2 \, d\mathbf{p}_3$$
$$\times \frac{1}{i(\omega - \omega_1) - \varepsilon_0(\mathbf{p} - \mathbf{p}_1) + \mu} \frac{1}{i(\omega_3 - \omega_1) - \varepsilon_0(\mathbf{p}_3 - \mathbf{p}_1) + \mu}$$
$$\times \frac{1}{i\omega_3 - \varepsilon_0(\mathbf{p}_3) + \mu} \frac{1}{i\omega_2 - \varepsilon_0(\mathbf{p}_2) + \mu} \frac{1}{i(\omega_1 + \omega_2) - \varepsilon_0(\mathbf{p}_1 + \mathbf{p}_2) + \mu}$$
$$\times [U(\mathbf{p}_1)]^2 U(\mathbf{p} - \mathbf{p}_3),$$

where

$$\omega_1 = 2n\pi T, \qquad \omega_2, \omega_3 = (2n + 1)\pi T.$$

Next, we consider the other version of the diagram technique for the case of two-particle interactions. To go over to the Fourier representation in the

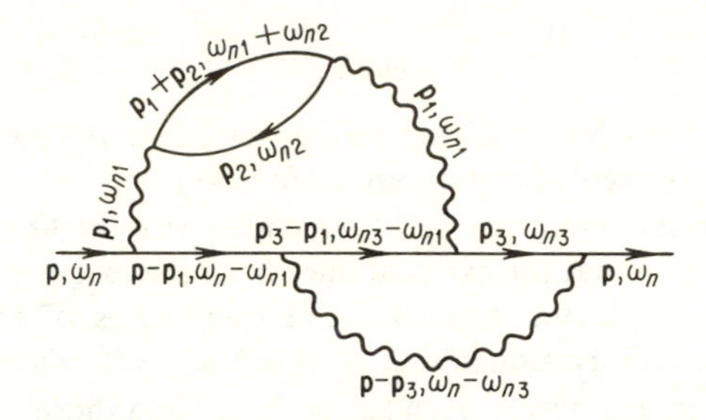

FIGURE 43

corresponding expressions of Sec. 13, it is convenient to use the following formal device: Earlier, we introduced the quantity $\mathscr{T}^{(0)}_{\gamma_1, \gamma_2, \gamma_3, \gamma_4}(z_1, z_2; z_3, z_4)$. This quantity depends on four "times" τ_i, where each τ_i varies in the interval from 0 to $1/T$. We continue the function $\mathscr{T}^{(0)}$ onto the interval from $-1/T$ to $1/T$ by using relations like (11.8) for the $\mathscr{G}$-function, i.e.,

$$\mathscr{T}^{(0)}(\tau_1 < 0, \tau_2; \tau_3, \tau_4) = \mp \mathscr{T}^{(0)}(\tau_1 + \frac{1}{T}, \tau_2; \tau_3, \tau_4),$$

and similarly for τ_2, τ_3, τ_4. Then we define the Fourier components with respect to the τ_i by

$$\frac{1}{16} \int_{-1/T}^{1/T} \cdots \int_{-1/T}^{1/T} d\tau_1 \cdots d\tau_4 \, e^{i(\omega_1\tau_1 + \omega_2\tau_2 - \omega_3\tau_3 - \omega_4\tau_4)} \mathscr{T}^{(0)}(\tau_1, \tau_2; \tau_3, \tau_4).$$

Obviously, all four frequencies are "odd" in the case of Fermi statistics, and "even" in case of Bose statistics.

Moreover, we note that since $\mathscr{T}^{(0)}(z_1, z_2; z_3, z_4)$ is by definition a function only of the coordinate and "time" differences, the Fourier components of $\mathscr{T}^{(0)}$ with respect to the space and time variables contain a δ-function $\delta(\mathbf{p}_1 + \mathbf{p}_2 - \mathbf{p}_3 - \mathbf{p}_4)$ involving a sum of momenta and a Kronecker delta $\delta_{\omega_1+\omega_2-\omega_3-\omega_4}$ involving a sum of frequencies. Therefore, we define the Fourier component of $\mathscr{T}^{(0)}$ directly as

$$\frac{(2\pi)^3}{T}\delta(\mathbf{p}_1 + \mathbf{p}_2 - \mathbf{p}_3 - \mathbf{p}_4)\delta_{\omega_1+\omega_2-\omega_3-\omega_4}\mathscr{T}^{(0)}(\mathbf{p}_1, \omega_1, \mathbf{p}_2, \omega_2; \mathbf{p}_3, \omega_3, \mathbf{p}_4, \omega_4)$$
$$= \frac{1}{16}\int_{-1/T}^{1/T}\cdots\int_{-1/T}^{1/T} d\tau_1\cdots d\tau_4 \int d\mathbf{r}_1\cdots d\mathbf{r}_4 \tag{14.11}$$
$$\times\, e^{-i(\mathbf{p}\cdot\mathbf{r}_1 + \mathbf{p}_2\cdot\mathbf{r}_2 - \mathbf{p}_3\cdot\mathbf{r}_3 - \mathbf{p}_4\cdot\mathbf{r}_4) + i(\omega_1\tau_1 + \omega_2\tau_2 - \omega_3\tau_3 - \omega_4\tau_4)}\mathscr{T}^{(0)}(z_1, z_2; z_3, z_4).$$

For example, suppose we carry out the Fourier transformation of the

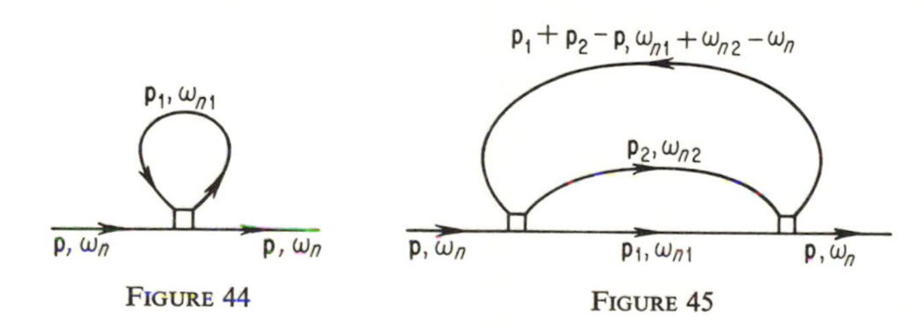

FIGURE 44 FIGURE 45

expression for the first-order correction corresponding to diagram (b) of Fig. 35 (see p. 116). Then, after simple calculations, we obtain

$$-\frac{1}{[i\omega - \varepsilon_0(\mathbf{p}) + \mu]^2}\frac{T}{(2\pi)^3}$$
$$\times \sum_{\omega_1}\int d\mathbf{p}_1\, \mathscr{T}^{(0)}_{\alpha\gamma,\gamma\beta}(\mathbf{p}, \omega, \mathbf{p}_1, \omega_1; \mathbf{p}_1, \omega_1, \mathbf{p}, \omega)\frac{1}{i\omega_1 - \varepsilon_0(\mathbf{p}_1) + \mu},$$

which corresponds to the diagram shown in Fig. 44. Similar calculations for the correction corresponding to the diagram shown in Fig. 45 lead to the formula

$$\frac{1}{2}\frac{1}{[i\omega - \varepsilon_0(\mathbf{p}) + \mu]^2}\frac{T^2}{(2\pi)^6}\sum_{\omega_1,\omega_2}\int d\mathbf{p}_1\, d\mathbf{p}_2\, \mathscr{T}^{(0)}_{\alpha\gamma_1,\gamma_2\gamma_3}(p, p_1 + p_2 - p; p_1, p_2)$$
$$\times \frac{1}{i\omega_1 - \varepsilon_0(\mathbf{p}_1) + \mu}\,\frac{1}{i\omega_2 - \varepsilon_0(\mathbf{p}_2) + \mu}\,\frac{1}{i(\omega_1 + \omega_2 - \omega) - \varepsilon_0(\mathbf{p}_1 + \mathbf{p}_2 - \mathbf{p}) + \mu}$$
$$\times\, \mathscr{T}^{(0)}_{\gamma_2\gamma_3,\gamma_1\beta}(p_1, p_2; p_1 + p_2 - p, p),$$

where we have used the four-dimensional notation $p = (\mathbf{p}, \omega_n)$.

The nth-order diagram for the $\mathscr{G}$-function contains n squares (vertices) and $2n + 1$ lines, where the $2n - 1$ internal lines obey the conservation laws $\Sigma\, \mathbf{p}' = 0, \Sigma\, \omega_n' = 0$ at the vertices. It is easy to see that there are a total of n independent integrations and summations over the momenta and frequencies of the internal lines. The general rules for calculating the contribution of an arbitrary diagram are as follows:

1. With each internal line associate a quantity

$$\frac{1}{i\omega' - \varepsilon_0(\mathbf{p}') + \mu}.$$

2. With each external line associate a quantity

$$\frac{1}{i\omega - \varepsilon_0(\mathbf{p}) + \mu}.$$

3. With each vertex associate a function

$$\mathscr{T}^{(0)}_{\alpha\beta,\gamma\delta}(p_1, p_2; p_3, p_1 + p_2 - p_3).$$

4. Integrate and sum over all the independent internal momenta and frequencies.
5. Sum over the indices $\alpha, \beta, \ldots$ of the quantities $\mathscr{T}^{(0)}$ joined by $\mathscr{G}^{(0)}$-lines.
6. Multiply the resulting expression by

$$\frac{T^n}{(2\pi)^{3n}} A_n,$$

where A_n is the same as in Rule 5, p. 117.

For example, the expression corresponding to Fig. 46 is

$$- \frac{1}{4} \frac{1}{[i\omega - \varepsilon_0(\mathbf{p}) + \mu]^2} \frac{T^3}{(2\pi)^9} \sum_{\omega_1,\omega_2,\omega_3} \int d\mathbf{p}_1\, d\mathbf{p}_2\, d\mathbf{p}_3$$

$$\times \mathscr{T}^{(0)}_{\alpha\gamma_1,\gamma_2\gamma_3}(p, p_1 + p_2 - p; p_1, p_2) \frac{1}{i\omega_1 - \varepsilon_0(\mathbf{p}_1) + \mu}$$

$$\times \frac{1}{i\omega_2 - \varepsilon_0(\mathbf{p}_2) + \mu} \mathscr{T}^{(0)}_{\gamma_2\gamma_3,\gamma_4\gamma_5}(p_1, p_2; p_3, p_1 + p_2 - p_3)$$

$$\times \frac{1}{i\omega_3 - \varepsilon_0(\mathbf{p}_3) + \mu} \frac{1}{i(\omega_1 + \omega_2 - \omega_3) - \varepsilon_0(\mathbf{p}_1 + \mathbf{p}_2 - \mathbf{p}_3) + \mu}$$

$$\times \mathscr{T}^{(0)}_{\gamma_4\gamma_5,\gamma_1\beta}(p_3, p_1 + p_2 - p_3; p_1 + p_2 - p, p)$$

$$\times \frac{1}{i(\omega_1 + \omega_2 - \omega) - \varepsilon_0(\mathbf{p}_1 + \mathbf{p}_2 - \mathbf{p}) + \mu}.$$

In the case of a point interaction, the function $\mathscr{T}^{(0)}$ does not depend on the momenta and frequencies.

B. Interaction between particles and phonons. In the case of interaction between particles and phonons, only the expressions corresponding to diagrams of even order are nonzero. An arbitrary diagram of order $2n$ has $3n + 1$ internal (electron and phonon) lines and $2n$ vertices, corresponding to

$$3n - 1 - (2n - 1) = n$$

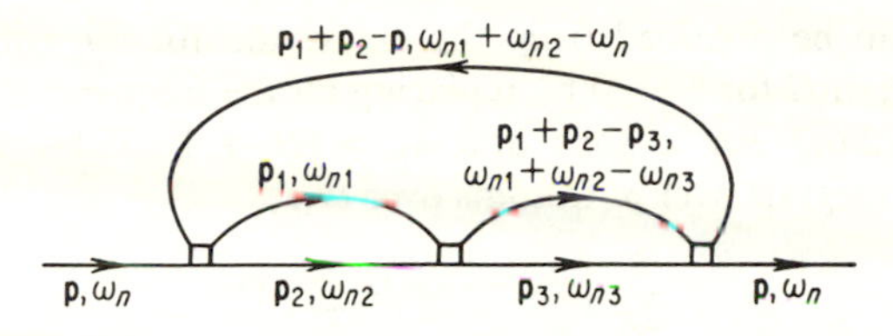

FIGURE 46

independent integrations. To calculate the contribution made by such a diagram, we use the following general rules:

1. With each solid internal line associate a quantity

$$\frac{1}{i\omega' - \varepsilon_0(\mathbf{p}') + \mu},$$

and with the two solid external lines (in the diagrams for corrections to the particle $\mathscr{G}$-function) associate a quantity

$$\frac{\delta_{\alpha\beta}}{[i\omega - \varepsilon_0(\mathbf{p}) + \mu]^2}.$$

2. With each phonon (dashed) line associate a quantity

$$-\frac{\omega_0^2(\mathbf{k})}{\omega^2 + \omega_0^2(\mathbf{k})}.$$

3. Multiply the result by

$$g^{2n}\frac{T^n}{(2\pi)^{3n}}(-1)^n(2s+1)^F(\mp 1)^F,$$

where g is the coupling constant, F is the number of closed loops, and s is the spin of the particles.

For example, the expression for the second-order correction to the phonon $\mathscr{D}$-function corresponding to the diagram shown in Fig. 47 equals

$$\pm\left[\frac{\omega_0^2(\mathbf{k})}{\omega^2 + \omega_0^2(\mathbf{k})}\right]^2 g^2(2s+1)\frac{T}{(2\pi)^3}$$
$$\times \sum_{\omega'}\int d\mathbf{p}'\,\frac{1}{i\omega' - \varepsilon_0(\mathbf{p}') + \mu}\,\frac{1}{i(\omega' - \omega) - \varepsilon_0(\mathbf{p}' - \mathbf{k}) + \mu}.$$

The rules given in this section stand in a very close relation to the corresponding rules for calculating corrections to the Green's function for $T = 0$. As is easily verified, the correction to the temperature Green's

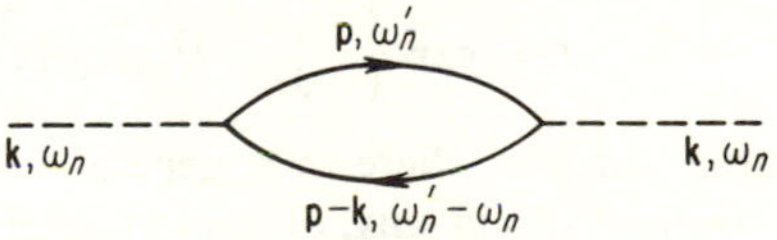

FIGURE 47

function $\mathscr{G}$ can be obtained from the expression for the correction to the Green's function G for $T = 0$ by replacing all the frequencies ω in G by $i\omega_n$ [where $\omega_n = 2n\pi T$ for bosons and $\omega_n = (2n + 1)\pi T$ for fermions] and changing all integrals over ω to sums over ω_n:

$$\frac{1}{2\pi} \int d\omega \cdots \to iT \sum_{\omega_n} \cdots$$

Finally, we examine how to make the passage to the limit $T = 0$ in the technique just developed. As $T \to 0$ the chief role in the sums over the frequencies ω_n is played by large values of n, and therefore these sums can be replaced by integrals. Observing that

$$\Delta\omega = \omega_{n+1} - \omega_n = 2\pi T,$$

we obtain

$$T \sum_{\omega_n} \cdots \to \frac{1}{2\pi} \int d\omega \cdots$$

We emphasize that $\mathscr{G}(\omega)$ evaluated at $T = 0$ does not coincide with $G(\omega)$. The relation between these two quantities will be established later on.

15. The Perturbation Series for the Thermodynamic Potential Ω

In certain cases, one finds that it is more convenient to calculate the thermodynamic potential Ω directly, instead of first finding the Green's function

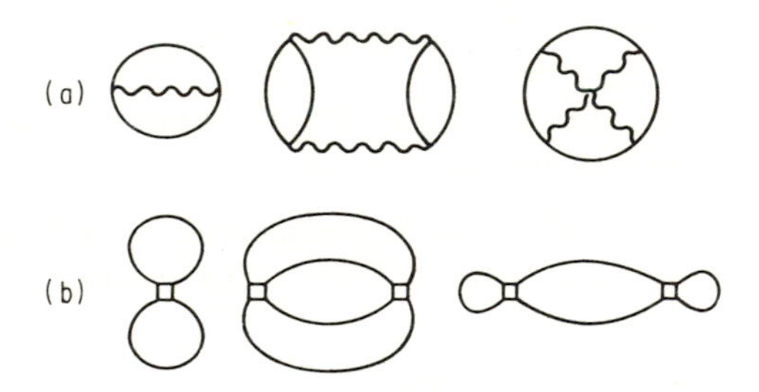

FIGURE 48

and then calculating thermodynamic quantities. In terms of the average value of the $\mathscr{S}$-matrix, the correction to the thermodynamic potential is given by the formula

$$\Delta\Omega = -T \ln \langle \mathscr{S} \rangle,$$
$$\mathscr{S} = \exp\left\{-\int_0^{1/T} \hat{H}_{\text{int}}(\tau)\, d\tau\right\} \tag{15.1}$$

[see (12.11)]. It turns out that there is a general way of evaluating the logarithm in (15.1), or more specifically, of constructing a diagram technique for finding the quantity Ω directly. The considerations given above make it clear that the diagrams describing the perturbation series for Ω have the form of closed loops. Typical diagrams are shown in Fig. 48 for the case of two-particle interactions and in Fig. 49 for the interaction of particles with phonons.[10] The expression corresponding to diagram (a) of Fig. 49 actually vanishes.

Among the diagrams corresponding to the perturbation-theory approximation of a given order, there are two types of diagrams, connected and disconnected. The latter contain two or more closed loops which are not connected to each other by any lines. Connected diagrams come about as follows: Suppose we use Wick's theorem to write an arbitrary term of the series for $\langle \mathscr{S} \rangle$:

$$\frac{(-1)^n}{n!} \int_0^{1/T} \cdots \int_0^{1/T} d\tau_1 \cdots d\tau_n \langle T_\tau(\hat{H}_{\text{int}}(\tau_1) \cdots \hat{H}_{\text{int}}(\tau_n)) \rangle. \tag{15.2}$$

Then a connected diagram is obtained if we start our pairing with any operator appearing in $\hat{H}_{\text{int}}(\tau_1)$ and ultimately come back to $\hat{H}_{\text{int}}(\tau_1)$ without skipping a single $\hat{H}_{\text{int}}$ in the process. In any other case, we get a disconnected diagram.

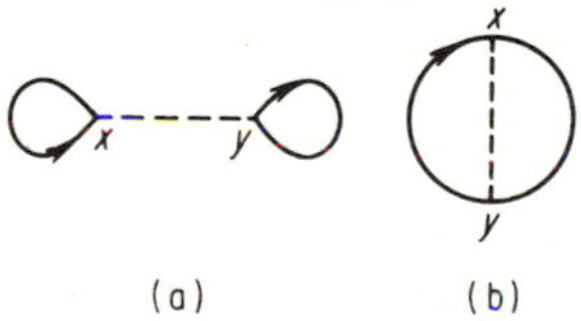

FIGURE 49

Suppose a disconnected diagram of order n consists of k closed loops. First consider the case where all k loops contain different numbers of vertices. Such a diagram leads to the expression

$$\begin{aligned} \frac{(-1)^n}{n!} & \int d\tau_1^{(1)} \cdots d\tau_{m_1}^{(1)} \langle T_\tau(\hat{H}_{\text{int}}(\tau_1^{(1)}) \cdots \hat{H}_{\text{int}}(\tau_{m_1}^{(1)})) \rangle_{\text{con}} \\ & \times \int d\tau_1^{(2)} \cdots d\tau_{m_2}^{(2)} \langle T_\tau(\hat{H}_{\text{int}}(\tau_1^{(2)}) \cdots \hat{H}_{\text{int}}(\tau_{m_2}^{(2)})) \rangle_{\text{con}} \cdots \\ & \times \int d\tau_1^{(k)} \cdots d\tau_{m_k}^{(k)} \langle T_\tau(\hat{H}_{\text{int}}(\tau_1^{(k)}) \cdots \hat{H}_{\text{int}}(\tau_{m_k}^{(k)})) \rangle_{\text{con}}, \end{aligned} \tag{15.3}$$

where

$$m_1 + m_2 + \cdots + m_k = n \; (m_1 \neq m_2 \neq \cdots \neq m_k),$$

and the symbol $\langle \cdots \rangle_{\text{con}}$ denotes an average which in each case corresponds to a definite connected diagram. We now sum all the topologically equivalent diagrams containing k loops of the type just chosen. Obviously, this can be done by merely multiplying (15.3) by F_k, the total number of such diagrams. This is just the number of ways of assigning n operators $\hat{H}_{\text{int}}$ to k different "cells," containing $m_1, m_2, \cdots, m_k$ places, respectively, i.e.,

$$F_k = \frac{n!}{m_1! m_2! \cdots m_k!}.$$

[10] Fig. 48(a) corresponds to the first version of the two-particle theory, and Fig. 48(b) corresponds to the second version.

As a result, we obtain

$$\frac{(-1)^{m_1}}{m_1!}\int d\tau_1^{(1)}\cdots d\tau_{m_1}^{(1)}\langle T_\tau(\hat{H}_{\text{int}}(\tau_1^{(1)})\cdots\hat{H}_{\text{int}}(\tau_{m_1}^{(1)}))\rangle_{\text{con}}$$
$$\times\frac{(-1)^{m_2}}{m_2!}\int d\tau_1^{(2)}\cdots d\tau_{m_2}^{(2)}\langle T_\tau(\hat{H}_{\text{int}}(\tau_1^{(2)})\cdots\hat{H}_{\text{int}}(\tau_{m_2}^{(2)}))\rangle_{\text{con}}\cdots \quad (15.4)$$
$$\times\frac{(-1)^{m_k}}{m_k!}\int d\tau_1^{(k)}\cdots d\tau_{m_k}^{(k)}\langle T_\tau(\hat{H}_{\text{int}}(\tau_1^{(k)})\cdots\hat{H}_{\text{int}}(\tau_{m_k}^{(k)}))\rangle_{\text{con}}.$$

Next, we observe that from now on the assumption that each average $\langle\cdots\rangle_{\text{con}}$ corresponds to a connected diagram of a given type can actually be dropped, and instead we can assume that $\langle\cdots\rangle_{\text{con}}$ represents the sum of *all* connected diagrams with a given number of vertices. In fact, we can conclude that the sum of all disconnected diagrams containing k closed loops with $m_1, m_2, \ldots, m_k$ vertices, respectively, has the form

$$\Xi_{m_1}\Xi_{m_2}\cdots\Xi_{m_k},$$

where

$$\Xi_m = \frac{(-1)^m}{m!}\int d\tau_1\cdots d\tau_m\langle T_\tau(\hat{H}_{\text{int}}(\tau_1)\cdots\hat{H}_{\text{int}}(\tau_m))\rangle_{\text{con}} \quad (15.5)$$

is just the sum of all connected diagrams for $\langle\mathscr{S}\rangle$ of order m. Moreover, it is clear that

$$1 + \Xi_1 + \Xi_2 + \cdots = \langle\mathscr{S}\rangle_{\text{con}}. \quad (15.6)$$

If some of the numbers $m_1, m_2, \ldots$ are the same, so that the diagram decomposes into $p_1 + p_2 + \cdots + p_k$ closed loops, where p_1 loops contain m_1 vertices, p_2 loops contain m_2 vertices, and so on ($m_1 \neq m_2 \neq \cdots \neq m_k$), then it can be shown that the expression (15.5) should be replaced by [11]

$$\frac{1}{p_1!}\Xi_{m_1}^{p_1}\frac{1}{p_2!}\Xi_{m_2}^{p_2}\cdots\frac{1}{p_k!}\Xi_{m_k}^{p_k}, \quad (15.7)$$

or equivalently by

$$\frac{1}{p_1!}\Xi_1^{p_1}\frac{1}{p_2!}\Xi_2^{p_2}\cdots\frac{1}{p_l!}\Xi_l^{p_l}, \quad (15.8)$$

where the numbers $p_l = 0, 1, 2, \ldots$ shows how many closed loops of order l are contained in the entire disconnected diagram. Summing (15.8) over all

---

[11] This can be seen by the following argument: In the case where some of the numbers $m_1, m_2, \ldots$ are the same, the number F_k referred to above equals the number of ways in which $p_1 m_1 + p_2 m_2 + \cdots + p_k m_k = n$ operators $\hat{H}_{\text{int}}$ can be assigned to $p_1 + p_2 + \cdots + p_k$ cells $\langle\cdots\rangle_{\text{con}}$, where p_1 cells each contain m_1 places, p_2 cells each contain m_2 places, and so on. Then the number F_k equals

$$F_k = \frac{n!}{p_1!(m_1!)^{p_1}p_2!(m_2!)^{p_2}\cdots p_k!(m_k!)^{p_k}}.$$

the p_l (the summations over different p_l are obviously independent), we find that

$$\begin{aligned}\langle \mathscr{S} \rangle &= \sum_{p_1, p_2, \ldots} \frac{1}{p_1!} \Xi_1^{p_1} \frac{1}{p_2!} \Xi_2^{p_2} \cdots \\ &= \sum_{p_1} \frac{1}{p_1!} \Xi_1^{p_1} \sum_{p_2} \frac{1}{p_2!} \Xi_2^{p_2} \cdots = e^{\Xi_1} e^{\Xi_2} \cdots \\ &= \exp(\Xi_1 + \Xi_2 + \cdots).\end{aligned} \tag{15.9}$$

Substituting (15.9) into (15.1), we finally obtain

$$\Delta\Omega = -T(\Xi_1 + \Xi_2 + \cdots) = -T(\langle \mathscr{S} \rangle_{\text{con}} - 1). \tag{15.10}$$

This proves the very important fact that to calculate the correction to the thermodynamic potential it is sufficient to calculate only the contribution to $\langle \mathscr{S} \rangle$ of the connected diagrams. As already noted, the diagrams for $\langle \mathscr{S} \rangle$ have the form of closed loops, and can be calculated by essentially the same rules as used to calculate $\mathscr{G}$-functions, except that the coefficient associated with a diagram is different.

In Sec. 12 we saw that the coefficient $1/n!$ in the perturbation series for the $\mathscr{G}$-function (12.13) is cancelled out if we take into account all topologically equivalent diagrams, of which there are precisely $n!$ We encounter a different situation in calculating $\langle \mathscr{S} \rangle_{\text{con}}$. In this case, the number of equivalent diagrams corresponding to the nth term of the series (12.12) equals $(n - 1)!$, so that the factor $1/n$ is associated with each diagram (if only topologically nonequivalent diagrams are considered to be nonzero).[12] The presence of a coefficient which depends in an essential way on the order n makes the perturbation series for Ω very inconvenient, especially in cases where it is not permissible to consider only a finite number of terms in the series and we have to sum infinite sequences of diagrams.

We now give some examples illustrating the calculation of the corrections ΔΩ, restricting ourselves (for brevity) to the case of interaction of particles and phonons. In the second-order approximation of perturbation theory, only the connected diagram shown in Fig. 49(b) makes a nonzero contribution. Using Wick's theorem to calculate this contribution, we find that

$$\Omega_2 = -T\Xi_2 = \pm \frac{1}{2} Tg^2 \int d^4x\, d^4y\, \mathscr{G}^{(0)}_{\alpha\beta}(x - y) \mathscr{G}^{(0)}_{\beta\alpha}(y - x) \mathscr{D}^{(0)}(x - y),$$

where we have used four-dimensional notation. The quantity Ω_2 is proportional to V, the volume of the system, as can easily be seen by introducing a

[12] The equivalent diagrams are obtained by making all possible permutations of $n - 1$ operators $\hat{H}_{\text{int}}$ in formula (12.12). One of the operators $\hat{H}_{\text{int}}$ has to be regarded as fixed. In calculating $\mathscr{G}$, the initial point and final point of the external lines [i.e., the operators $\psi_\alpha(\mathbf{r}_1, \tau_1)$ and $\bar{\psi}_\beta(\mathbf{r}_2, \tau_2)$ in (12.13)] were held fixed.

new variable of integration $x' = x - y$. The same holds true in any approximation, which is to be expected, since, as is well known, the potential Ω has the form

$$\Omega = -VP(\mu, T),$$

where $P(\mu, T)$ is the pressure expressed as a function of the chemical potential and temperature. Therefore, from now on, we shall write these formulas in terms of ΔP, where

$$P = P_0(\mu, T) + \Delta P,$$

and P_0 is the pressure corresponding to a system of free particles. Thus, for ΔP_2 we have

$$\Delta P_2 = \mp \frac{1}{2} g^2 \int d^4x\, \mathscr{G}^{(0)}_{\alpha\beta}(x)\mathscr{G}^{(0)}_{\beta\alpha}(-x)\mathscr{D}^{(0)}(x). \tag{15.11}$$

Transforming to the momentum representation, we obtain

$$\Delta P_2 = \mp \frac{1}{2} g^2 \frac{T^2}{(2\pi)^6} (2s + 1)$$
$$\times \sum_{\omega_1, \omega_2} \int d\mathbf{p}\, d\mathbf{k} \frac{1}{i\omega_1 - \varepsilon_0(\mathbf{p}) + \mu} \frac{1}{i(\omega_1 + \omega_2) - \varepsilon_0(\mathbf{p} + \mathbf{k}) + \mu} \frac{\omega_0^2(\mathbf{k})}{\omega_2^2 + \omega_0^2(\mathbf{k})}.$$

The corresponding diagram is shown in Fig. 50.

Next, consider an arbitrary diagram of order $2n$. Such a diagram contains $3n$ lines and $2n$ vertices. However, one of the $2n$ conservation laws is

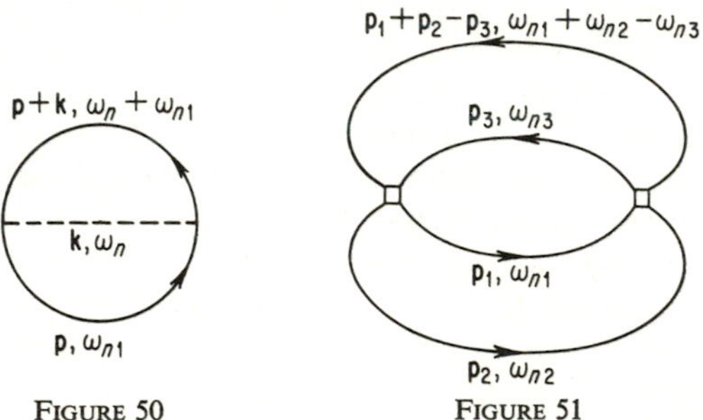

FIGURE 50 FIGURE 51

satisfied identically, if the other $2n - 1$ conservation laws hold. Thus, in a diagram of order $2n$ there are a total of $n + 1$ independent integrations. The superfluous conservation law leads to the appearance of an extra factor $\delta(\mathbf{p} = 0)$, proportional to the volume V of the system, in the diagram for $\langle \mathscr{S} \rangle$.[13] The rules used to associate Green's functions (and vertex parts for other interactions) with separate elements of a diagram remain the same as in

[13] By definition,

$$\delta(\mathbf{p} = 0) = \frac{1}{(2\pi)^3} \int d\mathbf{r} = \frac{V}{(2\pi)^3}.$$

the case of a diagram for $\mathscr{G}$. The coefficient associated with a diagram of order $2n$ for the correction ΔP equals

$$M_n = \frac{(-1)^{n+1}}{2n} g^{2n} \left[\frac{T}{(2\pi)^3}\right]^{2n} (\mp 1)^F (2s+1)^F,$$

where F is the number of closed loops formed by single $\mathscr{G}$-lines of particles.

We can also give expressions for ΔP in the case of two-particle interactions. For interactions of the form (13.7), the second-order correction to the pressure corresponding to the diagram shown in Fig. 51 has the form

$$-\frac{1}{4}\frac{T^3}{(2\pi)^9} \sum_{\omega_1,\omega_2,\omega_3} \int d\mathbf{p}_1\, d\mathbf{p}_2\, d\mathbf{p}_3 \frac{1}{i\omega_1 - \varepsilon_0(\mathbf{p}_1) + \mu} \frac{1}{i\omega_2 - \varepsilon_0(\mathbf{p}_2) + \mu}$$
$$\times \frac{1}{i\omega_3 - \varepsilon(\mathbf{p}_3) + \mu} \frac{1}{i(\omega_1 + \omega_2 - \omega_3) - \varepsilon_0(\mathbf{p}_1 + \mathbf{p}_2 - \mathbf{p}_3) + \mu}$$
$$\times \mathscr{T}^{(0)}_{\alpha\beta,\gamma\delta}(p_1, p_2; p_1 + p_2 - p_3, p_3) \mathscr{T}^{(0)}_{\gamma\delta,\beta\alpha}(p_1 + p_2 - p_3, p_3; p_2, p_1).$$

16. Dyson's Equation. Many-Particle Green's Functions

16.1. Dyson's equation. In statistical problems for $T \neq 0$, just as for $T = 0$, it is almost never possible to get along with just the first few terms of the perturbation series for the corrections to the Green's functions. In practically any physically well-posed problem, the formal parameter for making expansions in the diagram technique is the interaction Hamiltonian $\hat{H}_{\text{int}}$, which does not turn out to be small. As a result, some infinite sequence of terms of the perturbation series makes a first-order contribution to the quantity of interest.

In the preceding chapter, we saw that field-theory methods based on the diagram technique can be used to sum infinite series. In this technique, the sum of a series can be represented as a diagram whose elements (lines and vertices) in turn represent sums of infinite numbers of diagrams. Moreover, definite analytical expressions are associated with the elements of such a diagram according to the same rules as used for the diagrams dealt with in perturbation theory. This fact allows us to construct various equations for the Green's functions. In Chap. 2, we already encountered one such equation, *Dyson's equation*, which expresses the Green's function in terms of the mass operator.

In constructing such equations, two properties of the diagram technique are vital, i.e., the topological structure of the diagrams and the rules whereby a given expression is associated with a diagram. The diagrams in the technique for absolute zero and the diagrams in Matsubara's technique differ only in that integration over frequencies for $T = 0$ is replaced by summation over discrete "frequencies" $i\omega_n$ for $T \neq 0$. More precisely, any expression for a correction to a temperature Green's function $\mathscr{G}$, corresponding to a

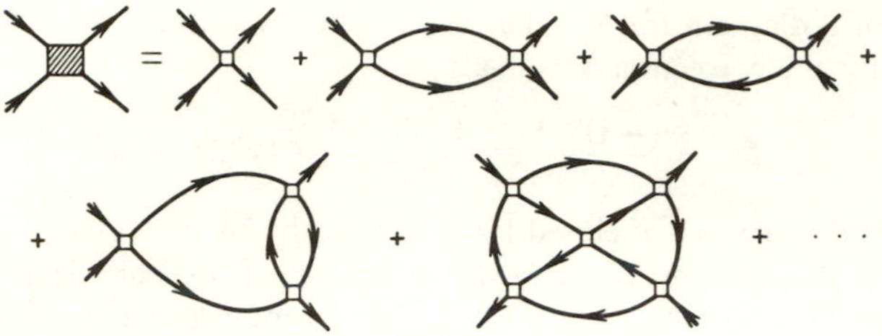

FIGURE 52

certain diagram, can be obtained from the expression for the Green's function G for $T = 0$, corresponding to the same diagram, if in the latter expression we replace ω by $i\omega_n$ and integration by summation, according to the rule

$$\frac{1}{2\pi}\int d\omega \cdots \to iT\sum_{\omega_n} \cdots$$

(see the end of Sec. 14). This fact allows us to immediately extend all the results of Sec. 10 to the case $T \neq 0$, provided we simply change the notation. In particular, Dyson's equation still holds in Matsubara's technique, and takes the form

$$\begin{aligned} \mathscr{G}_{\alpha\beta}(p) = \mathscr{G}^{(0)}_{\alpha\beta}(p) &- \frac{T}{(2\pi)^3}\sum_{\omega_1}\int d\mathbf{p}_1\, \mathscr{T}^{(0)}_{\gamma_1\gamma_2,\gamma_3\gamma_4}(p, p_1; p_1, p) \\ &\times \mathscr{G}^{(0)}_{\alpha\gamma_1}(p)\mathscr{G}_{\gamma_3\gamma_2}(p_1)\mathscr{G}_{\gamma_4\beta}(p) \\ &+ \frac{1}{2}\mathscr{G}^{(0)}_{\alpha\gamma_1}(p)\frac{T^2}{(2\pi)^6}\sum_{\omega_1,\omega_2}\int d\mathbf{p}_1\, d\mathbf{p}_2\, \mathscr{T}^{(0)}_{\gamma_1\gamma_2,\gamma_3\gamma_4}(p, p_1 + p_2 - p; p_1, p_2) \\ &\times \mathscr{G}_{\gamma_3\gamma_5}(p_1)\mathscr{G}_{\gamma_4\gamma_6}(p_2)\,\mathscr{G}_{\gamma_7\gamma_2}(p_1 + p_2 - p) \\ &\times \mathscr{T}_{\gamma_5\gamma_6,\gamma_7\gamma_8}(p_1, p_2; p_1 + p_2 - p, p)\mathscr{G}_{\gamma_8\beta}(p) \end{aligned} \tag{16.1}$$

for a system of particles with interactions between pairs of particles. Here $\mathscr{T}$ is the exact vertex part, and has the same meaning as in the technique for $T = 0$, i.e., it is the sum of all compact diagrams with four external lines, where each external line represents an exact Green's function $\mathscr{G}$ (examples of such diagrams are shown in Fig. 52). Dyson's equation (16.1) has the graphical interpretation shown in Fig. 53, just like the corresponding equation of Chap. 2. Here, a thick line indicates $\mathscr{G}$ and a thin line $\mathscr{G}^{(0)}$, while the vertex part $\mathscr{T}$ is indicated by a shaded square.

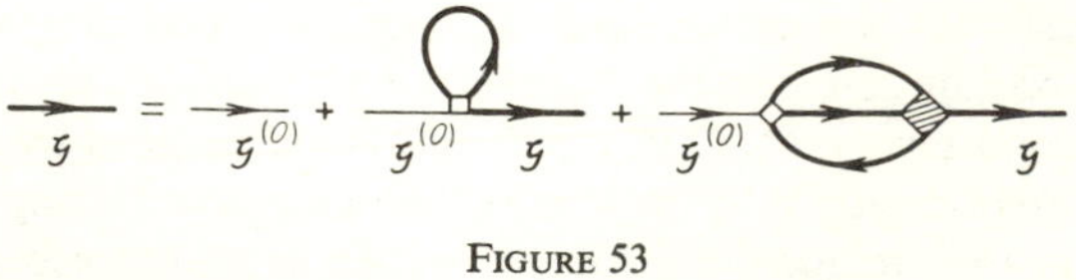

FIGURE 53

Introducing $\mathscr{G}_{\alpha\beta}^{-1}$, the inverse of the matrix $\mathscr{G}_{\alpha\beta}$, we can write (16.1) in the form

$$\begin{aligned}\mathscr{G}_{\alpha\beta}^{-1}(p) = {} & [i\omega - \varepsilon(\mathbf{p}) + \mu]\delta_{\alpha\beta} \\ & + \frac{T}{(2\pi)^3}\sum_{\omega_1}\int d\mathbf{p}_1\, \mathscr{T}^{(0)}_{\alpha\gamma_1,\gamma_2\beta}(p, p_1; p_1, p)\mathscr{G}_{\gamma_2\gamma_1}(p_1) \\ & - \frac{T^2}{2(2\pi)^6}\sum_{\omega_1,\omega_2}\int d\mathbf{p}_1\, d\mathbf{p}_2\, \mathscr{T}^{(0)}_{\alpha\gamma_1,\gamma_2\gamma_3}(p, p_1 + p_2 - p; p_1, p_2) \\ & \times \mathscr{G}_{\gamma_2\gamma_4}(p_1)\mathscr{G}_{\gamma_3\gamma_5}(p_2)\, \mathscr{G}_{\gamma_6\gamma_1}(p_1 + p_2 - p) \\ & \times \mathscr{T}_{\gamma_4\gamma_5,\gamma_6\beta}(p_1, p_2; p_1 + p_2 - p; p). \end{aligned} \tag{16.2}$$

Similarly, in the case of interaction of particles and phonons, the system of equations for $\mathscr{G}$ and $\mathscr{D}$ takes the form (see Fig. 54):

$$\begin{aligned}\mathscr{G}_{\alpha\beta}^{-1}(\mathbf{p}, \omega_n) = {} & [i\omega_n - \varepsilon(\mathbf{p}) + \mu]\delta_{\alpha\beta} \\ & + \frac{gT}{(2\pi)^3}\sum_{\omega_n}\int d\mathbf{p}'\, \mathscr{G}_{\alpha\beta}(\mathbf{p}', \omega_n')\mathscr{D}(\mathbf{p}' - \mathbf{p}, \omega_n' - \omega_n) \\ & \times \mathscr{T}(\mathbf{p}, \mathbf{p}'; \omega_n, \omega_n'), \end{aligned} \tag{16.3}$$

$$\begin{aligned}\mathscr{D}^{-1}(\mathbf{k}, \omega_n) = {} & -\omega_0^{-2}(\mathbf{k})[\omega_n^2 + \omega_0^2(\mathbf{k})] \\ & \mp \frac{gT}{(2\pi)^3}\sum_{\omega_n'}\int d\mathbf{p}'\, \mathscr{G}_{\alpha\beta}(\mathbf{p}', \omega_n')\mathscr{G}_{\beta\alpha}(\mathbf{p}' - \mathbf{k}, \omega_n' - \omega_n) \\ & \times \mathscr{T}(\mathbf{p}', \mathbf{p}' - \mathbf{k}; \omega_n', \omega_n' - \omega_n). \end{aligned}$$

In (16.3), $\mathscr{T}$ represents the exact vertex part, and describes the sum of all compact diagrams with two external particle lines and one external phonon line (see Fig. 55).

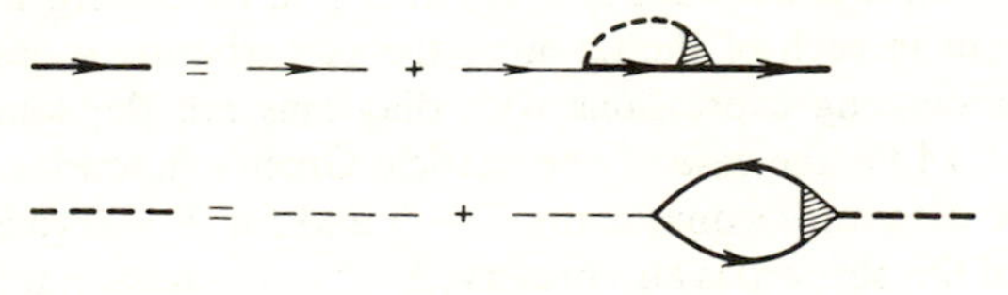

FIGURE 54

Just as at the absolute zero of temperature, the exact vertex parts for $T \neq 0$ are related by definite formulas to the many-particle temperature Green's functions. The latter are expressed by formulas (11.1)–(11.4) in Matsubara's technique, which are identical in appearance with the corresponding expressions for $T = 0$. Taking into account also that the formulation of Wick's theorem is identical in both cases, we immediately arrive at the conclusion that diagrams with many external lines can be used to calculate the many-particle Green's functions, with the same rules (in the space $\mathbf{r}$, τ) for associating expressions with diagrams as described in Sec. 14.

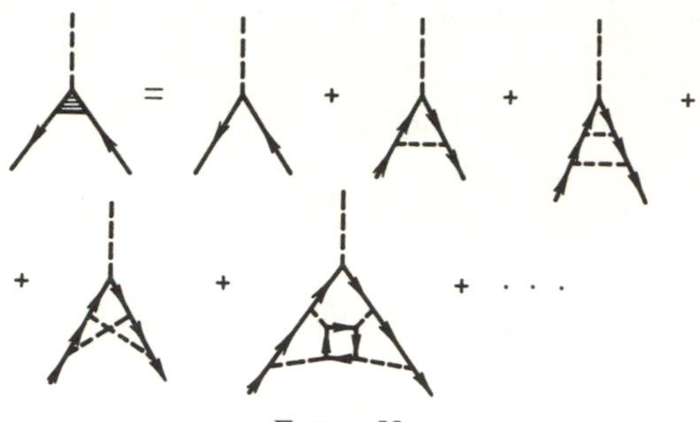

FIGURE 55

We now discuss the formal method used to go over to the momentum representation. For example, consider the two-particle Green's function

$$\mathscr{G}^{\mathrm{II}}_{\alpha\beta,\gamma\delta}(1, 2; 3, 4) = \langle T_\tau(\tilde{\psi}_\alpha(1)\tilde{\psi}_\beta(2)\tilde{\bar{\psi}}_\gamma(3)\tilde{\bar{\psi}}_\delta(4))\rangle, \qquad (16.4)$$

where the $\tilde{\psi}$ denote the "Heisenberg" operators (11.3). The expression (16.4) depends on four "times" $\tau_1, \tau_2, \tau_3, \tau_4$, each of which varies from 0 to $1/T$. Using (11.8) to give the relation between the values of $\mathscr{G}^{\mathrm{II}}(1, 2; 3, 4)$ for $\tau_1 < 0$ and its values for $\tau_1 > 0$, we extend (16.4) onto the interval $-1/T \leqslant \tau_1 \leqslant 0$, and we similarly extend the definition of (16.4) in the variables τ_2, τ_3 and τ_4. Next, using (14.2) to calculate the Fourier transformation of (16.4) with respect to each of the τ_i, we see at once that the frequency corresponding to each "Fermi" variable [i.e., to each Fermi operator in (16.4)] can only take "odd" values $(2n + 1)\pi T$, while the frequency corresponding to each "Bose" variable can only take "even" values $2n\pi T$. Moreover, any other many-particle Green's function can be treated in the same way.

As in Sec. 14, it is not hard to see that a Fourier transformation in τ can be carried out in each of the terms of the perturbation series, and that the rules for associating expressions with diagrams are the same as the rules given in Sec. 14 for the case of one-particle Green's functions. The relation between the diagram technique for $T = 0$ and for $T \neq 0$ (mentioned at the beginning of this section) is also preserved. The expression for the correction to $\mathscr{G}^{\mathrm{II}}(1, 2; 3, 4)$ is obtained from the expression for $G^{\mathrm{II}}(1, 2; 3, 4)$ corresponding to the same diagram by making the substitutions

$$\omega \to i\omega_n, \qquad \frac{1}{2\pi}\int d\omega \cdots \to iT\sum_{\omega_n} \cdots,$$

already alluded to. The existence of this connection between the cases $T = 0$ and $T \neq 0$ allows us to repeat word for word everything said in Sec. 10 concerning diagrams for many-particle functions. The perturbation series for $\mathscr{G}^{\mathrm{II}}(1, 2; 3, 4)$ can be reduced to the sum of all compact diagrams made up of only thick lines, corresponding to exact Green's functions $\mathscr{G}(1, 2)$. These diagrams coincide with the diagrams for the exact vertex part $\mathscr{T}$, which means that there must be a relation between $\mathscr{G}^{\mathrm{II}}(1, 2; 3, 4)$ and $\mathscr{T}$. It is not hard to see that this relation has the form[14]

$$\begin{aligned}
&\mathscr{G}^{\mathrm{II}}_{\alpha\beta,\gamma\delta}(p_1, p_2; p_3, p_4) \\
&\quad = \frac{(2\pi)^3}{T}\Big\{\frac{(2\pi)^3}{T}\Big[\mathscr{G}_{\alpha\delta}(p_1)\mathscr{G}_{\beta\gamma}(p_2)\delta_{\omega_1\omega_4}\delta(\mathbf{p}_1 - \mathbf{p}_4) \\
&\quad \mp \mathscr{G}_{\alpha\gamma}(p_1)\mathscr{G}_{\beta\delta}(p_2)\delta_{\omega_1\omega_3}\delta(\mathbf{p}_1 - \mathbf{p}_3)\Big] \qquad (16.5) \\
&\quad \pm \mathscr{G}_{\alpha\lambda}(p_1)\mathscr{G}_{\beta\mu}(p_2)\mathscr{T}_{\lambda\mu,\nu\tau}(p_1, p_2; p_3, p_4)\,\mathscr{G}_{\nu\gamma}(p_3)\mathscr{G}_{\tau\delta}(p_4)\Big\} \\
&\quad \times \delta_{\omega_1+\omega_2-\omega_3-\omega_4}\delta(\mathbf{p}_1 + \mathbf{p}_2 - \mathbf{p}_3 - \mathbf{p}_4).
\end{aligned}$$

In the case of interaction of particles with phonons, the vertex part $\mathscr{T}(p_1, p_2)$ is connected with the Fourier components of the Green's function

$$\mathscr{G}_{\alpha\beta}(1, 2; 3) = \langle T_\tau(\tilde{\psi}_\alpha(1)\bar{\tilde{\psi}}_\beta(2)\tilde{\varphi}(3))\rangle$$

by the relation

$$\begin{aligned}
\mathscr{G}_{\alpha\beta}(p_1, p_2; k) &= \frac{(2\pi)^3}{T}\,\mathscr{G}_{\alpha\gamma}(p_1)\mathscr{G}_{\gamma\beta}(p_2)\mathscr{D}(k) \qquad (16.6) \\
&\quad \times \mathscr{T}(p_1, p_1 + k)\delta(\mathbf{p}_1 - \mathbf{p}_2 + \mathbf{k})\delta_{\omega_1-\omega_2+\omega}.
\end{aligned}$$

As was to be expected, the relations (16.5) and (16.6) differ from the corresponding relations (10.17) and (10.21) only by numerical coefficients.

It should be emphasized that the method of graphical summation can be applied only to diagrams for $\mathscr{G}$-functions. In the perturbation series for the potential Ω (considered in Sec. 15), graphical summation cannot be carried out due to the presence of the coefficient $1/n$, associated with the nth-order diagrams. Obviously, such diagrams, unlike those for $\mathscr{G}$, can no longer be decomposed into separate blocks, and the result of summing an infinite sequence of diagrams no longer reduces to just replacing thin lines by thick lines. In particular, the graphical construction shown in Fig. 56 is completely illegitimate.

16.2. Relation between the Green's functions and the thermodynamic potential Ω. We now derive a series of relations between the temperature Green's functions and the thermodynamic potential Ω, beginning with the case of two-particle interactions. Instead of the potential $\mathscr{T}^{(0)}$, we introduce the potential $\lambda\mathscr{T}^{(0)}$, where $0 < \lambda < 1$, and we then differentiate the expression

$$\Omega = \Omega_0 - T\ln\langle\mathscr{S}\rangle$$

---

[14] The coefficients in (16.5) and (16.6) are most easily verified by calculating both sides of the equation to the first-order approximation of perturbation theory.

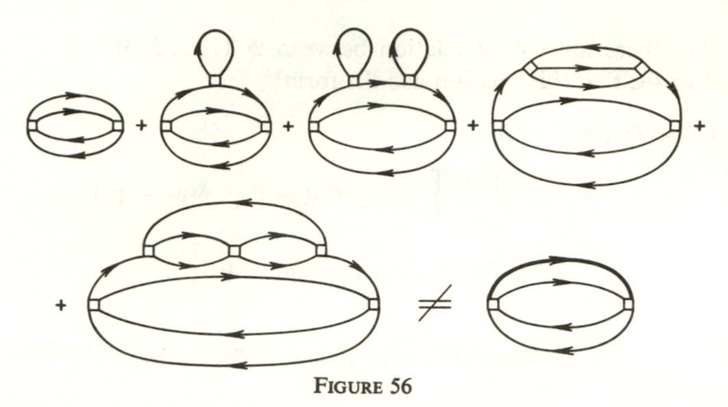

FIGURE 56

with respect to λ. This gives

$$\frac{\partial\Omega}{\partial\lambda} = \frac{T}{4}\int d^4x_1\, d^4x_2\, d^4x_3\, d^4x_4\, \mathcal{T}^{(0)}_{\alpha\beta,\gamma\delta}(x_1, x_2; x_3, x_4) \times \frac{\langle T_\tau(\bar{\psi}_\alpha(x_1)\bar{\psi}_\beta(x_2)\psi_\delta(x_4)\psi_\gamma(x_3)\mathcal{S}(\lambda))\rangle}{\langle\mathcal{S}(\lambda)\rangle},$$

or according to the definition (16.4),

$$\frac{\partial\Omega}{\partial\lambda} = \frac{T}{4}\int d^4x_1\, d^4x_2\, d^4x_3\, d^4x_4\, \mathcal{T}^{(0)}_{\alpha\beta,\gamma\delta}(x_1, x_2; x_3, x_4)\mathcal{G}^{\mathrm{II}}_{\delta\gamma,\alpha\beta}(x_4, x_3; x_1, x_2).$$

Transforming to the Fourier representation and using (16.5), we obtain

$$\begin{aligned}\frac{\partial\Omega}{\partial\lambda} = \frac{VT}{2\lambda(2\pi)^3}\sum_{\omega}\int d\mathbf{p}\Big[&\frac{T}{(2\pi)^3}\sum_{\omega_1}\int d\mathbf{p}_1\\ &\times \lambda\mathcal{T}^{(0)}_{\gamma_1\gamma_2,\gamma_3\gamma_4}(p, p_1; p, p_1)\mathcal{G}_{\gamma_3\gamma_1}(p, \lambda)\mathcal{G}_{\gamma_4\gamma_2}(p_1, \lambda)\\ &\pm \frac{T^2}{2(2\pi)^6}\sum_{\omega_1,\omega_2}\int d\mathbf{p}_1\, d\mathbf{p}_2\, \lambda\mathcal{T}^{(0)}_{\gamma_1\gamma_2,\gamma_3\gamma_4}(p, p_1 + p_2 - p; p_1, p_2)\\ &\times \mathcal{G}_{\gamma_3\gamma_6}(p_1, \lambda)\mathcal{G}_{\gamma_4\gamma_5}(p_2, \lambda)\mathcal{G}_{\gamma_8\gamma_2}(p_1 + p_2 - p, \lambda)\\ &\times \mathcal{T}_{\gamma_5\gamma_6,\gamma_7\gamma_8}(p_2, p_1; p, p_1 + p_2 - p)\mathcal{G}_{\gamma_7\gamma_1}(p, \lambda)\Big],\end{aligned} \tag{16.7}$$

where $\mathcal{G}(p, \lambda)$ is the Green's function for $\lambda \neq 1$. We note that the expression in brackets in (16.7) is the same as the right-hand side of Dyson's equation (16.1) with the interaction potential $\lambda\mathcal{T}^{(0)}$, i.e.,

$$\frac{\partial\Omega}{\partial\lambda} = \pm\frac{VT}{2\lambda(2\pi)^3}\sum_{\omega}\int d\mathbf{p}\, \mathcal{G}^{(0)-1}_{\alpha\beta}(p)[\mathcal{G}_{\beta\alpha}(p, \lambda) - \mathcal{G}^{(0)}_{\beta\alpha}(p)]. \tag{16.8}$$

The relation in which we are interested can be found by integrating (16.8) with respect to λ and taking into account the condition $\Omega(\lambda = 0) = \Omega_0$:

$$\Omega = \Omega_0 \pm \frac{1}{2}V\int_0^1 \frac{d\lambda}{\lambda}\frac{T}{(2\pi)^3}\sum_{\omega}\int d\mathbf{p}\, \mathcal{G}^{(0)-1}_{\alpha\beta}(p)[\mathcal{G}_{\beta\alpha}(p, \lambda) - \mathcal{G}^{(0)}_{\beta\alpha}(p)]. \tag{16.9}$$

In the case of interaction of particles with phonons, there exist similar relations expressing Ω in terms of an integral of the $\mathscr{G}$-function of the system with respect to the coupling constant g. Calculations identical with those used to derive (16.9) lead to the result

$$\Omega = \Omega_0 \pm V \int_0^g \frac{dg_1}{g_1} \frac{T}{(2\pi)^3} \sum_{\omega_n} \int d\mathbf{p}\, [i\omega_n - \varepsilon(\mathbf{p}) + \mu][\mathscr{G}_{\alpha\alpha}(\mathbf{p}, \omega_n) - \mathscr{G}^{(0)}_{\alpha\alpha}(\mathbf{p}, \omega_n)]$$

$$= \Omega_0 + V \int_0^g \frac{dg_1}{g_1} \frac{T}{(2\pi)^3} \sum_{\omega_n} \int d\mathbf{k}\, \frac{\omega_n^2 + \omega_0^2(\mathbf{k})}{\omega_0^2(\mathbf{k})} [\mathscr{D}(\mathbf{k}, \omega_n) - \mathscr{D}^{(0)}(\mathbf{k}, \omega_n)]. \tag{16.10}$$

Another useful formula can be obtained from the familiar relation between the derivative of Ω with respect to m, the mass of the particles, and the derivative of $\hat{H}$, the total Hamiltonian of the system, with respect to m (see e.g., L8):

$$\left(\frac{\partial \Omega}{\partial m}\right)_{T,V,\mu} = \left\langle \frac{\partial \hat{H}}{\partial m} \right\rangle.$$

Since

$$\frac{\partial \hat{H}}{\partial m} = \frac{1}{2m^2} \int d\mathbf{r}\, \psi_\alpha^+(\mathbf{r})\, \nabla^2 \psi_\alpha(\mathbf{r}),$$

it follows at once from the definition (11.1) of the $\mathscr{G}$-function that

$$\frac{\partial \Omega}{\partial m} = \pm \frac{1}{2m^2} \int d\mathbf{r}\, [\nabla_{\mathbf{r}}^2 \mathscr{G}_{\alpha\alpha}(\mathbf{r}, \mathbf{r}'; -0)]_{\mathbf{r}=\mathbf{r}'},$$

or in Fourier components,[15]

$$\frac{\partial \Omega}{\partial m} = \mp \frac{VT}{m(2\pi)^3} \sum_{\omega_n} \int d\mathbf{p}\, \frac{\mathbf{p}^2}{2m} \mathscr{G}_{\alpha\alpha}(\mathbf{p}, \omega_n) e^{i\omega_n \tau} \qquad (\tau \to +0). \tag{16.11}$$

Using the relations (16.9)–(16.11) and the relation (11.6), which in Fourier components takes the form

$$\frac{\partial \Omega}{\partial \mu} = -N = \mp \frac{VT}{(2\pi)^3} \sum_{\omega_n} \int d\mathbf{p}\, \mathscr{G}_{\alpha\alpha}(\mathbf{p}, \omega_n) e^{i\omega_n \tau} \qquad (\tau \to +0), \tag{16.12}$$

we can calculate the derivatives of Ω with respect to various parameters, thereby expressing the thermodynamic potential in terms of the Green's functions.

Finally, we derive one more formula expressing the potential Ω in terms of the exact Green's function $\mathscr{G}$. As already noted, the presence of the factor $1/n$ in the perturbation-theory diagrams for Ω makes it impossible to apply the method of summation of diagrams to the perturbation series for Ω (see Sec. 15). Nevertheless, there still exists the interesting possibility of carrying out partial summation and writing Ω in the form of a sum of infinitely many diagrams which consist only of thick lines representing exact $\mathscr{G}$-functions [see Luttinger and Ward (L17)]. To keep the analysis simple, we restrict

[15] In Sec. 17, we shall show how to calculate the sums over ω_n which appear here.

ourselves to the case of two-particle interactions, assuming that the system is not ferromagnetic, so that $\mathscr{G}_{\alpha\beta} = \delta_{\alpha\beta}\mathscr{G}$. From the perturbation-theory diagrams for the potential Ω, we choose all *compact diagrams*, i.e., all diagrams which have no internal parts representing corrections to the $\mathscr{G}$-function (see p. 93), and then in these diagrams we replace all thin lines by thick lines.[16] The sum of all diagrams obtained in this way (with a coefficient $1/n$ for each diagram!) will be denoted by Ω'.

By definition, the quantity Ω' is a functional of the exact Green's function $\mathscr{G}$. By varying each diagram for Ω' in turn with respect to $\delta\mathscr{G}(p)$, it is not hard to see that (except for a factor) we obtain a series representing corrections to the exact Green's function $\mathscr{G}(p)$, provided that in this series we substitute $\delta\mathscr{G}(p)$ for the product of the Green's functions $\mathscr{G}^{(0)}(p)$ and $\mathscr{G}(p)$, corresponding to the lines on the extreme left and on the extreme right, respectively. More precisely,

$$\delta\Omega' = \pm 2VT \sum_n \int \frac{d\mathbf{p}}{(2\pi)^3} \Sigma(p)\, \delta\mathscr{G}(p), \tag{16.13}$$

where $\Sigma(p)$ is the self-energy part of the exact Green's function $\mathscr{G}(p)$; $\Sigma(p)$ is defined just like the self-energy part in the diagram technique for $T = 0$ (see Sec. 10.1) and is related to $\mathscr{G}$ by the formula

$$\mathscr{G}^{-1} = (\mathscr{G}^{(0)})^{-1} - \Sigma. \tag{16.14}$$

If by Σ we mean the sum of the corresponding diagrams consisting of thick lines, then (16.14) is nothing other than Dyson's equation (16.2).

Next, we form the expression

$$\bar{\Omega} = \Omega_0 \pm 2VT \sum_n \int \frac{d\mathbf{p}}{(2\pi)^3} \{\ln[1 - \mathscr{G}^{(0)}(p)\Sigma(p)] + \Sigma(p)\mathscr{G}(p)\} + \Omega', \tag{16.15}$$

which, as we now show, equals Ω. To see this, we first note that the quantity $\bar{\Omega}$, regarded as a functional of $\mathscr{G}$ (or equivalently, as a functional of Σ) has the "stationarity property"

$$\frac{\delta\bar{\Omega}}{\delta\Sigma} = 0, \tag{16.16}$$

provided that $\mathscr{G}$ satisfies Dyson's equation (16.14). In fact, varying (16.15) and taking account of (16.13), we immediately find that

$$\begin{aligned}
\delta\bar{\Omega} &= 2VT \sum_n \int \frac{d\mathbf{p}}{(2\pi)^3} \left\{ -\frac{\delta\Sigma(p)}{(\mathscr{G}^{(0)})^{-1}(p) - \Sigma(p)} + \mathscr{G}(p)\,\delta\Sigma(p) + \Sigma(p)\,\delta\mathscr{G}(p) \right\} \\
&\quad \pm 2VT \sum_n \int \frac{d\mathbf{p}}{(2\pi)^3} \Sigma(p)\,\delta\mathscr{G}(p) \\
&= \pm 2VT \sum_n \int \frac{d\mathbf{p}}{(2\pi)^3} \left\{ \mathscr{G}(p) - \frac{1}{(\mathscr{G}^{(0)})^{-1}(p) - \Sigma(p)} \right\} \delta\Sigma(p),
\end{aligned}$$

---

[16] For example, only the first and second of the diagrams shown in Fig. 48(b), p. 130, should be selected.

which implies (16.16). Then, just as on p. 139, we consider the interaction potential $\lambda\mathcal{T}^{(0)}$ instead of the potential $\mathcal{T}^{(0)}$, and we calculate

$$\frac{\partial\bar{\Omega}(\lambda)}{\partial\lambda}.$$

In so doing, because of (16.16), we do not have to take account of the dependence of $\mathcal{G}$ and Σ on λ. Therefore

$$\frac{\partial\bar{\Omega}(\lambda)}{\partial\lambda} = \frac{\partial\Omega'(\lambda)}{\partial\lambda},$$

where, in calculating $\partial\Omega'(\lambda)/\partial\lambda$, $\mathcal{G}(\lambda)$ should not be differentiated. On the other hand, it follows from the structure of the diagrams that the functional $\Omega'(\lambda)$ coincides with Ω' for $\lambda = 1$, if in Ω' we substitute $\sqrt{\lambda}\mathcal{G}(\lambda)$ for $\mathcal{G}(\lambda = 1)$. Thus, using (16.13), we find that

$$\frac{\partial\Omega'(\lambda)}{\partial\lambda} = \pm 2VT\sum_n\int\frac{d\mathbf{p}}{(2\pi)^3}\frac{\Sigma(\lambda)}{\sqrt{\lambda}}\frac{\partial}{\partial\lambda}[\sqrt{\lambda}\mathcal{G}(\lambda)],$$

which, if we let $\partial\mathcal{G}(\lambda)/\partial\lambda \to 0$, implies

$$\frac{\partial\bar{\Omega}(\lambda)}{\partial\lambda} = \pm\frac{VT}{\lambda}\sum_n\int\frac{d\mathbf{p}}{(2\pi)^3}\Sigma(\lambda)\mathcal{G}(\lambda).$$

Comparing this expression with (16.8) (and recalling the definition of Σ), we conclude that

$$\frac{\partial\bar{\Omega}(\lambda)}{\partial\lambda} = \frac{\partial\Omega(\lambda)}{\partial\lambda}. \tag{16.17}$$

On the other hand, according to the definition (16.15),

$$\bar{\Omega}(0) = \Omega(0),$$

so that, integrating (16.17) and setting $\lambda = 1$, we find that $\bar{\Omega} = \Omega$, as asserted.

Actually, formula (16.15) for Ω is valid for arbitrary interactions between the particles (and not just for two-particle interactions). In fact, we need only take Ω' to be a functional with the property (16.13). The existence of such a functional can be proved by starting from the argument given in Sec. 19.4 for the case $T = 0$ and repeating it step by step for the case of finite temperatures. However, the construction of Ω' is now quite complicated, and will not be given here.

It should be noted that

$$\Omega_0 = \pm 2VT\sum_n\int\frac{d\mathbf{p}}{(2\pi)^3}[\ln\mathcal{G}^{(0)}(p)]e^{i\omega_n\tau} \qquad (\tau\to +0)$$

(as proved e.g., in L17), so that (16.15) can also be written in the form

$$\Omega = \pm 2VT\sum_n\int\frac{d\mathbf{p}}{(2\pi)^3}(\ln\mathcal{G} - \Sigma\,\mathcal{G})e^{i\omega_n\tau} + \Omega' \qquad (\tau\to +0). \tag{16.18}$$

17. Time-Dependent Green's Functions for T ≠ 0. Analytic Properties of the Green's Functions

The time-dependent Green's functions G introduced in Chap. 2 preserve their meaning for finite temperatures. Later on, we shall give various examples showing that the functions G determine the transport properties of the system, in particular, the electrical resistance and the complex dielectric constant ε as a function of the frequency of the field. These functions also describe the process of inelastic scattering of particles by condensed matter.

For nonzero temperatures, the one-particle Green's function should be defined as

$$G_{\alpha\beta}(\mathbf{r}_1 - \mathbf{r}_2, t_1 - t_2; E_n, N) = -i\langle E_n, N|T_t(\tilde{\psi}_\alpha(\mathbf{r}_1, t_1)\tilde{\psi}_\beta^+(\mathbf{r}_2, t_2))|E_n, N\rangle, \tag{17.1}$$

where $\tilde{\psi}$, $\tilde{\psi}^+$ are the Heisenberg operators of the system, and the average is over the state of the system with energy E_n and particle number N. The definition (17.1) includes as a special case the definition of G for $T = 0$, in which case the average is over the ground state of the system. The Green's function (17.1) depends on the total energy E of the system and on the number of particles in the system. In quantum statistics, it is more convenient to regard all quantities as functions of the temperature and chemical potential μ; this is equivalent to transforming from the microcanonical distribution to the canonical distribution (see e.g., L8). Averaging (17.1) over the Gibbs distribution, we obtain

$$\begin{aligned} G_{\alpha\beta}(\mathbf{r}_1 - \mathbf{r}_2, t_1 - t_2; T, \mu) &= \sum_{N,n} e^{(\Omega + \mu N - E_n)/T} G_{\alpha\beta}(\mathbf{r}_1 - \mathbf{r}_2, t_1 - t_2, E_n, N) \\ &= -i \operatorname{Sp}\{e^{(\Omega + \mu\hat{N} - \hat{H})/T} T_t(\tilde{\psi}_\alpha(\mathbf{r}_1, t_1)\tilde{\psi}_\beta^+(\mathbf{r}_2, t_2))\}. \end{aligned} \tag{17.2}$$

Similar formulas are used to define many-particle functions. The phonon Green's function has the form

$$D(1, 2) = -i \operatorname{Sp}\{e^{(\Omega - \hat{H})/T} T_t(\tilde{\varphi}(1)\tilde{\varphi}(2))\}, \tag{17.3}$$

and the two-particle Green's function is

$$G^{\text{II}}(1, 2; 3, 4) = \operatorname{Sp}\{e^{(\Omega + \mu\hat{N} - \hat{H})/T} T_t(\tilde{\psi}(1)\tilde{\psi}(2)\tilde{\psi}^+(3)\tilde{\psi}^+(4))\}. \tag{17.4}$$

The Fourier components $G(\mathbf{p}, \omega)$ of the Green's function satisfy a very important relation [see Landau (L5)]. To derive this relation, we first note that the time dependence of the Heisenberg operators $\tilde{\psi}$ and $\tilde{\psi}^+$ is given by the expressions

$$\begin{aligned} \tilde{\psi}_{nm}(\mathbf{r}, t) &= \psi_{nm}(\mathbf{r})e^{i\omega_{nm}t}, \\ \tilde{\psi}_{nm}^+(\mathbf{r}, t) &= \psi_{nm}^+(\mathbf{r})e^{i\omega_{nm}t}, \\ \omega_{nm} &= E_n - E_m - \mu(N_n - N_m), \end{aligned} \tag{17.5}$$

where N_n always equals $N_m \pm 1$. Moreover, in the case where the system under consideration is homogeneous and infinite, the spatial dependence of the matrix elements of the operators $\psi(\mathbf{r})$ and $\psi^+(\mathbf{r})$ has the form

$$\begin{aligned} \psi_{nm}(\mathbf{r}) &= \psi_{nm}(0)e^{-i\mathbf{p}_{nm}\cdot\mathbf{r}}, \\ \psi^+_{nm}(\mathbf{r}) &= \psi^+_{nm}(0)e^{-i\mathbf{p}_{nm}\cdot\mathbf{r}}, \\ \mathbf{p}_{nm} &= \mathbf{P}_n - \mathbf{P}_m, \end{aligned} \tag{17.6}$$

where $\mathbf{P}_n$, $\mathbf{P}_m$ are the momenta of the system in the states n, m, and the quantities $\psi_{nm}(0)$ and $\psi^+_{nm}(0)$ no longer depend on the coordinates (see footnote 6, p. 55). Substituting (17.5) and (17.6) into (17.1), we obtain

$$G_{\alpha\beta}(\mathbf{r}, t > 0) = -i\sum_{n,m} e^{(\Omega+\mu N_n - E_n)/T}e^{i\omega_{nm}t - i\mathbf{p}_{nm}\cdot\mathbf{r}}(\psi_\alpha(0))_{nm}(\psi^+_\beta(0))_{mn},$$

$$G_{\alpha\beta}(\mathbf{r}, t < 0) = \pm i\sum_{n,m} e^{(\Omega+\mu N_m - E_m)/T}e^{-i\omega_{mn}t + i\mathbf{p}_{mn}\cdot\mathbf{r}}(\psi_\alpha(0))_{nm}(\psi^+_\beta(0))_{mn}.$$

Next, we transform from the coordinate representation of the Green's function to its Fourier components:

$$G(\mathbf{p}, \omega) = \int G(\mathbf{r}, t)e^{-i\mathbf{p}\cdot\mathbf{r}+i\omega t}d\mathbf{r}\, dt.$$

The integration with respect to $\mathbf{r}$ gives a δ-function of $\mathbf{p} + \mathbf{p}_{nm}$. We integrate with respect to t separately over the intervals $-\infty$ to 0 and from 0 to ∞, using the formula

$$\int_0^\infty e^{i\alpha x}\, dx = \pi\delta(\alpha) + \frac{i}{\alpha}.$$

Integrating with respect to t from 0 to ∞, we obtain

$$(2\pi)^3 \sum_{n,m} e^{(\Omega+\mu N_n - E_n)/T}(\psi_\alpha(0))_{nm}(\psi^+_\beta(0))_{mn}\delta(\mathbf{p} + \mathbf{p}_{nm}) \times \left[\frac{1}{\omega + \omega_{nm}} - i\pi\delta\,(\omega + \omega_{nm})\right]$$

($N_n = N_m - 1$), whereas the integral from $-\infty$ to 0 gives

$$\pm\,(2\pi)^3 \sum_{n,m} e^{(\Omega+\mu N_m - E_m)/T}(\psi_\alpha(0))_{nm}(\psi^+_\beta(0))_{mn}\delta(\mathbf{p} - \mathbf{p}_{mn}) \times \left[\frac{1}{\omega - \omega_{mn}} + i\pi\delta(\omega - \omega_{mn})\right].$$

It follows that

$$G_{\alpha\beta}(\mathbf{p}, \omega) = -(2\pi)^3 \sum_{n,m} e^{(\Omega+\mu N_n - E_n)/T}(\psi_\alpha(0))_{nm}(\psi^+_\beta(0))_{mn} \times \delta(\mathbf{p} - \mathbf{p}_{mn})\left\{\frac{1}{\omega_{mn} - \omega}\,[1 \pm e^{-\omega_{mn}/T}] + i\pi\delta\,(\omega - \omega_{mn})\,[1 \mp e^{-\omega_{mn}/T}]\right\}. \tag{17.7}$$

The rest of the argument hinges on the dependence of $G_{\alpha\beta}$ on the spin variables. If the system is not ferromagnetic (and this is the only case which will be considered here), it follows from symmetry considerations that $G_{\alpha\beta}$ must be proportional to the unit tensor $\delta_{\alpha\beta}$, i.e.,

$$G_{\alpha\beta} = \delta_{\alpha\beta} G, \tag{17.8}$$

and

$$\begin{aligned} G(\mathbf{p}, \omega) &= \frac{1}{2s+1} G_{\alpha\alpha}(\mathbf{p}, \omega) \\ &= -(2\pi)^3 \sum_{n,m} e^{(\Omega + \mu N_n - E_n)/T} \frac{1}{2s+1} \sum_{\alpha} |(\psi_\alpha(0))_{nm}|^2 \delta(\mathbf{p} - \mathbf{p}_{mn}) \\ &\times \left[\frac{1}{\omega_{mn} - \omega}(1 \pm e^{-\omega_{mn}/T}) + i\pi\delta(\omega - \omega_{mn})(1 \mp e^{-\omega_{mn}/T})\right], \end{aligned} \tag{17.9}$$

where s is the spin of the particles. Comparing the two terms in brackets in (17.9), we see that there is a definite relation between the real and imaginary parts G' and G'' of the Green's function (see L5). In fact, we have

$$G'(\mathbf{p}, \omega) = \frac{\mathrm{P}}{\pi} \int_{-\infty}^{\infty} \frac{G''(\mathbf{p}, x)}{x - \omega} \coth \frac{x}{2T} dx \tag{17.10}$$

in the case of Fermi statistics (where P denotes the operation of taking the principal value of an integral), and

$$G'(\mathbf{p}, \omega) = \frac{\mathrm{P}}{\pi} \int_{-\infty}^{\infty} \frac{G''(\mathbf{p}, x)}{x - \omega} \tanh \frac{x}{2T} dx \tag{17.11}$$

in the case of Bose statistics. Moreover, it follows from (17.9) that for bosons G'' is always negative, whereas for fermions G'' changes sign when $\omega = 0$, being positive for $\omega < 0$ and negative for $\omega > 0$.

The relations (17.10) and (17.11) imply that G is not analytic, regarded as a function of the complex variable ω. However, the function G is connected by simple formulas with two functions G^R and G^A (see footnote 8, p. 57) which are analytic in the upper and lower half-planes (of the variable ω), respectively. In terms of the real part G' and the imaginary part G'' of G, the function G^R is

$$G^R(\mathbf{p}, \omega) = G'(\mathbf{p}, \omega) + iG''(\mathbf{p}, \omega) \coth \frac{\omega}{2T} \tag{17.12}$$

for fermions, and

$$G^R(\mathbf{p}, \omega) = G'(\mathbf{p}, \omega) + iG''(\mathbf{p}, \omega) \tanh \frac{\omega}{2T} \tag{17.13}$$

for bosons. Similarly, the function G^A is

$$G^A(\mathbf{p}, \omega) = G'(\mathbf{p}, \omega) - iG''(\mathbf{p}, \omega) \coth \frac{\omega}{2T}$$

$$G^A(\mathbf{p}, \omega) = G'(\mathbf{p}, \omega) - iG''(\mathbf{p}, \omega) \tanh \frac{\omega}{2T}.$$

The functions G^R and G^A satisfy the *dispersion relations*

$$\operatorname{Re} G^R(\omega) = \frac{\mathrm{P}}{\pi}\int_{-\infty}^{\infty} \frac{\operatorname{Im} G^R(x)}{x-\omega}\,dx,$$
$$\operatorname{Re} G^A(\omega) = -\frac{\mathrm{P}}{\pi}\int_{-\infty}^{\infty} \frac{\operatorname{Im} G^A(x)}{x-\omega}\,dx, \tag{17.14}$$

which imply that G^R and G^A are analytic, according to a well-known theorem from the theory of functions of a complex variable.

It is not hard to see that G^R is just the so-called *retarded* Green's function

$$G^R(1,2) = \begin{cases} -i \operatorname{Sp}\{e^{(\Omega+\mu\hat{N}-\hat{H})/T}[\tilde{\psi}(1)\tilde{\psi}^+(2) \pm \tilde{\psi}^+(2)\tilde{\psi}(1)]\} & \text{for } t_1 > t_2, \\ 0 & \text{for } t_1 < t_2, \end{cases} \tag{17.15}$$

while G^A is the *advanced* Green's function

$$G^A(1,2) = \begin{cases} 0 & \text{for } t_1 > t_2, \\ i \operatorname{Sp}\{e^{(\Omega+\mu\hat{N}-\hat{H})/T}[\tilde{\psi}(1)\tilde{\psi}^+(2) \pm \tilde{\psi}^+(2)\tilde{\psi}(1)]\} & \text{for } t_1 < t_2. \end{cases} \tag{17.16}$$

In fact, carrying out exactly the same argument for (17.15) as was made for G, we obtain

$$G^R(\mathbf{p},\omega) = -(2\pi)^3 \sum_{n,m} e^{(\Omega+\mu N_n - E_n)/T}|\psi_{nm}(0)|^2(1 \pm e^{-\omega_{mn}/T}) \times \delta(\mathbf{p}-\mathbf{p}_{mn})\left[i\pi\delta(\omega-\omega_{mn}) - \frac{1}{\omega-\omega_{mn}}\right], \tag{17.17}$$

which clearly satisfies the relations (17.14).

We can write formula (17.17) and the analogous formula for G^A in the somewhat different form

$$G^R(\mathbf{p},\omega) = \int_{-\infty}^{\infty} \frac{\rho(\mathbf{p},x)}{x-\omega-i\delta}\,dx,$$
$$G^A(\mathbf{p},\omega) = \int_{-\infty}^{\infty} \frac{\rho(\mathbf{p},x)}{x-\omega+i\delta}\,dx, \tag{17.18}$$

where $\delta \to +0$ and ρ is the real function

$$\rho(\mathbf{p},\omega) = -(2\pi)^3 \sum_{n,m} e^{(\Omega+\mu N_n - E_n)/T}|\psi_{nm}(0)|^2 \times (1 \pm e^{-\omega_{mn}/T})\delta(\mathbf{p}-\mathbf{p}_{mn})\delta(\omega-\omega_{mn}). \tag{17.19}$$

Expressions of the type (17.18) were first obtained by Lehmann (L11), as applied to the Green's functions of quantum electrodynamics.

In particular, using (17.18), we can deduce the behavior of G^R and G^A for large ω. In the first place, noting that the integral

$$\int_{-\infty}^{\infty} \rho\,dx$$

is finite, we find that

$$G^R \approx G^A \approx -\frac{1}{\omega}\int_{-\infty}^{\infty} \rho\, dx.$$

On the other hand, we can calculate the integral of ρ by using the definition (17.15) of G^R. In fact, because of the commutation rules for the Heisenberg operators at $t_1 = t_2$, we have

$$G^R(\mathbf{r}_1, \mathbf{r}_2; t_1 = t_2 + 0) = -i\delta(\mathbf{r}_1 - \mathbf{r}_2),$$
$$G^R(\mathbf{p}; t_1 = t_2 + 0) = -i.$$

Then, expressing $G^R(\mathbf{p}; t_1 = t_2 + 0)$ in terms of $G^R(\mathbf{p}, \omega)$, i.e.,

$$G^R(\mathbf{p}; t_1 = t_2 + 0) = \frac{1}{2\pi}\int_{-\infty}^{\infty} d\omega\, G^R(\mathbf{p}, \omega)e^{-i\omega\alpha} \qquad (\alpha \rightarrow +0),$$

and substituting the expression (17.18) for G^R into this formula, we find that

$$-i = \frac{1}{2\pi}\int_{-\infty}^{\infty} dx\, \rho(x, \mathbf{p})\int_{-\infty}^{\infty} d\omega\, \frac{e^{-i\omega\alpha}}{x - \omega - i\delta} = i\int_{-\infty}^{\infty} \rho(x, \mathbf{p})\, dx.$$

It follows that

$$\int_{-\infty}^{\infty} \rho(x, \mathbf{p})\, dx = -1, \tag{17.20}$$

so that for large ω the functions G^R and G^A behave like

$$G^R \approx G^A \approx \frac{1}{\omega}, \tag{17.21}$$

i.e., like the corresponding functions for noninteracting particles.

The retarded and advanced potentials satisfy an infinite system of coupled equations [see Bogoliubov and Tyablikov (B8), Zubarev (Z1)]. However, they cannot be calculated by a diagram technique like that used to calculate the temperature Green's functions $\mathscr{G}$. Therefore, it is of interest to establish the connection between G^R and $\mathscr{G}$. To this end, we construct an integral representation for $\mathscr{G}$, analogous to (17.18).

According to the definition (11.1), we can write

$$\mathscr{G}(\mathbf{r}, \tau) = -\sum_{n,m} e^{(\Omega + \mu N_n - E_n)/T} e^{\omega_{nm}\tau - i\mathbf{p}_{nm}\cdot\mathbf{r}}|\psi_{nm}(0)|^2 \tag{17.22}$$

for $\tau > 0$. Using the formula

$$\mathscr{G}(\mathbf{p}, \omega_k) = \int_0^{1/T} d\tau \int d\mathbf{r}\, e^{i\omega_k\tau - i\mathbf{p}\cdot\mathbf{r}}\mathscr{G}(\mathbf{r}, \tau)$$

to go over to Fourier components in (17.22), where $\omega_k = (2k + 1)\pi T$ for fermions and $\omega_k = 2k\pi T$ for bosons, we obtain

$$\mathscr{G}(\mathbf{p}, \omega_k) = -(2\pi)^3 \sum_{n,m} e^{(\Omega + \mu N_n - E_n)/T}|\psi_{nm}(0)|^2\delta(\mathbf{p} - \mathbf{p}_{mn})\, \frac{1 \pm e^{-\omega_{mn}/T}}{\omega_{mn} - i\omega_k}. \tag{17.23}$$

The function (17.23) can be represented in the form[17]

$$\mathscr{G}(\mathbf{p}, \omega_n) = \int_{-\infty}^{\infty} \frac{\rho(\mathbf{p}, x)}{x - i\omega_n} dx, \tag{17.24}$$

with the same ρ as in (17.19), from which we obtain a relation between $\mathscr{G}$ for $\omega_n > 0$ and $G^R(\omega)$:

$$\mathscr{G}(\omega_n) = G^R(i\omega_n) \quad \text{for} \quad \omega_n > 0. \tag{17.25}$$

On the other hand, it follows from (17.24) that

$$\mathscr{G}(\omega_n) = \mathscr{G}^*(-\omega_n). \tag{17.26}$$

Thus, from a knowledge of the function $G^R(\omega)$, which is analytic in the upper half-plane, we can use (17.25) and (17.26) to construct the temperature Green's function $\mathscr{G}$ for all "frequencies" ω_n. However, the inverse problem of constructing the function G^R from a knowledge of $\mathscr{G}$ is of much greater interest. Suppose we know $\mathscr{G}$ for all frequencies ω_n, and suppose we have succeeded in constructing a function $F(\omega)$ which is analytic in the upper half-plane and has the property

$$F(i\omega_n) = \mathscr{G}(\omega_n) \quad \text{for} \quad \omega_n > 0.$$

Then, using a familiar theorem from the theory of functions of a complex variable,[18] we see at once that $F(\omega)$ coincides with $G^R(\omega)$ everywhere in the upper half-plane. Thus, the problem of constructing the function $G^R(\omega)$ reduces to the problem of continuing $\mathscr{G}(\omega_n)$ analytically from a discrete set of points into the entire upper half-plane [see Abrikosov, Gorkov and Dzyaloshinski (A5), Fradkin (F2)]. Although this problem does not have a solution in general form, the analytic continuation can be carried out in various special cases. In later chapters, we shall encounter examples of this type.

From a knowledge of the functions $G^R(\omega)$ and $D^R(\omega)$, or of the Green's function $D^R_{\alpha\beta}(\omega)$ of the electromagnetic field (see Sec. 29), we can determine a number of transport properties of the system. Thus, finding the pole of the Green's function $G^R(\omega)$ for an electron in a metal allows us to calculate the mean free path of an electron as a function of its energy, while the pole of the phonon Green's function $D^R(\omega)$ gives the absorption coefficient of sound. The function $D^R_{\alpha\beta}(\omega)$ determines the complex dielectric constant $\varepsilon(\omega)$ of the system, and with it the low-frequency conductivity of the metal [by using the relation $\varepsilon(\omega) \to 4\pi i\sigma/\omega$ as $\omega \to 0$]. This makes it clear that the technique of temperature Green's functions, together with the method of analytic continuation, allows us to transcend the purely statistical problem

---

[17] It might seem that in the Bose case, the integrand in (17.24) has a singularity at the point $x = 0$ when $\omega_n = 0$. However, it follows from (17.19) that in this case $\rho \sim x$ for small x.

[18] The theorem in question states that two analytic functions coincide if they take the same values at an infinite sequence of points which have a limit point in the region of analyticity. In our case, the infinite sequence of points consists of the uniformly spaced points $i\omega_n$, and the limit point is the point at infinity.

of calculating the thermodynamic potential Ω; as a matter of fact, while calculating Ω, we can simultaneously find certain transport properties as well.

On the other hand, a variety of transport properties (e.g., viscosity, thermal conductivity, etc.) are connected with many-particle Green's functions, which, in principle, obey relations resembling (17.24). Since these functions depend on a large number of frequencies (e.g., on three frequencies, in the case of a two-particle function), the general relations in question turn out to be quite formidable. However, in most cases of practical importance, we only need Green's functions some of whose arguments coincide (examples of this kind will be encountered in the chapter on superconductivity). In such cases, the required relations do not differ from the corresponding relations for one-particle functions, and in particular, formulas connecting the relevant temperature functions and time-dependent functions can be derived without difficulty.

For example, consider the scattering of slow neutrons in a liquid, which for simplicity we assume to consist of bosons with zero spin. As is well known (see e.g., L7), the interaction between an incident slow neutron, with radius vector $\mathbf{r}$ and mass m_n, and an atom of the liquid, with radius vector $\mathbf{R}$ and mass m, is described by the potential energy

$$V(\mathbf{r} - \mathbf{R}) = 2\pi \frac{m + m_n}{mm_n} a\delta(\mathbf{r} - \mathbf{R}), \tag{17.27}$$

where a is the scattering amplitude. Summing (17.27) over all the atoms of the liquid, we find that the energy of interaction between a slow neutron and the liquid as a whole is

$$V(\mathbf{r}) = 2\pi \frac{m + m_n}{mm_n} a \sum_k \delta(\mathbf{r} - \mathbf{R}_k). \tag{17.28}$$

In the second-quantized representation for particles of the liquid, $V(\mathbf{r})$ has the form

$$V(\mathbf{r}) = 2\pi \frac{m + m_n}{mm_n} a\psi^+(\mathbf{r})\psi(\mathbf{r}),$$

where ψ and ψ^+ are the field operators of the particles of the liquid in the Schrödinger representation.

The matrix element for a transition involving scattering of the neutron accompanied by a transfer of momentum $\mathbf{q}$ is proportional to the quantity

$$a \int e^{-i\mathbf{q}\cdot\mathbf{r}} \langle i|\psi^+(\mathbf{r})\psi(\mathbf{r})|f\rangle \, d\mathbf{r},$$

where i denotes the initial state and f the final state of the liquid. It follows that the differential scattering cross-section is

$$d\sigma \sim a^2 \int d\mathbf{r}_1 \, d\mathbf{r}_2 \, e^{-i\mathbf{q}\cdot(\mathbf{r}_1 - \mathbf{r}_2)} \langle i|\psi^+(\mathbf{r}_1)\psi(\mathbf{r}_1)|f\rangle \times \langle f|\psi^+(\mathbf{r}_2)\psi(\mathbf{r}_2)|i\rangle \delta(E_i - E_f + \Delta),$$

where Δ is the energy transfer. Summing this expression over the final state f, and calculating the Gibbs average over the initial state i, we obtain

$$d\sigma \sim a^2 \sum_{i,f} \int d\mathbf{r}_1\, d\mathbf{r}_2\, e^{-i\mathbf{q}\cdot(\mathbf{r}_1-\mathbf{r}_2)} e^{(\Omega+\mu N_i-E_i)/T} \langle i|\psi^+(\mathbf{r}_1)\psi(\mathbf{r}_1)|f\rangle \times \langle f|\psi^+(\mathbf{r}_2)\psi(\mathbf{r}_2)|i\rangle \delta(E_f - E_i - \Delta). \tag{17.29}$$

Finally, substituting the expressions (17.6) for the operators $\psi(\mathbf{r})$ and $\psi^+(\mathbf{r})$ into (17.29), we obtain

$$d\sigma \sim a^2(2\pi)^3 V \sum_{i,f} e^{(\Omega+\mu N_i-E_i)/T} |\langle i|\psi^+(0)\psi(0)|f\rangle|^2 \delta(\mathbf{p}_{fi} - \mathbf{q})\delta(\omega_{fi} - \Delta), \tag{17.30}$$

where V is the volume of the system, $\omega_{fi} = E_f - E_i$ and $\mathbf{p}_{fi} = \mathbf{P}_{fi} - \mathbf{P}_i$.

It is easily seen that to within a constant factor, the expression (17.30) coincides with the imaginary part of the Fourier component of the function

$$K(\mathbf{r}_1 - \mathbf{r}_2, t_1 - t_2) = -i\,\mathrm{Sp}\,\{e^{(\Omega+\mu\hat{N}-\hat{H})/T} T_t(\tilde{\psi}^+(\mathbf{r}_1, t_1)\tilde{\psi}(\mathbf{r}_1, t_1)\tilde{\psi}^+(\mathbf{r}_2, t_2)\tilde{\psi}(\mathbf{r}_2, t_2))\},$$

i.e.,

$$d\sigma \propto -\, Va^2 \frac{\mathrm{Im}\, K(\mathbf{q}, \Delta)}{1 + e^{-\Delta/T}}. \tag{17.31}$$

The function K is a two-particle Green's function (multiplied by i) with two pairs of identical arguments, and its analytic properties are exactly the same as those of a one-particle Bose Green's function G. If we introduce functions K^R and $\mathscr{K}$, by analogy with G^R and $\mathscr{G}$, then all that was said about G, G^R and $\mathscr{G}$ applies word for word to the functions K, K^R and $\mathscr{K}$, if at the same time we replace the operators $\psi(1)$ and $\psi^+(2)$ by $\psi^+(1)\psi(1)$ and $\psi^+(2)\psi(2)$ in all the formulas from (17.1) to (17.21). Thus, to calculate $d\sigma$, it is sufficient to find the temperature Green's function $\mathscr{K}$ and then construct K^R, its analytic continuation into the upper half-plane. Afterwards, $d\sigma$ can be found by using the relation

$$d\sigma \propto -Va^2 \frac{\mathrm{Im}\, K^R(\mathbf{q}, \Delta)}{1 - e^{-\Delta/T}}. \tag{17.31'}$$

Sometimes formula (17.31′) is written in a different form, i.e.,

$$d\sigma = A\, \frac{S(\mathbf{q}, \Delta)}{1 - e^{-\Delta/T}},$$

where A is a constant, and $S(\mathbf{q}, \Delta)$ is the so-called *structure function*

$$S(\mathbf{q}, \Delta) = 2\, \mathrm{Im}\, K^R(\mathbf{q}, \Delta).$$

It is not hard to verify that the function $S(\mathbf{q}, \omega)$ is the Fourier component of the average of the commutator of the density operators:

$$\langle[\rho(0), \rho(x)]\rangle = \frac{1}{(2\pi)^4} \int S(\mathbf{q}, \omega) e^{i\mathbf{q}\cdot\mathbf{r} - i\omega t}\, d\mathbf{q}\, d\omega.$$

The quantity $S(\mathbf{q}, \omega)$ satisfies a useful relation [see Cohen and Feynman (C2), Nozières and Pines (N1)], valid both for bosons and for fermions:

$$\frac{1}{2\pi}\int_0^{\infty} \omega S(\mathbf{q}, \omega)\, d\omega = \frac{\bar{n}\mathbf{q}^2}{2m}. \tag{17.32}$$

Here, $\bar{n} = \langle \rho(x) \rangle$ is the density of particles in the system, and m is the mass of a free atom.

To derive (17.32), it is convenient to start from the equation of continuity obeyed by the density operator and the particle current-density operator, i.e.,

$$\frac{\partial \rho(x)}{\partial t} + \operatorname{div} \mathbf{j}(x) = 0, \tag{17.33}$$

where $\rho(x) = \psi^+(x)\psi(x)$, and the current-density operator has the form

$$\mathbf{j}(x) = -\frac{i}{2m}(\psi^+\nabla\psi - (\nabla\psi^+)\psi),$$

if the interaction between the particles in the liquid does not depend on their velocities. Equation (17.33) expresses the law of conservation of matter; applying it to the commutator $[\rho(1), \rho(2)]$ for $t_1 = t_2$, we have

$$\frac{\partial}{\partial t_1}[\rho(1), \rho(2)]_{t_1 = t_2} = -\operatorname{div}_1[\mathbf{j}(1), \rho(2)]_{t_1 = t_2}.$$

Then, using the commutation rules (6.2) for the operators ψ and ψ^+ with identical time arguments, we obtain

$$\frac{\partial}{\partial t_1}[\rho(1), \rho(2)]_{t_1 = t_2} = \frac{i}{m}\operatorname{div}_1\left(\rho(1)\nabla_1\delta(\mathbf{r}_1 - \mathbf{r}_2) - \frac{1}{2}\delta(\mathbf{r}_1 - \mathbf{r}_2)\nabla_1\rho(1)\right).$$

Taking the average of this operator equation, and using the spatial homogeneity of the system (which implies $\nabla\bar{n} = 0$), we find that

$$\frac{1}{2\pi}\int_{-\infty}^{\infty} \omega S(\mathbf{q}, \omega)\, d\omega = \frac{\bar{n}\mathbf{q}^2}{m},$$

in terms of Fourier components. But

$$S(\mathbf{q}, -\omega) = -S(\mathbf{q}, \omega),$$

which proves (17.32).

It also follows from (17.30) that

$$\frac{1}{2\pi}\int_{-\infty}^{\infty} \frac{S(\mathbf{q}, \omega)}{1 - e^{-\omega/T}}\, d\omega = e^{-i\mathbf{q}\cdot\mathbf{r}}\langle \rho(0)\rho(\mathbf{r})\rangle\, d\mathbf{r} = \bar{n}S(\mathbf{q}),$$

where $S(\mathbf{q})$ is the correlation function introduced on p. 9 (for $\mathbf{q} \neq 0$). It is convenient to transform this last relation into the form

$$\frac{1}{2\pi}\int_0^{\infty} S(\mathbf{q}, \omega) \coth\frac{\omega}{2T}\, d\omega = \bar{n}S(\mathbf{q}).$$

We conclude this chapter by showing how to evaluate sums over ω_n of the type (16.12). First we note that for large ω_n, the function $\mathscr{G}$ has the form

$$\mathscr{G} \approx \frac{1}{i\omega_n} \tag{17.34}$$

[this is an immediate consequence of (17.24) and (17.20)], and hence the sum

$$T\sum_{\omega_n} \mathscr{G}(\omega_n)e^{-i\omega_n\tau} \tag{17.35}$$

diverges for $\tau = 0$. Actually, this means that (17.35) has a discontinuity for $\tau = 0$, as is clear from the definition of $\mathscr{G}(\tau)$. If we assume that τ is arbitrarily small but finite, the series (17.35) converges. Using (17.26), we can write (17.35) in the form

$$2T \sum_{\omega_n \geqslant 0}' \cos(\omega_n\tau) \operatorname{Re} \mathscr{G}(\omega_n) + 2T \sum_{\omega_n \geqslant 0}' \sin(\omega_n\tau) \operatorname{Im} \mathscr{G}(\omega_n),$$

where the prime on the summation sign means that the term $\omega_n = 0$ is taken with weight $\frac{1}{2}$. Since $\operatorname{Re} \mathscr{G}$ converges to zero faster than $1/\omega_n$ as $\omega_n \to \infty$, because of (17.34), we can simply set $\tau = 0$ in the first sum. To evaluate the second sum, we observe that the chief contribution to the sum is obtained for $\omega_n\tau \sim 1$, i.e., corresponds to large values of ω_n when $\tau \to 0$. Therefore, the sum over ω_n can be replaced by an integral, i.e.,

$$T\sum \cdots \to \frac{1}{2\pi}\int \cdots,$$

where, of course, we must use the asymptotic value (17.34) for $\mathscr{G}(\omega_n)$. This gives

$$\lim_{\tau \to 0} 2T \sum_{\omega_n \geqslant 0}' \sin(\omega_n\tau) \operatorname{Im} \mathscr{G}(\omega_n) = -\frac{1}{\pi}\int_0^\infty \frac{\sin x\tau}{x}\,dx = -\frac{1}{2}\operatorname{sgn}\tau,$$

and hence we have the following rule for evaluating the sum (17.35):

$$\lim_{\tau \to 0} T\sum_{\omega_n} \mathscr{G}(\omega_n)e^{-i\omega_n\tau} = 2T\sum_{\omega_n \geqslant 0}' \operatorname{Re} \mathscr{G}(\omega_n) - \frac{1}{2}\operatorname{sgn}\tau.. \tag{17.36}$$

4

THEORY OF THE FERMI LIQUID

18. Properties of the Vertex Part for Small Momentum Transfer. Zero Sound.[1]

Our primary aim in this chapter is to show how the methods of quantum field theory can be used to establish the foundations of a theory of the Fermi liquid. To this end, we consider a system of fermions at $T = 0$, with arbitrary short-range interaction forces. The properties of the Green's function in this case were considered in Sec. 7, where, in particular, it was shown that excitations of the "particle" type correspond to a pole of the function G^R in the lower half-plane near the positive real half-line of the complex variable ε,[2, 3] while the "holes" correspond to a pole of the function G^A in the upper half-plane near the half-line $\varepsilon < 0$. Since both of these functions are obtained by analytic continuation of the G-function from different real half-lines (of the variable ε), it follows that the G-function has the form

$$G(\mathbf{p}, \varepsilon) = \frac{a}{\varepsilon - v(|\mathbf{p}| - p_0) + i\delta \operatorname{sgn}(|\mathbf{p}| - p_0)}, \tag{18.1}$$

near $|\mathbf{p}| = p_0, \varepsilon = 0$. Here, $\delta \to +0$, a is a coefficient whose meaning was explained in Sec. 7 [see (7.40)], p_0 is the *Fermi momentum* defined by the relation $\varepsilon(p_0) = \mu$, and $v(|\mathbf{p}| - p_0)$ is the difference $\varepsilon(\mathbf{p}) - \mu$ expanded in the

---

[1] This section is in large measure based on results obtained by Landau (L6).

[2] For convenience, in this chapter we use ε to denote the frequency variable in the Green's function.

[3] For the Green's function with fixed N (defined in Sec. 7), this corresponds to $\varepsilon > \mu$.

neighborhood of $|\mathbf{p}| = p_0$. The expansion coefficient v is the velocity of the excitations at the Fermi surface, equal to p_0/m^*, where m^* is the effective mass of the excitations.

We now consider the properties of the vertex part Γ, which, together with G, plays an essential role in the theory of the Fermi liquid. Let us examine the behavior of the vertex part for p_1 near p_3 and p_2 near p_4. We introduce the notation

$$\Gamma(p_1, p_2; p_1 + k, p_2 - k) \equiv \Gamma(p_1, p_2; k), \tag{18.2}$$

where the energy-momentum transfer $k = (\mathbf{k}, \omega)$ is a small four-vector, i.e., $|\mathbf{k}| \ll p_0$, $|\omega| \ll \mu$. Consider the simplest diagrams for Γ, shown in Fig. 57. The expressions corresponding to these diagrams contain integrals of two Green's functions. Although for diagrams (a) and (b) there is nothing special about the case $k = 0$, for diagram (c) the poles of the two Green's

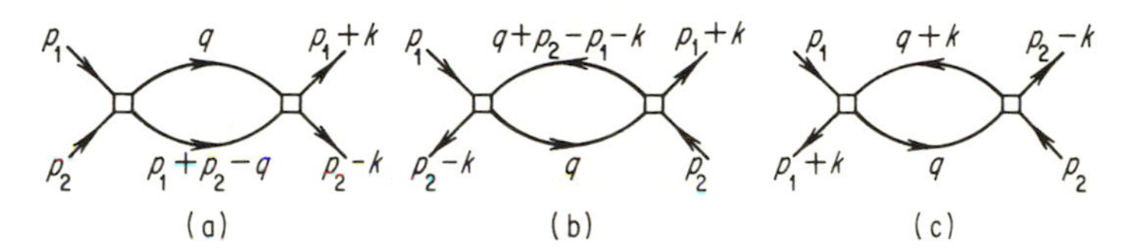

FIGURE 57

functions come together as $k \to 0$. As we shall see below, this leads to the appearance of singularities in Γ. It should be noted that although the diagrams in Fig. 57 formally pertain to the case of two-particle interaction forces, diagram (c) will still exhibit special behavior even in the case of arbitrary interaction forces.

Let $\Gamma^{(1)}$ denote the set of all possible diagrams for Γ which do not contain "anomalous elements," i.e., $G(p)G(p + k)$ lines. It is not hard to see that the exact Γ is obtained by summing over the "ladder diagrams" shown in Fig. 58, where the vertices are $\Gamma^{(1)}$-functions and all the lines are "anomalous." This sum can be expressed as an integral equation

$$\begin{aligned}\Gamma_{\alpha\beta,\gamma\delta}(p_1, p_2; k) = \Gamma^{(1)}_{\alpha\beta,\gamma\delta}(p_1, p_2) - i \int \Gamma^{(1)}_{\alpha\xi,\gamma\eta}(p_1, q)G(q) \\ \times G(q + k)\Gamma_{\eta\beta,\xi\delta}(q, p_2; k)\frac{d^4q}{(2\pi)^4}\end{aligned} \tag{18.3}$$

where we have set $k = 0$ in the function $\Gamma^{(1)}$, since it has no singularities at $k = 0$ (the forces are short-range).

Next, we consider the integral in (18.3), which consists of a contribution from regions far from the point $|\mathbf{p}| = p_0$, $\varepsilon = 0$ and a contribution from the neighborhood of this point. The latter integral determines the singularities of the whole expression. If k is small, we can take the neighborhood of $|\mathbf{p}| = p_0$, $\varepsilon = 0$ to be small, and the only important contribution to the corresponding integral will be from the part of the contour going around the poles of the G-functions. Since the arguments of the two G-functions are close together, we can assume that all other quantities in the integrand are slowly varying with respect to q. Then there is a contribution from the poles only when the poles lie on different sides of the real axis. This requires that $|\mathbf{q}| < p_0$, $|\mathbf{q} + \mathbf{k}| > p_0$ or $|\mathbf{q}| > p_0$, $|\mathbf{q} + \mathbf{k}| < p_0$. Bearing in mind that k is small, we easily see that this implies that $|\mathbf{q}| \approx p_0$ and $\varepsilon \approx 0$. Thus, in

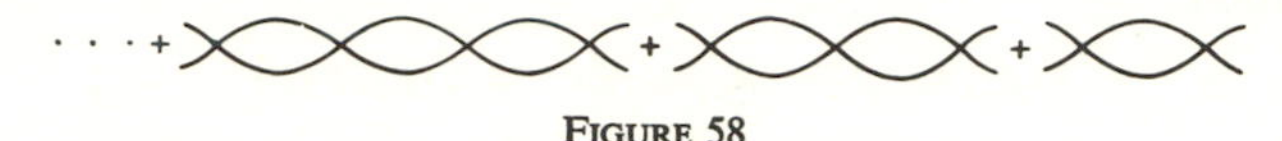

FIGURE 58

the part of the integral with respect to q which comes from the detours around the poles, the product $G(q)G(q + k)$ can be replaced by $A\delta(\varepsilon)\delta(|\mathbf{q}| - p_0)$.

The coefficient A can be determined by integrating $G(q)G(q + k)$ with respect to ε and $|\mathbf{q}|$, and is found to be

$$A = \frac{2\pi i a^2}{v} \frac{\mathbf{v}\cdot\mathbf{k}}{\omega - \mathbf{v}\cdot\mathbf{k}},$$

where $\mathbf{v}$ is a vector directed along $\mathbf{q}$ and equal to v in absolute value. Thus, the product $G(q)G(q + k)$ can be written in the form

$$G(q)G(q + k) = \frac{2\pi i a^2}{v} \frac{\mathbf{v}\cdot\mathbf{k}}{\omega - \mathbf{v}\cdot\mathbf{k}} \delta(\varepsilon)\delta(|\mathbf{q}| - p_0) + \varphi(q), \tag{18.4}$$

where $\varphi(q)$ represents the regular part of the product $G(q)G(q + k)$, which is important only in the integral over the "distant" region [hence we set $k = 0$ in $\varphi(q)$].

The limit of the expression (18.4) as $\mathbf{k}, \omega \to 0$ depends in an essential way on the ratio of ω to $|\mathbf{k}|$, and the same applies to Γ. We first examine Γ in the limit $\omega \to 0$, $|\mathbf{k}|/\omega \to 0$. Using (18.3) and (18.4), and denoting this limit by Γ^ω, we obtain

$$\Gamma^{\omega}_{\alpha\beta,\gamma\delta}(p_1, p_2) = \Gamma^{(1)}_{\alpha\beta,\gamma\delta}(p_1, p_2) - i \int \Gamma^{(1)}_{\alpha\xi,\gamma\eta}(p_1, q)\varphi(q)\Gamma^{\omega}_{\eta\beta,\xi\delta}(q, p_2) \frac{d^4q}{(2\pi)^4}. \tag{18.5}$$

With a view to eliminating $\Gamma^{(1)}$ from (18.3) and (18.5), we temporarily write these equations in operator form

$$\begin{aligned} \Gamma^{\omega} &= \Gamma^{(1)} - i\Gamma^{(1)}\varphi\Gamma^{\omega}, \\ \Gamma &= \Gamma^{(1)} - i\Gamma^{(1)}(i\Phi + \varphi)\Gamma, \end{aligned} \tag{18.6}$$

where products are interpreted as integrals, and $i\Phi$ denotes the first term in (18.4). The first of the equations (18.6) gives

$$\Gamma^{\omega} = (1 + i\Gamma^{(1)}\varphi)^{-1}\Gamma^{(1)}.$$

In the second equation, we transpose the term involving φ to the left-hand side, and then apply the operator $(1 + i\Gamma^{(1)}\varphi)^{-1}$, obtaining

$$\Gamma = \Gamma^{\omega} + \Gamma^{\omega}\Phi\Gamma.$$

Writing this equation in explicit form, we find that

$$\Gamma_{\alpha\beta,\gamma\delta}(p_1, p_2; k) = \Gamma^{\omega}_{\alpha\beta,\gamma\delta}(p_1, p_2) + \frac{a^2p_0^2}{(2\pi)^3 v}\int \Gamma^{\omega}_{\alpha\xi,\gamma\eta}(p_1, q)\Gamma_{\eta\beta,\xi\delta}(q, p_2)\frac{\mathbf{v}\cdot\mathbf{k}}{\omega - \mathbf{v}\cdot\mathbf{k}}\,d\Omega. \tag{18.7}$$

Next, we examine Γ in the other limiting case, i.e., $|\mathbf{k}| \to 0$, $\omega/|\mathbf{k}| \to 0$, this time denoting the limit by Γ^k. Using (18.7), we find the following relation between Γ^k and Γ^{ω}:

$$\Gamma^{k}_{\alpha\beta,\gamma\delta}(p_1, p_2) = \Gamma^{\omega}_{\alpha\beta,\gamma\delta}(p_1, p_2) - \frac{p_0^2a^2}{v(2\pi)^3}\int \Gamma^{\omega}_{\alpha\xi,\gamma\eta}(p_1, q)\Gamma^{k}_{\eta\beta,\xi\delta}(q, p_2)\,d\Omega. \tag{18.8}$$

We now investigate the poles of the function $\Gamma(p_1, p_2; k)$ for small $\mathbf{k}$ and ω. Since

$$\Gamma(p_1, p_2; k) \gg \Gamma^{\omega}(p_1, p_2)$$

in the neighborhood of the pole, we can neglect the term Γ^{ω} in the right-hand side of equation (18.7). It should also be noted that the variable p_2 and the indices β, δ act as parameters in (18.7). Therefore, near its pole the function Γ can be represented as a product $\chi_{\alpha\gamma}(p_1; k)\chi'_{\beta\delta}(p_2; k)$ of two functions. We can then cancel the common factor $\chi'_{\beta\delta}(p_2; k)$ appearing in both sides of (18.7). Writing

$$\nu_{\alpha\gamma}(n) = \frac{\mathbf{n}\cdot\mathbf{k}}{\omega - v\mathbf{n}\cdot\mathbf{k}}\chi_{\alpha\gamma}(p_1; k),$$

where $\mathbf{n}$ is a unit vector in the direction of $\mathbf{p}_1$, we obtain the following relation for $\nu_{\alpha\gamma}$:

$$(\omega - v\mathbf{n}\cdot\mathbf{k})\nu_{\alpha\gamma}(\mathbf{n}) = \mathbf{n}\cdot\mathbf{k}\frac{p_0^2a^2}{(2\pi)^3}\int \Gamma^{\omega}_{\alpha\xi,\gamma\eta}(\mathbf{n}, l)\nu_{\eta\xi}(l)\,d\Omega. \tag{18.9}$$

Equation (18.9) has the same form as the equation for zero sound and spin waves [see (2.24), p. 25]. In the next section, we shall show that this is natural, since the poles of Γ determine the acoustic excitation spectrum of the Fermi liquid. The quantity $a^2\Gamma^{\omega}$ plays the same role in equation (18.9) as the function f introduced in our previous treatment of the theory of the Fermi liquid (see Sec. 2). The quantity $a^2\Gamma^{\omega}$ has no direct physical meaning in its own right. However, equation (18.8) relates $a^2\Gamma^{\omega}$ to $a^2\Gamma^k$, and as we shall now show, this latter quantity can be interpreted (except for a constant factor) as the forward scattering amplitude of two quasi-particles with $|\mathbf{p}_1| = |\mathbf{p}_2| = p_0$.

We begin by considering the auxiliary problem of the scattering of two particles in the vacuum. Suppose that at time $t = -\infty$, the wave function of the system is $a^+_{\mathbf{p}_3\gamma}a^+_{\mathbf{p}_4\delta}\Phi_0$, where Φ_0 is the vacuum wave function. Then at time $t = \infty$, the system goes into the state $S(\infty)a^+_{\mathbf{p}_3\gamma}a^+_{\mathbf{p}_4\delta}\Phi_0$. The scattering amplitude for a transition into the state $\mathbf{p}_1\alpha$, $\mathbf{p}_2\beta$ is proportional to

$$-i\langle a_{\mathbf{p}_1\alpha}a_{\mathbf{p}_2\beta}S(\infty)a^+_{\mathbf{p}_3\gamma}a^+_{\mathbf{p}_4\delta}\rangle_0, \tag{18.10}$$

where $\langle\cdots\rangle_0$ denotes the vacuum average. The operator $S(\infty)$ is defined by equation (8.8), i.e., when $S(\infty)$ is expanded in a power series in H_{int}, it represents a sum of integrals of T-products of ψ-operators.

According to formula (8.10), each of these T-products can be written in terms of a set of N-products. It is clear that the only terms that matter in our matrix element are those containing N-products of four ψ-operators multiplied by all possible pairings. Since the pairings are c-numbers, they do not participate in the averages. In the present case, there are two particles, and hence

$$N(\psi^+(x_3)\psi^+(x_4)\psi(x_1)\psi(x_2)) = \psi^+(x_3)\psi^+(x_4)\psi(x_1)\psi(x_2).$$

In calculating the vacuum average (18.10), this N-product leads to the factor

$$\exp\{i(\mathbf{r}_1\cdot\mathbf{p}_1 + \mathbf{r}_2\cdot\mathbf{p}_2 - \mathbf{r}_3\cdot\mathbf{p}_3 - \mathbf{r}_4\cdot\mathbf{p}_4) \\ - i[\varepsilon_0(\mathbf{p}_1)t_1 + \varepsilon_0(\mathbf{p}_2)t_2 - \varepsilon_0(\mathbf{p}_3)t_3 - \varepsilon_0(\mathbf{p}_4)t_4]\}.$$

In (18.10), this factor and an expression containing only pairings is integrated over the coordinates. As a result, we obtain the set of all diagrams with four vertices, where the energy and momentum at each external point are connected by the relation $\varepsilon = \varepsilon_0(\mathbf{p})$. This quantity corresponds to the last term of the previously introduced two-particle Green's function (10.17), which has no external $G^{(0)}$-functions. In the present problem, involving the scattering of two particles in the vacuum, these $G^{(0)}$-functions are just exact G-functions. In fact, according to formula (7.3), we have

$$G^{(0)}(\mathbf{r}, t) = 0 \quad \text{for} \quad t < 0.$$

Moreover, in any diagram for corrections to the G-function, we can always find at least one pair of lines with opposite directions[4] (i.e., one $G^{(0)}$ with $t > 0$ and one $G^{(0)}$ with $t < 0$), as a result of which any correction to $G^{(0)}$ vanishes. For the same reason, the only diagrams for the vertex part which are left are those such that all the $G^{(0)}$-lines go in the same direction, i.e., diagrams of the type shown in Fig. 57(a).

It follows from all these considerations that the scattering amplitude (18.10) equals[5]

$$\Gamma'_{\alpha\beta,\gamma\delta}(p_1, p_2; p_3, p_4)|_{\varepsilon_i = \varepsilon_0(p_i)},$$

where Γ' is the vertex part for the problem being discussed. In the present case, formula (9.17), expressing the relation between the two-particle Green's function and the vertex part, takes the form

$$G^{\text{II}'}_{\alpha\beta,\gamma\delta}(p_1, p_2; p_3, p_4) = G^{(0)}(p_1)\, G^{(0)}(p_2)(2\pi)^4 \\ \times [\delta(p_1 - p_3)\delta_{\alpha\gamma}\delta_{\beta\delta} - \delta(p_1 - p_4)\delta_{\beta\gamma}\delta_{\alpha\delta}] \qquad (18.11) \\ + iG^{(0)}(p_1)\, G^{(0)}(p_2)\, G^{(0)}(p_3)\, G^{(0)}(p_4)\, \Gamma'_{\alpha\beta,\gamma\delta}(p_1, p_2; p_3, p_4),$$

---

[4] Diagrams of the type shown in Fig. 4(a), p. 69 are an exception, but, as already noted in Sec. 8, in such diagrams $G^{(0)}(0)$ must be regarded as the limit $\lim_{t\to+0} G^{(0)}(-t)$, which in the present case equals zero.

[5] In all the formulas, we omit a constant factor, which equals $4\pi/m$, as shown by comparison with perturbation theory (see Chap. 1).

where the momenta obey appropriate conservation laws. The quantity $G^{\text{II}'}$ can be regarded as a two-particle Green's function (which explains the notation). The first term in (18.11) corresponds to free motion of the particles, and the second term corresponds to scattering of the particles by each other.

Returning to the case of the Fermi liquid, we compare formulas (10.17) and (18.11). According to (18.1), in the region of small ε and values of $|\mathbf{p}|$ near p_0, the Green's functions have a form very close to that of free-particle Green's functions. We have to divide the function G^{II} by a^2 before we can regard it as a Green's function for two interacting quasi-particles. Then the free term will have exactly the same normalization as in the case of real particles with energy $\varepsilon(\mathbf{p})$. The second term in (10.17) corresponds to scattering of quasi-particles. Comparing it with the similar expression for real particles, we arrive at the conclusion that the quantity

$$a^2\Gamma_{\alpha\beta,\gamma\delta}(p_1, p_2; p_3, p_4)|_{\varepsilon_i = \varepsilon(p_i) - \mu} \tag{18.12}$$

acts as an amplitude for scattering of quasi-particles. In particular, all the $\varepsilon_i = 0$ if $|\mathbf{p}_1| = |\mathbf{p}_2| = p_0$, and then the scattering amplitude for small momentum transfer equals $a^2\Gamma(p_1, p_2; k)$ with $\omega = 0$, while the forward scattering amplitude equals $a^2\Gamma^k$.

The relation (18.8) connecting $a^2\Gamma^k$ with the function $f = a^2\Gamma^\omega$ can be solved if we assume that the spin-dependent part of the particle interactions is due purely to exchange. Then we can write $a^2\Gamma^k$ in the form

$$\frac{p_0^2}{\pi^2 v} a^2\Gamma^k = A(\mathbf{n}_1, \mathbf{n}_2) + B(\mathbf{n}_1, \mathbf{n}_2)\boldsymbol{\sigma}_1 \cdot \boldsymbol{\sigma}_2, \tag{18.13}$$

and, according to (2.23),

$$\frac{p_0^2}{\pi^2 v} a^2\Gamma^\omega = F.$$

The equations for Φ and Z [cf. (2.28)] separate, i.e.,

$$\begin{aligned} A(\mathbf{n}_1, \mathbf{n}_2) &= \Phi(\mathbf{n}_1, \mathbf{n}_2) - \int \Phi(\mathbf{n}_1, \mathbf{n}')A(\mathbf{n}', \mathbf{n}_2)\frac{d\Omega}{4\pi}, \\ B(\mathbf{n}_1, \mathbf{n}_2) &= Z(\mathbf{n}_1, \mathbf{n}_2) - \int Z(\mathbf{n}_1, \mathbf{n}')B(\mathbf{n}', \mathbf{n}_2)\frac{d\Omega}{4\pi}. \end{aligned} \tag{18.14}$$

In an isotropic liquid, all the quantities at the Fermi surface depend only on $\cos \angle(\mathbf{n}_1, \mathbf{n}_2) = \cos\chi$. Suppose we expand these quantities in Legendre polynomials, e.g.,

$$A(\chi) = \sum_l A_l P_l(\cos\chi).$$

Then we immediately obtain the following relations between the expansion coefficients:

$$A_l = \frac{\Phi_l}{1 + \dfrac{\Phi_l}{2l+1}}, \qquad B_l = \frac{Z_l}{1 + \dfrac{Z_l}{2l+1}}. \tag{18.15}$$

19. Effective Mass. Relation Between the Fermi Momentum and the Particle Number.[6] Excitations of the Bose Type. Heat Capacity.

19.1. Some useful relations. We begin by deriving some relations involving the Green's function. Suppose our system is in an infinitely small field $\delta U(t)$, which is homogeneous in space and slowly varying in time. The corresponding interaction Hamiltonian has the form

$$H_{\text{int}} = \int \psi_\alpha^+ (\mathbf{r})\, \delta U(t) \psi_\alpha(\mathbf{r})\, d\mathbf{r}.$$

Going over to the interaction representation with respect to H_{int}, and expanding the Green's function in a power series in δU up to terms of the first order, we find that

$$\delta G_{\alpha\beta}(x, x') = -\int d^4y\, \delta U(t_y)[\langle T(\tilde{\psi}_\alpha(x)\tilde{\psi}_\gamma^+(y)\tilde{\psi}_\gamma(y)\tilde{\psi}_\beta^+(x'))\rangle - \langle T(\tilde{\psi}_\alpha(x)\tilde{\psi}_\beta^+(x'))\rangle\langle\tilde{\psi}_\gamma^+(y)\tilde{\psi}_\gamma(y)\rangle],$$

where the $\tilde{\psi}$ are the Heisenberg operators of the interacting particles in the absence of the field δU. Using formula (10.17), we obtain

$$\delta G_{\alpha\beta}(x, x') = \delta_{\alpha\beta} \int \delta U(t_y) G(x - y) G(y - x')\, d^4y - i \int \delta U(t_y) G(x - x_1) G(y - x_2)\, G(x_3 - x') G(x_4 - y) \times \Gamma_{\alpha\gamma,\beta\gamma}(x_1, x_2; x_3, x_4),\, d^4y\, d^4x_1\, d^4x_2\, d^4x_3\, d^4x_4$$

or, transforming to Fourier components,

$$\delta G_{\alpha\beta} = \delta_{\alpha\beta} G(p)\, \delta U(\omega) G(p + k_1) - iG(p)G(p + k_1) \int \Gamma_{\alpha\gamma,\beta\gamma}(p, q; k_1) G(q)\, \delta U(\omega) G(q + k_1) \frac{d^4q}{(2\pi)^4},$$

where $k_1 = (0, \omega)$.

Since the field δU has no effect on the spin of the particles, $\delta G_{\alpha\beta}$ must be proportional to $\delta_{\alpha\beta}$. Performing the operation $\frac{1}{2}$ Sp, we obtain

$$\delta G = G(p)\, \delta U G(p + k_1) - iG(p)G(p + k_1) \times \frac{1}{2} \int \Gamma_{\alpha\beta,\alpha\beta}(p, q: k_1) G(q)\, \delta U(\omega) G(q + k_1) \frac{d^4q}{(2\pi)^4}.$$

On the other hand, if we add the term

$$\delta U(t) \int \psi_\alpha^+(\mathbf{r}) \psi_\alpha(\mathbf{r})\, d\mathbf{r} = \delta U(t) \hat{N}$$

to the Hamiltonian, we find that in the limit $\delta U \to$ const, the function G is

[6] These results were obtained by Landau and Pitayevski, and were partially published in P6.

simply multiplied by $e^{-i\delta U(t-t')}$, which corresponds to adding the term $-\delta U$ to ε. Thus,

$$\frac{\delta G}{\delta U} \to -\frac{\partial G}{\partial \varepsilon}$$

in the limit $\omega \to 0$, and we have

$$\frac{\partial G}{\partial \varepsilon} = -\{G^2(p)\}_\omega\left[1 - \frac{i}{2}\int \Gamma^{\omega}_{\alpha\beta,\,\alpha\beta}(p, q)\{G^2(q)\}_\omega \frac{d^4q}{(2\pi)^4}\right],$$

where $\{G^2(p)\}_\omega = \varphi$ [see (18.4)] denotes the limit of $G(p)G(p + k_1)$ as $\omega \to 0$. Examining this formula near the pole of $G(p)$, we can write $G(p)$ in the form (18.1). Then, dividing by $-\{G^2(p)\}_\omega$, we obtain our first formula

$$\frac{\partial G^{-1}(p)}{\partial \varepsilon} = \frac{1}{a} = 1 - \frac{i}{2}\int \Gamma^{\omega}_{\alpha\beta,\,\alpha\beta}(p, q)\{G^2(q)\}_\omega \frac{d^4q}{(2\pi)^4}. \tag{19.1}$$

To obtain another important relation, we argue as follows: Suppose the particles have infinitely small charge δe, and suppose the system is in a magnetic field which is weakly homogeneous in space and constant in time. Then the term $-\mathbf{A}\delta e$ should be added to the momentum operator in the Hamiltonian. If the charge δe is very small, the change in the Hamiltonian is given by the expression

$$-\frac{\delta e}{m}\int \psi^+_\alpha(\mathbf{r})(\hat{\mathbf{p}}\cdot\mathbf{A}(\mathbf{r}))\psi_\alpha(\mathbf{r})\, dV,$$

where $\mathbf{p}$ is the momentum operator. The resulting change in the Green's function is obtained in the same way as before, and is found to be

$$\delta G = -G(p)\frac{\delta e}{m}(\mathbf{p}\cdot\mathbf{A})G(p + k_2) + \frac{i}{2}G(p)G(p + k_2)$$
$$\times \int \Gamma_{\alpha\beta,\,\alpha\beta}(p, q; k_2)G(q)\frac{\delta e}{m}(\mathbf{q}\cdot\mathbf{A})G(q + k_2)\frac{d^4q}{(2\pi)^4},$$

where $k_2 = (\mathbf{k}, 0)$ [here $\mathbf{k}$ is assumed to be small]. On the other hand, in the limit $\mathbf{k} \to 0$, it follows from the gauge invariance that all functions depending on momenta must go over into functions of $\mathbf{p} - \mathbf{A}\delta e$, so that

$$\frac{\delta G}{\mathbf{A}\delta e} = -\frac{\partial G}{\partial p},$$

in the limit $\mathbf{k} \to 0$. Therefore, in the limit $\delta e \to 0$, $\mathbf{k} \to 0$, we obtain a second relation for $G(p)$ near the pole:

$$\begin{aligned}\nabla_{\mathbf{p}} G^{-1} &= -\frac{\mathbf{v}}{a} = -\frac{\mathbf{p}}{m^*a} \\ &= -\frac{\mathbf{p}}{m} + \frac{i}{2}\int \Gamma^{k}_{\alpha\beta,\,\alpha\beta}(p, q)\frac{\mathbf{q}}{m}\{G^2(q)\}_k \frac{d^4q}{(2\pi)^4}.\end{aligned} \tag{19.2}$$

A third relation can be obtained by considering the change of the Green's function when the system as a whole moves with a small, slowly varying velocity $\delta\mathbf{u}(t)$. In this case, the Hamiltonian of the system is changed by addition of the term

$$-\delta\mathbf{u}\cdot\hat{\mathbf{P}} = -\delta\mathbf{u}\cdot\int\psi_\alpha^+(\mathbf{r})\hat{\mathbf{p}}\psi_\alpha(\mathbf{r})\,d\mathbf{r},$$

where $\hat{\mathbf{P}}$ is the total momentum operator of the system. The corresponding change in the Green's function is

$$\delta G = -G(p)(\mathbf{p}\cdot\delta\mathbf{u})(\omega)G(p+k_1)$$
$$+\frac{i}{2}G(p)G(p+k_1)\int\Gamma_{\alpha\beta,\alpha\beta}(p,q;k_1)(\mathbf{q}\cdot\delta\mathbf{u})(\omega)G(q)G(q+k_1)\frac{d^4q}{(2\pi)^4},$$

where $k_1 = (0, \omega)$. On the other hand, for $\omega = 0$ this reduces to a transformation to a coordinate system moving with constant velocity $\delta\mathbf{u}$, and then the energy changes by the quantity $-\mathbf{P}\cdot\delta\mathbf{u}$ (according to the usual Galilean formula). Moreover, the frequency ε becomes $\varepsilon + \mathbf{p}\cdot\delta\mathbf{u}$, so that the Green's function is changed by the quantity $(\partial G/\partial\varepsilon)\mathbf{p}\cdot\delta\mathbf{u}$. Thus, in the limit $\omega \to 0$, $\delta\mathbf{u} \to 0$, we find that

$$\mathbf{p}\frac{\partial G^{-1}}{\partial\varepsilon} = \frac{\mathbf{p}}{a} = \mathbf{p} - \frac{i}{2}\int\Gamma^\omega_{\alpha\beta,\alpha\beta}(p,q)\mathbf{q}\{G^2(q)\}_\omega\frac{d^4q}{(2\pi)^4} \tag{19.3}$$

near the pole.

Finally, we derive one last relation by considering the change of the G-function under the influence of a small field $\delta U(\mathbf{r})$ which is constant in time and weakly inhomogeneous in space. In this case, the change in G equals

$$\delta G = G(p)\,\delta U(\mathbf{k})G(p+k_2)$$
$$-\frac{i}{2}G(p)G(p+k_2)\int\Gamma_{\alpha\beta,\alpha\beta}(p,q;k_2)G(q)\,\delta U(\mathbf{k})G(q+k_2)\frac{d^4q}{(2\pi)^4},$$

where $k_2 = (\mathbf{k}, 0)$. On the other hand, the equilibrium condition

$$\mu + \delta U = \text{const}$$

must be satisfied in a constant external field. In the limit $\mathbf{k} \to 0$, the chemical potential changes by the small constant $-\delta U$. Thus, in the limit $\mathbf{k} \to 0$, $\delta U \to 0$, we have

$$\frac{\partial G^{-1}}{\partial\mu} = 1 - \frac{i}{2}\int\Gamma^k_{\alpha\beta,\alpha\beta}(p,q)\{G^2(q)\}_k\frac{d^4q}{(2\pi)^4}, \tag{19.4}$$

a formula which is valid for arbitrary momenta.

19.2. Basic relations of the theory of the Fermi liquid. We are now in a position to derive the basic relations of the theory of the Fermi liquid, by using formulas (19.1)–(19.4) and formula (18.8) connecting Γ^k and Γ^ω. We note in passing that formula (18.8) holds for arbitrary momenta p_1 and p_2, which are by no means obliged to lie near the Fermi surface.

First of all, we substitute (18.8) into (19.2), obtaining

$$-\frac{\mathbf{p}}{m^*a}+\frac{\mathbf{p}}{m}=\frac{i}{2}\int\Gamma^{\omega}_{\alpha\beta,\,\alpha\beta}(p,q)\,\frac{\mathbf{q}}{m}\{G^2(q)\}_k\,\frac{d^4q}{(2\pi^4)}$$
$$-\frac{1}{2}\,\frac{p_0^2a^2}{(2\pi)^3v}\int\Gamma^{\omega}_{\alpha\beta,\,\alpha\beta}(p,q)\Big(-\frac{\mathbf{q}}{m^*a}+\frac{\mathbf{q}}{m}\Big)\,d\Omega.$$

It follows from formula (18.4) that

$$\{G^2(p)\}_k=\{G^2(p)\}_\omega-\frac{2\pi ia^2}{v}\,\delta(\varepsilon)\,\delta(|\mathbf{p}|-p_0). \tag{19.5}$$

We substitute this expression into the first integral in the preceding equation, and use formula (19.3). Then, after some cancellation, we find that

$$\frac{1}{m}=\frac{1}{m^*}+\frac{p_0}{2(2\pi)^3}\int a^2\Gamma^{\omega}_{\alpha\beta,\,\alpha\beta}(\chi)\cos\chi\,d\Omega. \tag{19.6}$$

for $|\mathbf{p}|=p_0,\ \varepsilon=0$. It is easy to see that this formula coincides with (2.12), if

$$a^2\Gamma^{\omega}_{\alpha\beta,\,\alpha\beta}=\mathrm{Sp}_\sigma\,\mathrm{Sp}_{\sigma'}\,f(\chi,\boldsymbol{\sigma},\boldsymbol{\sigma}').$$

Next, we prove formula (2.1). Consider the expression (18.1) for the Green's function G near the pole, i.e., for $|\mathbf{p}|\to p_0,\ \varepsilon\to 0$. The coefficients a and v in (18.1), as well as the momentum p_0, depend on the chemical potential μ. Differentiating G with respect to μ, we easily see that the terms coming from differentiation of a and v are small near the pole $[\sim(|\mathbf{p}|-p_0)/\mu$ or $\varepsilon/\mu]$ compared to the term coming from differentiation of p_0. Thus we have

$$\frac{\partial G}{\partial\mu}\approx -G^2\,\frac{v}{a}\,\frac{dp_0}{d\mu},$$

which implies

$$v\,\frac{dp_0}{d\mu}=a\Big(\frac{\partial G^{-1}}{\partial\mu}\Big)_{\substack{\varepsilon=0\\|\mathbf{p}|=p_0}}$$

Substituting (19.4) into this formula, and using (18.8) to express Γ^k, we obtain

$$\frac{v}{a}\,\frac{dp_0}{d\mu}=1-\frac{i}{2}\int\Gamma^{\omega}_{\alpha\beta,\,\alpha\beta}(p,q)\{G^2(q)\}_k\,\frac{d^4q}{(2\pi)^4}$$
$$-\frac{1}{2}\,\frac{p_0^2a^2}{(2\pi)^3v}\int\Gamma^{\omega}_{\alpha\beta,\,\alpha\beta}(p,q)\Big(\frac{v}{a}\,\frac{dp_0}{d\mu}-1\Big)\,d\Omega.$$

Then, substituting the expression (19.5) for $\{G^2\}_k$ into this last equation, and using formula (19.1), we find after some cancellation that

$$v\,\frac{dp_0}{d\mu}=\Big[1+\frac{p_0^2}{2(2\pi)^3v}\int a^2\Gamma^{\omega}_{\alpha\beta,\,\alpha\beta}(p,q)\,d\Omega\Big]^{-1} \tag{19.7}$$

The total number of particles in the system is given by (7.37). Differentiation of this formula with respect to μ gives[7]

$$\frac{d(N/V)}{d\mu} = -2i\int\frac{\partial G(p)}{\partial\mu}\frac{d^4p}{(2\pi)^4} = 2i\int\frac{\partial G^{-1}(p)}{\partial\mu}\{G^2(p)\}_k\frac{d^4p}{(2\pi)^4}.$$

Substituting (19.4) into this formula, and using (18.8) to express Γ^k in terms of Γ^ω, we obtain

$$\begin{aligned}\frac{d(N/V)}{d\mu} &= 2i\int\{G^2(p)\}_k\frac{d^4p}{(2\pi)^4}\\ &+ \int\{G^2(p)\}_k\Gamma^\omega_{\alpha\beta,\,\alpha\beta}(p,q)\{G^2(q)\}_k\frac{d^4q}{(2\pi)^4}\frac{d^4p}{(2\pi)^4}\\ &- \frac{ip_0^2a^2}{(2\pi)^3v}\int\{G^2(p)\}_k\Gamma^\omega_{\alpha\beta,\,\alpha\beta}(p,q)\left(\frac{v}{a}\frac{dp_0}{d\mu}-1\right)\frac{d^4p}{(2\pi)^4}\,d\Omega.\end{aligned}$$

Then, substituting (19.5) into this last equation, and using (19.1), we find that

$$\begin{aligned}\frac{d(N/V)}{d\mu} &= 2i\int\{G^2(p)\}_\omega\frac{d^4p}{(2\pi)^4}\\ &+ \int\{G^2(p)\}_\omega\Gamma^\omega_{\alpha\beta,\,\alpha\beta}(p,q)\{G^2(q)\}_\omega\frac{d^4q}{(2\pi)^4}\frac{d^4p}{(2\pi)^4}\\ &+ \frac{8\pi ap_0^2}{(2\pi)^3v} - \frac{8\pi p_0^2(a-1)}{(2\pi)^3v}v\frac{dp_0}{d\mu}\\ &- 8\pi\left[\frac{p_0^2a^2}{(2\pi)^3v}\right]^2\frac{1}{2}\int\Gamma^\omega_{\alpha\beta,\,\alpha\beta}(p,q)\frac{v}{a}\frac{dp_0}{d\mu}\,d\Omega.\end{aligned}$$

According to (19.1), the first two terms in the right-hand side are just

$$-2i\int\frac{\partial G}{\partial\varepsilon}\frac{d^4p}{(2\pi)^4} = -2i\int[G(\varepsilon=\infty)-G(\varepsilon=-\infty)]\frac{d\mathbf{p}}{(2\pi)^4},$$

but this expression equals zero, since, according to (7.11), G vanishes as $\varepsilon\to\pm\infty$. (This is also obvious from the fact that the expression in question equals the change in the number of particles in the system if we change the origin from which the energy is measured.) In the last of the remaining terms, we use (19.7) to express $\int a^2\Gamma^\omega\,d\Omega$, obtaining

$$\frac{d(N/V)}{d\mu} = \frac{8\pi p_0^2}{(2\pi)^3}\frac{dp_0}{d\mu}. \tag{19.8}$$

Integrating with respect to μ, we finally find that

$$\frac{N}{V} = \frac{8\pi}{3}\frac{p_0^3}{(2\pi)^3}.$$

Using (19.7), we can verify the expression (2.19) for the velocity of sound. We need only observe that because of (19.8), formulas (19.7) and (2.18) coincide.

[7] Here, for brevity, we omit the factor $e^{i\varepsilon t}$ $(t\to+0)$ in the integrand.

19.3. Excitations of the Bose type. We now consider the case of acoustic excitations. To this end, we carry out an analysis similar to formulas (7.32)–(7.33), examining the time variation of the state which at the time $t = t'$ is described by the wave function

$$\Psi_0(t') = \frac{1}{V}\sum_{\mathbf{p},\alpha} \psi_{\mathbf{p}\alpha}(t')\psi^+_{\mathbf{p}+\mathbf{k}\alpha}(t')\Phi_i(t'). \tag{19.9}$$

Here $|\mathbf{k}| \ll p_0$, and the $\psi_{\mathbf{p}\alpha}$ are the operators of a particle with momentum $\mathbf{p}$ in the interaction representation. Handling all the operators in the same sequence as in (7.32)–(7.33), we obtain the probability amplitude

$$\begin{aligned}
\langle\Psi_0^*(t)\Psi(t)\rangle &= \frac{1}{V^2}\sum_{\mathbf{p}_1,\mathbf{p}_2,\alpha,\beta} \langle\tilde{\psi}_{\mathbf{p}_2\beta}(t)\tilde{\psi}_{\mathbf{p}_2-\mathbf{k},\beta}(t)\tilde{\psi}_{\mathbf{p}_1\alpha}(t')\tilde{\psi}^+_{\mathbf{p}_1+\mathbf{k},\alpha}(t')\rangle \\
&= -\int G^{\mathrm{II}}_{\alpha\beta,\alpha\beta}(\mathbf{p}_1, t', \mathbf{p}_2, t; \mathbf{p}_1 + \mathbf{k}, t', \mathbf{p}_2 - \mathbf{k}, t)\frac{d\mathbf{p}_1}{(2\pi)^3}\frac{d\mathbf{p}_2}{(2\pi)^3} \\
&= -\int G^{\mathrm{II}}_{\alpha\beta,\alpha\beta}(p_1, p_2; p_1 + k, p_2 - k)\frac{d^4p_1}{(2\pi)^4}\frac{d^4p_2}{(2\pi)^4}e^{-i\omega(t-t')}\frac{d\omega}{2\pi},
\end{aligned}$$

where $k = (\mathbf{k}, \omega)$. Substituting formula (10.17) for G^{II} into this expression, we find that

$$\begin{aligned}
\langle\Psi_0^*(t)\Psi(t)\rangle = \int\Big[2\int & G(p)G(p+k)\frac{d^4p}{(2\pi)^4} \\
&-i\int G(p_1)G(p_1+k)\Gamma_{\alpha\beta,\alpha\beta}(p_1, p_2; k) \\
&\times G(p_2)G(p_2-k)\frac{d^4p_1}{(2\pi)^4}\frac{d^4p_2}{(2\pi)^4}\Big]e^{-i\omega(t-t')}\frac{d\omega}{2\pi}.
\end{aligned} \tag{19.10}$$

The expression in brackets in (19.10), which we denote by $ia\Pi$, can be transformed by using formulas (18.4), (18.7) and (19.1), with the aim of eliminating terms containing integrations remote from the point $|\mathbf{p}| = p_0$, $\varepsilon = 0$. This is done in the same spirit as in all the previous derivations given in this section. Using (18.7) to express the vertex part Γ in terms of Γ^ω, and using (19.1) to replace the integral of the type $\int\Gamma^\omega\varphi$, we find that the function Π equals

$$\begin{aligned}
\Pi &= \frac{2p_0^2}{(2\pi)^3v}\int\frac{\mathbf{v}\cdot\mathbf{k}}{\omega - \mathbf{v}\cdot\mathbf{k}}\,d\Omega \\
&+ \left(\frac{p_0^2}{(2\pi)^3v}\right)^2\int\frac{\mathbf{v}_1\cdot\mathbf{k}}{\omega - \mathbf{v}_1\cdot\mathbf{k}}\,a^2\Gamma_{\alpha\beta,\alpha\beta}(p_1, p_2; k)\frac{\mathbf{v}_2\cdot\mathbf{k}}{\omega - \mathbf{v}_2\cdot\mathbf{k}}\,d\Omega_1\,d\Omega_2.
\end{aligned}$$

We can easily verify directly that Π can be written in the form

$$\frac{p_0^2}{(2\pi)^3v}\int\Pi_{1\alpha\alpha}\,d\Omega,$$

where $\Pi_{1\alpha\gamma}$ satisfies the equation

$$\begin{aligned}
&(\omega - \mathbf{v}\cdot\mathbf{k})\Pi_{1\alpha\gamma}(\mathbf{k}, \mathbf{n}, \omega) \\
&\quad - \frac{p_0^2\mathbf{v}\cdot\mathbf{k}}{(2\pi)^3v}\frac{1}{2}\int a^2\Gamma^\omega_{\gamma\eta,\alpha\xi}(\mathbf{n}, \mathbf{n}')\Pi_{1\eta\xi}(\mathbf{k}, \mathbf{n}', \omega)\,d\Omega' = (\mathbf{v}\cdot\mathbf{k})\delta_{\alpha\gamma}.
\end{aligned} \tag{19.11}$$

As in Sec. 7, it follows from formula (19.10) that the value of $\langle\Psi_0^*(t)\Psi(t)\rangle$ as $t \to \infty$ is determined by the poles of the function Π in the lower half-plane (regarded as a function of the complex variable ω). The equation for the poles is obtained from (19.11) by replacing the right-hand side by zero. Comparing the resulting equation with (18.9), we see that the quantity Π_1 corresponds to ν. Since we are interested in the quantity $\mathrm{Sp}_\sigma \int \Pi_1 \, d\Omega$, and not in the quantity Π_1 itself, we only consider the solutions of (18.9) which are isotropic in the plane perpendicular to the vector **k**. In the same way, by choosing other functions $\Psi_0(t)$, we can obtain equations for all the components of the quantity $\nu_{\alpha\gamma}$, with respect to both angles and spins.

Thus, we have shown that in a Fermi liquid there can exist excitations whose spectrum is determined by the poles of the function Γ, i.e., by equation (18.9). These excitations obey Bose statistics, since they correspond to operators which are bilinear in the Fermi operators [see (19.9)]. As shown in Sec. 2, such excitations correspond to various branches of the zero-sound spectrum. This explains the physical meaning of the poles of the function Γ in the region of small energy and momentum transfer, and proves the identity of equations (18.9) and (2.24). It follows from formula (19.9) that an acoustic excitation can be regarded as a bound pair consisting of a quasi-particle and a hole with neighboring values of the momentum.

19.4. Another derivation of the relation between the Fermi momentum and the particle number.[8] In order to give another derivation of formula (2.1), we begin by writing formula (7.37) for the density in the form

$$\frac{N}{V} = 2i \int \left[\frac{\partial}{\partial \varepsilon} \ln G(p) - G(p) \frac{\partial}{\partial \varepsilon} \Sigma(p)\right] e^{i\varepsilon t} \frac{d^4p}{(2\pi)^4}, \ (t \to +0), \quad (19.12)$$

where Σ is the exact irreducible self-energy part $[G = (\varepsilon - \xi - \Sigma)^{-1}]$. In the expression on the right, the integral of the second term in brackets vanishes. To see this, we first prove that $\Sigma(p)$ can be regarded as the variational derivative with respect to $G(p)$ of some functional X:

$$\delta X = \int \Sigma(p) \, \delta G(p) \frac{d^4p}{(2\pi)^4}. \quad (19.13)$$

Suppose we calculate the variational derivative

$$\frac{\delta \Sigma(p)}{\delta G(q)}.$$

To this end, we consider the diagrams making up Σ, and in each such diagram we vary all the G-lines in turn. For example, consider the self-energy part shown in Fig. 10(c), p. 76. If we direct our attention to each of the three G-lines in turn, each time denoting the corresponding energy and momentum by ε_q and q, it is not hard to see that the result will equal the G-function in

---

[8] The derivation given in this section is mainly due to Luttinger and Ward (L17).

question multiplied by a sum of two diagrams, which simply represent the second approximation to the function $\Gamma^{(1)}(p, q)$ introduced in Sec. 18, i.e., to the part of the function $\Gamma(p, q)$ which does not contain "anomalous elements" with two identical G-functions. Applying this procedure to all the diagrams making up Σ, we obtain

$$\delta\Sigma(p) = -\frac{i}{2}\int \Gamma^{(1)}_{\alpha\beta,\,\alpha\beta}(p, q)\,\delta G(q)\,\frac{d^4q}{(2\pi)^4},$$

which implies

$$\frac{\delta\Sigma(p)}{\delta G(q)} = -\frac{i}{2}\,\Gamma^{(1)}_{\alpha\beta,\,\alpha\beta}(p, q). \tag{19.14}$$

This quantity is symmetric with respect to the permutation $p \rightleftarrows q$, which, as is well known, is a sufficient condition for the existence of the functional X.

The functional X can be represented as a set of diagrams containing only exact G-lines. It is clear from formula (19.13) that these diagrams are obtained from the diagrams for Σ (skeleton diagrams with exact G-lines) if we use an exact G-line to "close" every such diagram. To obtain formula (19.13) with the correct normalization, we have to introduce a numerical coefficient, depending on the type of diagram (for example, if there is only one type of interaction, this coefficient equals $1/n$, where n is the number of vertices[9]). The diagrams making up X do not change if all the frequencies in the G-lines are shifted by a small amount $\delta\varepsilon^{(0)}$, since the limits of the integrations over frequency are $-\infty$ and ∞, and the δ-functions assigned to the vertices contain just as many ε_i with plus signs as with minus signs. It follows that

$$\frac{\delta X}{\delta\varepsilon^{(0)}} = \int \Sigma(p)\frac{\partial G(p)}{\partial\varepsilon}\frac{d^4p}{(2\pi)^4} = 0.$$

Returning to the expression (19.12), we can write the integral of the second term in brackets in the form

$$-2i\int G(p)\frac{\partial}{\partial\varepsilon}\Sigma(p)e^{i\varepsilon t}\frac{d^4p}{(2\pi)^4}$$
$$= -2i\int G(p)\Sigma(p)\frac{d\mathbf{p}}{(2\pi)^4}\Big|_{-\infty}^{\infty} + 2i\int\Sigma(p)\frac{\partial G(p)}{\partial\varepsilon}\frac{d^4p}{(2\pi)^4},$$

where the second term on the right vanishes. As for the first term, according to formula (7.21)

$$G(\mathbf{p}, \varepsilon) \approx G^{(0)}(\mathbf{p}, \varepsilon) \approx \frac{1}{\varepsilon},$$

as $\varepsilon \to \infty$, i.e., $\Sigma(p)$ cannot grow in proportion to ε, and $G(p)\Sigma(p) \to 0$ as $\varepsilon \to \infty$. Thus, we have finally proved that the integral of the second term in brackets in (19.12) vanishes, and hence

$$\frac{N}{V} = 2i\int\frac{\partial}{\partial\varepsilon}\ln G(p)e^{i\varepsilon t}\frac{d^4p}{(2\pi)^4}. \tag{19.15}$$

[9] Compare with the quantity Ω' on p. 142.

As already noted in Sec. 7, the G-function is not analytic, but the function $G_R(\varepsilon)$, which equals $G(\varepsilon)$ for $\varepsilon > 0$ and $G^*(\varepsilon)$ for $\varepsilon < 0$, is analytic in the upper half-plane. It can also be shown that the function G_R has no zeros in the upper half-plane.[10] It follows that

$$\frac{N}{V} = 2i \int_{-\infty}^{\infty} \frac{\partial}{\partial \varepsilon} \ln G_R(p) e^{i\varepsilon t} \frac{d^4p}{(2\pi)^4} + 2i \int_{-\infty}^{0} \frac{d\varepsilon}{2\pi} \int \frac{d\mathbf{p}}{(2\pi)^3} \frac{\partial}{\partial \varepsilon} \ln \frac{G(p)}{G^*(p)}$$

$$= 2i \int_{-\infty}^{0} \frac{d\varepsilon}{2\pi} \int \frac{d\mathbf{p}}{(2\pi)^3} \frac{\partial}{\partial \varepsilon} \ln \frac{G(p)}{G^*(p)} = \frac{i}{\pi} \int \frac{d\mathbf{p}}{(2\pi)^3} \ln \frac{G(p)}{G^*(p)} \Bigg|_{-\infty}^{0}$$

(in the integral involving G_R, the contour can be shifted to the region $\operatorname{Im} \varepsilon = \infty$, and then the integral vanishes).

Letting φ denote the phase of the G-function, we have

$$\frac{N}{V} = -\frac{2}{\pi} \int [\varphi(0) - \varphi(-\infty)] \frac{d\mathbf{p}}{(2\pi)^3}.$$

We now examine how the phase of the G-function changes in going from $\varepsilon = 0$ to $\varepsilon = -\infty$. As we know from Sec. 7, $\operatorname{Im} G > 0$ for $\varepsilon < 0$, and $\operatorname{Im} G = 0$ at the point $\varepsilon = 0$. Moreover, as $\varepsilon \to -\infty$, $\operatorname{Im} G$ falls off more rapidly than $\operatorname{Re} G$, with $\operatorname{Re} G \approx 1/\varepsilon < 0$. Since $\operatorname{Im} G$ does not change sign, the point $(\operatorname{Re} G, \operatorname{Im} G)$ in the complex G-plane (for a given ε) moves around only in the upper half-plane, i.e., the phase can only vary from 0 to π. Since

$$\frac{\operatorname{Im} G}{\operatorname{Re} G} \to -0$$

as $\varepsilon \to -\infty$, we have $\varphi(-\infty) = \pi$. The value of the phase for $\varepsilon = 0$ is determined by the sign of $\operatorname{Re} G(\mathbf{p}, 0) \equiv G(\mathbf{p}, 0)$, i.e., $\varphi(0) = 0$ if $G(\mathbf{p}, 0) > 0$ and $\varphi(0) = \pi$ if $G(\mathbf{p}, 0) < 0$.

Thus, from formula (19.15) we obtain

$$\frac{N}{V} = 2 \int_{G(\mathbf{p},0)>0} \frac{d\mathbf{p}}{(2\pi)^3}. \tag{19.16}$$

The region $G(\mathbf{p}, 0) > 0$ is bounded by a surface on which the function G either vanishes or becomes infinite. The case where $G(\mathbf{p}, 0)$ vanishes $(\Sigma \to \infty)$ probably corresponds to superconductivity (see Sec. 34), whereas $G(\mathbf{p}, 0)$ can become infinite at the Fermi surface for an ordinary Fermi liquid. In the neighborhood of the Fermi surface (the Fermi sphere $|\mathbf{p}| = p_0$, in the present case),

$$G(\mathbf{p}, 0) \approx -\frac{a}{\xi},$$

where $a > 0$, i.e., the region $G(\mathbf{p}, 0) > 0$ corresponds to $\xi < 0$ (the interior of the Fermi sphere). Finally, evaluating the integral in (19.16), we obtain formula (2.1).

---

[10] This is proved in the same way as it is proved that the complex dielectric constant $\varepsilon(\omega)$ has no zeros in the upper half-plane of the variable ω (see L9, Sec. 62).

19.5. Heat capacity. So far, we have considered the Fermi liquid at $T = 0$. However, it is also of interest to investigate the properties of the Fermi liquid at nonzero temperatures. In the case of low temperatures, it is quite natural to expect that all quantities will be determined by the values of certain basic characteristics of the Fermi liquid at $T = 0$. As an example illustrating this point, we now calculate the heat capacity, using a method of isolating the "temperature correction" which can also be helpful in making other calculations.

Our starting point is the expression (16.12) for the total number of particles in the system, as a function of μ and T:

$$\frac{N(\mu, T)}{V} = 2T \sum_{\varepsilon} \int_{\tau \to +0} \frac{d\mathbf{p}}{(2\pi)^3} \mathscr{G}(\mathbf{p}, \varepsilon) e^{i\varepsilon\tau}. \tag{19.17}$$

From a knowledge of this function, we can determine the entropy by using the thermodynamic relation

$$\left(\frac{\partial N}{\partial T}\right)_{\mu} = \left(\frac{\partial S}{\partial \mu}\right)_{T}. \tag{19.18}$$

Consider the $\mathscr{G}$-function appearing in formula (19.17). If we bear in mind that it is an exact $\mathscr{G}$-function, obtained by summation of all possible diagrams which contain sums over frequencies, it becomes clear that the dependence of $\mathscr{G}(\mathbf{p}, \varepsilon)$ on temperature involves more than its dependence on the discrete variable $\varepsilon = (2n + 1)\pi T$. Therefore, we write $\mathscr{G}(T; \mathbf{p}, \varepsilon)$ instead of $\mathscr{G}(\mathbf{p}, \varepsilon)$. As we know from Chap. 3, the formula

$$\mathscr{G}^{-1}(T; \mathbf{p}, \varepsilon) = (\mathscr{G}^{(0)})^{-1}(\mathbf{p}, \varepsilon) - \Sigma(T; \mathbf{p}, \varepsilon) \tag{19.19}$$

connects the $\mathscr{G}$-function and the self-energy part Σ. It should be noted that the function $\mathscr{G}^{(0)}(\mathbf{p}, \varepsilon)$ depends on the temperature only through the variable ε. If we let $T \to 0$, while holding $\varepsilon = \text{const}$, the relation (19.19) takes the form

$$\mathscr{G}^{-1}(0; \mathbf{p}, \varepsilon) = (\mathscr{G}^{(0)})^{-1}(\mathbf{p}, \varepsilon) - \Sigma(0; \mathbf{p}, \varepsilon). \tag{19.20}$$

From (19.19) and (19.20), we find that

$$\mathscr{G}^{-1}(T; \mathbf{p}, \varepsilon) = \mathscr{G}^{-1}(0; \mathbf{p}, \varepsilon) - [\Sigma(T; \mathbf{p}, \varepsilon) - \Sigma(0; \mathbf{p}, \varepsilon)], \tag{19.21}$$

where the quantity $\Sigma(0; \mathbf{p}, \varepsilon)$ differs from $\Sigma(T; \mathbf{p}, \varepsilon)$ in that all sums over frequencies are replaced by integrals, according to the prescription

$$T \sum_{\varepsilon} \cdots \to \frac{1}{2\pi} \int_{-\infty}^{\infty} d\varepsilon \cdots$$

The calculation of the difference $\Sigma(T) - \Sigma(0)$ at low temperatures goes as follows: We consider two identical diagrams for $\Sigma(T)$ and $\Sigma(0)$, say $\Sigma_1(T)$ and $\Sigma_1(0)$, regarding these diagrams as made up of exact $\mathscr{G}$-lines. Suppose all sums are replaced by integrals in $\Sigma_1(T)$. This diagram already differs from $\Sigma_1(0)$ because of the presence of the functions $\mathscr{G}(T)$ which differ from the functions $\mathscr{G}(0)$ figuring in $\Sigma_1(0)$, and to a first approximation, we need only take account of the difference of one of the functions $\mathscr{G}(T)$ from $\mathscr{G}(0)$, setting all the others equal to $\mathscr{G}(0)$. A second contribution to the difference $\Sigma_1(T) - \Sigma_1(0)$ is due to the presence of sums instead of integrals. The difference between a sum and an integral matters only if the corresponding frequency is small, of order T. The frequency regions which are important to a first approximation are those such that one of the $\mathscr{G}$-lines has a frequency of order T while all the others have much larger frequencies. Therefore, we can single out the $\mathscr{G}$-line with the small frequency, replacing sums by integrals and functions $\mathscr{G}(T)$ by functions $\mathscr{G}(0)$ everywhere else.

Thus, we have to vary the diagram with respect to all the $\mathscr{G}$-lines appearing in it. The required calculation is completely analogous to that made in Sec. 19.4. Applying this procedure to all the diagrams making up Σ, we obtain

$$\begin{aligned}\Sigma(T; \mathbf{p}, \varepsilon) &- \Sigma(0; \mathbf{p}, \varepsilon)\\ &= \frac{1}{2}\int \frac{d\mathbf{q}}{(2\pi)^3}\frac{d\varepsilon}{2\pi}\mathscr{T}^{(1)}_{\alpha\beta,\,\alpha\beta}(0; \mathbf{p}, \varepsilon; \mathbf{q}, \varepsilon_1)\,[\mathscr{G}(T; \mathbf{q}, \varepsilon_1) - \mathscr{G}(0; \mathbf{q}, \varepsilon_1)]\\ &\quad + \frac{1}{2}\left[T\sum_{\varepsilon_1} - \frac{1}{2\pi}\int d\varepsilon_1\right]\int \frac{d\mathbf{q}}{(2\pi)^3}\mathscr{T}^{(1)}_{\alpha\beta,\,\alpha\beta}(0; \mathbf{p}, \varepsilon; \mathbf{q}, \varepsilon_1)\mathscr{G}(0; \mathbf{q}, \varepsilon_1).\end{aligned}$$

Using formula (19.21) and solving the resulting equation for $\Sigma(T) - \Sigma(0)$ [to do this, it suffices to form an equation for $\mathscr{T}$ similar to (18.3)], we find that

$$\begin{aligned}\Sigma(T; \mathbf{p}, \varepsilon) &- \Sigma(0; \mathbf{p}, \varepsilon)\\ &\approx \frac{1}{2}\left[T\sum_{\varepsilon_1} - \frac{1}{2\pi}\int d\varepsilon_1\right]\int \frac{d\mathbf{q}}{(2\pi)^3}\mathscr{T}_{\alpha\beta,\,\alpha\beta}(0; \mathbf{p}, \varepsilon; \mathbf{q}, \varepsilon_1)\mathscr{G}(0; \mathbf{q}, \varepsilon_1),\end{aligned} \tag{19.22}$$

which, together with (19.21), gives the first approximation to $\mathscr{G}(T)$:

$$\begin{aligned}\mathscr{G}(T; \mathbf{p}, \varepsilon) &= \mathscr{G}(0; \mathbf{p}, \varepsilon) + \frac{1}{2}\mathscr{G}^2(0; \mathbf{p}, \varepsilon)\\ &\quad\times\left[T\sum_{\varepsilon_1} - \frac{1}{2\pi}\int d\varepsilon_1\right]\int \frac{d\mathbf{q}}{(2\pi)^3}\mathscr{T}_{\alpha\beta,\,\alpha\beta}(0; \mathbf{p}, \varepsilon; \mathbf{q}, \varepsilon_1)\mathscr{G}(0; \mathbf{q}, \varepsilon_1).\end{aligned}$$

Substituting this expression into (19.19) and subtracting $N(\mu, 0)/V$ from the result, we can obviously write

$$\begin{aligned}&\frac{1}{V}[N(\mu, T) - N(\mu, 0)]\\ &= 2\left[T\sum_{\varepsilon} - \frac{1}{2\pi}\int d\varepsilon\right]\mathscr{G}(0; \mathbf{p}, \varepsilon) + \frac{1}{2\pi}\int d\varepsilon\frac{d\mathbf{p}}{(2\pi)^3}\mathscr{G}^2(0; \mathbf{p}, \varepsilon)\left[T\sum_{\varepsilon_1} - \frac{1}{2\pi}\int d\varepsilon_1\right]\\ &\quad\times\int \frac{d\mathbf{q}}{(2\pi)^3}\mathscr{T}_{\alpha\beta,\,\alpha\beta}(0; \mathbf{p}, \varepsilon; \mathbf{q}, \varepsilon_1)\mathscr{G}(0; \mathbf{q}, \varepsilon_1) = 2\left[T\sum_{\varepsilon} - \frac{1}{2\pi}\int d\varepsilon\right]\\ &\quad\times\int \frac{d\mathbf{p}}{(2\pi)^3}\mathscr{G}(0; \mathbf{p}, \varepsilon)\left[1 + \frac{1}{2}\int\frac{d\varepsilon_1}{2\pi}\int \frac{d\mathbf{q}}{(2\pi)^3}\mathscr{T}_{\alpha\beta,\,\alpha\beta}(0; \mathbf{p}, \varepsilon; \mathbf{q}, \varepsilon_1)\mathscr{G}^2(0; \mathbf{q}, \varepsilon_1)\right]\end{aligned} \tag{19.23}$$

to within terms of the first order. In the last equality, we have used the symmetry property

$$\mathscr{T}_{\alpha\beta,\,\alpha\beta}(\mathbf{p}, \varepsilon; \mathbf{q}, \varepsilon_1) = \mathscr{T}_{\alpha\beta,\,\alpha\beta}(\mathbf{q}, \varepsilon_1; \mathbf{p}, \varepsilon).$$

This expression can be written in a somewhat different form. In just the same way as we derived formula (19.4), we can deduce a corresponding formula in the "temperature technique":

$$\frac{\partial}{\partial\mu}\mathscr{G}^{-1}(T;\mathbf{p},\varepsilon) = 1 + \frac{1}{2}T\sum_{\varepsilon_1}\int\frac{d\mathbf{q}}{(2\pi)^3}\mathscr{T}_{\alpha\beta,\alpha\beta}(T;\mathbf{p},\varepsilon;\mathbf{q},\varepsilon_1)\mathscr{G}^2(T;\mathbf{q},\varepsilon_1).$$

In the limit $T \to 0$, with $\varepsilon = \text{const}$, we obtain

$$\frac{\partial}{\partial\mu}\mathscr{G}^{-1}(0;\mathbf{p},\varepsilon) = 1 + \frac{1}{2}\int\frac{d\varepsilon_1}{2\pi}\int\frac{d\mathbf{q}}{(2\pi)^3}\mathscr{T}_{\alpha\beta,\alpha\beta}(0;\mathbf{p},\varepsilon;\mathbf{q},\varepsilon_1)\mathscr{G}^2(0;\mathbf{q},\varepsilon_1). \tag{19.24}$$

Substitution of (19.24) into (19.23) gives

$$\begin{aligned}\frac{1}{V}[N(\mu,T) - N(\mu,0)] &= 2\left[T\sum_{\varepsilon} - \frac{1}{2\pi}\int d\varepsilon\right]\int\frac{d\mathbf{p}}{(2\pi)^3}\mathscr{G}(0;\mathbf{p},\varepsilon)\frac{\partial}{\partial\mu}\mathscr{G}^{-1}(0;\mathbf{p},\varepsilon)\\ &= -2\left[T\sum_{\varepsilon} - \frac{1}{2\pi}\int d\varepsilon\right]\int\frac{d\mathbf{p}}{(2\pi)^3}\frac{\partial}{\partial\mu}\ln\mathscr{G}(0;\mathbf{p},\varepsilon).\end{aligned} \tag{19.25}$$

Differentiating (19.25) with respect to the temperature, while holding μ constant, and comparing the result with (19.18), we find that

$$\frac{S}{V} = -2\frac{\partial}{\partial T}\left[T\sum_{\varepsilon}\int\frac{d\mathbf{p}}{(2\pi)^3}\ln\mathscr{G}(0;\mathbf{p},\varepsilon)\right]. \tag{19.26}$$

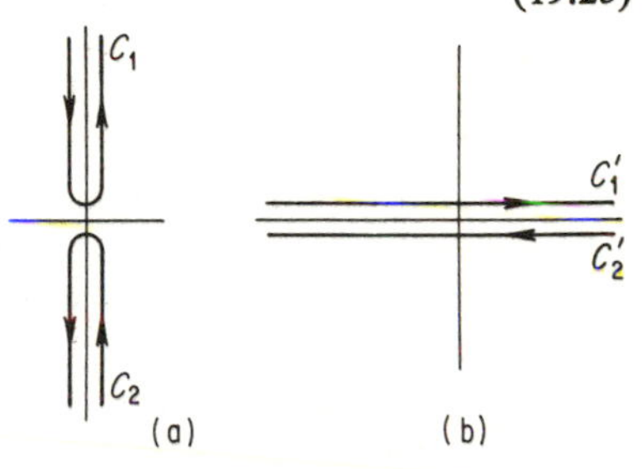

FIGURE 59

The rest of the calculation goes as follows: Using the relation between the temperature Green's function and the retarded and advanced Green's functions, we write the expression (19.26) for the entropy as a sum of two contour integrals

$$\begin{aligned}\frac{S}{V} &= -2\frac{\partial}{\partial T}\left\{\int\frac{d\mathbf{p}}{(2\pi)^3}\left(\frac{1}{4\pi i}\int_{C_1}\left[\tanh\frac{\varepsilon}{2T}\right]\ln G_R(\mathbf{p},\varepsilon)\,d\varepsilon\right.\right.\\ &\qquad\left.\left.+\frac{1}{4\pi i}\int\left[\tanh\frac{\varepsilon}{2T}\right]\ln G_A(\mathbf{p},\varepsilon)\,d\varepsilon\right)\right\}\\ &= 2\int\frac{d\mathbf{p}}{(2\pi)^3}\frac{1}{2\pi iT}\left(\int_{C_1}\varepsilon\left[-\frac{\partial n_F(\varepsilon)}{\partial\varepsilon}\right]\ln G_R(\mathbf{p},\varepsilon)\,d\varepsilon\right.\\ &\qquad\left.+\int_{C_2}\varepsilon\left[-\frac{\partial n_F}{\partial\varepsilon}\right]\ln G_A(\mathbf{p},\varepsilon)\,d\varepsilon\right),\end{aligned}$$

where n_F is the Fermi distribution function, and the contours C_1 and C_2 are shown in Fig. 59(a). The function G_R has no zeros in the upper half-plane, and G_A has no zeros in the lower half-plane (see footnote 10, p. 168). Using this property, as well as the analyticity of G_R and G_A in the appropriate half-planes and the fact that $\partial n_F/\partial\varepsilon$ falls off rapidly as $\varepsilon \to \pm\infty$, we can "unfold" the contours C_1 and C_2 along the real axis, as shown in Fig. 59(b), thereby obtaining

$$\frac{S}{V} = 2\int \frac{d\mathbf{p}}{(2\pi)^3}\frac{1}{2\pi i T}\int_{-\infty}^{\infty} \varepsilon\left[-\frac{\partial n_F(\varepsilon)}{\partial \varepsilon}\right][\ln G_R(\mathbf{p}, \varepsilon) - \ln G_A(\mathbf{p}, \varepsilon)]\, d\varepsilon.$$

Applying the usual rule for evaluating integrals involving the Fermi distribution,[11] we find that

$$\begin{aligned}\frac{S}{V} &= \frac{2\pi^2 T}{3}\frac{1}{2\pi i}\int \frac{d\mathbf{p}}{(2\pi)^3}\left[G_R^{-1}\frac{\partial G_R}{\partial \varepsilon} - G_A^{-1}\frac{\partial G_A}{\partial \varepsilon}\right]_{\varepsilon=0} \qquad (19.27)\\ &= \frac{2\pi^2 T}{3}\frac{1}{2\pi}\int \frac{d\mathbf{p}}{(2\pi)^3}\, 2\,\mathrm{Im}\left[G_R^{-1}\frac{\partial G_R}{\partial \varepsilon}\right]_{\varepsilon=0}\end{aligned}$$

The last equality stems from the fact that $G_R^* = G_A$ on the real axis. It is not hard to see that the integral with respect to $\mathbf{p}$ comes mostly from the region near the Fermi surface. Making the substitution

$$G_R \approx \frac{a}{\varepsilon - \xi + i\delta},$$

we obtain

$$\frac{S}{V} = \frac{p_0 m^*}{3} T. \qquad (19.28)$$

It is clear that here the heat capacity equals the entropy.

We observe that in our calculation of the temperature correction, all that mattered was the neighborhood of the point $\xi = 0$, $\varepsilon = 0$, i.e., the real poles of the function G_R (or G) for $T = 0$. It seems that this is the general state of affairs for arbitrary temperature corrections. In other words, to a first approximation, the temperature corrections must always be determined by the poles of the functions G (or Γ) for $T = 0$, i.e., by the spectrum of the elementary excitations.

20. Singularities of the Vertex Part When the Total Momentum of the Colliding Particles is Small[12]

In addition to the singularity when the energy-momentum transfer is small, the vertex part has another singularity, which, as we shall see later in Chap. 7, is of interest in the theory of superconductivity. Suppose the energies ε_1, ε_2 and the sum of the momenta $\mathbf{p}_1 + \mathbf{p}_2 = \mathbf{s}$ are small. Then it is not hard to see that of the three diagrams shown in Fig. 57, diagram (a) represents a special case in that the two G-functions in the integrand have poles which lie close together. We deal with this vertex part in the same way as in Sec. 18.

---

[11] $$\int_{-\infty}^{\infty} f(\varepsilon)\frac{\partial n_F(\varepsilon)}{\partial \varepsilon}\, d\varepsilon = -f(0) - \frac{\pi^2}{6}T^2\left[\frac{\partial^2 f}{\partial \varepsilon^2}\right]_{\varepsilon=0} + \cdots$$

[12] This section is based on unpublished work of Abrikosov, Gorkov, Khalatnikov and Landau.

We write

$$\Gamma_{\alpha\beta,\gamma\delta}(p_1, p_3; s) = \Gamma_{\alpha\beta,\gamma\delta}(p_1, -p_1 + s; p_3, -p_3 + s),$$

and we let $\Gamma^{(2)}$ denote the sum of all "nonanomalous" diagrams. In $\Gamma^{(2)}$, s can be set equal to zero. To obtain the complete Γ, we have to sum all "ladder diagrams" of the type shown in Fig. 58. Before doing this, we differentiate Γ with respect to the fourth component of s, which we denote by λ. Each ladder diagram then leads to a sum of terms, where one "rung" is differentiated in each term. If the differentiated rung is held fixed, it is easily seen that all diagrams can be divided into two independent ladders, one going to the left and the other to the right, where the sum of each ladder is the exact vertex part. In this way, we obtain the equation

$$\frac{\partial}{\partial\lambda}\Gamma_{\alpha\beta,\gamma\delta}(p_1, p_3; s) \tag{20.1}$$
$$= \frac{i}{2}\int \Gamma_{\alpha\beta,\xi\eta}(p_1, q; s)G(q)\left[\frac{\partial}{\partial\lambda}G(-q + s)\right]\Gamma_{\xi\eta,\gamma\delta}(q, p_3; s)\frac{d^4q}{(2\pi)^4}.$$

Near $|\mathbf{q}| = p_0$, $\varepsilon = 0$, the expression

$$G(q)\frac{\partial}{\partial\lambda}G(-q + s)$$

in the integrand of (20.1) has the form

$$\begin{aligned} &-\frac{a^2}{[\varepsilon - v(|\mathbf{q}| - p_0) + i\delta \operatorname{sgn}(|\mathbf{q}| - p_0)]} \\ &\times \frac{1}{[\varepsilon - \lambda + v(|\mathbf{q} - \mathbf{s}| - p_0) - i\delta \operatorname{sgn}(|\mathbf{q} - \mathbf{s}| - p_0)]^2} \end{aligned} \tag{20.2}$$

It is clear from (20.2) that the integral with respect to

$$\frac{d^4q}{(2\pi)^4} \to \frac{p_0^2 d|\mathbf{q}|\,d\Omega\,d\varepsilon}{(2\pi)^4}$$

comes mostly from the neighborhood of the point $|\mathbf{q}| = p_0, \varepsilon = 0$. Assuming that the function Γ in the integrand is not rapidly varying in this region, we can regard Γ as a constant when integrating with respect to $d|\mathbf{q}|$ and $d\varepsilon$. Then we need only integrate the expression (20.2), obtaining

$$\begin{aligned} &\frac{\partial}{\partial\lambda}\Gamma_{\alpha\beta,\gamma\delta}(p_1, p_3; s) \\ &= \frac{a^2p_0^2}{2(2\pi)^3 v}\int \Gamma_{\alpha\beta,\xi\eta}(p_1, q; s)\Gamma_{\xi\eta,\gamma\delta}(q, p_3; s)\frac{\lambda}{\lambda^2 - (\mathbf{v}\cdot\mathbf{s})^2 + i\delta}\,d\Omega. \end{aligned} \tag{20.3}$$

We now take the gradient of Γ with respect to $\mathbf{s}$. In differentiating the second of the G-functions, it must not be forgotten that $\mathbf{v}\cdot\mathbf{s}$ not only appears directly in the denominator, but also determines the sign of the imaginary part. Therefore, near $|\mathbf{q}| = p_0$, $\varepsilon = 0$, the expression

$$G(q)\nabla_s G(-q + s)$$

has the form

$$-\frac{a^2\mathbf{v}}{[\varepsilon - v(|\mathbf{q}| - p_0) + i\delta\,\mathrm{sgn}\,(|\mathbf{q}| - p_0)]}$$
$$\times \frac{1}{\left\{\varepsilon - \lambda + v(|\mathbf{q}| - p_0) - \mathbf{v}\cdot\mathbf{s} - i\delta\,\mathrm{sgn}\left[|\mathbf{q}| - p_0 - \dfrac{(\mathbf{v}\cdot\mathbf{s})}{v}\right]\right\}^2}$$
$$+ \frac{2\pi i a^2\delta(\varepsilon - \lambda)\delta[\mathbf{v}\cdot\mathbf{s} - v(|\mathbf{q}| - p_0)]\mathbf{v}}{\lambda - \mathbf{v}\cdot\mathbf{s} + i\delta\,\mathrm{sgn}\,(\mathbf{v}\cdot\mathbf{s})}.$$

The integral of this expression with respect to $d|\mathbf{q}|d\varepsilon$ leads to the following relation involving Γ:

$$\nabla_s\Gamma_{\alpha\beta,\gamma\delta}(p_1, p_3; s) = -\frac{a^2p_0^2}{2(2\pi)^3v}\int \Gamma_{\alpha\beta,\xi\eta}(p_1, q; s)\Gamma_{\xi\eta,\gamma\delta}(q, p_3; s)\frac{\mathbf{v}(\mathbf{v}\cdot\mathbf{s})}{\lambda^2 - (\mathbf{v}\cdot\mathbf{s})^2 + i\delta}\,d\Omega. \tag{20.4}$$

Combining equations (20.3) and (20.4), we obtain

$$\left(\lambda\frac{\partial}{\partial\lambda} + \mathbf{s}\cdot\nabla_s\right)\Gamma_{\alpha\beta,\gamma\delta}(p_1, p_3; s) = \frac{a^2p_0^2}{2(2\pi)^3v}\int \Gamma_{\alpha\beta,\xi\eta}(p_1, q; s)\Gamma_{\xi\eta,\gamma\delta}(q, p_3; s)\,d\Omega. \tag{20.5}$$

If we neglect magnetic interactions, the forces between the particles depend only on the relative orientation of the spins. Then, if we assume that the total spin is conserved when the particles interact, equation (20.5) separates into two independent equations. One of these equations corresponds to interaction between two particles with opposite spins (e.g., $\alpha = \frac{1}{2}$, $\beta = -\frac{1}{2}$), and the other to interaction between two particles with parallel spins. These equations are completely identical, the only difference coming from the initial conditions (i.e., from $\Gamma^{(2)}$). Therefore, from now on we write just Γ, with the understanding that Γ denotes either of the two different components.

We now make the assumption (to be justified later) that $\Gamma(p_1, p_2; s)$ does not depend on the angles $\angle(\mathbf{p}_1, \mathbf{s})$ and $\angle(\mathbf{p}_3, \mathbf{s})$. Then this quantity can be expanded in Legendre polynomials depending on $\cos\theta$, where θ is the angle between $\mathbf{p}_1$ and $\mathbf{p}_3$:

$$\Gamma(p_1, p_3; s) = \sum_l \Gamma_l P_l\,(\cos\theta). \tag{20.6}$$

We note that $\Gamma_{\alpha\beta,\xi\eta}(p_1, p_3; s)$ is antisymmetric or symmetric with respect to the spins depending on whether the spins are antiparallel or parallel. Permuting the momenta of the original particles corresponds to replacing $\cos\theta$ by $-\cos\theta$. Since $\Gamma_{\alpha\beta,\xi\eta}(p_1, p_2; p_3, p_4)$ must be antisymmetric with respect to the permutation $p_1, \alpha \rightleftarrows p_2, \beta$, only odd harmonics appear in the expansion (20.6) when the spins are parallel, and only even harmonics appear when the spins are antiparallel. The equations (20.5) for the individual Γ_l separate and take the form

$$\left(\lambda \frac{\partial}{\partial \lambda} + \mathbf{s}\cdot\nabla_{\mathbf{s}}\right)\Gamma_l = \frac{4\pi a^2 p_0^2}{(2\pi)^3 v}\frac{\Gamma_l^2}{2l+1} \tag{20.7}$$

[the factor of 2 comes from taking the sum over the spins in (20.5)]. The solution of equation (20.7) is

$$\Gamma_l(\lambda, |\mathbf{s}|) = -\frac{(2\pi)^3 v(2l+1)}{2\pi a^2 p_0^2}\frac{1}{\ln(\lambda|\mathbf{s}|) + f_l(\lambda/|\mathbf{s}|)}, \tag{20.8}$$

where $f_l(x)$ is an arbitrary function.

Using (20.3) and (20.4), we can obtain limiting expressions for $f_l(x)$ as $x \to 0$ and $x \to \infty$. Consider equation (20.3) in the limit $|\mathbf{s}| \to 0$. Expanding Γ in spherical harmonics, we find that

$$\lambda \frac{d\Gamma_l}{d\lambda} = \frac{4\pi a^2 p_0^2}{(2\pi)^3 v(2l+1)}\Gamma_l^2. \tag{20.9}$$

The solution of this equation is

$$\Gamma_l(\lambda, 0) = -\frac{(2\pi)^3 v(2l+1)}{4\pi a^2 p_0^2}\frac{1}{\ln(\lambda/c_1^l)}, \tag{20.10}$$

where c_1 is a constant. Thus,

$$f_l(x) \to \ln \frac{x}{(c_1^l)^2}$$

as $x \to \infty$. Similarly, letting $\lambda \to 0$ in equation (20.4), we obtain

$$\Gamma_l(0, |\mathbf{s}|) = -\frac{(2\pi)^3 v(2l+1)}{4\pi a^2 p_0^2}\cdot\frac{1}{\ln(v|\mathbf{s}|/c_2^l)}, \tag{20.11}$$

and hence

$$f_l(x) \to \ln \frac{v^2}{(c_2^l)^2 x}$$

as $x \to 0$. The constants c_1^l and c_2^l have the dimension of energy and can be complex. For example, if the constant c_1^l is of the order of the Fermi energy, it follows from (20.10) that $\Gamma_l(\lambda, 0)$ goes to zero like $[\ln(c_1^l/\lambda)]^{-1}$ as $\lambda \to 0$. However, the special case where this constant is small can also occur, and then $\Gamma_l(\lambda, 0)$ has a pole for some value of λ (usually complex). The significance of this fact is intimately connected with the phenomenon of superconductivity, and will be explained in Chap. 7.

With this investigation of Γ in the case where the total momentum of the colliding particles is small, we conclude our microscopic analysis of the isotropic Fermi liquid with arbitrary short-range interactions between the particles. At present, only one isotropic Fermi liquid is known, i.e., liquid He^3. The class of anisotropic Fermi liquids, i.e., electrons in metals, is much larger. However, in addition to its anisotropy, the "electron liquid" in metals has such specific features as long-range Coulomb forces, interaction with lattice vibrations, etc. Some of these complications will be considered in the next two sections, where we consider first the isotropic model of electron-phonon interactions and then the degenerate plasma.

There is no question that a general study of the electron Fermi liquid is one of the most important problems which ought to be solved by using the methods of quantum field theory. In particular, a deeper study of the phenomenon of superconductivity is needed, since so far superconductivity has only been studied by using the simplest model (see Chap. 7).

21. Electron-Phonon Interactions[13]

We now use the isotropic model to study the interaction of electrons and phonons in a metal. In this model, superconductivity will occur at sufficiently low temperatures, i.e., at temperatures below a *critical temperature* T_c (see Chap. 7). We assume that the temperature T of the metal lies in the range

$$T_c \ll T \ll \omega_D, \tag{21.1}$$

where ω_D is the Debye frequency of the phonons. Of course, our model differs drastically from the actual state of affairs in a normal metal, since not only are real metals anisotropic, but there is also Coulomb interaction between the electrons. At the end of the section, we shall discuss the qualitative changes that arise because of this Coulomb interaction.

21.1. The vertex part. In Sec. 16 we derived Dyson's equations for the electron and phonon Green's functions in the temperature technique. We now begin by considering the vertex part $\mathscr{T}$ which appears in these equations. It will be shown that $\mathscr{T}$ differs from its zeroth-order approximation, equal to g, by small quantities of order $\sqrt{m/M}$, where m and M are the masses of the electron and nucleus, respectively. Suppose we calculate the first-order correction to $\mathscr{T}$ shown in Fig. 55, p. 138, assuming that an electron has momentum of order p_0, while the momentum k and energy ω of a phonon are such that[14]

$$k \lesssim k_D \sim p_0, \qquad \omega \lesssim \omega_D.$$

These are just the values of the energy and momenta which will be most important in what follows.

According to Sec. 14, the contribution of the first-order diagram equals

$$\mathscr{T}^{(1)}(p+q, p; q) = -g^3 T \sum_{\varepsilon_1} \int \frac{d\mathbf{p}_1}{(2\pi)^3} \mathscr{D}^{(0)}(p - p_1)\mathscr{G}^{(0)}(p_1 + q)\mathscr{G}^{(0)}(p_1), \tag{21.2}$$

[13] Our treatment is based on the work of Migdal (M6) and Eliashberg (E4).

[14] In this section, the phonon four-momentum in the temperature technique is denoted by $q = (\mathbf{k}, \omega)$, and the quantity $|\mathbf{k}|$ is written simply as k.

where $\mathscr{G}^{(0)}$ and $\mathscr{D}^{(0)}$ are given by the formulas

$$\mathscr{G}^{(0)}(p) = \frac{1}{i\varepsilon - \varepsilon_0(\mathbf{p}) + \mu}, \tag{21.3}$$

with

$$\varepsilon = (2n+1)\pi T, \quad \varepsilon_0(\mathbf{p}) = \frac{\mathbf{p}^2}{2m}, \quad \mu = \frac{p_0^2}{2m},$$

and

$$\mathscr{D}^{(0)}(q) = -\frac{\omega_0^2(\mathbf{k})}{\omega^2 + \omega_0^2(\mathbf{k})} \tag{21.4}$$

We shall assume that the phonon spectrum is bounded, i.e., that $k < k_D \sim p_0$, and that $\mathscr{D}^{(0)}$ vanishes for larger momenta. Physically, this restriction means that there are no vibrations with wavelengths less than the interatomic distances.

For $|\mathbf{p}_1 - \mathbf{p}| \sim p_0$ and $|\varepsilon - \varepsilon_1| \ll \omega_D$, the function $\mathscr{D}^{(0)}$ in the integrand of (21.2) can be replaced by a constant:

$$\mathscr{D}^{(0)} = -1.$$

In the region $|\varepsilon - \varepsilon_1| \gg \omega_D$, the function $\mathscr{D}^{(0)}$ falls off like $(\varepsilon - \varepsilon_1)^{-2}$. This allows us to evaluate the integral for $\mathscr{T}^{(1)}$:

$$\mathscr{T}^{(1)} \approx -g^3 T \sum_{|\varepsilon - \varepsilon_1| < \omega_D} \int_{|\mathbf{p} - \mathbf{p}_1| < k_D} \frac{d\mathbf{p}_1}{(2\pi)^3} \mathscr{G}^{(0)}(p_1 + q)\mathscr{G}^{(0)}(p_1). \tag{21.5}$$

The summation over ε_1 gives a factor

$$\omega_D \sim up_0 \sim \mu \sqrt{\frac{m}{M}},$$

where u is the velocity of sound. As a result, $\mathscr{T}^{(1)}$ is of order

$$\mathscr{T}^{(1)} \sim g^3 \frac{p_0^2}{v} \sqrt{\frac{m}{M}} \sim g\eta \sqrt{\frac{m}{M}}$$

(see Sec. 9), where $v = p_0/m$ and $\eta \sim 1$, provided that the integration with respect to p_1 does not introduce factors of order ω_D^{-1}. In this regard, the only suspicious case is when the energy-momentum transfer is small, i.e., when $k \ll p_0$ and $\omega \ll \omega_D$. In this case, the poles of the two $\mathscr{G}$-functions in the integral for $\mathscr{T}^{(1)}$ approach each other, and it is easy to see that the integral over the momenta comes mainly from the region near $|\mathbf{p}| = p_0$, while the summation over the energies comes mainly from the region $\varepsilon \ll \mu$. Then $\mathscr{G}^{(0)}$ can be replaced by

$$\mathscr{G}^{(0)} = \frac{1}{i\varepsilon - v(|\mathbf{p}| - p_0)}. \tag{21.6}$$

The situation here is very reminiscent of the situation considered in Sec. 18, but there is an important difference. As in Sec. 18, the product of the two $\mathscr{G}^{(0)}$-functions has a sharp maximum near $|\mathbf{p}_1| = p_0$ for small ε_1. The sum and integral of the product of the two $\mathscr{G}$-functions taken between infinite limits is formally divergent, and hence depends in an essential way on the order of integration. In Sec. 18, we integrated first with respect to frequencies and then with respect to $\xi = v(|\mathbf{p}| - p_0)$. This was done because the integral with respect to ε actually extends over the infinite interval, while the integral with respect to ξ is essentially confined to the region $|\xi| \ll \mu$. In the present case, according to (21.3), the presence of the $\mathscr{D}^{(0)}$-function in the integral makes the integral convergent. Therefore, we can first sum over frequencies and afterwards integrate with respect to ξ_1, or vice versa; actually, it is more convenient to first integrate with respect to ξ_1 and then sum over ε_1. As a result, we obtain

$$-\frac{p_0^2}{(2\pi)^3 v}\int d\Omega\,\frac{i\omega}{\mathbf{v}\cdot\mathbf{k} - i\omega}.$$

This quantity fails to be small only when $\omega \gtrsim vk$, and as we shall see below, this case is not important for our subsequent considerations.

The estimate just made is not changed by taking account of higher-order diagrams, and hence

$$\mathscr{T} = g\left[1 + O\left(\sqrt{\frac{m}{M}}\right)\right]. \tag{21.7}$$

21.2. The phonon Green's function. We now find the phonon Green's function, using the second of Dyson's equations (16.3). Let

$$\mathscr{K}(q) = 2g^2T\sum_{\varepsilon}\int\frac{d\mathbf{p}}{(2\pi)^3}\,\mathscr{G}(p)\mathscr{G}(p-q) \tag{21.8}$$

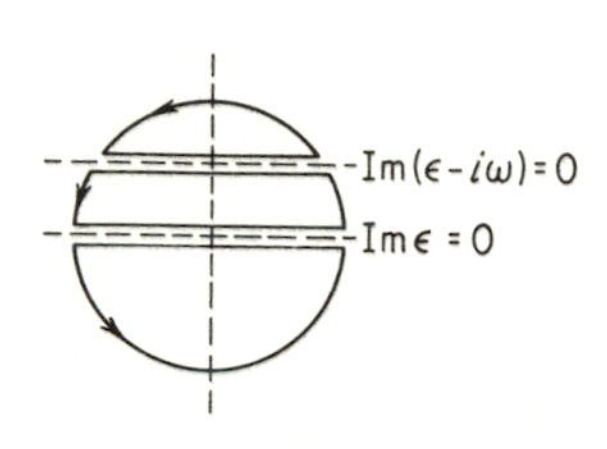

FIGURE 60

be the irreducible self-energy part appearing in this equation. We now make an analytic continuation of $\mathscr{K}$ into the region of real frequencies. Suppose the phonon frequency $\omega = 2n\pi T$ is positive. Then the sum over ε in (21.8) can be represented in the form of a contour integral

$$\mathscr{K}(q) = \frac{g^2}{2\pi i}\int_C d\varepsilon\,\tanh\frac{\varepsilon}{2T}\int\frac{d\mathbf{p}}{(2\pi)^3}\,G_{RA}(\mathbf{p},\varepsilon)G_{RA}(\mathbf{p}-\mathbf{k},\varepsilon - i\omega),$$

where the contour of integration is shown in Fig. 60. The Green's functions have singularities on the horizontal lines

$$\operatorname{Im}\varepsilon = 0, \qquad \operatorname{Im}(\varepsilon - i\omega) = 0.$$

Therefore, the function $G_{RA}(\varepsilon)$ in the integrand should be interpreted as $G_R(\varepsilon)$ for $\operatorname{Im}\varepsilon > 0$ and as $G_A(\varepsilon)$ for $\operatorname{Im}\varepsilon < 0$. Since the integral along the large circle vanishes, only the integrals along the lines $\operatorname{Im}\varepsilon = 0$ and $\operatorname{Im}\varepsilon = \omega$ remain. In the latter integral, we make the change of variables

$$\varepsilon - i\omega \to \varepsilon,$$

using the fact that

$$\tanh\left(\frac{\varepsilon}{2T} + n\pi i\right) = \tanh\frac{\varepsilon}{2T}.$$

Moreover, we observe that

$$G_R(\varepsilon) = G_A^*(\varepsilon).$$

As a result, we obtain

$$\mathscr{II} = \frac{g^2}{\pi(2\pi)^3}\int_{-\infty}^{\infty} d\varepsilon \int d\mathbf{p}\, [\operatorname{Im} G_R(\mathbf{p}, \varepsilon)G_A(\mathbf{p} - \mathbf{k}, \varepsilon - i\omega) + \operatorname{Im} G_R(\mathbf{p} - \mathbf{k}, \varepsilon)\, G_R(\mathbf{p}, \varepsilon + i\omega)] \tanh\frac{\varepsilon}{2T}. \tag{21.9}$$

We can now easily carry out the analytic continuation into the region of real ω, thereby obtaining the retarded function Π_R. Separating the real and imaginary parts of Π_R, we find that

$$\begin{aligned} \operatorname{Re}\Pi_R &= \frac{2g^2}{(2\pi)^4}\int d^4p\, [\operatorname{Im} G_R(\mathbf{p}, \varepsilon)\operatorname{Re} G_R(\mathbf{p} - \mathbf{k}, \varepsilon - \omega) \\ &\qquad + \operatorname{Im} G_R(\mathbf{p} - \mathbf{k}, \varepsilon)\operatorname{Re} G_R(\mathbf{p}, \varepsilon + \omega)] \tanh\frac{\varepsilon}{2T}, \\ \operatorname{Im}\Pi_R &= -\frac{2g^2}{(2\pi)^4}\int d^4p\, \operatorname{Im} G_R(\mathbf{p}, \varepsilon)\operatorname{Im} G_R(\mathbf{p} - \mathbf{k}, \varepsilon - \omega) \\ &\qquad \times\left(\tanh\frac{\varepsilon}{2T} - \tanh\frac{\varepsilon - \omega}{2T}\right). \end{aligned} \tag{21.10}$$

Below, we shall show that the function G_R has the form

$$G_R = \frac{1}{\varepsilon - \xi(\mathbf{p}) - \Sigma_R(\varepsilon)}, \tag{21.11}$$

where Σ depends only on ε. For $\varepsilon < \omega_D$,

$$\Sigma(\varepsilon) \approx -b\varepsilon,$$

where b is a constant of order unity, while for $\varepsilon \gg \omega_D$,

$$\Sigma(\varepsilon) \approx \text{const} \sim \omega_D.$$

We begin by considering the case of long-wavelength phonons, where $k \ll p_0$. In this case, we can make the approximation

$$\xi(\mathbf{p} - \mathbf{k}) \approx \xi(\mathbf{p}) - \mathbf{v}\cdot\mathbf{k}.$$

We can also set

$$\xi(\mathbf{p}) = \varepsilon_0(\mathbf{p}) - \mu \approx v(|\mathbf{p}| - p_0),$$

since the important contribution to the integral comes from the region $\xi \sim vk$. (This approximation will not be used in the case of phonons with $k \sim p_0$, to be considered later.) Then, after substituting for the function G_R, we can carry out the integration with respect to ξ in the equations (21.10). The integral for $\operatorname{Re}\Pi_R$ is formally divergent, and therefore we make use of the fact that the integration with respect to ξ is between finite limits, which we denote by $-L_1$ and L_2. These limits are of order $\mu \sim vp_0$, and hence are much larger than Σ. It follows that

$$\operatorname{Re}\Pi_R = -\frac{p_0^2}{(2\pi)^3 v}\int d\Omega\, \mathrm{P}\int_{-\infty}^{\infty} d\varepsilon \tanh\frac{\varepsilon}{2T} \times\left\{\frac{\theta(L_2 - \varepsilon)\theta(\varepsilon + L_1)}{-\omega + \mathbf{v}\cdot\mathbf{k} - \Sigma(\varepsilon - \omega) + \Sigma(\varepsilon)} + \frac{\theta(L_2 - \mathbf{v}\cdot\mathbf{k} - \varepsilon)\theta(\varepsilon + L_1 + \mathbf{v}\cdot\mathbf{k})}{\omega - \mathbf{v}\cdot\mathbf{k} - \Sigma(\varepsilon + \omega) + \Sigma(\varepsilon)}\right\},$$

where, as usual, P denotes the operation of taking the principal value of an integral, and

$$\theta(x) = \begin{cases} 1 & \text{for} \quad x > 0, \\ 0 & \text{for} \quad x < 0. \end{cases}$$

Since $\Sigma(\varepsilon) \lesssim \omega_D$, the regions making the main contribution to the integral with respect to ε are the neighborhoods of $\varepsilon = -L_1$ and $\varepsilon = L_2$, where $\Sigma(\varepsilon \pm \omega_D) \approx \Sigma(\varepsilon)$. Taking account of this fact and the condition $\omega \sim uk \ll vk$, we obtain

$$\operatorname{Re} \Pi_R = -\frac{2g^2p_0^2}{(2\pi)^3 v}\int d\Omega \frac{\mathbf{v}\cdot\mathbf{k}}{\mathbf{v}\cdot\mathbf{k} - \omega} \approx -\frac{g^2 m p_0}{\pi^2}, \tag{21.12}$$

The integral (21.10) for $\operatorname{Im} \Pi_R$ is convergent, and hence we can regard the limits of integration with respect to ξ as being infinite. As a result, we find that

$$\operatorname{Im} \Pi_R = -\frac{\pi g^2 p_0^2}{(2\pi)^3 v}\int d\Omega \int_{-\infty}^{\infty} d\varepsilon\, \delta[\mathbf{v}\cdot\mathbf{k} - \omega + \Sigma(\varepsilon) - \Sigma(\varepsilon - \omega)]$$
$$\times \left(\tanh\frac{\varepsilon}{2T} - \tanh\frac{\varepsilon - \omega}{2T}\right).$$

The main contribution to this integral comes from the values $\varepsilon \sim T$ if $\omega \ll T$, and from the values $\varepsilon \sim \omega$ if $\omega \gg T$. In the case under consideration,

$$\omega \sim uk \ll \omega_D \qquad (k \ll p_0 \sim k_D),$$

so that $\varepsilon \ll \omega_D$ under all circumstances. It follows that

$$\Sigma(\varepsilon) = -b\varepsilon$$

in the argument of the δ-function, where b is a constant (see below), and therefore, the δ-function becomes

$$\delta[\mathbf{v}\cdot\mathbf{k} - (1 + b)\omega].$$

The integration with respect to ε gives 2ω, and hence

$$\operatorname{Im} \Pi_R = -\frac{g^2 p_0 m}{2\pi}\frac{\omega}{vk}. \tag{21.13}$$

We note that both $\operatorname{Re} \Pi_R$ and $\operatorname{Im} \Pi_R$ turn out to be the same as in the first approximation of perturbation theory, i.e., as in the case where $\mathcal{G}$ is replaced by $\mathcal{G}^{(0)}$ in (21.8). In the case of $\operatorname{Re} \Pi_R$, this is explained by the fact that the region of integration corresponds to large frequencies, where the self-energy Σ does not depend on frequency; in the case of $\operatorname{Im} \Pi_R$, this is due to the fact that $\Sigma(\varepsilon)$ is a linear function for $\varepsilon \ll \omega_D$.

Next, we consider the case $k \sim k_D \sim p_0$, where $vk \gg \omega_D$ always holds. Since $\Sigma \lesssim \omega_D$, the G-functions in (21.10) can be replaced by free $G^{(0)}$-functions. Moreover, since $\omega \sim \omega_D \gg T$, we can substitute $\operatorname{sgn} \varepsilon$ for $\tanh(\varepsilon/2T)$, and we can set $\omega = 0$ in the integral for $\operatorname{Re} \Pi_R$. However, the substitution

$$\xi = \frac{\mathbf{p}^2}{2m} - \frac{p_0^2}{2m}$$

is not permissible. As a result, we obtain

$$\operatorname{Re}\Pi_R = -\frac{2g^2\pi}{(2\pi)^4}\int d\mathbf{p}\Big\{\operatorname{Re} G_R^{(0)}[\mathbf{p}-\mathbf{k}, \xi(\mathbf{p})] \operatorname{sgn} \xi(\mathbf{p}) + \operatorname{Re} G^{(0)}[\mathbf{p}, \xi(\mathbf{p}-\mathbf{k})] \operatorname{sgn} \xi(\mathbf{p}-\mathbf{k})]\Big\}$$

$$= -\frac{4g^2}{(2\pi)^3}\int\limits_{\substack{\xi(\mathbf{p}+\frac{1}{2}\mathbf{k})>0 \\ \xi(\mathbf{p}-\frac{1}{2}\mathbf{k})<0}} d\mathbf{p}\,\frac{1}{\xi(\mathbf{p}+\frac{1}{2}\mathbf{k})-\xi(\mathbf{p}-\frac{1}{2}\mathbf{k})} \tag{21.14}$$

$$= -\frac{g^2 m p_0}{2\pi^2}\, h\Big(\frac{k}{2p_0}\Big),$$

where

$$h(x) = 1 + \frac{1-x^2}{2x}\ln\left|\frac{1+x}{1-x}\right|.$$

As for $\operatorname{Im}\Pi_R$, we retain only the ω in $\tanh[(\varepsilon-\omega)/2T]$, thereby obtaining

$$\operatorname{Im}\Pi_R = -\frac{g^2\omega}{(2\pi)^2}\int d\mathbf{p}\,\delta[\xi(\mathbf{p})-\xi(\mathbf{p}-\mathbf{k})] = -\frac{g^2 p_0 m}{2\pi}\frac{\omega}{vk}\,\theta(2p_0-k). \tag{21.15}$$

The expressions just found include formulas (21.12) and (21.13) [valid for $k \ll p_0$] as special cases, and therefore can be regarded as universal. Making the analytic continuation of the temperature Dyson's equation, we find that

$$D_R^{-1}(\mathbf{k},\omega) = (D_R^{(0)})^{-1} - \Pi_R$$
$$= \frac{1}{\omega_0^2(k)}\Big\{\omega^2 - \omega_0^2(k)\Big[1 - \eta h\Big(\frac{k}{2p_0}\Big) - \eta\,\frac{i\pi m\omega}{p_0 k}\,\theta(2p_0-k)\Big]\Big\}, \tag{21.16}$$

where the constant

$$\eta = \frac{g^2 p_0 m}{2\pi^2} \sim 1$$

has been introduced instead of g^2.

The pole of the function D_R determines the true energy and attenuation of the phonons, i.e.,

$$\omega(k) = \omega_0(k)\sqrt{1 - \eta h\Big(\frac{k}{2p_0}\Big)} \tag{21.17}$$

and

$$\gamma_1(k) = \frac{\pi}{2}\frac{\omega_0^2(k)m}{p_0 k}\,\theta(2p_0-k). \tag{21.18}$$

According to (21.17),

$$\omega(k) \approx \omega_0(k)\sqrt{1-2\eta}$$

for $k \ll p_0$. At $k = 2p_0$, the derivative $d\omega(k)/dk$ has a logarithmic singularity

$$\frac{d\omega(k)}{dk} = \frac{\omega(2p_0)}{4p_0}\frac{\eta}{1-\eta}\ln\frac{4p_0}{|2p_0-k|}, \tag{21.19}$$

and at the same point the attenuation drops discontinuously to zero. It should be noted that this behavior of $\omega(k)$ and $\gamma_1(k)$ is a result of the approximation made in deriving the formula for Π. For example, these singularities disappear if we do not neglect ω in the $G^{(0)}$-functions appearing in (21.10), and in particular,

$$\gamma_1 \sim \arctan \frac{m\omega}{p_0(k - 2p_0)} \qquad (k > 2p_0),$$

i.e., γ_1 drops to zero gradually and not discontinuously.[15] Obviously, the corrections to the formulas derived above become important in the region

$$|k - 2p_0| \sim \frac{\omega}{v} \sim \frac{\omega_D}{v} \ll p_0.$$

To make an exact calculation of the behavior of $\omega(k)$ and $\gamma_1(k)$ in this region, not only must we take account of ω in the formula (21.10), but we must also not replace G by $G^{(0)}$. It should be noted that here the difference between $\mathcal{T}$ and g becomes important. As a result of all this, the calculation becomes very complicated.

It is clear from (21.18) and (21.17) that the attenuation $\gamma_1(k)$ is relatively weak in the case where η is not too close to $\frac{1}{2}$, and in fact

$$\frac{\gamma_1(k)}{\omega(k)} = \frac{\pi}{2} \frac{\omega_0^2(k)}{p_0 k \omega(k)} \sim \frac{\pi}{2} \frac{u}{v} \frac{\eta}{1 - 2\eta} \sim \sqrt{\frac{m}{M}} (1 - 2\eta)^{-1}. \qquad (21.20)$$

It should be noted that $D_R(\mathbf{k}, \omega)$, the phonon function appearing in (21.16), can be written in the following form:

$$D_R(\mathbf{k}, \omega) = \frac{\omega_0^2(k)}{2\omega(k)} \left[\frac{1}{\omega - \omega(k) + i\gamma_1(k)} - \frac{1}{\omega + \omega(k) + i\gamma_1(k)} \right]. \qquad (21.21)$$

This shows that the function $D_R(\mathbf{k}, \omega)$ differs from $D_R^{(0)}(\mathbf{k}, \omega)$ by a constant factor and a change of proper frequency.

21.3. The electron Green's function. We now turn our attention to the electron Green's function. Dyson's equation (16.3) for $\mathcal{G}$ involves the self-energy Σ, which satisfies the equation

$$\Sigma(\mathbf{p}, \varepsilon) = -\frac{g^2 T}{(2\pi)^3} \sum_{\varepsilon_1} \int d\mathbf{p}_1\, \mathcal{G}(\mathbf{p}_1, \varepsilon_1) \mathcal{D}(\mathbf{p} - \mathbf{p}_1, \varepsilon - \varepsilon_1). \qquad (21.22)$$

Just as in Sec. 21.2, we carry out an analytic continuation into the region of real frequencies. To do this, we first write (21.22) in the form of a contour integral

$$\begin{aligned}\Sigma(\mathbf{p}, \varepsilon) = \frac{ig^2 T}{2(2\pi)^4} \int d\mathbf{p}_1 \int_C d\varepsilon_1\, G_{RA}(\mathbf{p}_1, \varepsilon_1) D_{RA}(\mathbf{p} - \mathbf{p}_1, i\varepsilon - \varepsilon_1) \tanh \frac{\varepsilon_1}{2T} \\ - \frac{g^2 T}{(2\pi)^3} \int d\mathbf{p}_1\, \mathcal{G}(\mathbf{p}, \varepsilon) \mathcal{D}(\mathbf{p} - \mathbf{p}_1, 0),\end{aligned} \qquad (21.23)$$

---

[15] The maximum value attained by $d\omega/dk$ is of order v.

where the contour of integration is similar to Fig. 60, and G_{RA}, D_{RA} have their previous meaning. The second term in (21.23) is due to the fact that the contour integral does not include the term corresponding to $\varepsilon_1 = \varepsilon$ in (21.22). We recall that the frequencies appearing in the $\mathscr{D}$-function are "even" (i.e., of the form $2n\pi T$).

As before, the integral along the large circle vanishes, and only the integrals along the horizontal lines $\operatorname{Im} \varepsilon_1 = 0$ and $\operatorname{Im}(\varepsilon_1 - i\varepsilon) = 0$ remain. Bearing in mind that $G_A = G_R^*$ and

$$\tanh\left[\frac{\varepsilon + (2n+1)\pi iT}{2T}\right] = \coth\frac{\varepsilon}{2T},$$

we transform (21.23) in the same way as in the case of the function $\mathscr{T}$. Here the integral around the point $\varepsilon_1 = i\varepsilon$ gives a term which cancels the second term in (21.23). Then it is not hard to make the analytic continuation into the upper half-plane of the variable ε. As a result, we obtain

$$\begin{aligned}\Sigma_R(\mathbf{p}, \varepsilon) = &-\frac{g^2}{(2\pi)^4}\int d\mathbf{p}_1 \int_{-\infty}^{\infty} d\varepsilon_1 [\operatorname{Im} G_R(\mathbf{p}_1, \varepsilon_1)] D_R(\mathbf{p} - \mathbf{p}_1, \varepsilon - \varepsilon_1) \tanh\frac{\varepsilon_1}{2T}\\ &-\frac{g^2}{(2\pi)^4}\int d\mathbf{p}_1 \int_{-\infty}^{\infty} d\varepsilon_1 G_R(\mathbf{p}_1, \varepsilon - \varepsilon_1)[\operatorname{Im} D_R(\mathbf{p} - \mathbf{p}_1, \varepsilon_1)] \coth\frac{\varepsilon_1}{2T}.\end{aligned} \tag{21.24}$$

According to formulas (17.14) and (17.18),

$$G_R(\mathbf{p}, \varepsilon) = \frac{1}{\pi}\int_{-\infty}^{\infty} \frac{\operatorname{Im} G_R(\mathbf{p}, \varepsilon_1)}{\varepsilon_1 - \varepsilon - i\delta}\, d\varepsilon_1, \tag{21.25}$$

and similarly for D_R. Substituting (21.25) into (21.24), and making a change of variables, we obtain

$$\begin{aligned}\Sigma_R(\mathbf{p}, \varepsilon) = &-\frac{g^2}{(2\pi)^4\pi}\int d\mathbf{p}_1 \int_{-\infty}^{\infty} d\omega \int_{-\infty}^{\infty} d\varepsilon_1\\ &\times \frac{\operatorname{Im} G_R(\mathbf{p}_1, \varepsilon_1) \operatorname{Im} D_R(\mathbf{p} - \mathbf{p}_1, \omega)}{\omega + \varepsilon_1 - \varepsilon - i\delta}\left(\tanh\frac{\varepsilon_1}{2T} + \coth\frac{\omega}{2T}\right).\end{aligned} \tag{21.26}$$

Introducing the new variables

$$\xi_1 = \frac{\mathbf{p}_1^2}{2m} - \frac{p_0^2}{2m}, \qquad k = |\mathbf{p} - \mathbf{p}_1|,$$

we first evaluate the integral with respect to $\mathbf{p}_1$ in (21.26):

$$\int_0^{\infty} |\mathbf{p}_1|^2 d|\mathbf{p}_1| \int_{-1}^{1} dx = \frac{m}{|\mathbf{p}|}\int_0^{k_D} k\, dk \int_{\xi(|\mathbf{p}|-k)}^{\xi(|\mathbf{p}|+k)} d\xi_1. \tag{21.27}$$

Next, we look for an expression for G_R of the form (21.11). The imaginary part of (21.11) can be written as

$$\operatorname{Im} G_R(\mathbf{p}_1, \varepsilon_1) = \frac{\operatorname{Im}\Sigma_R(\varepsilon_1)}{[\varepsilon_1 - \xi_1 - \operatorname{Re}\Sigma_R(\varepsilon_1)]^2 + [\operatorname{Im}\Sigma_R(\varepsilon_1)]^2}. \tag{21.28}$$

If we assume that the external momentum is near p_0, then, according to

(21.27), ξ_1 varies over a region depending on k. In the case where the values $k \sim k_D$ make the main contribution to the integral, it is safe to assume that ξ_1 varies between the limits $-\infty$ and ∞. Below, we shall also encounter the case where the momentum values $k \ll p_0$ are important, but then the scale of all energies appearing in the integrals will be of order $\omega(\mathbf{k}) \ll \omega_D$. In both cases, $\operatorname{Im}\Sigma(\varepsilon) \ll \varepsilon$, and consequently

$$\operatorname{Im} G_R(\mathbf{p}_1, \varepsilon_1) \approx \pi\delta[\varepsilon_1 - \operatorname{Re}\Sigma_R(\varepsilon_1) - \xi_1] \operatorname{sgn} \operatorname{Im}\Sigma_R(\varepsilon_1) = -\pi\delta[\varepsilon_1 - \operatorname{Re}\Sigma_R(\varepsilon_1) - \xi_1].$$

Substituting this expression into (21.24), we obtain

$$\Sigma_R(\mathbf{p}, \varepsilon) = \frac{g^2}{(2\pi)^3} \frac{m}{|\mathbf{p}|} \int_0^{k_D} k\, dk \int_{\varepsilon'}^{\varepsilon''} d\varepsilon_1 \times \int_{-\infty}^{\infty} d\omega \frac{\operatorname{Im} D_R(k, \omega)}{\omega + \varepsilon_1 - \varepsilon - i\delta} \left(\tanh\frac{\varepsilon_1}{2T} + \coth\frac{\omega}{2T}\right), \tag{21.29}$$

where

$$\varepsilon'' - \operatorname{Re}\Sigma_R(\varepsilon'') = \xi(|\mathbf{p}| + k), \qquad \varepsilon' - \operatorname{Re}\Sigma_R(\varepsilon') = \xi(|\mathbf{p}| - k).$$

As we shall see below, it turns out that the values $\varepsilon_1 \lesssim \omega_D$ play the main role. In order for these small values to fall in the region over which ε_1 varies, it is necessary that k be less than $2p_0$. Thus, the upper limit of the integral with respect to k is essentially

$$k_1 = \min\{k_D, 2p_0\}.$$

We now consider $\operatorname{Re}\Sigma_R$, making the assumption (to be justified later on) that the range of values $k \sim k_D \sim p_0$ is important in the integral with respect to k. Then the integral with respect to ε_1 can be carried out between infinite limits. Going from $\int_{-\infty}^{\infty} d\omega \cdots$ to $\int_0^{\infty} d\omega \cdots$, we obtain

$$\operatorname{Re}\Sigma_R = \frac{g^2}{(2\pi)^3} \frac{m}{p_0} \int_0^{k_1} k\, dk \int_{-\infty}^{\infty} d\varepsilon_1 \int_0^{\infty} d\omega \times \frac{\operatorname{Im} D_R(k, \omega)}{\omega + \varepsilon_1} \left(\tanh\frac{\varepsilon + \varepsilon_1}{2T} + \tanh\frac{\varepsilon - \varepsilon_1}{2T}\right),$$

where we have used the antisymmetry property

$$\operatorname{Im} D_R(\omega) = -\operatorname{Im} D_R(-\omega).$$

Because of the relation between $\operatorname{Re} D_R$ and $\operatorname{Im} D_R$ [see (21.25)] and the antisymmetry of $\operatorname{Im} D_R$, the formula for $\operatorname{Re}\Sigma_R$ can be written in the form

$$\operatorname{Re}\Sigma_R = \frac{g^2}{16\pi^2} \frac{m}{p_0} \int_0^{k_D} k\, dk\, \mathrm{P} \int_{-\infty}^{\infty} d\omega \operatorname{Re} D_R(k, \omega) \left(\tanh\frac{\varepsilon + \omega}{2T} + \tanh\frac{\varepsilon - \omega}{2T}\right). \tag{21.30}$$

Substituting (21.21) into (21.30), we obtain

$$\operatorname{Re}\Sigma_R = \frac{g^2}{16\pi^2} \frac{m}{p_0} \int_0^{k_1} k\, dk\, \mathrm{P} \int_{-\infty}^{\infty} d\omega \frac{\omega_0^2(k)}{\omega^2 - \omega^2(k)} \left(\tanh\frac{\varepsilon + \omega}{2T} + \tanh\frac{\varepsilon - \omega}{2T}\right).$$

It follows at once that the values $k \sim k_1$ are important, which confirms the assumption made above. As for ω, the main contribution is made by the values $\omega \sim \varepsilon$ or $\omega \sim \omega_D$ if $\varepsilon \gg T$, and by the values $\omega \sim T \ll \omega_D$ if $\varepsilon \ll T$. In all cases, we can write

$$\operatorname{Re}\Sigma_R = \frac{g^2}{8\pi^2}\frac{m}{p_0}\int_0^{k_1} k\,dk\,\mathrm{P}\int_{-\varepsilon}^{\varepsilon} d\omega\,\frac{\omega_0^2(k)}{\omega^2 - \omega^2(k)}. \tag{21.31}$$

Integration with respect to ω gives

$$\operatorname{Re}\Sigma_R = -\frac{g^2}{8\pi^2}\frac{m}{p_0}\int_0^{k_1} k\,dk\,\frac{\omega_0^2(k)}{\omega(k)}\ln\left|\frac{\varepsilon + \omega(k)}{\varepsilon - \omega(k)}\right|, \tag{21.32}$$

where the main contribution to this integral comes from the values $k \sim k_1$, i.e., $\omega(k) \sim \omega_D$. If $\varepsilon \ll \omega_D$, then

$$\operatorname{Re}\Sigma_R \approx -\frac{g^2\varepsilon}{4\pi^2}\frac{m}{p_0}\int_0^{k_1} k\,dk\,\frac{\omega_0^2(k)}{\omega^2(k)} = -b\varepsilon, \tag{21.32'}$$

where b is a positive constant of order 1. In the other limit $\omega \gg \omega_D$, we obtain

$$\operatorname{Re}\Sigma_R \approx -\frac{g^2}{4\pi^2\varepsilon}\frac{m}{p_0}\int_0^{k_1} k\,dk\,\omega_0^2(k) \sim -\frac{\omega_D^2}{\varepsilon} \to 0.$$

Next, we consider $\operatorname{Im}\Sigma_R$, which is obviously obtained by going around the pole in the denominator of (21.29):

$$\operatorname{Im}\Sigma_R(\mathbf{p}, \varepsilon) = \frac{g^2\pi}{(2\pi)^3}\frac{m}{p_0}\int_0^{k_1} k\,dk\int_{\varepsilon'}^{\varepsilon''} d\varepsilon_1\,\operatorname{Im} D_R(k, \varepsilon - \varepsilon_1) \times\left(\tanh\frac{\varepsilon_1}{2T} + \coth\frac{\varepsilon - \varepsilon_1}{2T}\right). \tag{21.33}$$

According to (21.21), we have

$$\operatorname{Im} D_R(k, \omega) = -\frac{2\omega_0^2(k)\omega\gamma_1(k)}{[\omega^2 - \omega^2(k)]^2 + 4\omega^2\gamma_1^2(k)}. \tag{21.34}$$

If (21.34) is substituted into (21.33), two cases are possible in principle: If ω varies in an interval which is large compared to that of $\gamma_1(k)$, then

$$\operatorname{Im} D_R(k, \omega) \approx -\pi\omega_0^2(k)\delta[\omega^2 - \omega^2(k)]\operatorname{sgn}\omega. \tag{21.35}$$

However, if ω varies in an interval which is much smaller than that of $\gamma_1(k)$, we can neglect ω in the denominator of (21.34) in comparison with the other quantities, and then

$$\operatorname{Im} D_R(k, \omega) \approx -\frac{2\omega_0^2(k)\omega\gamma_1(k)}{\omega^4(k)}. \tag{21.36}$$

Suppose we substitute (21.35) into (21.33), making the change of variable $\varepsilon_1 - \varepsilon = \omega$. Then, because of the tanh and coth factors, the region $\omega \sim \max(\varepsilon, T)$ is important. On the other hand, the δ-function requires that $\omega \sim \omega(k) \sim uk$, so that the region $k \sim \varepsilon/u$, i.e., $vk \gg \varepsilon$, is important. This means that in the present case, the limits of integration with respect to ω can be regarded as infinite, and consequently

$$\operatorname{Im} \Sigma_R (\mathbf{p}, \varepsilon)$$
$$= \frac{g^2 m}{16\pi p_0} \int_0^{k_1} \frac{k\, dk\, \omega_0^2(k)}{\omega(k)} \left[\tanh \frac{\omega(k) + \varepsilon}{2T} - \tanh \frac{\varepsilon - \omega(k)}{2T} - 2 \coth \frac{\omega(k)}{2T} \right]. \tag{21.37}$$

If $\max \{\varepsilon, T\} \ll \omega_D$, we obtain

$$\operatorname{Im} \Sigma_R (\varepsilon) = -\frac{\pi \eta T^3}{4(1 - 2\eta) u^2 p_0^2} f_1\left(\frac{\varepsilon}{T}\right), \tag{21.38}$$

where

$$u = \left. \frac{d\omega(k)}{dk} \right|_{k=0}$$

is the velocity of sound, and

$$f_1(x) = \int_0^\infty z^2\, dz \left(\coth \frac{z}{2} - \frac{1}{2} \tanh \frac{z + x}{2} - \frac{1}{2} \tanh \frac{z - x}{2} \right). \tag{21.39}$$

If $\varepsilon \ll T \ll \omega_D$, this gives

$$\operatorname{Im} \Sigma_R (\varepsilon) = -\frac{7\pi\zeta(3)}{8} \frac{\eta}{1 - 2\eta} \frac{T^3}{u^2 p_0^2} \tag{21.40}$$

(where ζ denotes the Riemann zeta function), and if $T \ll \varepsilon \ll \omega_D$, we have

$$\operatorname{Im} \Sigma_R (\varepsilon) = -\frac{\pi}{12} \frac{\eta}{1 - 2\eta} \frac{|\varepsilon|^3}{u^2 p_0^2}. \tag{21.41}$$

In the case where $\varepsilon \gg \omega_D$ (by hypothesis $T \ll \omega_D$), we obtain

$$\operatorname{Im} \Sigma_R (\varepsilon) = \text{const} \sim \omega_D. \tag{21.42}$$

In making this derivation, the values $\omega \sim \max \{\varepsilon, T\}$ are important. Therefore, according to the foregoing, the replacement of (21.34) by (21.35) is legitimate only if

$$\max \{\varepsilon, T\} \gg \max \gamma_1(k) \sim \omega_D \sqrt{\frac{m}{M}}.$$

In the opposite limiting case, we must use formula (21.36). Then the values $k \sim k_1$ are important in the integral, so that

$$\operatorname{Im} \Sigma_R (\varepsilon) = -\frac{g^2 m}{2\pi^2 p_0} \int_0^{k_1} k\, dk\, \frac{\omega_0^2(k) \gamma_1(k)}{\omega^4(k)}$$
$$\times \int_0^\infty \omega\, d\omega \left(\coth \frac{\omega}{2T} - \frac{1}{2} \tanh \frac{\omega + \varepsilon}{2T} - \frac{1}{2} \tanh \frac{\omega - \varepsilon}{2T} \right). \tag{21.43}$$

Evaluating the integral with respect to ω, we find that

$$\operatorname{Im} \Sigma_R (\varepsilon) = -\frac{A}{v p_0} (\pi^2 T^2 + \varepsilon^2), \tag{21.44}$$

if $\max \{\varepsilon, T\} \ll \omega_D \sqrt{m/M}$, where

$$A = \frac{g^2 p_0}{4\pi^2} \int_0^{k_1} k\, dk\, \frac{\omega_0^2(k) \gamma_1(k)}{\omega^4(k)} = \text{const} \sim 1.$$

Formulas (21.38) and (21.44) correspond to different attenuation mechanisms. The first describes the attenuation due to radiation and absorption of phonons by electrons. However, in the case where the energy of the quasi-particles is very close to the Fermi surface, i.e., $\varepsilon \ll \omega_D\sqrt{m/M}$, the interaction between electrons due to phonon exchange becomes important, which guarantees that the attenuation is described by (21.44). It follows from what has been said earlier (see Sec. 2) that the attenuation due to interaction between fermions must have just this form.

The energy of the electron excitations can be determined from the real part of the pole of the function G_R. According to (21.11), we have $\varepsilon - \Sigma_R(\varepsilon) = \xi$. In the case $\varepsilon \ll \omega_D$, using (21.32), we obtain

$$\varepsilon = \frac{v}{1+b}(|\mathbf{p}| - p_0). \tag{21.45}$$

Thus, the velocity of the quasi-particles decreases at the Fermi surface $(b > 0)$. Moreover, near its pole the G-function takes the form (18.1), where

$$a = \frac{1}{1+b} < 1.$$

It follows from (21.41) that for $\varepsilon \sim \omega_D$ the attenuation of the quasi-particles is comparable to their energy. However, it is not hard to see [cf. (21.42)] that as the energy of the excitations increases further, the attenuation ceases to increase and again becomes less than the energy of the quasi-particles. Thus, there are two regions in which the concept of quasi-particles is meaningful, i.e., $|\varepsilon| \ll \omega_D$ and $|\varepsilon| \gg \omega_D$. In both regions, the energy of the electrons is of the form $v(|\mathbf{p}| - p_0)$, but the velocities v are different.

We now indicate qualitatively what is found if the direct Coulomb interaction between electrons is taken into account. Because of the screening of the Coulomb forces at distances of the order of the lattice period (i.e., of order $1/p_0$), these forces can also be regarded as short-range in the present case. Taking them into account leads to changes in the velocity at the Fermi surface and in the coefficient a in the G-function near its pole. Moreover, a qualitative difference appears in the amount of attenuation. The Coulomb interaction leads to an attenuation proportional to $(vp_0)^{-1}\max\{\varepsilon^2, T^2\}$ (see Sec. 2). In the region $\varepsilon \ll \omega_D\sqrt{m/M}, T \ll \omega_D\sqrt{m/M}$, this attenuation is added to the attenuation (21.44), which has the same structure and order of magnitude. However, in the region $\varepsilon \gg \omega_D\sqrt{m/M}, T \gg \omega_D\sqrt{m/M}$, the phonon attenuation becomes predominant. As already noted, for $\varepsilon \gg \omega_D$ the attenuation becomes a constant of order ω_D. Coulomb interaction is also present in this region, and it begins to dominate when

$$|\varepsilon| \gg \sqrt{vp_0\omega_D} \sim \omega_D\sqrt[4]{M/m} > \omega_D.$$

It follows that whereas the excitation spectrum in the region $|\varepsilon| \gg \omega_D$ is determined by Coulomb interactions between the electrons, the phonon attenuation continues to dominate the electron attenuation for a while.

21.4. Correction to the linear term in the electronic heat capacity. Using the results obtained above, we can draw an interesting conclusion about the electronic heat capacity. At first glance, it appears that the correction to the linear term in (19.27) must be of relative order $(T/\mu)^2$. However, the electron-phonon interaction actually leads to a larger correction [see Eliashberg (E5)].

Consider the general formula (21.32) for $\operatorname{Re}\Sigma_R$. If $\varepsilon \ll \omega_D$ we obtain the expression (21.32′) as a first approximation. It is not hard to see that by expanding the logarithm in the integrand of (21.32) up to terms of order ε^3, we obtain a logarithmically divergent integral with respect to k. To within logarithmic accuracy, we have[16]

$$\delta \operatorname{Re}\Sigma_R = -\frac{1}{6}\frac{\eta}{1-2\eta}\frac{1}{u^2p_0^2}\varepsilon^3 \ln\frac{\omega_D}{|\varepsilon|}.$$

Taking this correction into account in calculating the entropy (see Sec. 19.5) leads to the formula

$$\delta\left(\frac{S}{V}\right) = 2\int\frac{d\mathbf{p}}{(2\pi)^3}\frac{1}{T}\int_{-\infty}^{\infty}\varepsilon\left[-\frac{\partial n_F(\varepsilon)}{\partial\varepsilon}\right][\operatorname{Im} G_R(\varepsilon)]\,\delta\operatorname{Re}\Sigma_R(\varepsilon)\,d\varepsilon.$$

Setting

$$\operatorname{Im}\Sigma_R = -\pi\delta[\xi - \varepsilon - \Sigma_R(\varepsilon)],$$

and evaluating the integral with respect to the Fermi distribution function, we obtain

$$\delta\left(\frac{S}{V}\right) = \frac{7\pi^2}{180}\frac{\eta}{1-2\eta}\frac{m}{u^2p_0}T^3\ln\frac{\omega_D}{T},$$

and then differentiation with respect to T gives

$$\delta\left(\frac{C}{V}\right) = T\frac{\partial}{\partial T}\left[\delta\frac{S}{V}\right] = \frac{7\pi^2}{60}\frac{\eta}{1-2\eta}\frac{m}{u^2p_0}T^3\ln\frac{\omega_D}{T}. \qquad (21.46)$$

Thus, the contribution to the linear term in the heat capacity C turns out to be of relative order

$$\frac{\eta}{1-2\eta}\frac{T^2}{\omega_D^2}\ln\frac{\omega_D}{T}.$$

It is interesting to compare this correction with the cubic term due to the lattice vibrations. The order of magnitude of the lattice heat capacity is T^3/u^3. It follows that the correction just found is

$$\frac{\eta}{1-2\eta}\frac{u}{v}\ln\frac{\omega_D}{T} \qquad (21.47)$$

times the size of the lattice heat capacity, and therefore, as a rule, can be

---

[16] We note that $\delta\operatorname{Re}\Sigma_R$ and the attenuation (21.41) are the real and imaginary parts of the analytic function

$$\delta\Sigma_R = -\frac{1}{12}\frac{\eta}{1-2\eta}\frac{1}{u^2p_0^2}\varepsilon^3\ln\left[-\frac{\omega_D^2}{(\varepsilon+i\delta)^2}\right].$$

considered to be small. However, since (21.47) is generally of the order 10^{-1}, the possibility is not excluded that this term makes an appreciable contribution to the heat capacity of certain metals.[17]

22. Some Properties of a Degenerate Plasma

22.1. Statement of the problem. As an example of a system with Coulomb interactions, we now consider a *plasma*, i.e., a mixture of an electron gas and an ion gas. In this case, the interaction Hamiltonian has the form

$$\begin{aligned} H_{\text{int}} = \frac{e^2}{2} \int \psi_\alpha^+(\mathbf{r})\psi_\beta^+(\mathbf{r}') \frac{1}{|\mathbf{r} - \mathbf{r}'|} \psi_\beta(\mathbf{r}')\psi_\alpha(\mathbf{r})\, d\mathbf{r}\, d\mathbf{r}' \\ - Ze^2 \int \psi_\alpha^+(\mathbf{r})\Phi^+(\mathbf{r}') \frac{1}{|\mathbf{r} - \mathbf{r}'|} \Phi(\mathbf{r}')\psi_\alpha(\mathbf{r})\, d\mathbf{r}\, d\mathbf{r}' \\ + \frac{Z^2e^2}{2} \int \Phi^+(\mathbf{r})\Phi^+(\mathbf{r}') \frac{1}{|\mathbf{r} - \mathbf{r}'|} \Phi(\mathbf{r}')\Phi(\mathbf{r})\, d\mathbf{r}\, d\mathbf{r}', \end{aligned} \tag{22.1}$$

where ψ_α and Φ are the operators of the electron and ion fields, respectively. We assume that the electron gas is degenerate and that the ion gas has a Boltzmann distribution. This requires that the temperature satisfy the inequalities

$$\frac{1}{m}\left(\frac{N}{V}\right)^{2/3} \gg T \gg \frac{1}{M}\left(\frac{N}{V}\right)^{2/3}. \tag{22.2}$$

In making calculations, we can provisionally regard the ions as a Fermi gas, since the Boltzmann limit is the same for both kinds of statistics.

Next, we assume that the effect of the Coulomb interaction is small, which requires that

$$\frac{e^2}{\bar{E}\bar{r}} \ll 1,$$

where $\bar{E}$ is some average energy and $\bar{r}$ is the average distance between particles. Since $\bar{E} \sim T$ for ions and $\bar{E} \sim p_0^2/2m$ for electrons, our condition is equivalent to the requirement that

$$\frac{T}{e^2} \gg \left(\frac{N}{V}\right)^{1/3} \gg e^2 m. \tag{22.3}$$

It is not hard to see that (22.3) is compatible with the assumption that the electron gas is degenerate. If (22.3) is satisfied, the Coulomb interaction will almost always have a small effect on the properties of the plasma. The

[17] The presence of this term can be observed if the metal makes the transition to the superconducting state, occurring at the critical temperature T_c (see Chap. 7). This transition does not affect the lattice heat capacity, but it leads to an exponential decrease in the electronic heat capacity as $T \to 0$. Comparing the cubic term in the heat capacity above T_c with the same term as $T \to 0$, we can detect the appearance of the electronic correction (21.46) to the cubic term above T_c.

case where the collisions between the particles involve small momentum transfer is exceptional. In fact, the Fourier components of the Coulomb potential have the form[18]

$$U(k) = \frac{4\pi e^2}{k^2}, \tag{22.4}$$

and hence the role of collisions with small momentum transfer is very important.

Before proceeding, we have to consider an apparent difficulty connected with the Coulomb interaction. The diagrams for the G-function in Sec. 8 contain diagrams of the type shown in Fig. 5(b), p. 70, containing $U(0)$, and similar diagrams are encountered in the case of the temperature $\mathscr{G}$-function. According to (22.4), the analytical expressions corresponding to such diagrams are infinite. To avoid this trouble, we go from the case where the chemical potentials are given to the case where the particle numbers are given, and we replace the Coulomb potential by the potential

$$U(r) = e^2 \frac{\exp(-\alpha r)}{r},$$

where it is assumed that the quantity α is small (we shall ultimately set $\alpha = 0$). Moreover, to avoid further difficulties due to the fact that the particle numbers are given, we shall use $(\mathbf{r}, \tau)$ space in our subsequent analysis.

Consider, for example, an electron line $\mathscr{G}_e$. We sum all irreducible self-energy parts of the type shown Figs. 4(a), 8(e) and 8(f), pp. 69, 73, which are joined to the basic $\mathscr{G}$-line by one wavy line. It is not hard to see that the sum of all such diagrams gives

$$\begin{aligned}\Sigma'(\mathbf{r} - \mathbf{r}', \tau - \tau') &= 2e^2 \int d\mathbf{r}[\mathscr{G}_e(0, -0) - Z\mathscr{G}_i(0, -0)] \\ &\qquad \times \frac{\exp(-\alpha|\mathbf{r} - \mathbf{r}_1|)}{|\mathbf{r} - \mathbf{r}_1|} \delta(\mathbf{r} - \mathbf{r}')\, \delta(\tau - \tau') \qquad (22.5) \\ &= \frac{N_e - ZN_i}{V} \frac{4\pi e^2}{\alpha^2} \delta(\mathbf{r} - \mathbf{r}')\, \delta(\tau - \tau') \equiv 0,\end{aligned}$$

where, in the last equation, we have used the condition for electrical neutrality, i.e., $N_e = ZN_i$. In the same way, the analogous corrections for the $\mathscr{G}_i$-lines vanish. This means that all diagrams containing volume integrals of the Coulomb potential (the zeroth Fourier component) should simply be set equal to zero.

Afterwards, we can go back to the representation with a given μ by using the following formal device: All the $\mathscr{G}^{(0)}$-functions appearing in the diagrams are written in the form

$$\mathscr{G}_N^{(0)} = \mathscr{G}_\mu^{(0)} e^{-\mu(\tau_1 - \tau_2)}. \tag{22.6}$$

[18] Since four-vectors are not used in this section, we denote magnitudes of three-dimensional vectors by lightface Latin letters.

Then it is not hard to see that the resulting $\mathscr{G}$-function will simply be represented by diagrams containing $\mathscr{G}_{\mu}^{(0)}$, with every diagram multiplied by the identical factor $e^{-\mu(\tau_1-\tau_2)}$, so that, dividing by this factor, we arrive at $\mathscr{G}_{\mu}$. Thus, it is apparent that the prescription for dealing with a Coulomb potential merely consists in discarding all diagrams involving $U(0)$. However, we must bear in mind that the results obtained in this way are correct only for a choice of chemical potentials satisfying the condition

$$N_e(\mu_e, \mu_i) = ZN_i(\mu_e, \mu_i),$$

or equivalently

$$\frac{\partial \Omega}{\partial \mu_e} = Z\frac{\partial \Omega}{\partial \mu_i}. \tag{22.7}$$

22.2. The vertex part for small momentum transfer. We begin by considering the vertex part for small momentum transfer. Since in this case the ions are very important, we use the temperature technique. The first-order correction to the expression (22.4) is represented by two diagrams of the type shown in Fig. 61(a), where the loop represents electrons in one diagram and

FIGURE 61

ions in the other. Although this correction contains an extra factor of e^2, it also contains $(1/k^2)^2$. Therefore, the correction will be important when the momentum transfer is small, and we have to sum the chain of "bubble" diagrams shown in Fig. 61(b) with an arbitrary number of electron and ion loops.[19] As a result, it turns out that all the vertices (electron-electron, ion-ion and ion-electron) are multiplied by the same factor, i.e.,

$$\begin{aligned}\mathscr{T}_{12}(\mathbf{k}, \omega_m) &= \frac{4\pi e^2 Z_1 Z_2}{k^2\left\{1 - \dfrac{4\pi e^2}{k^2}\left[\mathscr{H}_e(\mathbf{k}, \omega_m) + Z^2\mathscr{H}_i(\mathbf{k}, \omega_m)\right]\right\}} \\ &= \frac{4\pi e^2 Z_1 Z_2}{k^2 - 4\pi e^2[\mathscr{H}_e(\mathbf{k}, \omega_m) + Z^2\mathscr{H}_i(\mathbf{k}, \omega_m)]},\end{aligned} \tag{22.8}$$

where $\omega_m = 2m\pi T$ (m is an integer), $\mathscr{H}_e$ corresponds to an electron loop and $\mathscr{H}_i$ corresponds to an ion loop.

As already noted, the ions can be regarded provisionally as a Fermi gas. Therefore, the first step of the calculation is the same for both $\mathscr{H}_e$ and $\mathscr{H}_i$:

---

[19] Such a summation for a system with Coulomb interactions was first carried out by Gell-Mann and Brueckner (G3).

$$\begin{aligned}
\mathscr{T} &= 2T \sum_n \int \frac{d\mathbf{p}}{(2\pi)^3} \frac{1}{[i\varepsilon_n - \varepsilon_0(\mathbf{p}) + \mu][i\varepsilon_n + i\omega_m - \varepsilon_0(\mathbf{p}+\mathbf{k}) + \mu]} \\
&= 2T \int \frac{d\mathbf{p}}{(2\pi)^3} \frac{1}{i\omega_m - \varepsilon_0(\mathbf{p}+\mathbf{k}) + \varepsilon_0(\mathbf{p})} \sum_n \left\{ \frac{1}{i\varepsilon_n - \varepsilon_0(\mathbf{p}) + \mu} \right. \\
&\qquad \left. - \frac{1}{i(\varepsilon_n + \omega_m) - \varepsilon_0(\mathbf{p}+\mathbf{k}) + \mu} \right\} \\
&= -2T \int \frac{d\mathbf{p}}{(2\pi)^3} \frac{1}{i\omega_m - \varepsilon_0(\mathbf{p}+\mathbf{k}) + \varepsilon_0(\mathbf{p})} \sum_{n>0} \left\{ \frac{2[\varepsilon_0(\mathbf{p}) - \mu]}{(2n+1)^2\pi^2T^2 + [\varepsilon_0(\mathbf{p}) - \mu]^2} \right. \\
&\qquad \left. - \frac{2[\varepsilon_0(\mathbf{p}+\mathbf{k}) - \mu]}{(2n+1)^2\pi^2T^2 + [\varepsilon_0(\mathbf{p}+\mathbf{k}) - \mu]^2} \right\} \\
&= -\int \frac{d\mathbf{p}}{(2\pi)^3} \frac{1}{i\omega_m - \varepsilon_0(\mathbf{p}+\mathbf{k}) + \varepsilon_0(\mathbf{p})} \left\{ \tanh\left[\frac{\varepsilon_0(\mathbf{p}) - \mu}{2T}\right] \right. \\
&\qquad \left. - \tanh\left[\frac{\varepsilon_0(\mathbf{p}+\mathbf{k}) - \mu}{2T}\right] \right\} \\
&= -2 \int \frac{d\mathbf{p}}{(2\pi)^3} \frac{n(\mathbf{p}+\mathbf{k}) - n(\mathbf{p})}{i\omega_m - \varepsilon_0(\mathbf{p}+\mathbf{k}) + \varepsilon_0(\mathbf{p})}. \qquad (22.9)
\end{aligned}$$

Here

$$n(\mathbf{p}) = \frac{1}{e^{[\varepsilon_0(\mathbf{p}) - \mu]/T} + 1}$$

for the electrons, and

$$2n(\mathbf{p}) = e^{[\mu - \varepsilon_0(\mathbf{p})]/T}$$

for the ions, and we have used the formula

$$\sum_{n>0} \frac{1}{(2n+1)^2 + x^2} = \frac{\pi}{4x} \tanh \frac{\pi x}{2}. \qquad (22.10)$$

We shall assume that $|\mathbf{k}| \ll p_0$. The average momentum of an ion is of order $\sqrt{MT}$, which, according to (22.2), is much larger than p_0. Therefore, expanding the expression (22.9) with respect to $\mathbf{k}$, we can write it in the simplified form

$$\mathscr{T}(\mathbf{k}, \omega_m) = -2 \int \frac{d\mathbf{p}}{(2\pi)^3} \frac{\partial n}{\partial \varepsilon} \frac{\mathbf{v}\cdot\mathbf{k}}{i\omega_m - \mathbf{v}\cdot\mathbf{k}}. \qquad (22.11)$$

The vertex part under consideration here depends only on $\mathbf{k}$ and ω_m, and can therefore be associated with a $\mathscr{D}$-function expressing the electromagnetic interaction between the particles. In fact, consider for example the quantity

$$\langle T(\tilde{\psi}_\alpha(\mathbf{r}_1, \tau_1)\tilde{\psi}_\beta(\mathbf{r}_2, \tau_2)\bar{\tilde{\psi}}_\alpha(\mathbf{r}_1, \tau_1)\bar{\tilde{\psi}}_\beta(\mathbf{r}_2, \tau_2)\rangle,$$

where $\langle \cdots \rangle$ is meant in the sense of an ordinary temperature average. It is not hard to see that the Fourier transform of this quantity with respect to the variables $\mathbf{r}_1 - \mathbf{r}_2$ and $\tau_1 - \tau_2$ has all the properties of a Bose temperature Green's function. On the other hand, this Fourier transform obviously equals

$$\mathscr{T}_e(\mathbf{k}, \omega_m)\mathscr{F}_{ee}(\mathbf{k}, \omega_m)\mathscr{T}_e(\mathbf{k}, \omega_m),$$

and the same applies to other vertices with small momentum transfer.

It is now clear how to go from temperature functions to time-dependent functions. As we already know, to make this change in the case of a Green's function, we need only find a function which is analytic in the upper half-plane of the variable ω and which coincides with the temperature Green's function at the points $i\omega_m = i2m\pi T$. This is how the retarded function is defined. The actual Green's function equals the retarded function for $\omega > 0$ and its complex conjugate for $\omega < 0$.

This procedure can easily be carried out for the functions $\mathscr{T}_e$, $\mathscr{T}_i$ and we denote the resulting functions by Π_e, Π_i. As can be seen from the integral (22.11), to accomplish this we must change $i\omega_m$ to $\omega + i\delta \operatorname{sgn} \omega$. Since in the diagrams making up the average written above, only the factors $\mathscr{T}^n$ depend on ω_m, we can obtain the corresponding actual Green's function in the same way. Moreover, the same obviously applies to the functions $\Gamma(\mathbf{k}, \omega)$.[20] Thus, the time-dependent vertex parts $\Gamma(\mathbf{k}, \omega)$ are given by the same formulas (22.8) and (22.9) as before, where, however, $i\omega_m$ has been changed to $\omega + i\delta$ $\operatorname{sgn} \omega$. The appearance of a correction to k^2 in the denominator of Γ is just due to the Debye screening of the Coulomb interaction, which in general makes the interaction retarded (i.e., Γ depends on ω).

Next, we examine the behavior of Γ as a function of the ratio between ω and k. In the case $\omega \ll v_i k$, formula (22.11) gives

$$\mathscr{T}_e = \Pi_e = -\frac{1}{V}\frac{\partial N_e}{\partial \mu_e}, \qquad \mathscr{T}_i = \Pi_i = -\frac{1}{V}\frac{\partial N_i}{\partial \mu_i}, \tag{22.12}$$

where

$$\frac{1}{V}\frac{\partial N}{\partial \mu} \sim \frac{N}{V\mu}$$

in order of magnitude. For electrons we have $\mu_e \sim p_0^2/m$ and for ions $\mu_i \sim T \ln T$, so that $\Pi_i \gg \Pi_e$. From formulas (22.8) and (22.11) we obtain

$$\Gamma_{12}(\mathbf{k}, \omega) = \frac{4\pi e^2 Z_1 Z_2}{k^2 + \varkappa_i^2}, \tag{22.13}$$

where

$$\varkappa_i = \sqrt{\frac{4\pi Z^2 e^2 N_i}{VT}}$$

is the reciprocal Debye radius of the ions.

The next region is

$$v_e k \gg |\omega| \gg v_i k,$$

and here we have

$$\Pi_i = \frac{k^2}{\omega^2}\frac{N}{MV}, \qquad \Pi_e = -\frac{1}{V}\frac{\partial N_e}{\partial \mu} = -\frac{p_0 m}{\pi^2}. \tag{22.14}$$

Substitution of (22.14) into (22.8) gives

$$\Gamma_{12} = \frac{4\pi e^2 Z_1 Z_2 \omega^2}{(k^2 + \varkappa_e^2)\omega^2 - k^2\omega_{p1}^2}, \tag{22.15}$$

---

[20] This procedure, which is correct for the separate terms of a series, may not be applicable to its sum. However, in the present case, it can be seen that the procedure leads to a correct result.

where

$$\varkappa_e = \sqrt{\frac{4p_0me^2}{\pi}}$$

is the reciprocal Debye radius of the electrons, and

$$\omega_{p1} = \sqrt{\frac{4\pi N_i Z^2 e^2}{MV}}. \tag{22.16}$$

The function Γ_{12} has a pole for

$$\omega = \frac{\omega_{p1} k}{\sqrt{k^2 + \varkappa_e^2}}. \tag{22.17}$$

For $k \ll \varkappa_e$, the function is linear, corresponding to "ionic sound," with velocity

$$\frac{\omega}{k} = \frac{\omega_{p1}}{\varkappa_e} = p_0\sqrt{\frac{Z}{3Mm}}. \tag{22.18}$$

For $k \gg \varkappa_e$, ω approaches a constant $\omega \approx \omega_{p1}$. The pole of Γ_{12} stays real right up to $k \sim \omega_{p1}/v_i$, and therefore ω achieves the values ω_{p1} in the case where $\omega_{p1} \gg v_i\varkappa_e$. This condition is satisfied because of the first inequality in (22.2). The attenuation of the oscillations is determined by the imaginary parts of Π_i and Π_e. Substituting the equilibrium distributions for electrons and ions into formula (22.11), and letting $i\omega_m \to \omega + i\delta \operatorname{sgn} \omega$, we obtain

$$\operatorname{Im} \Pi_i = -\frac{|\omega|}{k}\frac{N_i}{V}\sqrt{\frac{2\pi M}{T^3}}\, e^{-M\omega^2/2k^2T}, \qquad \operatorname{Im} \Pi_e = -\frac{|\omega|}{k}\frac{m^2}{2\pi}. \tag{22.19}$$

The argument of the exponential in the first of the formulas (22.19) is of order p_0^2/mT or less, while the ratio of the other factor in $\operatorname{Im} \Pi_i$ to $\operatorname{Im} \Pi_e$ is of order

$$\left(\frac{p_0^2}{mT}\right)^{3/2}\left(\frac{M}{m}\right)^{1/2}$$

Therefore, either of these quantities can turn out to make the larger contribution. Taking the imaginary part of the pole of Γ_{12}, we find that the attenuation of the oscillations is

$$\gamma = \frac{\omega^4}{k^3}\left[\left(\frac{\pi}{8}\right)^{1/2}\left(\frac{M}{T}\right)^{3/2} e^{-M\omega^2/2k^2T} + \frac{m^2e^2}{\omega_{p1}^2}\right], \tag{22.20}$$

where ω is given by (22.17).

For $v_e k \gg \omega \gg \omega_{p1}$, we have

$$\Gamma_{12} = \frac{4\pi e^2 Z_1 Z_2}{k^2 + \varkappa_e^2}, \tag{22.21}$$

according to (22.15). In the case where $\omega \sim v_e k$, the electron loop dominates the ion loop. According to (22.11), the complete expression for Π_e equals

$$\Pi_e = -\frac{p_0^2}{\pi^2 v_e}\left[1 - \frac{\omega}{2v_e k}\ln\left|\frac{\omega + v_e k}{\omega - v_e k}\right| + \frac{i\pi|\omega|}{2v_e k}\,\theta(v_e k - |\omega|)\right]. \tag{22.22}$$

Using (22.22), we find that

$$\Pi_e = \frac{p_0^3}{\pi^2 m}\frac{k^2}{3\omega^2}\left(1 + \frac{3}{5}\frac{v_e^2 k^2}{\omega^2}\right)$$

in the region $\omega \gg v_e k$. Substituting this result into (22.8), we obtain

$$\Gamma_{12}(k, \omega) = \frac{4\pi e^2 Z_1 Z_2}{k^2\left[1 - \frac{4\pi e^2 N_e}{m\omega^2 V}\left(1 + \frac{3}{5}\frac{v_e^2 k^2}{\omega^2}\right)\right]}, \tag{22.23}$$

which has a pole at the point

$$\omega^2 = \omega_{p2}^2 + \frac{3}{5} v_e^2 k^2, \tag{22.24}$$

where

$$\omega_{p2}^2 = \frac{4\pi e^2 N_e}{mV} \qquad (v_e k \ll \omega_{p2}). \tag{22.25}$$

This pole corresponds to "plasma oscillations of the electrons." The dispersion of the oscillations consists of a small correction. The attenuation of the oscillations can be obtained in the same way as before, by taking account of the exponentially small contribution coming from the detour around the pole when the integral (22.11) is transformed. This contribution turns out to be proportional to

$$e^{-m\omega_{p2}^2/2k^2T}.$$

For very low temperatures, this expression is no longer correct, since the attenuation then contains larger terms coming from the higher-order approximations in e^2.

It follows from formula (22.23) that $\omega \to 0$, $\Gamma \to -\infty$ in the limit $k/\omega \to 0$. Thus, it is clear from the present example that Γ^ω contains an infinite constant in the case of Coulomb interactions.

22.3. The electron spectrum. We now find the electron Green's function.[21] The first-order correction to the self-energy part is shown in Fig. 62. The analytical expression for this correction has the form

$$\Sigma_1 = -4\pi e^2 T \lim_{t \to +0} \sum_{\varepsilon_1} \int \frac{d\mathbf{p}_1}{(2\pi)^3} e^{i\varepsilon_1 t} \mathscr{G}^{(0)}(\mathbf{p}_1, \varepsilon_1) \frac{1}{(\mathbf{p} - \mathbf{p}_1)^2}. \tag{22.26}$$

Using the rules (17.36) and (22.10) to evaluate the sum over ε_1, we obtain

$$\Sigma_1 = -4\pi e^2 \int \frac{d\mathbf{p}_1}{(2\pi)^3} n_F(\mathbf{p}_1) \frac{1}{(\mathbf{p} - \mathbf{p}_1)^2},$$

where n_F is the Fermi distribution function. We now add and subtract the same expression with $n_F(T = 0)$ instead of $n_F(\mathbf{p}_1)$, and use the fact that the

[21] Sec. 22.3 is based on the paper A1, and on a calculation made by Kochkin for $T \neq 0$.

difference $n_F(\mathbf{p}, T) - n_F(\mathbf{p}, 0)$ falls off rapidly to zero as we move away from the Fermi surface. This enables us to change

$$\xi_1 = \frac{\mathbf{p}_1^2}{2m} - \frac{p_0^2}{2m}$$

to $v(p_1 - p_0)$ in the corresponding integral.[22] Then, after integrating over the angles, we obtain

$$\Sigma_1 = -\frac{e^2}{\pi p}\int_0^{p_0} p_1\, dp_1 \ln\left|\frac{p_1 + p}{p_1 - p}\right| - \frac{e^2}{\pi v}\int_0^\infty \frac{d\xi_1}{e^{\xi_1/T} + 1} \ln\frac{\xi_1 + \xi}{|\xi_1 - \xi|}.$$

Evaluating the first integral, and making some elementary transformations, we find that

$$\Sigma_1 = \frac{e^2 m}{\pi p}\left\{\xi\left[\ln\frac{(p + p_0)^2}{2mT} - 1\right] - Tf\left(\frac{\xi}{T}\right)\right\} - \frac{e^2}{\pi} p_0, \quad (22.27)$$

where

$$\xi = \frac{\mathbf{p}^2}{2m} - \frac{p_0^2}{2m},$$

$$f(x) = \frac{1}{2}\int_0^\infty dz \ln z\left(\tanh\frac{z + x}{2} - \tanh\frac{z - x}{2}\right).$$

FIGURE 62

Since (22.27) does not depend on ε, Σ_1 represents a correction to the energy of the quasi-particles. Moreover, the Fermi momentum is not affected by the interaction and is connected with the chemical potential by the relation $\varepsilon(p_0) = \mu$. Therefore, for $p = p_0$ the expression Σ_1 must be regarded as the change in the chemical potential, i.e.,

$$\Delta\mu = -\frac{e^2}{\pi} p_0. \quad (22.28)$$

For $|p - p_0| \ll p_0$,

$$\Sigma_1 - \Delta\mu = \frac{e^2}{\pi v}\left[\xi\left(\ln\frac{2vp_0}{T} - 1\right) - Tf\left(\frac{\xi}{T}\right)\right], \quad (22.29)$$

from which we obtain

$$\begin{aligned} \Sigma_1 - \Delta\mu &= \frac{e^2}{\pi v}\xi \ln\frac{2p_0 v}{\xi} && \text{for} \quad \xi \ll T, \\ \Sigma_1 - \Delta\mu &= \frac{e^2}{\pi v}\xi\left(\ln\frac{4\gamma v p_0}{\pi T} - 1\right) && \text{for} \quad \xi \ll T, \end{aligned} \quad (22.30)$$

where $\gamma = e^c \approx 1.78$ (c is Euler's constant).

It follows from (22.30) that the expression (22.27) is not completely correct for small $\xi(\mathbf{p})$ and T. As $T \to 0$, $\xi \to 0$, the correction to the velocity $\nabla_{\mathbf{p}}\Sigma$ of the quasi-particles near the Fermi surface goes to infinity like a multiple of

---

[22] Here we use v to denote p_0/m, the electron velocity at the Fermi surface.

$\ln (vp_0/T)$ or $\ln (vp_0/\xi)$. This result is connected with the fact that as $p \to p_0$, the case where the momentum transfer at the vertices is small becomes important in the integral (22.26). Then we have to take account of all loops strung along the basic wavy line, i.e., we have to replace $4\pi e^2/(\mathbf{p}_1 - \mathbf{p})^2$ by the expression for $\Gamma_{ee}(\mathbf{p}_1 - \mathbf{p}, \varepsilon_1 - \varepsilon)$ corresponding to (22.8). Instead of calculating the whole quantity Σ, it will be more convenient to calculate the difference between Σ and the part Σ_1 already calculated. This difference is given by the expression (22.26) with $4\pi e^2/(\mathbf{p}_1 - \mathbf{p})^2$ replaced by

$$\Gamma_{ee} - \frac{4\pi e^2}{(\mathbf{p}_1 - \mathbf{p})^2}.$$

We note that taking account of higher-order approximations in the $\mathscr{G}$-function appearing in this integral does not introduce significant corrections, since in the corrected $\mathscr{G}$-function, p_0 corresponds to the Fermi surface, as before.

It was shown above that as a function of frequency, Γ_{ee} has all the properties of a Green's function for bosons. The same is true of the difference between Γ_{ee} and its zeroth-order approximation. It follows that the problem greatly resembles the calculation of the self-energy of an electron interacting with phonons, which was carried out in Sec. 21. However, in the case under discussion, small values of the momentum k are important. If we write

$$D(k, \omega) = \Gamma_{ee}(k, \omega) - \frac{4\pi e^2}{k^2},$$

then formula (21.29) is completely applicable to the present problem (with g^2 changed to 1). Bearing in mind that $\Sigma(\varepsilon) \ll \varepsilon$, and also that the important values of k are $\ll p_0$, we obtain

$$\Sigma_R = \frac{1}{(2\pi)^3 v} \int_{-\infty}^{\infty} d\beta \int_0^{\infty} k\, dk \int_{\xi - vk - \varepsilon}^{\xi + vk - \varepsilon} d\omega \frac{\operatorname{Im} D_R(k, \beta)}{\beta + \omega - i\delta} \times \left(\tanh \frac{\omega + \varepsilon}{2T} + \coth \frac{\beta}{2T}\right). \tag{22.31}$$

Since we are interested in the spectrum, which to a first approximation is given by the formula $\varepsilon = \xi$, it can be assumed that the integral with respect to ω is evaluated between the limits $-vk$ and vk.

The real part of Σ_R equals the principal value of the integral (22.31). After making various transformations, while taking into account the antisymmetry of $\operatorname{Im} D_R(k, \beta)$ and the relation between $\operatorname{Im} D_R(k, \beta)$ and $\operatorname{Re} D_R(k, \beta)$, we find that

$$\operatorname{Re} \Sigma_R = \frac{\pi}{(2\pi)^3 v} \int_0^{\infty} d\omega \int_{\omega/v}^{\infty} k\, dk \operatorname{Re} D_R(k, \omega) \times \left(\tanh \frac{\varepsilon + \omega}{2T} + \tanh \frac{\varepsilon - \omega}{2T}\right). \tag{22.32}$$

Similarly, the detour around the pole in (22.31) gives

$$\operatorname{Im}\Sigma_R = \frac{\pi}{(2\pi)^3 v}\int_0^\infty d\omega \int_{\omega/v}^\infty k\,dk\, \operatorname{Im} D_R(k,\omega) \times \left(2\coth\frac{\omega}{k} - \tanh\frac{\omega+\varepsilon}{2T} - \tanh\frac{\omega-\varepsilon}{2T}\right). \tag{22.33}$$

Just as before, the factor in parentheses involving the tanh and coth terms shows that only the larger of the two variables ε and T is important, except for the case where the pole of D_R is important (see below). The value of D_R is determined by the region in which ω lies, where we have the following three cases:

(a) $\omega \ll v_i k$,

(b) $v_i k \ll \omega \ll v_e k$,

(c) $\omega \sim v_e k$.

It is not hard to see that the corresponding regions of values of max $\{\varepsilon, T\}$ are (a) less than ω_{p1}, (b) between ω_{p1} and ω_{p2}, and (c) greater than ω_{p2}. However, according to the inequalities (22.2) and (22.3), we have $T \gg \omega_{p1}$, and consequently it is only necessary to consider cases (b) and (c).

We first consider the case $\omega_{p1} \ll \max\{\varepsilon, T\} \ll \omega_{p2}$. In the expression for $\operatorname{Re}\Sigma_R$ we divide the integral with respect to k into two regions

$$\frac{\omega}{v_e} \lesssim k \lesssim \frac{\omega}{v_i}, \qquad \frac{\omega}{v_i} \lesssim k < \infty.$$

Substituting

$$D_R = \frac{(4\pi e^2)^2 (Z^2\Pi_i + \Pi_e)}{k^2[k^2 - 4\pi e^2(Z^2\Pi_i + \Pi_e)]}, \tag{22.34}$$

we find that the contribution from the region $\omega/v_i \lesssim k < \infty$ is of order

$$\frac{e^2}{v}\,\omega_{p1}\tanh\frac{\varepsilon}{2T},$$

and hence is small compared to Σ_1. The other region $\omega/v_e \lesssim k \lesssim \omega/v_i$ makes an important contribution to Σ. Bearing in mind that the important values of ω satisfy the relation $\omega \sim \max\{\varepsilon, T\} \gg \omega_{p1}$, we can assume that

$$\Pi_i \ll \Pi_e, \qquad 4\pi e^2 \Pi_e = -\varkappa_e^2,$$

which implies

$$\begin{aligned}\operatorname{Re}\Sigma_R &= -\frac{e^2}{2\pi v}\int_0^\infty d\omega \int_{\omega/v}^\infty \frac{\varkappa_e^2\,dk}{k(k^2+\varkappa_e^2)}\left(\tanh\frac{\varepsilon+\omega}{2T} + \tanh\frac{\varepsilon-\omega}{2T}\right) \\ &= -\frac{e^2}{\pi v}\left[\varepsilon\ln\frac{v\varkappa_e}{T} - Tf\left(\frac{\varepsilon}{T}\right)\right].\end{aligned} \tag{22.35}$$

The same decomposition into domains of the variable k can be carried out for $\operatorname{Im}\Sigma_R$. In the region $\omega/v_e \lesssim k \lesssim \omega/v_i$, the function D_R has a pole, corresponding to "ionic sound," and the detour around this pole makes the following contribution to $\operatorname{Im}\Sigma_R$:

$$-\frac{\varepsilon^2\omega_{p1}}{4v}\int_0^\infty d\omega\int_{\omega/v}^{\sim\omega/v_i}\frac{k^2dk}{(k^2+\varkappa_e^2)^{3/2}}\,\delta\left(\omega-\omega_{p1}\frac{k}{\sqrt{k^2+\varkappa_e^2}}\right)$$
$$\times\left(2\coth\frac{\omega}{2T}-\tanh\frac{\omega+\varepsilon}{2T}-\tanh\frac{\omega-\varepsilon}{2T}\right)=-\frac{e^2T}{v}\ln\frac{\varkappa_i}{\varkappa_e}. \tag{22.36}$$

(This result is valid with logarithmic accuracy.) At the same time, the imaginary components of Π_i and Π_e may be important. Substituting these into the expression for $\operatorname{Im}\Sigma_R$ gives

$$\frac{e^2}{2\pi v}\int_0^\infty d\omega\int_{\omega/v}^\infty\frac{k\,dk\,4\pi e^2\operatorname{Im}(\Pi_e+Z^2\Pi_i)}{[k^2-4\pi e^2(\Pi_e+Z^2\Pi_i)]^2}$$
$$\times\left(2\coth\frac{\omega}{2T}-\tanh\frac{\omega+\varepsilon}{2T}-\tanh\frac{\omega-\varepsilon}{2T}\right).$$

Because of the exponential character of $\operatorname{Im}\Pi_i$, the important region in the corresponding integral is $k \gtrsim \omega/v_i$. An estimate shows that this integral makes a contribution to $\operatorname{Im}\Sigma_R$ of order $(e^2/v)T$, and hence with logarithmic accuracy is small compared to (22.36). The second imaginary component $\operatorname{Im}\Pi_e$ gives an integral over the region $\omega \sim \max\{\varepsilon, T\}$, $k \sim \varkappa_e$, and therefore the lower limit with respect to k can be set equal to zero:

$$-\frac{e^2}{4v^2}\int_0^\infty\omega\,d\omega\left(2\coth\frac{\omega}{2T}-\tanh\frac{\omega+\varepsilon}{2T}-\tanh\frac{\omega-\varepsilon}{2T}\right)\int_0^\infty\frac{\varkappa_e^2\,dk}{(k^2+\varkappa_e^2)^2}$$
$$=-\frac{\pi e^2}{16v^2\varkappa_e}[\varepsilon^2+(\pi T)^2]. \tag{22.37}$$

The second term in this expression is small compared to (22.36), but the term proportional to ε^2 can be larger than the ion attenuation.

We now turn our attention to the region $\omega_{p2} \ll \max\{\varepsilon, T\}$. By examining the integral (22.35) for $\operatorname{Re}D_R$, it is not hard to see that it makes an unimportant contribution of order

$$\frac{e^2}{v}\omega_{p2}\tanh\frac{\varepsilon}{2T}.$$

Therefore, the real part of Σ_R is essentially equal to Σ_{R1}. As for the imaginary part, in this case the pole of D_R corresponding to ionic sound makes the same contribution (22.36), and similarly, the term coming from $\operatorname{Im}\Pi_i$ is again unimportant.

The electron attenuation coming from $\operatorname{Im}\Pi_e$ is an integral in which the most important region is $\omega \sim v\varkappa_e \sim \omega_{p2}$, $k \sim \varkappa_e$. If $T \gg \omega_{p2}$ this integral is of order $(e^2/v)T$, and can therefore be neglected. On the other hand, if $T \ll \omega_{p2}$ we can set $T = 0$ in the integral. Since in this case $\omega \sim vk$, we must substitute the whole expression (22.22) for Π_e. First of all, we use the fact that the expressions in parentheses in (22.32) and (22.33), which give the temperature dependence, coincide for $T = 0$, taking the same value 2. Combining these two formulas, we obtain

$$(\Sigma_R - \Sigma_{R1})\Big|_{\substack{T=0 \\ \varepsilon \gg \omega_{p2}}} = \frac{1}{(2\pi)^2 v} \int_0^\infty d\omega \int_{\omega/v}^\infty k\, dk\, D_R(k, \omega).$$

If we substitute (22.34) and (22.22) into this formula, the right-hand side becomes

$$-e^2 \operatorname{Re}(\beta_1 + i\beta_2), \tag{22.38}$$

where β_1 and β_2 are constants, equal to the real and imaginary parts of the integral

$$\frac{1}{2}\int_0^1 du\left[1 - u\left(\operatorname{arc\,tanh} u - \frac{i\pi}{2}\right)\right]^{1/2}. \tag{22.39}$$

(Here we take the value of the square root which has positive imaginary part.) The imaginary part of (22.38) gives the electron attenuation.

Using the equation

$$\varepsilon - \xi - \Sigma + \Delta\mu = 0,$$

we can now find the electron spectrum and the attenuation of the electron excitations in various regions:

(a) $$\varepsilon(\mathbf{p}) = \xi(\mathbf{p})\left[1 + \frac{e^2}{\pi v}\left(\ln\frac{2p_0}{\varkappa_e} - 1\right)\right],$$

$$\gamma(\mathbf{p}) = \frac{e^2}{v}\left[T \ln\frac{\varkappa_i}{\varkappa_e} + \frac{\pi\xi^2(\mathbf{p})}{16 v \varkappa_e}\right] \quad \text{for} \quad \omega_{p1} \ll \max\{\xi, T\} \ll \omega_{p2}; \tag{22.40}$$

(b) $$\varepsilon(\mathbf{p}) = \xi(\mathbf{p})\left\{1 + \frac{e^2 m}{\pi p}\left[\ln\frac{(p + p_0)^2}{2mT} - 1\right]\right\} - \frac{e^2 m T}{\pi p} f\left[\frac{\xi(\mathbf{p})}{T}\right],$$

$$\gamma(\mathbf{p}) = \frac{e^2}{v} T \ln\frac{\varkappa_i}{\varkappa_e} + e^2 \varkappa_e \beta_2 \quad \text{for} \quad \omega_{p2} \ll \max\{\xi, T\}.$$

In particular, we have

$$\varepsilon(\mathbf{p}) = \xi(\mathbf{p})\left[1 + \frac{e^2}{\pi v}\left(\ln\frac{4\gamma v p_0}{\pi T} - 1\right)\right] \quad \text{for} \quad \frac{p_0^2}{2m} \gg T \gg \max\{\xi(\mathbf{p}), \omega_{p2}\},$$

$$\varepsilon(\mathbf{p}) = \xi(\mathbf{p})\left[1 + \frac{e^2 m}{\pi p} \ln\left|\frac{p + p_0}{p - p_0}\right|\right] \quad \text{for } \xi(\mathbf{p}) \gg \max\{T, \omega_{p2}\}. \tag{22.41}$$

22.4. Thermodynamic functions. We conclude this section by considering the thermodynamic functions of a degenerate plasma [calculated by Vedenov (V1)]. According to formula (10.22), we have

$$\Delta\Omega = \Omega - \Omega_0 = \frac{1}{2}\int_0^{e^2} d(e^2) \int d\mathbf{r}\, d^4x' \frac{\delta(\tau - \tau')}{|\mathbf{r} - \mathbf{r}'|}$$

$$\times \left[\langle \bar{\tilde{\psi}}_\alpha(x) \bar{\tilde{\psi}}_\beta(x') \tilde{\psi}_\beta(x') \tilde{\psi}_\alpha(x)\rangle - 2Z\langle \bar{\tilde{\psi}}_\alpha(x) \bar{\tilde{\Phi}}(x') \tilde{\Phi}(x') \tilde{\psi}_\alpha(x)\rangle\right.$$

$$\left. + Z^2 \langle \bar{\tilde{\Phi}}(x) \bar{\tilde{\Phi}}(x') \tilde{\Phi}(x') \tilde{\Phi}(x)\rangle\right]. \tag{22.42}$$

The expressions in angular brackets $\langle\cdots\rangle$ can be written in terms of the functions $\mathscr{G}$ and $\mathscr{T}$, e.g.,

$$\langle\tilde{\psi}_\alpha(x)\tilde{\psi}_\beta(x')\tilde{\psi}_\beta(x')\tilde{\psi}_\alpha(x)\rangle = -2\mathscr{G}_e^{(0)}(x - x')\mathscr{G}_e^{(0)}(x' - x) + \left(\frac{N_e}{V}\right)^2$$
$$- \int d^4x_1\, d^4x_2\, d^4x_3\, d^4x_4\, \mathscr{G}_e^{(0)}(x - x_1)\mathscr{G}_e^{(0)}(x' - x_2)$$
$$\times\, \mathscr{G}_e^{(0)}(x_3 - x)\mathscr{G}_e^{(0)}(x_4 - x')\mathscr{T}^{ee}_{\alpha\beta\alpha\beta}(x_1, x_2; x_3, x_4).$$

The term $(N_e/V)^2$ can be omitted, since in (22.34) it will cancel a similar term coming from electron-ion and ion-ion interactions (because of the electrical neutrality of the plasma). Moreover, when writing down the averages involving four field operators, coming from the electron-ion and ion-ion interactions, we can leave out the term involving two $\mathscr{G}$-functions. The point is that this term is due to exchange, and exchange of ions with electrons is impossible, whereas exchange of ions with ions gives rise to a very small effect, since the ions form a Boltzmann gas.

Thus, the resulting expression consists of two terms. The first term comes from the product of two electron $\mathscr{G}$-functions and represents the exchange energy of the electrons, while the second term corresponds to a sum of terms with different $\mathscr{G}$-functions. We begin by considering the first term, which equals

$$\frac{\Delta\Omega_1}{V} = -\lim_{\tau_1,\tau_2\to+0}\sum_{\varepsilon_1,\varepsilon_2}\int\frac{d\mathbf{p}_1\, d\mathbf{p}_2}{(2\pi)^6}\frac{1}{(\mathbf{p}_1 - \mathbf{p}_2)^2}$$
$$\times\frac{e^{i\varepsilon_1\tau_1}}{i\varepsilon_1 - \varepsilon_0(\mathbf{p}_1) + \mu}\frac{e^{i\varepsilon_2\tau_2}}{i\varepsilon_2 - \varepsilon_0(\mathbf{p}_2) + \mu}$$

in the momentum representation. The factors $e^{i\varepsilon_m\tau}$ take account of the order of the ψ-operators in the Hamiltonian (22.1), and the sums over ε_1 and ε_2 are independent. Recalling the definition of the Fourier components of the function $\mathscr{G}$, we obtain

$$\lim_{\tau\to+0} T\sum_{\varepsilon}\frac{e^{i\varepsilon\tau}}{i\varepsilon - \varepsilon_0(\mathbf{p}) + \mu} = \lim_{\tau\to+0}\mathscr{G}_e^{(0)}(\mathbf{p}, -\tau) = n(\mathbf{p}),$$

which leads to

$$\frac{\Delta\Omega_1}{V} = -4\pi e^2\int\frac{d\mathbf{p}_1\, d\mathbf{p}_2}{(2\pi)^6}\frac{n(\mathbf{p}_1)n(\mathbf{p}_2)}{(\mathbf{p}_1 - \mathbf{p}_2)^2}. \tag{22.43}$$

For a Boltzmann gas, (22.43) is small since it involves occupation numbers $n \ll 1$. This justifies neglecting the ion-exchange term.

As for the second term, it equals

$$\frac{\Delta\Omega_2}{V} = -\frac{4\pi}{2}\int_0^{e^2} d(e^2)T^3\sum_{\varepsilon_1,\varepsilon_2,\omega}\int\frac{d\mathbf{k}\, d\mathbf{p}_1\, d\mathbf{p}_2}{(2\pi)^9}\frac{1}{k^2}$$
$$\times[4\mathscr{G}^e(\mathbf{p}_1, \varepsilon_1)\mathscr{G}^e(\mathbf{p}_1 + \mathbf{k}, \varepsilon_1 + \omega)\mathscr{G}^e(\mathbf{p}_2, \varepsilon_2)\mathscr{G}^e(\mathbf{p}_2 + \mathbf{k}, \varepsilon_2 + \omega)\mathscr{T}_{ee}(\mathbf{k}, \omega)$$
$$- 8Z\mathscr{G}^i(\mathbf{p}_1, \varepsilon_1)\mathscr{G}^i(\mathbf{p}_1 + \mathbf{k}, \varepsilon_1 + \omega)\mathscr{G}^e(\mathbf{p}_2, \varepsilon_2)\mathscr{G}^e(\mathbf{p}_2 + \mathbf{k}, \varepsilon_2 + \omega)\mathscr{T}_{ei}(\mathbf{k}, \omega)$$
$$+ 4Z^2\mathscr{G}^i(\mathbf{p}_1, \varepsilon_1)\mathscr{G}^i(\mathbf{p}_1 + \mathbf{k}, \varepsilon_1 + \omega)\mathscr{G}^i(\mathbf{p}_2, \varepsilon_2)\mathscr{G}^i(\mathbf{p}_2 + \mathbf{k}, \varepsilon_2 + \omega)\mathscr{T}_{ii}(\mathbf{k}, \omega)] \tag{22.44}$$

in the momentum representation. Since the vertex parts $\mathcal{T}$ are themselves of order e^2, this expression is formally of order e^4. However, here it is important that small k and m make the main contribution to the integral over k and the sum over m. Comparing (22.44) with (22.9), we find that

$$\frac{\Delta\Omega_2}{V} = -2\pi \int_0^{e^2} d(e^2)T \sum_{\omega} \int \frac{d\mathbf{k}}{(2\pi)^3} \frac{1}{k^2}$$

$$\times [\mathcal{K}_e^2(\mathbf{k}, \omega)\mathcal{T}_{ee}(\mathbf{k}, \omega) + 2Z\mathcal{K}_i(\mathbf{k}, \omega)\mathcal{K}_e(\mathbf{k}, \omega)\mathcal{T}_{ei}(\mathbf{k}, \omega)$$

$$+ Z^2\mathcal{K}_i^2(\mathbf{k}, \omega)\mathcal{T}_{ii}(\mathbf{k}, \omega)]$$

$$= -8\pi^2 \int_0^{e^2} e^2\, d(e^2)T \sum_{\omega} \int \frac{d\mathbf{k}}{(2\pi)^3} \frac{1}{k^2} \frac{\mathcal{K}_e^2 + 2Z^2\mathcal{K}_i\mathcal{K}_e + Z^4\mathcal{K}_i^2}{k^2 - 4\pi e^2(\mathcal{K}_e + Z^2\mathcal{K}_i)} \tag{22.45}$$

$$= -\frac{1}{\pi} \int_0^{e^2} e^2\, d(e^2)T \sum_{\omega} \int \frac{d\mathbf{k}}{(2\pi)^3} \frac{1}{k^2} \frac{(\mathcal{K}_e + Z^2\mathcal{K}_i)^2}{k^2 - 4\pi e^2(\mathcal{K}_e + Z^2\mathcal{K}_i)},$$

where the important values of k^2 are of order $4\pi e^2(\mathcal{K}_e + Z^2\mathcal{K}_i)$. We must determine the relation between the values of v_ek, v_ik and $\omega = 2m\pi T$. It is not hard to see that the integral with respect to $\mathbf{k}$ in (22.45) increases as $4\pi e^2(\mathcal{K}_e + Z^2\mathcal{K}_i)$ is increased. Now consider formula (22.11) for $\mathcal{K}$. If we assume that $vk \ll T$, the largest value of $\mathcal{K}(\mathbf{k}, \omega)$ is obtained for $\omega = 0$. Then the most important ion loop is the one for which

$$\mathcal{K}_i \sim \frac{N_i}{VT},$$

and hence

$$4\pi e^2(\mathcal{K}_e + Z^2\mathcal{K}_i) \sim \frac{e^2}{V}\frac{N_iZ^2}{T}.$$

It follows that

$$(v_ik)^2 \sim \frac{T}{M}\frac{N_i}{V}\frac{e^2}{T} \sim \frac{e^2N_i}{VM} \ll T^2.$$

Thus, our assumption is justified, and in (22.45) we can just take the term $m = 0$, setting

$$\mathcal{K}_e + Z^2\mathcal{K}_i \approx -\frac{Z^2}{V}\frac{\partial N_i}{\partial \mu}.$$

After integrating with respect to k^2, this gives

$$\frac{\Delta\Omega_2}{V} = -\frac{2\sqrt{\pi}}{3}(Ze)^3T\left(\frac{1}{V}\frac{\partial N_i}{\partial\mu}\right)^{3/2} = -\frac{2\sqrt{\pi}}{3}(Ze)^3T^{1/2}\left(\frac{N_i}{V}\right)^{3/2}. \tag{22.46}$$

With our hypotheses (22.2) and (22.3), this term is small compared to (22.43). However, this is the only term of order e^3, since the correction to the term $\Delta\Omega_1$ must be of order e^4. If we do not impose the strong degeneracy condition on the electron gas, $\Delta\Omega_2$ can be of the same order as $\Delta\Omega_1$, but then we must take account of the electron loop. Thus, the final result is

$$\frac{\Delta\Omega}{V} = -4\pi e^2 \int \frac{d\mathbf{p}_1\, d\mathbf{p}_2}{(2\pi)^6} \frac{n(\mathbf{p}_1)n(\mathbf{p}_2)}{(\mathbf{p}_1 - \mathbf{p}_2)^2} - \frac{2\sqrt{\pi}}{3} e^3 T \left(\frac{Z^2}{V} \frac{\partial N_i}{\partial \mu} + \frac{1}{V} \frac{\partial N_e}{\partial \mu} \right)^{3/2},$$

subject to the conditions

$$e^2 m \left(\frac{V}{N} \right)^{1/3} \ll 1, \quad T \gg \max \left\{ e^2 \left(\frac{N}{V} \right)^{1/3}, \quad \frac{1}{M} \left(\frac{N}{V} \right)^{2/3} \right\}. \qquad (22.47)$$

5

SYSTEMS OF INTERACTING BOSONS

23. Application of Field Theory Methods to a System of Interacting Bosons for $T = 0$

The generalization of the methods of quantum field theory to the case of a system of interacting bosons at temperatures below the temperature of "Bose condensation" entails great difficulties. Nevertheless, the appropriate formalism has been developed [see Belyaev (B2)], and this chapter will be devoted to its study. As always, we first consider the case $T = 0$, i.e., the absolute zero of temperature.

When constructing the diagram technique in previous chapters, we consistently made use of the fact that the average of a product of several noninteracting operators ψ, ψ^+ can be reduced to products of averages of pairs of the operators. This is a consequence of Wick's theorem, which states that the average of a chronological product of any number of field operators decomposes into a sum of normal products with all possible pairings. For a system of fermions, the ground state or "vacuum" is such that the averages of the normal products can be made to vanish by properly changing the definition of the annihilation and creation operators. The situation is utterly different in the case of a system of bosons. In fact, in a Bose gas at low temperatures, an arbitrarily large number of particles can be "concentrated" in the state with zero momentum. In an ideal Bose gas at $T = 0$, the number of particles in the lowest level is simply equal to the number of particles in the system. Thus, a characteristic feature of the state of Bose condensation is that the density of the particles in the ground state which have zero momentum approaches a finite limit as N, the total number of particles, and V, the volume of the system, tend to infinity. Therefore, averages of normal products of the form $(a_0^+)^n a_0^n$, are not only nonvanishing, but can actually be arbitrarily large.

Consider a system of interacting bosons at the absolute zero of temperature. As already noted, in an ideal Bose gas all the particles will be in the state with zero momentum. Suppose that from the operators $\psi(x)$, $\psi^+(x)$ in the interaction representation we separate the operators corresponding to annihilation and creation of particles in the state with $p = 0$:

$$\psi(x) = \xi_0 + \psi'(x), \qquad \psi^+(x) = \xi_0^+ + \psi'^+(x). \tag{23.1}$$

(Below, we shall write $\xi_0 = a_0/\sqrt{V}$ and $\xi_0^+ = a_0^+/\sqrt{V}$.) The total number of particles $N = V\xi_0^+\xi_0$ becomes infinite as $V \to \infty$. Therefore, if we neglect the right-hand side in the commutation relation

$$\xi_0\xi_0^+ - \xi_0^+\xi_0 = \frac{1}{V}$$

the operators ξ_0 and ξ_0^+ can be regarded as numbers to a first approximation, just as was done in Sec. 4. However, we shall see later that this makes sense only for an infinitesimal interaction.

The total Hamiltonian of the system can be written as

$$H = H_0 + H_{\text{int}},$$

where

$$H_0 = \frac{1}{2m}\int \nabla\psi^+(x)\cdot\nabla\psi(x)\,d\mathbf{r},$$

and H_{int} is the interaction Hamiltonian, whose form we do not specify for the time being. All the familiar field theory formulas, which use the S-matrix to connect operators in the Heisenberg representation with operators in the interaction representation, remain the same, and so does the definition of the S-matrix itself:

$$S = T\exp\left\{-i\int H_{\text{int}}(x)\,d^4x\right\}. \tag{23.2}$$

The one-particle Green's function $G(x, x')$ is defined by

$$G(x - x') = -i\langle T(\tilde{\psi}(x)\tilde{\psi}^+(x'))\rangle \tag{23.3}$$

in terms of operators in the Heisenberg representation, and by

$$G(x - x') = -\frac{i\langle T(\psi(x)\psi^+(x')S)\rangle}{\langle S\rangle} \tag{23.3'}$$

in the interaction representation. [It is understood that the average $\langle\cdots\rangle$ is over a ground state containing N interacting particles in (23.3), and over a ground state containing N noninteracting particles in (23.3').] Instead of (23.3), it is more convenient to divide the Green's function $G(x - x')$ into two parts, i.e.,

$$G'(x - x') = -i\langle T(\tilde{\psi}'(x)\tilde{\psi}'^+(x'))\rangle = -\frac{i\langle T(\psi'(x)\psi'^+(x')S)\rangle}{\langle S\rangle} \tag{23.4}$$

and

$$G_0(t - t') = -i\langle T(\xi_0(t)\xi_0^+(t'))\rangle = -\frac{i\langle T(\xi_0(t)\xi_0^+(t')S)\rangle}{\langle S\rangle}, \tag{23.5}$$

where $G(x - x')$ is the Green's function of the particles not in the condensate,[1] and $G_0(t - t')$ is the Green's function of the particles in the condensate. Obviously, $G_0(t - t')$ does not depend on the difference $\mathbf{r} - \mathbf{r}'$, and can therefore be defined as the zero-momentum Fourier component of the exact Green's function:

$$G_0(t - t') = \int G(\mathbf{r} - \mathbf{r}', t - t')\, d\mathbf{r}'.$$

The density of the particles in the condensate is

$$n_0 = iG_0(t - t') \qquad (t' = t + 0),$$

and the density of the total number of particles is

$$n = n' + n_0 = i[G'(0, t - t') + G_0(t - t')] \qquad (t' = t + 0). \tag{23.6}$$

Once again we call attention to the fact (already noted in Sec. 4) that the number of particles in the condensate is different from the total number of particles when interaction is present.

Next, turning to the perturbation series for interacting particles, we construct the appropriate version of the diagram technique. In view of the special role of the particles in the condensate, we assume that the substitution (23.1) has been made in the expression for the Hamiltonian H_{int}, and that H_{int} has been brought into a form in which the operators ξ_0 and ξ_0^+, ψ' and ψ'^+ appear separately. Moreover, we assume that H_{int} is already written in this form in the definition of the S-matrix (23.2). [Our argument is applicable to a Hamiltonian $H_{\text{int}}(x)$ which is the product of any number of operators, with arbitrary interactions between particles.] After making this substitution, each of the operations of chronological ordering (the T-product) and averaging over a ground state of noninteracting particles can be represented as two consecutive operations, which act separately on the particles in the condensate and those not in the condensate, i.e.,

$$T = T^0 T', \quad \langle\cdots\rangle = \langle\langle\cdots\rangle'\rangle^0, \tag{23.7}$$

where T^0 and $\langle\cdots\rangle^0$ act on the operators ξ_0 and ξ_0^+, while T' and $\langle\cdots\rangle'$ act on the operators ψ' and ψ'^+.

Each term of the expansion of the S-matrix in powers of the interaction Hamiltonian H_{int} contains various products of the operators ξ_0, ξ_0^+, ψ' and ψ'^+. The free operators ψ' and ψ'^+ can be handled in the ordinary way by using Wick's theorem, since normal products involving particles not in the

---

[1] Here and elsewhere, the phrase "not in the condensate" has been chosen as the safest compromise for the suggestive Russian adjective надконденсатный (literally ≈ "over the condensate"). (*Translator*)

condensate have zero averages. The averages of chronological products of the form $\psi'\psi'^+$, i.e., the quantities

$$G^{(0)}(x - x') = -i\langle T'(\psi'(x)\psi'^+(x'))\rangle' \equiv -i\langle T(\psi'(x)\psi'^+(x'))\rangle \quad (23.8)$$

are nonzero, and the corresponding Fourier components have the form

$$G^{(0)}(x - x') = (2\pi)^{-4} \int G^{(0)}(p) e^{ip(x-x')}\, d^4p,$$
$$G^{(0)}(p) = \frac{1}{\omega - (\mathbf{p}^2/2m) + i\delta} \quad (23.9)$$

Therefore, if we regard the operators ξ_0 and ξ_0^+ as (numerical) parameters, they play the role of an external field at various vertices of the diagrams.

Next, we consider the problem of calculating the Green's function when the number of particles not in the condensate is arbitrary. The appropriate Green's function has the form

$$G_n(x_1, \ldots, x_n; x_1', \ldots, x_n') = \frac{(-i)^n \langle T(\psi'(x_1)\cdots\psi'(x_n)\psi'^+(x_1')\cdots\psi'^+(x_n')S)\rangle}{\langle S\rangle}. \quad (23.10)$$

Separating the operations T and $\langle\cdots\rangle$ into operations T', T^0, $\langle\cdots\rangle'$ and $\langle\cdots\rangle^0$, according to (23.7), we first study the perturbation series for the quantity

$$\bar{G}_n(x_1, \ldots, x_n; x_1', \ldots, x_n') = (-i)^n \langle T'(\psi'(x_1)\cdots\psi'(x_n)\psi'^+(x_1')\cdots\psi'^+(x_n')S)\rangle'. \quad (23.11)$$

Since the operations T' and $\langle\cdots\rangle$ have no effect on the operators ξ_0 and ξ_0^+, the latter act as parameters with respect to these operations, and have no effect whatsoever on the time-ordering and averaging of various products of operators of particles not in the condensate. Therefore, the corresponding matrix element can be written down by using the ordinary rules for constructing Feynman diagrams, and contains products of the chronological averages (23.8) and powers of the operators ξ_0, ξ_0^+.

The number of operators ξ_0, ξ_0^+ that appear to a given order in an expansion of the S-matrix in powers of the interaction Hamiltonian H_{int} depends on the form of H_{int} and on the terms in H_{int} which are chosen after making the substitution (23.1). For example, after the substitution

$$\psi \to \xi_0 + \psi', \qquad \psi^+ \to \xi_0^+ + \psi'^+,$$

the interaction Hamiltonian

$$H_{\text{int}} = \frac{1}{2}\int \psi^+(x)\psi^+(x')U(\mathbf{r} - \mathbf{r}')\psi(x')\psi(x)\, d\mathbf{r}\, d\mathbf{r}' \quad (23.12)$$

(see Sec. 25) decomposes into eight terms, beginning with a term

$$\frac{1}{2}(\xi_0^+)^2\xi_0^2 \int U(\mathbf{r})\, d\mathbf{r},$$

of degree four in ξ_0, ξ_0^+, and ending with the term

$$\frac{1}{2}\int \psi'^+(x)\psi'^+(x')U(\mathbf{r}-\mathbf{r}')\psi'(x')\psi'(x)\,d\mathbf{r}\,d\mathbf{r}'.$$

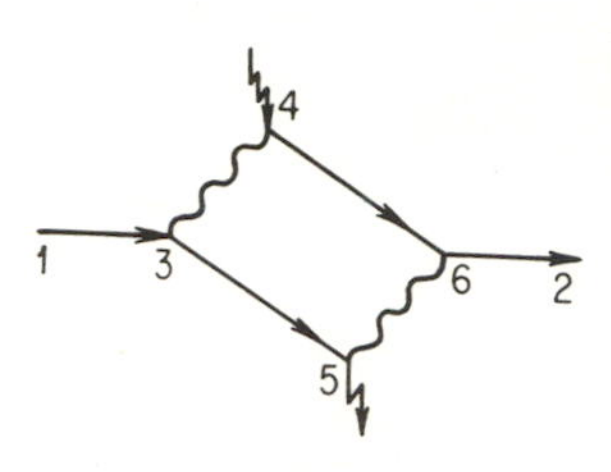

FIGURE 63

In Fig. 63 we illustrate one of the second-order diagrams for the function $\bar{G}(x-x')$. In this diagram, a solid line corresponds to the function $G^{(0)}(x-x')$ given by (23.9), a wavy line corresponds to the interaction potential $U(\mathbf{r}-\mathbf{r}')$, and a zigzag "free" line corresponds to one of the operators ξ_0, ξ_0^+ (ξ_0^+ is represented by a line directed towards a vertex and ξ_0 by a line directed away from a vertex). The matrix element of this diagram is

$$\begin{aligned}\bar{G}(x_1-x_2) = i\int G^{(0)}(x_1-x_3)G^{(0)}(x_3-x_5)\xi_0(t_5)U(\mathbf{r}_3-\mathbf{r}_4)\\ \times\xi_0^+(t_4)G^{(0)}(x_4-x_6)U(\mathbf{r}_6-\mathbf{r}_5)G^{(0)}(x_6-x_2)d^4x_3\cdots d^4x_6.\end{aligned}\tag{23.13}$$

In the general case (23.10), the mth-order matrix element in the function $\bar{G}_n(x_1,\ldots,x_n;x_1',\ldots,x_n')$ contains a product of any number of operators ξ_0 and ξ_0^+. However, it should be noted that the degrees of ξ_0 and ξ_0^+ must be the same, since the interaction H_{int} preserves the total number of particles. Therefore, if the numbers of operators ξ_0 and ξ_0^+ were unequal, the numbers of operators ψ' and ψ'^+ would also be unequal in the average $\langle\cdots\rangle'$, which would then vanish.

In (23.11) let $M_n(x_1,\ldots,x_n;x_1',\ldots,x_n')$ be a connected diagram with $2m$ vertices, corresponding to m operators ξ_0 and ξ_0^+. As usual, by a *connected diagram* we mean a diagram which does not decompose into separate parts not connected by at least one line. Together with M_n, we consider all diagrams differing from M_n by the presence of "vacuum" loops, i.e., various disconnected diagrams. As is well known from field theory, the effect of the whole set of such diagrams is to multiply each matrix element by the average value of the S-matrix. In our case, this means that M_n is multiplied by $\langle S\rangle'$.

Having discussed the calculation of $\bar{G}_n(x_1,\ldots,x_n;x_1',\ldots,x_n')$, we again turn our attention to the function

$$G_n(x_1,\ldots,x_n;x_1',\ldots,x_n')\equiv\frac{\langle T^0\bar{G}_n(x_1,\ldots,x_n;x_1',\ldots,x_n')\rangle^0}{\langle S\rangle}.$$

So far, we have been able to ignore the operator character of ξ_0 and ξ_0^+, since the operations T' and $\langle\cdots\rangle'$ do not affect ξ_0 and ξ_0^+, which commute with ψ' and ψ'^+. At this stage, however, the operator character of ξ_0 and ξ_0^+ becomes important. Each matrix element M_n in $\bar{G}_n$, like (23.13), contains in its integrand a definite number of operators ξ_0 and ξ_0^+, multiplied by averages of the form (23.8). Suppose M_n contains

$$\xi_0(t_1)\cdots\xi_0(t_m)\xi_0^+(t_1')\cdots\xi_0^+(t_m').$$

Then, to finally obtain $G_n(x_1, \ldots, x_n; x'_1, \ldots, x'_n)$, we have to calculate averages of the form

$$\frac{\langle T^0(\xi_0(t) \cdots \xi_0(t_m)\, \xi_0^+(t'_1) \cdots \xi_0^+(t'_m)\langle S\rangle')\rangle^0}{\langle S\rangle}.$$

Since the operations T^0 and $\langle \cdots \rangle^0$ do not affect particles which are not in the condensate, we see that the required averages are m-particle Green's functions of the particles in the condensate:

$$G_{0m}(t_1, \ldots, t_m; t'_1, \ldots, t'_m) = \frac{\langle T(\xi_0(t_1) \cdots \xi_0(t_m)\xi_0^+(t'_1) \cdots \xi_0^+(t'_m)S)\rangle}{\langle S\rangle}. \tag{23.14}$$

Therefore, to determine the Green's functions of the particles not in the condensate in terms of a perturbation series, we have to know the exact m-particle Green's functions of the particles in the condensate.

The complexity of directly calculating the functions G_{0m} from expressions of the form (23.14) in terms of n, the density of particles in the condensate in the absence of interaction, is associated with the fact that for products of the operators ξ_0, ξ_0^+ the expansion in normal products given by Wick's theorem is meaningless. This is because the average of such a normal product [of the form $N(a_0 \cdots a_0^+ \cdots)$] over the ground state is not only nonzero but actually very large. Moreover, we cannot neglect the fact that the operators ξ_0 and ξ_0^+ in (23.14) do not commute. In fact, $\langle S\rangle'$ can be written in the form[2]

$$\langle S\rangle' = \exp \sigma, \tag{23.15}$$

where σ, the sum of all connected[3] "vacuum" loops, is a functional of ξ_0 and ξ_0^+. This sum is proportional to the volume ($n_0 = \xi_0^+\xi_0$, the density of particles in the condensate, is a finite quantity). If in (23.15), $\langle S\rangle'$ is expanded formally as a power series in σ, arbitrary powers of V will appear. Therefore, we are not justified in neglecting the right-hand sides of the commutation relations

$$\xi_0\xi_0^+ - \xi_0^+\xi_0 = \frac{1}{V},$$

even though they are of order $1/V$, since the smallness of the quantity $1/V$ may be compensated by the corresponding power of V in the expansion (23.15).

Because of this, another approach is more convenient. Our starting point

---

[2] In field theory, the possibility of representing $\langle S\rangle'$ in the form (23.15) is proved under the assumption that ξ_0 and ξ_0^+ are external parameters, with no operator properties. However, we see that in (23.14), $\langle S\rangle'$ is subjected to the operation of time ordering with respect to the operators ξ_0 and ξ_0^+. Since Bose operators appearing inside a T-product commute with each other, by the very meaning of the T-product, it follows that the field theory assumption is satisfied here.

[3] I.e., not decomposable into independent parts.

is the observation that the expressions (23.14) can be written directly in terms of Heisenberg operators, i.e.,

$$G_{0m}(t_1,\ldots,t_m;t_1',\ldots,t_m') = \langle T(\xi_0(t_1)\cdots\xi_0(t_m)\xi_0^+(t_1')\cdots\xi_0^+(t_m'))\rangle, \quad (23.16)$$

where the average of the product is taken over a ground state of interacting particles. We first consider the average

$$V\langle\xi_0^+\xi_0\rangle,$$

which represents the exact number of particles with zero momentum. In an ideal gas at $T = 0$, this number is just the total number of particles N, since all the particles will be in the state with $\mathbf{p} = 0$, but interaction between the particles will decrease the number of particles in this state. (The interaction must be repulsive at sufficiently small distances, since attraction at all distances would make the system unstable.) However, as already emphasized in Sec. 4, the condensate will not disappear, i.e., the average number of particles with zero momentum tends to infinity as the number of particles in the system becomes infinite (as $V \to \infty$, the density n_0 of particles in the condensate remains finite for any interaction between particles). Physically, this fact is quite obvious, but of course we cannot exclude the logical possibility that n_0 might vanish for some interaction. We shall not attempt to prove this assertion here, especially since in the present instance, the unique physical object under consideration is helium. The interested reader can find a proof in Belyaev's paper B2. The fact that the total number of particles changes because of the interaction is precisely the reason we were previously unable to regard the free operators ξ_0 and ξ_0^+ as c-numbers.

The perturbation series for the Green's functions of the particles not in the condensate (constructed above) contains averages of the exact Heisenberg operators ξ_0 and ξ_0^+. If the condensate does not disappear, then, as far as their effect on the ground state of interacting particles is concerned, the operators ξ_0 and ξ_0^+ can in turn be regarded as c-numbers, to a first approximation. This fact can be used to simplify the expressions for the Green's functions of the particles in the condensate. However, it should not be forgotten that the operator ξ_0 annihilates a particle, whereas the operator ξ_0^+ creates a particle. Therefore, the essential point is actually that in all our subsequent considerations, the only important matrix elements of ξ_0, ξ_0^+ are those corresponding to transitions from the ground state of a system with N particles ($N \to \infty$) to the ground state of a system with $N \pm 1$ particles. From a physical point of view, by adding one particle to (or subtracting one particle from) the infinite number of particles in the Bose condensate, we do not change the ground state of the system, except to the extent of increasing its energy by an amount equal to the chemical potential μ. This fact will always be kept in mind when we refer to the operators ξ_0 and ξ_0^+ as numbers.

We now examine these matters in more detail, as applied to the one-particle function

$$G_0(t - t') = -i\langle T(\xi_0(t)\xi_0^+(t'))\rangle \approx -i\langle \xi_0(t)\xi_0^+(t')\rangle.$$

We write the last expression as a sum of products of matrix elements over intermediate states, i.e.,

$$\langle \Phi_N^* | \xi_0(t)\xi_0^+(t') | \Phi_N\rangle = \langle \Phi_N^* | \xi_0(t) | \Phi_{N+1}\rangle\langle \Phi_{N+1}^* | \xi_0(t') | \Phi_N\rangle$$
$$+ \sum_s \langle \Phi_N^* | \xi_0(t) | \Phi_{N+1}^s\rangle\langle \Phi_{N+1}^{s*} | \xi_0^+(t') | \Phi_N\rangle,$$

where Φ_N and Φ_{N+1} are the ground states of systems of N and $N + 1$ interacting particles, respectively, and the Φ_{N+1}^s are the states other than the ground state of a system of $N + 1$ particles. In this expression, the term containing the sum is small, since $\xi_0^+\Phi_N \approx \Phi_{N+1}$ (for example), while Φ_{N+1} and Φ_{N+1}^s are orthogonal. Moreover, the time dependence of the matrix elements for transitions from one ground state to the other can be found by using the ordinary quantum-mechanical formula

$$-i\frac{\partial}{\partial t}\langle \Phi_N^* | \xi_0(t) | \Phi_{N+1}\rangle = \langle \Phi_N^* | [\hat{H}, \xi_0(t)] | \Phi_{N+1}\rangle$$

or

$$\xi_0(t) = \xi_0(0)e^{-i(E_{N+1} - E_N)t}.$$

Using the definition $\mu = \partial E/\partial N$ of the chemical potential, and replacing $\xi_0(0)$ by $\sqrt{n_0}$, we find that

$$iG_0(t - t') = n_0 e^{-i\mu(t-t')}. \tag{23.17}$$

Thus, the function $G_0(t - t')$ has been separated into a product of two independent functions, where $\xi_0(t)$ has been associated with the factor $\sqrt{n_0}\,e^{-i\mu t}$ and $\xi_0^+(t')$ with the factor $\sqrt{n_0}\,e^{i\mu t'}$. Obviously, the situation is the same for an arbitrary Green's function of the particles in the condensate. In other words, in replacing the operators $\xi_0(t)$, $\xi_0^+(t)$ by numbers, each operator should be associated with a factor of the type indicated. Thus, we have reduced the diagram technique for calculating Green's functions of the particles not in the condensate to the ordinary diagram technique, with the operators ξ_0 and ξ_0^+ playing the role of an external field:

$$\xi_0(t) = \sqrt{n_0}\,e^{-i\mu t}, \quad \xi_0^+(t) = \sqrt{n_0}\,e^{i\mu t}. \tag{23.18}$$

As usual, in writing the perturbation series, it is only necessary to take account of the connected diagrams. In fact, as we have shown, taking account of the disconnected diagrams amounts to replacing the density of the particles in the condensate of an ideal gas by the exact density of the particles in the condensate of a gas with interaction between the particles; it also leads to the appearance of the frequency factors (23.18). Except for this, all the diagrams are the same as if we regard the operators ξ_0 and ξ_0^+ in the interaction representation as external parameters [after substituting the expressions

(23.1) for the operators appearing in the interaction Hamiltonian H_{int}], and then, in evaluating the expressions (23.10), carry out the average $\langle \cdots \rangle'$ and the time-ordering T' only with respect to particles not in the condensate (confining ourselves to connected diagrams). To obtain the final expressions, we need only use (23.18) to replace ξ_0 and ξ_0^+. We point out once again that n_0, the density of particles in the condensate, has different values for an ideal gas and for a gas of interacting particles.

Using the technique of Feynman diagrams just developed, we find that the Green's functions of the particles not in the condensate are expressions involving two parameters, the quantity n_0 and the chemical potential μ. Instead of using perturbation theory to calculate the dependence of n_0 and μ on n, the density of the total number of particles, we can use relations of a general character. An obvious relation connecting n with n_0 and μ is

$$n = n_0 + iG'(x - x') \qquad (\mathbf{r} = \mathbf{r}', t' = t + 0). \tag{23.19}$$

A second relation follows from the condition that the ground-state energy should be a minimum with respect to n_0. Using the above scheme to calculate the ground-state energy $E = \langle \hat{H} \rangle$, we obtain an expression for E which depends on two parameters n_0 and μ. Varying E with respect to n_0 with n (the density of the total number of particles) held fixed, we obtain a second relation

$$\left(\frac{\partial E}{\partial n_0}\right)_n = 0. \tag{23.20}$$

In principle, the relations (23.19) and (23.20) solve the problem, but for practical calculations it turns out to be more convenient to use the relation (24.17) [see below] instead of (23.20).

We conclude this section with a brief analysis of the problem of choosing thermodynamic variables. Until now, we have used the total number of particles in the system as the independent variable, since, to set up our perturbation-theory formalism, we had to start from the characteristics of an ideal Bose gas, and in such a gas there is no Bose condensation when the chemical potential is finite. In fact, as is well known, the chemical potential of an ideal Bose gas is identically zero in the whole interval from absolute zero to the condensation temperature T_0. However, for a system of interacting particles, the chemical potential μ is nonzero, and hence is a thermodynamic variable on the same footing as the total number of particles. As usual, the value of μ can be found from the condition that the average number of particles in the system be equal to the given actual number of particles, and this is essentially the content of the relation (23.19). Going over to the chemical potential μ as the independent variable has the formal convenience of allowing us to get rid of extra time dependence in the formulas (23.18). Otherwise, this time dependence, coming from the vertices with $\xi_0(t)$ and $\xi_0^+(t)$, would appear in our matrix elements.

Actually, as we have seen repeatedly, the transformation from the variable

N to the variable μ can be accomplished by replacing H, the total Hamiltonian of the system, by $H - \mu N$. Since the commutation relations involving the total number of particles N and the operators ψ and ψ^+ are such that

$$N\psi - \psi N = -\psi, \qquad N\psi^+ - \psi^+ N = \psi^+,$$

changing the Hamiltonian leads to additional time dependence in the operators ψ and ψ^+:

$$\psi \to e^{i\mu t}\psi, \qquad \psi^+ \to e^{-i\mu t}\psi^+. \tag{23.21}$$

At the same time, the Green's functions change, e.g.,

$$G(x - x') \to e^{i\mu(t-t')}G(x - x'), \tag{23.22}$$

in the case of the exact one-particle Green's function. As for the Fourier components, this transformation amounts to replacing all the frequencies ω in the old expressions by $\omega + \mu$. Therefore, after making the transformation (23.21), the Green's functions of the particles in the condensate, written in the new thermodynamic variables, are also independent of time. This allows us to drop the time factors (23.18) at the corresponding vertices of our diagrams. The reader can obtain the same result if he uses (23.22) to redefine the Green's functions and then studies the perturbation series directly, taking account of the transformation (23.21) and (23.22). From now on, we shall assume everywhere that μ has been chosen as the independent thermodynamic variable.

24. The Green's Functions

24.1. Structure of the equations. We now study in somewhat more detail the structure of the perturbation series for the one-particle Green's function of the particles not in the condensate. A diagram of arbitrary order can be divided into a number of irreducible parts, joined to each other by single lines corresponding to functions $G^{(0)}(x - x')$. Thus, any diagram for the Green's function consists of a chain of self-energy diagrams connected by zeroth-order Green's functions. Some examples are shown in Fig. 64, where the shaded circles represent irreducible self-energy parts, whose structure will not be made more explicit.

The presence of the condensate leads to the appearance of self-energy diagrams of a new type, not encountered in any of the problems considered in the preceding chapters. These diagrams stem from the interaction of particles not in the condensate with particles in the condensate, and they contain the operators ξ_0 and ξ_0^+ at certain vertices. According to Sec. 23, these operators act as a kind of external field, i.e., $\xi_0, \xi_0^+ \to \sqrt{n_0}$. As is easily seen from Fig. 64, the total number of lines "entering" any irreducible self-energy diagram always equals the total number of lines "leaving" such a diagram (we include the zigzag lines corresponding to the operators annihilating and creating particles of the condensate). Moreover, all the self-energy

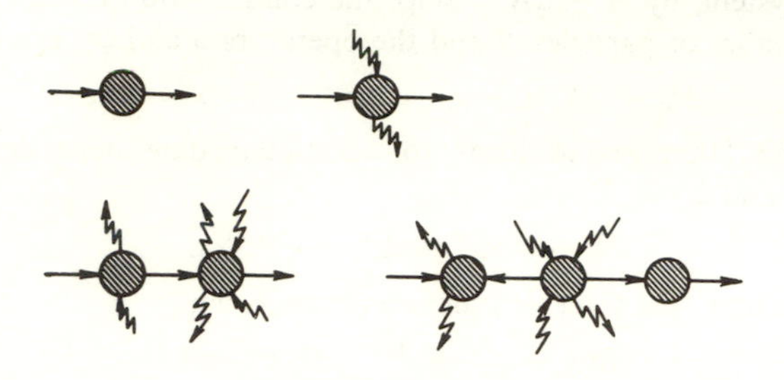

FIGURE 64

diagrams are joined by single straight lines, i.e., two particles not in the condensate "enter" or "leave" each self-energy diagram. It follows that we can make the following classification of all the irreducible self-energy diagrams:

1. Diagrams with one ingoing and one outgoing straight line, corresponding to particles not in the condensate. In such a diagram, the number of ingoing zigzag lines (equal to the degree of the operators $\tilde{\xi}_0^+$) is the same as the number of outgoing zigzag lines ($\tilde{\xi}_0$). We represent the sum of all self-energy diagrams of this type by a shaded circle, as in Fig. 65(a). The corresponding sum of matrix elements (in the coordinate representation) is denoted by $\Sigma_{11}(x - x')$.
2. Diagrams with two outgoing straight lines, corresponding to particles not in the condensate. In such a diagram, the number of ingoing zigzag lines exceeds by two the number of outgoing zigzag lines. We represent the sum of all self-energy diagrams of this type by a shaded circle with two ingoing zigzag lines, as in Fig. 65(b). The corresponding sum of matrix elements is denoted by $\Sigma_{02}(x - x')$.
3. Diagrams with two ingoing straight lines, corresponding to particles not in the condensate. In such a diagram, the number of outgoing zigzag lines exceeds by two the number of ingoing zigzag lines. We represent the sum of all self-energy diagrams of this type by a shaded circle with two outgoing zigzag lines, as in Fig. 65(c). The corresponding sum of matrix elements is denoted by $\Sigma_{20}(x - x')$.

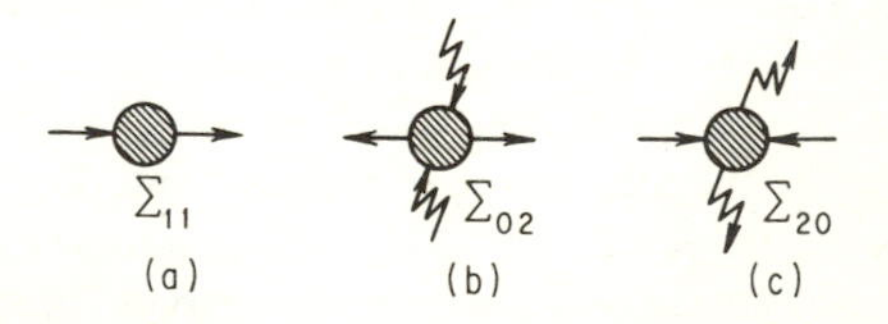

FIGURE 65

All three types of irreducible self-energy parts can be combined in any order in the diagrams for the Green's function $G'(x - x')$, subject to the obvious condition that the matrix elements Σ_{02} and Σ_{20} must appear the same number of times in any diagram. In Fig. 66 we show some examples of diagrams for the Green's function of the particles not in the condensate.

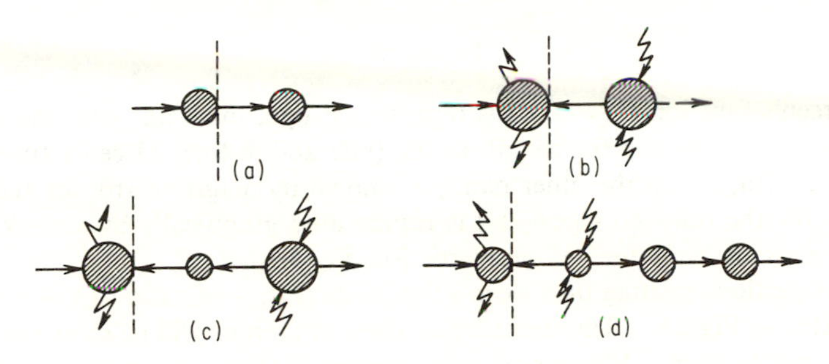

FIGURE 66

We are now in a position to write the analog of Dyson's equation for the Green's function of the particles not in the condensate. First we carry out the derivation graphically. To this end, we isolate the first irreducible self-energy part encountered in a diagram as we move from left to right along the chain of linked self-energy parts. Unlike the cases considered in the previous chapters, these irreducible parts can be of two types, i.e., Σ_{11} and Σ_{20}. In Fig. 66, we use a dashed line to indicate schematically how an arbitrary diagram is separated into two parts. To the right of the dashed line in Fig. 66(a), there appears a chain of linked self-energy parts whose sum is again the exact Green's function $G'(x - x')$. However, when the structures coming after the self-energy part Σ_{20} [like those appearing to the right of the dashed lines in diagrams (b), (c) and (d) of Fig. 66] are summed over all diagrams, a new function is obtained, which we denote by $\hat{G}(x - x')$. From a graphical point of view, this function is characterized by the fact that it is represented by diagrams whose external lines (corresponding to particles not in the condensate) point outwards.

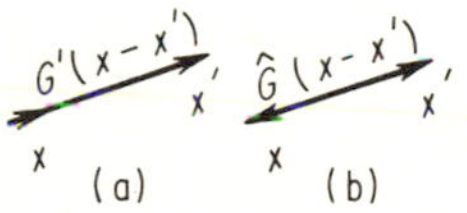

FIGURE 67

For convenience, we now introduce extra arrows (actually, arrowheads) along each line joining two points x and x' in a given diagram. These arrows show whether the line enters or leaves the points x and x'. For example, the Green's function $G^{(0)}(x - x')$ for noninteracting particles is by definition the average in the interaction representation of the T-product of the operators $\psi'(x)\psi'^{+}(x')$ [see (23.8)]. Therefore, to represent $G^{(0)}(x - x')$ graphically, at the point x along the line joining x to x' we attach an arrow pointing away from x [corresponding to the operator $\psi'(x)$], and at the point x' we attach an arrow pointing towards x' [corresponding to the operator $\psi'^{+}(x)$]. Clearly,

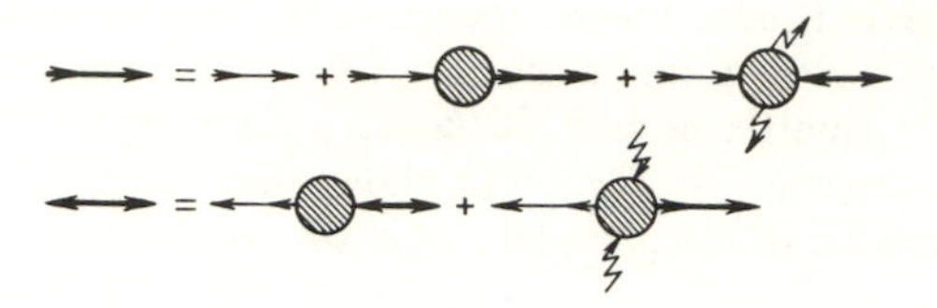

FIGURE 68

the Green's function $G'(x - x')$ is represented by a thick line with the same kind of arrows as for the zeroth-order (interaction-free) Green's function [see Fig. 67(a)]. On the other hand, as shown by diagrams (b), (c) and (d) of Fig. 66, the function $\hat{G}(x - x')$ is represented graphically by a thick line whose ends are both directed outwards [see Fig. 67(b)].

The equations relating the Green's functions $G'(x - x')$ and $\hat{G}(x - x')$ are illustrated in Fig. 68. The structure of these equations will be clear without further explanation. We merely note once again that the appearance of the function $\hat{G}(x - x')$ in this theory is due to interaction of the particles not in the condensate with particles in the condensate, and hence has no analog for noninteracting particles. As usual, the self-energy parts Σ_{11}, Σ_{20} and Σ_{02}

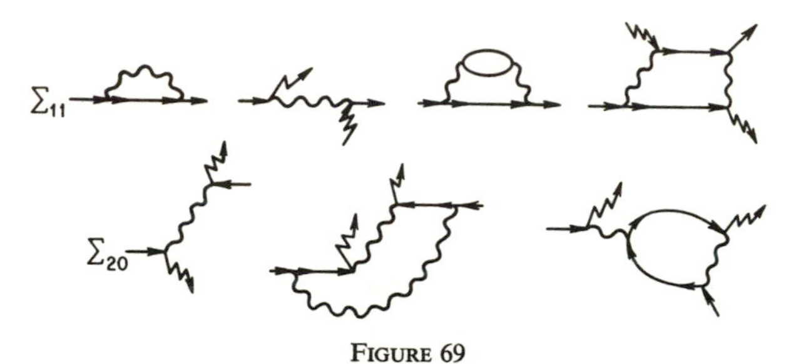

FIGURE 69

cannot be written in closed form in terms of the functions G' and $\hat{G}$. For these functions the technique of Feynman diagrams gives series expansions each term of which can be associated with a given diagram. Some diagrams of low order for Σ_{11} and Σ_{20} are shown in Fig. 69, for the case of the interaction Hamiltonian (23.12).

We now write down the equations corresponding to Fig. 68:[4]

$$\begin{aligned} G'(x - x') &= G^{(0)}(x - x') + \int G^{(0)}(x - y)[\Sigma_{11}(y - z)G'(z - x') \\ &\qquad + \Sigma_{20}(y - z)\hat{G}(z - x')]\, d^4z\, d^4y, \\ \hat{G}(x - x') &= \int G^{(0)}(y - x)[\Sigma_{11}(z - y)\hat{G}(z - x') \\ &\qquad + \Sigma_{02}(y - z)G'(z - x')]\, d^4z\, d^4y. \end{aligned} \tag{24.1}$$

[4] The choice of coefficients in (24.1) implies a corresponding definition of the quantities Σ_{ik} (see Sec. 25).

Taking Fourier components of all functions in (24.1), we obtain

$$\begin{aligned} G'(p) &= G^{(0)}(p) + G^{(0)}(p)\Sigma_{11}(p)G'(p) + G^{(0)}(p)\Sigma_{20}(p)\hat{G}(p), \\ \hat{G}(p) &= G^{(0)}(-p)\Sigma_{11}(-p)\hat{G}(p) + G^{(0)}(-p)\Sigma_{02}(p)G'(p). \end{aligned} \tag{24.2}$$

Using the expression (23.9) for $G^{(0)}(p)$, the Green's function for noninteracting particles, we can write (24.1) in the more convenient form

$$\begin{aligned} [\omega - \varepsilon_0(\mathbf{p}) + \mu - \Sigma_{11}(p)]G'(p) - \Sigma_{20}(p)\hat{G}(p) &= 1, \\ [-\omega - \varepsilon_0(\mathbf{p}) + \mu - \Sigma_{11}(-p)]\hat{G}(p) - \Sigma_{02}(p)G'(p) &= 0, \end{aligned} \tag{24.3}$$

where $\varepsilon_0(\mathbf{p}) = \mathbf{p}^2/2m$. Finally, introducing the notation

$$S(p) = \frac{\Sigma_{11}(p) + \Sigma_{11}(-p)}{2},$$

$$A(p) = \frac{\Sigma_{11}(p) - \Sigma_{11}(-p)}{2}$$

and using (24.3) to express $G'(p)$ and $\hat{G}(p)$ in terms of the quantities Σ_{11}, Σ_{02} and Σ_{20}, we find that

$$G'(p) = \frac{\omega + \varepsilon_0(\mathbf{p}) + S(p) + A(p) - \mu}{[\omega - A(p)]^2 - [\varepsilon_0(\mathbf{p}) + S(p) - \mu]^2 + \Sigma_{20}(p)\Sigma_{02}(p)}, \tag{24.4}$$

and

$$\hat{G}(p) = -\frac{\Sigma_{02}(p)}{[\omega - A(p)]^2 - [\varepsilon_0(\mathbf{p}) + S(p) - \mu]^2 + \Sigma_{20}(p)\Sigma_{02}(p)}. \tag{24.5}$$

These formulas for $G'(p)$ and $\hat{G}(p)$ generalize the usual expression for the one-particle function in terms of its self-energy part.

24.2. Analytic properties of the Green's functions. So far, the function $\hat{G}(p)$ has come under consideration as a result of summing certain diagrams. We now define $\hat{G}(p)$ in terms of the operators ψ'^{+}. To do this, we consider the quantity

$$-i\langle T(\tilde{\xi}_0\tilde{\xi}_0\tilde{\psi}'^{+}(x)\tilde{\psi}'^{+}(x'))\rangle,$$

and prove that its expansion in perturbation series is the same as that of the function $\hat{G}(x - x')$. Assuming that all operators are defined with factors $e^{i\mu t}$ or $e^{-i\mu t}$, as was done at the end of Sec. 23 [see (23.21)], we go over to the interaction representation:

$$\frac{-i\langle T(\xi_0\xi_0\psi'^{+}(x)\psi'^{+}(x'))S\rangle}{\langle S\rangle}.$$

Separating the operations T and $\langle\cdots\rangle$ into $T = T^0T'$ and $\langle\langle\cdots\rangle'\rangle^0$, and regarding ξ_0 and ξ_0^+ as external parameters, i.e., averaging over the particles not in the condensate, we find that the diagrams for the last expression are the same as those for the function $\hat{G}(x - x')$, whereas the matrix elements differ by the presence of two extra operators ξ_0. As already shown, averaging over the particles of the condensate amounts to replacing the operators

ξ_0, ξ_0^+ in the interaction representation by the Heisenberg operators $\tilde{\xi}_0$, $\tilde{\xi}_0^+$, which are then in turn replaced by numbers:

$$\tilde{\xi}_0 \to \sqrt{n_0}, \qquad \tilde{\xi}_0^+ \to \sqrt{n_0}.$$

Thus, for the function $\hat{G}(x - x')$ we can use two equivalent definitions

$$\hat{G}(x - x') = \frac{-i}{n_0} \langle T(\tilde{\xi}_0 \tilde{\xi}_0 \tilde{\psi}'^+(x) \tilde{\psi}'^+(x')) \rangle \tag{24.6}$$

or

$$\hat{G}(x - x') = -i \langle N + 2 | T(\tilde{\psi}'^+(x) \tilde{\psi}'^+(x')) | N \rangle. \tag{24.7}$$

In the last formula, $\hat{G}(x - x')$ is expressed in terms of the matrix element of $T(\psi'^+(x)\psi'^+(x'))$ with respect to ground states of a system consisting of $N + 2$ and N particles, respectively.

We now study the properties of the Green's functions $G'(x - x')$ and $\hat{G}(x - x')$. Using the definition (23.4), we represent the function $G'(x - x')$ as a sum of matrix elements over intermediate states (just as was done in Chap. 2), i.e.,

$$G'(x - x') = -i \sum_m \langle N | \tilde{\psi}'(x) | m \rangle \langle m | \tilde{\psi}'^+(x') | N \rangle \quad \text{for} \quad t > t',$$

and

$$G'(x - x') = -i \sum_n \langle N | \tilde{\psi}'^+(x') | n \rangle \langle n | \tilde{\psi}'(x) | N \rangle \quad \text{for} \quad t < t'.$$

Separating the space and time dependence of the matrix elements in the usual way, we obtain

$$G'(x - x') = \begin{cases} -i \sum\limits_m |\psi_{Nm}|^2 \, e^{i\mathbf{p}_m \cdot (\mathbf{r} - \mathbf{r}') - i\omega_{mN}(t - t') + i\mu(t - t')} & \text{for} \quad t > t', \\ -i \sum\limits_n |\psi_{Nn}|^2 \, e^{i\mathbf{p}_n \cdot (\mathbf{r}' - \mathbf{r}) - i\omega_{nN}(t' - t) - i\mu(t - t')} & \text{for} \quad t < t', \end{cases} \tag{24.8}$$

where $\mathbf{p}_m$, $\mathbf{p}_n$ are the momenta of the system in the states m, n,

$$\omega_{mN} = E_m - E_{N0}, \qquad \omega_{nN} = E_n - E_{N0},$$

E_m, E_n are the energies of the system in the states m, n, and E_{N0} is the ground-state energy of a system of N particles. According to the properties of the operators $\tilde{\psi}'$ and $\tilde{\psi}'^+$, the system has $N + 1$ particles in the states indexed by m, and $N - 1$ particles in the states indexed by n. [The appearance of the factors $e^{\pm i\mu t}$ in (24.8) is connected with this fact.] Using the definition

$$\mu \approx E_{N+1,0} - E_{N0},$$

we represent (24.8) in the form

$$G'(x - x') = \begin{cases} -i \sum\limits_m |\psi_{Nm}|^2 \, e^{i\mathbf{p}_m \cdot (\mathbf{r} - \mathbf{r}') - i(E_m - E_{N+1,0})(t - t')} & \text{for} \quad t > t', \\ -i \sum\limits_n |\psi_{Nn}|^2 \, e^{-i\mathbf{p}_n \cdot (\mathbf{r} - \mathbf{r}') + i(E_n - E_{N-1,0})(t - t')} & \text{for} \quad t < t'. \end{cases} \tag{24.9}$$

The energy differences $E_m - E_{N+1,0}$ and $E_n - E_{N-1,0}$ represent the spectra (or excitation energies) for systems consisting of $N + 1$ and $N - 1$ particles, respectively. When the number of particles is large, the spectra of these systems are the same to within terms of order $1/N$. Taking Fourier components of (24.9) with respect to the coordinate and time differences, we find that the Green's function in the momentum representation is

$$G'(p) = (2\pi)^3 \left[\sum_m \frac{\delta(\mathbf{p} - \mathbf{p}_m)|\psi_{Nm}|^2}{\omega - (E_m - E_{N+1,0}) + i\delta} - \sum_n \frac{\delta(\mathbf{p} + \mathbf{p}_n)|\psi_{Nn}|^2}{\omega + (E_n - E_{N-1,0}) - i\delta}\right]. \tag{24.10}$$

The poles of the function $G'(p)$ correspond to the values

$$\omega = \pm(E_m - E_0).$$

Thus, as always, the poles determine the spectrum of the system (except for sign), and their position relative to the ω-axis is clear from the rules for going around the contour associated with (24.10).

Next, we carry out a similar expansion with respect to intermediate states for the function $\hat{G}(x - x')$, using its representation in the form (24.7), i.e.,

$$\hat{G}(x - x') = \begin{cases} -i\sum_m \langle N + 2|\tilde{\psi}'^+(x)|m\rangle\langle m|\tilde{\psi}'^+(x')|N\rangle & \text{for} \quad t > t', \\ -i\sum_m \langle N + 2|\tilde{\psi}'^+(x')|m\rangle\langle m|\tilde{\psi}'^+(x)|N\rangle & \text{for} \quad t < t', \end{cases}$$

or

$$\hat{G}(x - x') = \begin{cases} -i\sum_m \psi^+_{N+2,m}\psi^+_{mN}e^{i\mathbf{p}_m\cdot(\mathbf{r}-\mathbf{r}') - i(E_m - E_{N+2,0} + \mu)t + i(E_m - E_{N0} - \mu)t'} \\ \qquad \text{for} \quad t > t', \\ -i\sum_m \psi^+_{N+2,m}\psi^+_{mN}e^{i\mathbf{p}_m\cdot(\mathbf{r}'-\mathbf{r}) - i(E_m - E_{N+2,0} + \mu)t' + i(E_m - E_{N0} - \mu)t} \\ \qquad \text{for} \quad t < t. \end{cases} \tag{24.11}$$

The states indexed by m correspond to states of a system consisting of $N + 1$ particles, with ground-state energy $E_{N+1,0}$. Again using the definition of the chemical potential, we transform (24.11) into

$$\hat{G}(x - x') = \begin{cases} -i\sum_m \psi^+_{N+2,m}\psi^+_{mN}e^{i\mathbf{p}_m\cdot(\mathbf{r}-\mathbf{r}') - i(E_m - E_{N+1,0})(t-t')} & \text{for} \quad t > t', \\ -i\sum_m \psi^+_{N+2,m}\psi^+_{mN}e^{i\mathbf{p}_m\cdot(\mathbf{r}'-\mathbf{r}) - i(E_m - E_{N+1,0})(t'-t)} & \text{for} \quad t < t'. \end{cases}$$

The Fourier components of the function $\hat{G}(x - x')$ are

$$\hat{G}(p) = (2\pi)^3 \sum_m \psi^+_{N+2,m}\psi^+_{mN} \times \left[\frac{\delta(\mathbf{p} - \mathbf{p}_m)}{\omega - (E_m - E_{N+1,0}) + i\delta} - \frac{\delta(\mathbf{p} + \mathbf{p}_m)}{\omega + (E_m - E_{N+1,0}) - i\delta}\right]. \tag{24.12}$$

Comparing (24.10) and (24.12), we conclude that the poles of the Green's function $G'(x - x')$ and $\hat{G}(x - x')$ coincide. In particular, returning to the representations (24.4) and (24.5) of $G'(p)$ and $\hat{G}(p)$ in terms of the irreducible self-energy parts, we see that the spectrum $\omega = \varepsilon(\mathbf{p})$ of the system is given by the equation

$$[\varepsilon(\mathbf{p}) - A(p)]^2 - [\varepsilon_0(\mathbf{p}) + S(p) - \mu]^2 + \Sigma_{20}(p)\Sigma_{02}(p) = 0,$$

where $p \equiv (\mathbf{p}, \varepsilon(\mathbf{p}))$.

In addition to the function $\hat{G}(x - x')$ defined by (24.6) and (24.7), it makes sense to introduce the function

$$\check{G}(x - x') = -\frac{i}{n_0}\langle T(\tilde{\Psi}'(x)\tilde{\Psi}'(x')\xi_0^+\xi_0^+)\rangle \\ \equiv -i\langle N|T(\tilde{\Psi}'(x)\tilde{\Psi}'(x'))|N + 2\rangle. \tag{24.13}$$

Carrying out an expansion with respect to intermediate states in (24.13), just as was done for the function $\hat{G}(x - x')$, we find the following expression, analogous to (24.12), for the Fourier components $\check{G}(p)$:

$$\check{G}(p) = (2\pi)^3 \sum_m \psi_{Nm}\psi_{m,\,N+2} \\ \times \left[\frac{\delta(\mathbf{p} - \mathbf{p}_m)}{\omega - (E_m - E_{N+1,0}) + i\delta} - \frac{\delta(\mathbf{p} + \mathbf{p}_m)}{\omega + (E_m - E_{N+1,0}) - i\delta}\right]. \tag{24.14}$$

Thus, the function $\check{G}(p)$ has poles (and detours around these poles) which are the same as those for the functions $\hat{G}(p)$ and $G'(p)$. As for the coefficients, i.e., the residues at these poles, they are real for $G'(p)$, whereas the residues at identical poles of $\hat{G}(p)$ and $\check{G}(p)$ are complex conjugates of each other.

The function $\check{G}(x - x')$ is represented graphically by a thick line with two arrows pointing towards each other. The equations connecting $\check{G}(x - x')$ with the ordinary Green's function are shown schematically in Fig. 70. These equations involve the Green's function $G'(x' - x)$, represented by a thick line with both arrows pointing towards the left. Taking the Fourier transforms of the equations corresponding to Fig. 70, we obtain

$$\check{G}(p) = G^{(0)}(p)\Sigma_{11}(p)\check{G}(p) + G^{(0)}(p)\Sigma_{20}(p)G'(-p),$$
$$G'(-p) = G^{(0)}(-p) + G_0(-p)[\Sigma_{11}(-p)G'(-p) + \Sigma_{02}(p)G(p)].$$

Then, solving these equations for $\check{G}(p)$, we find that

$$\check{G}(p) = -\frac{\Sigma_{20}(p)}{[\omega - A(p)]^2 - [\varepsilon_0(\mathbf{p}) + S(p) - \mu]^2 + \Sigma_{20}(p)\Sigma_{02}(p)}. \tag{24.15}$$

The only difference between the expressions (24.5) and (24.15) for $\hat{G}(p)$ and $\check{G}(p)$ is in the numerator, where one has $\Sigma_{02}(p)$ and the other has $\Sigma_{20}(p)$.

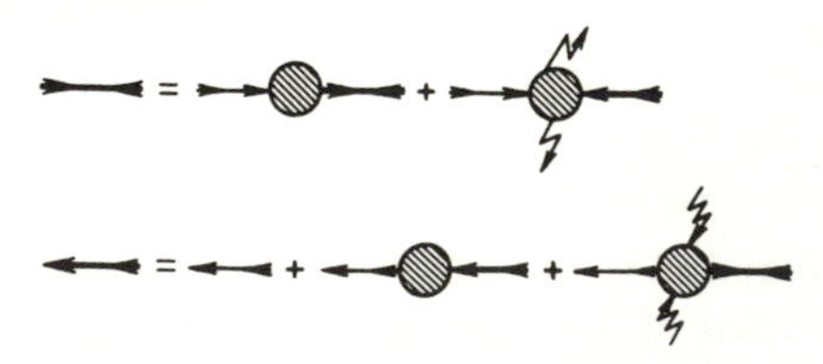

FIGURE 70

24.3. Behavior of the Green's functions for small momenta. We conclude this section by making some observations of a general nature concerning the results obtained above. Because of the homogeneity of space, all quantities depend on the magnitude of the vector $\mathbf{p}$. It follows from (24.12) and (24.14) that the functions $\hat{G}(p)$ and $\check{G}(p)$ are even functions of the frequency ω, from which it is easily seen that $\Sigma_{20}(p) = \Sigma_{02}(p)$. In fact, since the interaction Hamiltonian preserves the total number of particles, it is symmetric in the operators ψ and ψ^+. Therefore, with any diagram for Σ_{20} we can associate exactly the same diagram for Σ_{02}, obtained by replacing all ingoing lines in the diagram for Σ_{20} by outgoing lines, and *vice versa*. At the same time, this reverses the direction in which all internal lines are traversed. However, we can reverse the direction in which all internal lines are traversed by changing p to $-p$ in the matrix element corresponding to the given diagram for $\Sigma_{20}(p)$. Since, according to (24.15), Σ_{20} is an even function, it follows that

$$\Sigma_{20}(p) = \Sigma_{02}(p), \qquad \hat{G}(p) = \check{G}(p).$$

Now consider the equation

$$[\omega - A(p)]^2 - [\varepsilon_0(\mathbf{p}) + S(p) - \mu]^2 + \Sigma_{02}^2(p) = 0, \tag{24.16}$$

determining the poles of the Green's functions. It is clear from physical considerations that this equation must have solutions for arbitrarily small $\mathbf{p}$ and ω. In fact, the class of possible solutions for the energy spectrum of the excitations for small $\mathbf{p}$ must contain the acoustic spectrum $\omega = c|\mathbf{p}|$, i.e., the spectrum corresponding to long-wavelength oscillations. Therefore, suppose we set $\mathbf{p}$ and ω equal to zero in equation (24.16). Then we obtain the condition relating the chemical potential μ to the quantities $\Sigma_{11}(0)$, $\Sigma_{20}(0)$ and $\Sigma_{02}(0)$:

$$[\mu - \Sigma_{11}(0)]^2 = \Sigma_{02}^2(0).$$

From the two roots of this equation, we have to choose

$$\mu = \Sigma_{11}(0) - \Sigma_{02}(0), \tag{24.17}$$

as will be shown in Sec. 25.2. In order to find the form of the Green's functions in the region of small $\mathbf{p}$ and ω, we expand the denominators of (24.4), (24.5) and (24.15), confining ourselves everywhere to second-order terms in $\mathbf{p}$ and ω. Using (24.17), we obtain

$$\begin{aligned} G'(p) &= \frac{\Sigma_{11}(0) - \mu}{B(\omega^2 - c^2|\mathbf{p}|^2)} = \frac{\Sigma_{20}(0)}{B(\omega^2 - c^2|\mathbf{p}|^2)}, \\ \hat{G}(p) &= \check{G}(p) = -\frac{\Sigma_{20}(0)}{B(\omega^2 - c^2|\mathbf{p}|^2)}, \end{aligned} \tag{24.18}$$

where

$$B = \left[1 - \frac{\partial \Sigma_{11}(0)}{\partial \omega}\right]^2 - \frac{\partial^2 \Sigma_{11}(0)}{\partial \omega^2}\Sigma_{20}(0) + \frac{1}{2}\frac{\partial^2}{\partial \omega^2}\Sigma_{20}^2(0),$$

$$Bc^2 = 2\Sigma_{20}(0)\left\{\frac{1}{2m} + \frac{\partial \Sigma_{11}(0)}{\partial |\mathbf{p}|^2} - \frac{\partial \Sigma_{20}}{\partial |\mathbf{p}|^2}\right\}.$$

The quantity c is obviously the velocity of sound. As must be the case, c vanishes if $\Sigma_{20}(0)$ vanishes, since the velocity of sound is zero for an ideal Bose gas.

Comparing the expressions (24.18) with the general expansions (24.10), (24.12) and (24.14) for the Green's functions, we find that the ratio $\Sigma_{20}(0)/B$ is real and positive. Thus, for small $\mathbf{p}$ and ω ($\omega \sim c|\mathbf{p}|$), all the Green's functions $G'(p)$, $\hat{G}(p)$ and $\check{G}(p)$ have the same form

$$G(p) = \frac{\text{const}}{\omega^2 - c^2|\mathbf{p}|^2}. \tag{24.19}$$

25. The Dilute Nonideal Bose Gas

25.1. The diagram technique. To illustrate the methods developed in Secs. 23 and 24, we now consider in more detail the case where the interaction between the particles reduces to forces acting between pairs of particles [Belyaev (B4)]. Then the interaction Hamiltonian equals

$$H_{\text{int}} = \frac{1}{2}\int \psi^+(\mathbf{r})\psi^+(\mathbf{r}')U(\mathbf{r}-\mathbf{r}')\psi(\mathbf{r}')\,\psi(\mathbf{r})\,d\mathbf{r}\,d\mathbf{r}'. \tag{25.1}$$

If in (25.1) we use (23.1) to explicitly exhibit the operators ξ_0 and ξ_0^+ of the particles in the condensate, we obtain the following eight terms, whose sum equals H_{int}:

$$\begin{aligned}
H_a &= \frac{1}{2}\int \psi'^+(\mathbf{r})\psi'^+(\mathbf{r}')U(\mathbf{r}-\mathbf{r}')\psi'(\mathbf{r}')\psi'(\mathbf{r})\,d\mathbf{r}\,d\mathbf{r}',\\
H_b &= \frac{1}{2}V(\xi_0^+)^2\xi_0^2\int U(\mathbf{R})\,d\mathbf{R},\\
H_c &= \frac{1}{2}\int [\xi_0^+\psi'^+(\mathbf{r}') + \psi'^+(\mathbf{r})\xi_0^+]\psi'(\mathbf{r})\psi'(\mathbf{r}')U(\mathbf{r}-\mathbf{r}')\,d\mathbf{r}\,d\mathbf{r}',\\
H_d &= \frac{1}{2}\int \psi'^+(\mathbf{r})\psi'^+(\mathbf{r}')[\psi'(\mathbf{r}')\xi_0 + \xi_0\psi'(\mathbf{r})]U(\mathbf{r}-\mathbf{r}')\,d\mathbf{r}\,d\mathbf{r}',\\
H_e &= \frac{1}{2}\int [\xi_0^+\psi'^+(\mathbf{r}')\xi_0\psi'(\mathbf{r}) + \xi_0^+\psi'^+(\mathbf{r})\xi_0\psi'(\mathbf{r}')]U(\mathbf{r}-\mathbf{r}')\,d\mathbf{r}\,d\mathbf{r}', \qquad (25.2)\\
H_f &= \frac{1}{2}\int [\xi_0^+\psi'^+(\mathbf{r})\xi_0\psi'(\mathbf{r}) + \xi_0^+\psi'^+(\mathbf{r}')\xi_0\psi'(\mathbf{r}')]U(\mathbf{r}-\mathbf{r}')\,d\mathbf{r}\,d\mathbf{r}',\\
H_g &= \frac{1}{2}\int \xi_0^+\xi_0^+\psi'(\mathbf{r})\psi'(\mathbf{r}')U(\mathbf{r}-\mathbf{r}')\,d\mathbf{r}\,d\mathbf{r}',\\
H_h &= \frac{1}{2}\int \xi_0\xi_0\psi'^+(\mathbf{r})\psi'^+(\mathbf{r}')U(\mathbf{r}-\mathbf{r}')\,d\mathbf{r}\,d\mathbf{r}'.
\end{aligned}$$

In Fig. 71 we show the elementary processes corresponding to each of these different terms. To construct any matrix element, we proceed in the usual

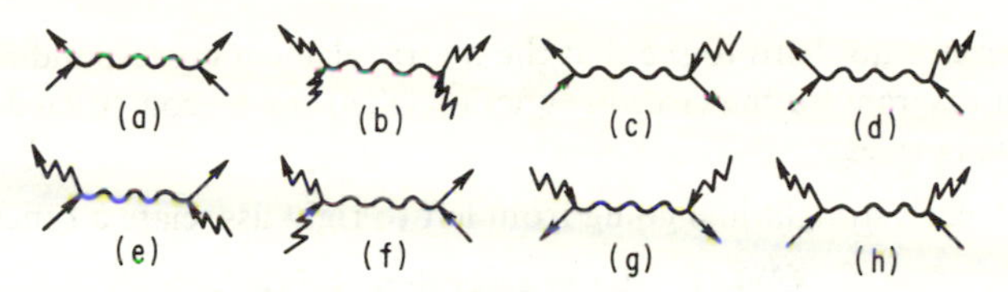

FIGURE 71

way, applying Wick's theorem to the operators of the particles not in the condensate. According to the results obtained above, we need only take account of connected diagrams for a given process, regarding the operators ξ_0, ξ_0^+ everywhere as external parameters and making the substitution $\xi_0, \xi_0^+ \to \sqrt{n_0}$ (provided that the frequencies of all particles participating in the process are measured from the value of the chemical potential). In deriving rules for establishing a one-to-one correspondence between matrix elements and diagrams (in the momentum representation), we shall confine ourselves to one-particle Green's functions, but otherwise our treatment will be quite general.

Consider an arbitrary diagram associated with the mth-order term in the perturbation series for one of the Green's functions, say $G'(x - x')$, i.e., a diagram whose analytical expression is contained in

$$(-i)\frac{(-i)^m}{m!}\langle T(\psi'(x)\int H_{\text{int}}(t_1)\cdots H_{\text{int}}(t_m)\psi'^+(x')\,dt_1\cdots dt_m)\rangle.$$

Suppose the diagram has s ingoing and s outgoing lines, corresponding to particles in the condensate. (As already noted several times, the total number of ingoing lines in any diagram equals the total number of outgoing lines.) There are $m!$ possible permutations of the m Hamiltonians $H_{\text{int}}(t_i)$ which do not destroy the order of the pairings determined by the given diagram. Moreover, there are obviously $2m - s + 1$ operators ψ' (and an equal number of operators ψ'^+) associated with the diagram.

By the definition of the Green's function $G^{(0)}$, for each pairing of ψ' and ψ'^+ we have to include a factor of $-i$. With each straight line we associate a Green's function $G^{(0)}$, and we introduce a wavy line corresponding to the potential

$$V(x - x') = U(\mathbf{r} - \mathbf{r}')\delta(t - t').$$

According to formula (25.1) and Fig. 71, the total number of "triple vertices" (i.e., vertices to which a wavy line is attached) is twice as large as the order of the perturbation. Suppose we take the Fourier components of all the quantities

$$G'(x - x') = \frac{1}{(2\pi)^4}\int G'(p)e^{ip(x-x')}\,d^4p,$$

$$V(x - x') = \frac{1}{(2\pi)^4}\int U(\mathbf{q})e^{iq(x-x')}\,d^4q,$$

etc. Then it is not hard to see that the matrix element corresponding to any mth-order diagram for the Green's function $G'(p)$ can be constructed by using the following rules:

1. With each straight line going from left to right associate a function
$$G^{(0)}(p) = [\omega - \varepsilon_0(\mathbf{p}) + \mu + i\delta]^{-1},$$
and with each straight line going in the opposite direction associate a function $G^{(0)}(-p)$.
2. With each wavy line of momentum $\mathbf{q}$ associate the Fourier component $U(\mathbf{q})$ of the interaction potential.
3. With each ingoing or outgoing line corresponding to a particle in the condensate associate a factor $\sqrt{n_0}$.
4. At each triple vertex let the momentum $\mathbf{q}$ of the wavy line equal the difference between the momenta of the particle lines.
5. Integrate over the momenta which are not determined by the conservation laws, introducing a factor $(2\pi)^{-4}$ for each such integration.
6. Multiply the whole matrix element by
$$A_{sm}(-i)^{s-m},$$
where A_{sm} depends on which terms of (25.2) participate in the diagram.

For the Green's function $\hat{G}(p)$ and $\check{G}(p)$, the above rules remain the same, provided we take s to mean the number of factors n_0 appearing in the given mth-order diagram. For example, if in one of the diagrams for $\hat{G}$, the number of ingoing lines corresponding to particles in the condensate equals l, then, according to the definition (24.7) of $\hat{G}$, the number of outgoing lines equals $l + 2$. The number of pairings of the operators ψ' and ψ'^{+} (the number of functions $G^{(0)}$) obviously equals
$$2m - l.$$
Since by definition each Green's function is accompanied by a factor of $-i$, the factor which must multiply the whole matrix element is
$$(-i)^{l+1-m}.$$
But $l + 1$ is just the degree s of the factors n_0 coming from the lines corresponding to particles in the condensate.

25.2. Relation between the chemical potential and the self-energy parts of the one-particle Green's functions. We now prove formula (24.17) for the chemical potential μ. Consider the operator $\xi_0(t)$ in the Heisenberg representation (it is assumed that the term $-\mu\hat{N}$ is included in the total Hamiltonian). The time dependence of $\xi_0(t)$ is given by the usual quantum-mechanical operator equation
$$\frac{i\partial\xi_0(t)}{\partial t} = [\xi_0(t), H] = -\mu\xi_0(t) - [H_{\text{int}}, \xi_0(t)].$$

Using this equation, we find the following equation for the Green's function $G_0(t - t')$ of the particles in the condensate:

$$\frac{\partial G_0(t-t')}{\partial t} = i\mu G_0(t-t') - \langle T([\xi_0(t), H_{\text{int}}]\xi_0^+(t'))\rangle.$$

But, according to the results of the preceding sections, $G_0(t - t')$ does not depend on time and is simply $-in_0$. It follows that

$$\mu n_0 = -\langle T([H_{\text{int}}, \xi_0(t)]\xi_0^+(t'))\rangle. \tag{25.3}$$

Next, we calculate the average in the right-hand side of (25.3). Going over to the interaction representation

$$\langle T([H_{\text{int}}, \xi_0(t)]\xi_0^+(t'))\rangle = \frac{\langle T([H_{\text{int}}, \xi_0(t)]\xi_0^+(t')S)\rangle}{\langle S\rangle},$$

we briefly reproduce the argument given in Sec. 23. In the operations T and $\langle\cdots\rangle$, we first carry out averaging and time-ordering with respect to the particles not in the condensate. According to our general prescription, in doing this we need only take account of all connected diagrams, regarding ξ_0 and ξ_0^+ as external parameters. In the present case, the connected diagrams are the different vacuum loops. The result of subjecting H_{int} to this averaging, which only affects the operators of particles not in the condensate, will be denoted by $\bar{H}_{\text{int}}^{\text{con}}$. The quantity $\bar{H}_{\text{int}}^{\text{con}}$ depends on both ξ_0 and ξ_0^+ as if they were parameters. To obtain the final result, we have to replace all the operators ξ_0 and ξ_0^+ appearing in $\bar{H}_{\text{int}}^{\text{con}}$ by the exact Heisenberg operators ξ_0 and ξ_0^+. As a result, (25.3) becomes

$$\mu n_0 = -\frac{\langle T^0([\bar{H}_{\text{int}}^{\text{con}}, \xi_0]\xi_0^+\langle S\rangle')\rangle^0}{\langle S\rangle} = -\langle T([\tilde{\bar{H}}_{\text{int}}^{\text{con}}, \xi_0]\xi_0^+)\rangle. \tag{25.4}$$

The commutator $[\bar{H}_{\text{int}}^{\text{con}}, \xi_0]$ contains commutators of ξ_0 with different products of ξ_0 and ξ_0^+ in vacuum averages of $\bar{H}_{\text{int}}^{\text{con}}$. To calculate $\bar{H}_{\text{int}}^{\text{con}}$, we use Wick's theorem to carry out the usual averaging of the operators of particles not in the condensate. Since this averaging consists of averaging pairs of operators ψ' and ψ'^+, the number of operators ξ_0 equals the number of operators ξ_0^+.

For example, consider a vacuum loop of order m containing s operators ξ_0 and s operators ξ_0^+. The result of commuting ξ_0 with one operator ξ_0^+ is

$$[\xi_0^+, \xi_0] = -\frac{1}{V},$$

but ξ_0 can be commuted with all s of the operators ξ_0^+. Let the correction to the ground-state energy corresponding to the given mth-order vacuum loop involving n_0^s be denoted by $\langle H_{\text{int}}\rangle_{ms}^{\text{con}}$. (This quantity is obtained from $\bar{H}_{\text{int}}^{\text{con}}$ by replacing the operators ξ_0 and ξ_0^+ by $\sqrt{n_0}$.) Then it is easily seen that the expression (25.4) is

$$\mu n_0 = \sum_{m,s} \frac{s}{V}\langle H_{\text{int}}\rangle_{ms}^{\text{con}},$$

or

$$\mu = \sum_{m,s} \frac{\partial}{\partial n_0} \frac{\langle H_{\text{int}} \rangle_{ms}^{\text{con}}}{V} = \frac{\partial}{\partial n_0} \frac{\langle H_{\text{int}} \rangle}{V}. \tag{25.5}$$

The vacuum average $\langle H_{\text{int}} \rangle$ is a function of the parameters μ and n_0, and hence in the right-hand side of (25.5) we have in mind a partial derivative with respect to n_0 with μ held fixed.

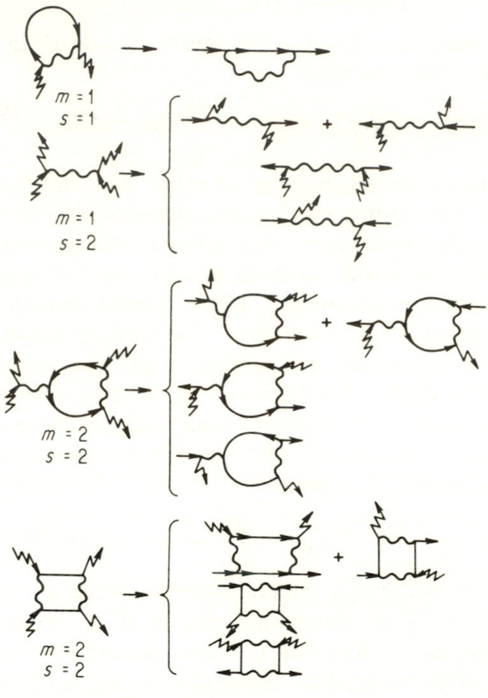

FIGURE 72

The idea of the rest of the proof [see Hugenholtz and Pines (H4)] is based on the fact that the operators ξ_0, ψ' and ξ_0^+, ψ'^+ enter the interaction Hamiltonian symmetrically. Therefore, with each vacuum loop of $\langle H_{\text{int}} \rangle$ which has a certain number of zigzag lines (corresponding to particles in the condensate) we can formally associate diagrams for the irreducible self-energy parts $\Sigma_{11}(0)$ and $\Sigma_{20}(0)$ by replacing the necessary number of ingoing and outgoing zigzag lines (representing the operators ξ_0^+ and ξ_0) by ingoing and outgoing straight lines (representing the operators ψ'^+ and ψ'). In Fig. 72, we show some simple examples of how this is done for diagrams corresponding to low-order terms in the perturbation series.[5] Since the

[5] The expressions associated with some of these diagrams vanish (see below), but for the purposes of illustrating our argument, this does not matter.

ingoing and outgoing zigzag lines "carry" zero four-momentum ($p = 0$), the indicated correspondence also holds for the matrix elements of $\Sigma_{11}(p)$ and $\Sigma_{20}(p) = \Sigma_{02}(p)$ when $p = 0$.

The matrix element $\langle H_{\text{int}}\rangle_{ms}^{\text{con}}$ of an arbitrary irreducible diagram is constructed by the same rules as those given above for the Green's functions. The only difference (as is easily verified) is that the numerical factor which should multiply the whole integral is now

$$(-i)^{s-m-2},$$

where m is the order of the term of the perturbation series, and s is the degree of n_0 in the given diagram. The corresponding factor in the matrix elements for the self-energy parts is

$$(-i)^{s-m}.$$

When a vacuum loop is differentiated with respect to n_0, the degree n_0 is reduced by 1. Consider all possible mth-order diagrams for $\Sigma_{11}(0)$, containing $s-1$ factors n_0 (and hence $s-1$ ingoing and $s-1$ outgoing zigzag lines, corresponding to particles in the condensate). All these diagrams can be obtained from the vacuum loop $(\overline{H}_{\text{int}}^{\text{con}})_{m-1,s}$ by replacing one of the s operators ξ_0^+ by an ingoing straight line and one of the s operators ξ_0 by an outgoing straight line, i.e., these diagrams can be obtained in s^2 different ways:

$$(\Sigma_{11}(0))_{m,s-1} = \frac{s^2}{n_0 V}\langle H_{\text{int}}\rangle_{m-1,s}^{\text{con}}.$$

As for the diagrams for $\Sigma_{20}(0)$, they can be obtained from $(\overline{H}_{\text{int}}^{\text{con}})_{m-1,s}$ by replacing two ingoing zigzag lines by two ingoing straight lines. Since this can be done in $s(s-1)$ different ways, we find that

$$(\Sigma_{20}(0))_{m,s-1} = \frac{s^2-s}{n_0 V}\langle H_{\text{int}}\rangle_{m-1,s}^{\text{con}}.$$

Comparing the difference $\Sigma_{11}(0) - \Sigma_{20}(0)$ with the expression (25.5), we immediately obtain

$$\mu = \Sigma_{11}(0) - \Sigma_{20}(0). \tag{25.6}$$

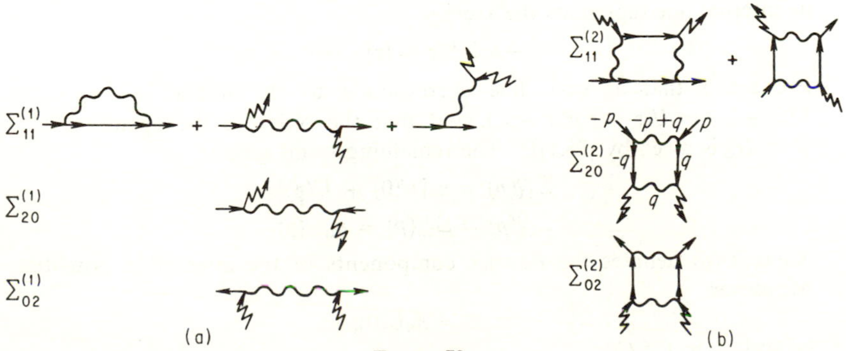

FIGURE 73

Of course, as noted in the preceding section, the validity of this relation is not confined to the present case, where the interaction forces act only between pairs of particles.

Equation (25.6), together with equation (23.19) relating the chemical potential to the density of the total number of particles in the system, constitute a set of two equations determining the values of the parameters μ and n_0. The proof of the equivalence of the conditions (23.20) and (25.6) will not be given here. We merely note that for perturbation-theory calculations (i.e., for the case of a gas of weakly interacting particles), it is more convenient to use (25.6), since this condition expresses μ directly in terms of quantities which are known for an ideal Bose gas.

25.3. The low-density approximation. The methods developed above will now be applied to the case of a gas of interacting bosons, with the Hamiltonian (25.1). We have already considered this example in Chap. 1, with the assumption that the interaction forces are weak. Because of this assumption, the expression (4.11) for the excitation spectrum contains the Fourier component of the interaction potential, which in the Born approximation is proportional to the scattering amplitude of the colliding particles. We now show that this result is actually valid in the more general case where the interaction is not assumed to be weak, provided the density of the gas is assumed to be low. This means that the dimensions of the particles of the gas are small compared to the average distance between the particles; if we characterize the dimensions of the colliding particles by their s-wave scattering amplitude f_0, then the quantity $f_0 n^{1/3}$ must be $\ll 1$.

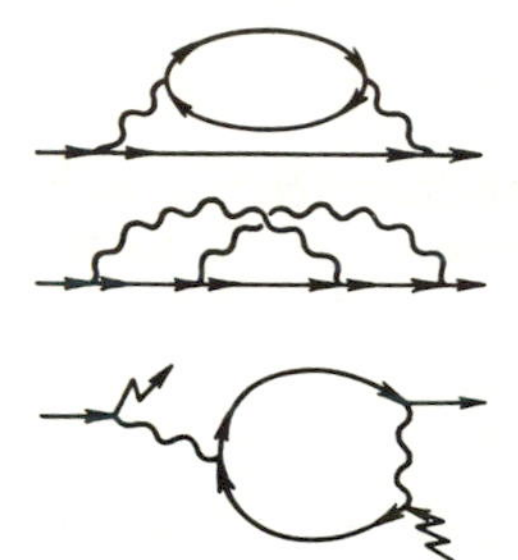

FIGURE 74

Consider the first-order diagrams for $\Sigma_{11}(p)$, $\Sigma_{20}(p)$ and $\Sigma_{02}(p)$ shown in Fig. 73(a). Of the three diagrams for Σ_{11}, the first corresponds to averaging the term H_a, given by (25.2), in the interaction Hamiltonian (25.1). This diagram equals zero, since its internal line represents the average

$$-i\langle\psi'^{+}(\mathbf{r}')\psi'(\mathbf{r})\rangle \equiv 0$$

[we recall that a wavy line corresponds to the interaction potential $V(x - x') = U(\mathbf{r} - \mathbf{r}')\delta(t - t')$, and that the order of the operators ψ'^{+}, ψ' in H_a is given by (25.1)]. The remaining terms give

$$\Sigma_{11}^{(1)}(p) = n_0[U(0) + U(\mathbf{p})],$$

$$\Sigma_{20}^{(1)}(p) = \Sigma_{02}^{(1)}(p) = n_0 U(\mathbf{p}),$$

where $U(\mathbf{p})$ denotes the Fourier components of the interaction potential. Moreover,

$$\mu = n_0 U(0),$$

according to (25.6).

Of all the second-order diagrams, only the diagrams shown in Fig. 73(b) are nonzero. For example, the diagrams for $\Sigma_{11}^{(2)}(p)$ shown in Fig. 74 are all zero, since each of them contains a product of Green's functions of the form

$$G^{(0)}(\mathbf{r} - \mathbf{r}', t_1 - t_2)G^{(0)}(\mathbf{r}'' - \mathbf{r}''', t_2 - t_1)$$

(in the coordinate representation), and, as we know,

$$G^{(0)}(\mathbf{r}_1 - \mathbf{r}_2, t_1 - t_2) \equiv 0$$

for $t_1 < t_2$.

To estimate the diagrams shown in Fig. 73(b), we assume for simplicity that the Fourier components of the interaction potential have the form

$$U(\mathbf{p}) = \begin{cases} U_0 & \text{for} \quad |\mathbf{p}| \ll \dfrac{1}{a}, \\ 0 & \text{for} \quad |\mathbf{p}| \gg \dfrac{1}{a}, \end{cases}$$

where $a \sim f_0$ is the radius of the particles, in order of magnitude. The estimate of any diagram may involve various parameters of the problem, i.e., the mass m, the parameters U_0 and a characterizing the interaction, and n_0, the density of particles in the condensate. From these parameters we can form two dimensionless combinations

$$\zeta \sim \frac{mU_0}{a}, \qquad \beta = \sqrt{n_0 a^3}.$$

The quantity ζ is a perturbation-theory parameter (for expansion in a series of successive Born approximations), while β is a "gas" parameter. From a formal standpoint, the perturbation series is an expansion in powers of $\zeta \ll 1$, but later we shall only assume that $\beta \ll 1$.

To keep things as simple as possible, we now consider the diagram for $\Sigma_{20}^{(2)}(p)$ shown in Fig. 73(b). In this case we have

$$\Sigma_{20}^{(2)}(p) \sim n_0 \int G^{(0)}(q)G^{(0)}(-q)U(\mathbf{q})U(\mathbf{p} - \mathbf{q})\, d\mathbf{q}\, d\omega.$$

Using (23.9) to substitute for the Green's function $G^{(0)}$ and integrating with respect to ω, we obtain

$$\Sigma_{20}^{(2)}(p) \sim n_0 U_0^2 \int \frac{d\mathbf{q}}{\mu - \varepsilon_0(\mathbf{q})}.$$

The main contribution to the last integral comes from large values of $|\mathbf{q}| \sim 1/a$, for which

$$\frac{\mu}{\varepsilon_0(\mathbf{q})} \sim mn_0 U_0 a^2 = \zeta\beta^2 \ll 1,$$

and hence

$$\Sigma_{20}^{(2)} \sim \frac{mn_0 U_0^2}{a} \sim \Sigma_{20}^{(1)}\zeta.$$

A similar estimate of the diagrams for $\Sigma_{11}^{(2)}$ shows that $\Sigma_{11}^{(2)} \sim \Sigma_{11}^{(1)}\zeta$.

Now consider the third-order diagram for $\Sigma_{20}^{(3)}$ shown in Fig. 75(a). In this case we have

$$\Sigma_{20}^{(3)} \sim n_0^2 \int G^{(0)}(-q)[G^{(0)}(q)]^2[U(\mathbf{q})]^2 U(\mathbf{p}+\mathbf{q})\, d\mathbf{q}\, d\omega \sim n_0^2 U_0^3 \int \frac{d\mathbf{q}}{[\mu - \varepsilon_0(\mathbf{q})]^2}.$$

The last integral (unlike the one just considered) converges at the upper limit, and the main contribution to the integral comes from the region

$$|\mathbf{q}| \sim \sqrt{m\mu} \sim \sqrt{n_0 U_0 m},$$

so that

$$\Sigma_{20}^{(3a)} \sim \frac{n_0^2 U_0^3 m^{3/2}}{\mu^{1/2}} \sim \Sigma_{20}^{(1)} \zeta^{3/2} \beta. \quad (25.7)$$

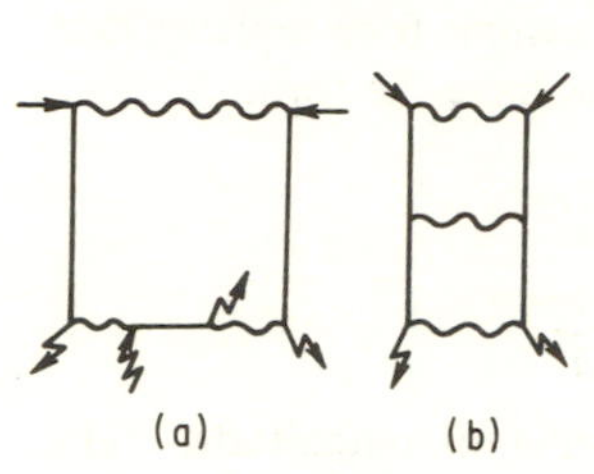

FIGURE 75

Moreover, the third-order diagram shown in Fig. 75(b) is of order

$$\Sigma_{20}^{(3b)} \sim \Sigma_{20}^{(1)} \zeta^2. \quad (25.8)$$

It is clear from (25.7) and (25.8) that

$$\Sigma_{20}^{(3a)} \sim \beta \zeta^{-1/2} \Sigma_{20}^{(3b)}.$$

This result is a consequence of the fact that $\Sigma_{20}^{(3b)}$ contains two integrals of products of two $G^{(0)}$-functions, each of which is formally divergent at the upper limit, whereas $\Sigma_{20}^{(3a)}$ contains an integral involving three $G^{(0)}$-functions, which converges without introducing a cutoff and is determined by the values of the integrand in the momentum region $|\mathbf{q}| \sim \sqrt{m\mu}$. In the diagrams, this difference consists of the number of solid lines in the closed loop (formed by straight lines and wavy lines).

Thus, each loop in Σ_{ik} with more than two straight lines introduces the small parameter β, while the loops with two straight lines do not contain β. Therefore, in the lowest approximation with respect to β we need only take account of diagrams of the second kind. Formally, this means that among all the diagrams for Σ_{11} and Σ_{20} we need only choose those which contain two ingoing or outgoing zigzag lines (corresponding to particles in the condensate), i.e., diagrams which are of the first order in n_0. In fact, it is clear from dimensionality considerations that all diagrams containing a larger power of n_0 involve extra dependence on the small parameter β. All the required diagrams are of the "ladder type," shown in Fig. 76.

Let $\Gamma^{(0)}(p_1, p_2; p_3, p_4)$ denote the set of all diagrams shown in Fig. 77 (in the momentum representation). Then the first approximation with respect to β differs from the first approximation of perturbation theory by having $\Gamma^{(0)}(p_1, p_2; p_1 - q, p_2 + q)$ in place of the Fourier components $U(\mathbf{q})$ of the interaction potential (the first "rung" of the ladder). Of course, summation over the "ladder loops" contained in the more complicated diagrams also leads to the appearance of $\Gamma^{(0)}$ in these diagrams (however, for our purposes there is no need to consider these diagrams, since, as already mentioned, they

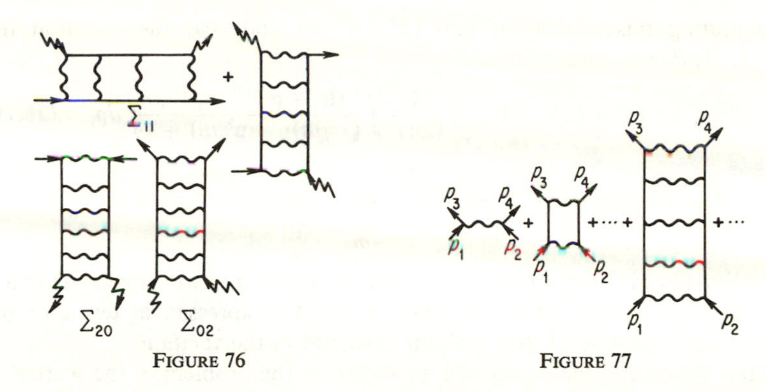

FIGURE 76

FIGURE 77

contribute terms of higher order in β). Thus, the potential $U(\mathbf{q})$ is eliminated from the problem, and its role is taken over by the effective potential $\Gamma^{(0)}$. From the structure of the diagrams shown in Fig. 77, we immediately deduce the following integral equation for the quantity $\Gamma^{(0)}(p_1, p_2; p_3, p_4)$:

$$\begin{aligned}\Gamma^{(0)}(p_1, p_2; p_3, p_4) = U(\mathbf{p}_3 - \mathbf{p}_1) + \frac{i}{(2\pi)^4}\int U(\mathbf{p}_1 - \mathbf{k})G^{(0)}(k) \\ \times G^{(0)}(p_1 + p_2 - k)\Gamma^{(0)}(k, p_1 + p_2 - k; p_3, p_4)\, d^4k.\end{aligned} \tag{25.9}$$

25.4. The effective interaction potential. Next, we investigate the integral equation (25.9). Introducing the total and relative momenta

$$p_1 + p_2 = p_3 + p_4 = P, \quad p_1 - p_2 = 2k, \quad p_3 - p_4 = 2k',$$

we can write the equation (25.9) for

$$\Gamma^{(0)}(p_1, p_2; p_3, p_4) \equiv \Gamma^{(0)}(k, k'; P)$$

in the form

$$\begin{aligned}\Gamma^{(0)}(k, k'; P) = U(\mathbf{k} - \mathbf{k}') + \frac{i}{(2\pi)^4}\int U(\mathbf{k} - \mathbf{p})G^{(0)}(\tfrac{1}{2}P + p) \\ \times G^{(0)}(\tfrac{1}{2}P - p)\Gamma^{(0)}(p, k'; P)\, d^4p.\end{aligned} \tag{25.10}$$

The interaction potential $V(x - x')$ contains no retardation effects, i.e.,

$$V(x - x') = U(\mathbf{r} - \mathbf{r}')\delta(t - t'),$$

and hence its Fourier components $V(q) \equiv U(\mathbf{q})$ do not depend on the fourth component of the four-vector q. It follows that $\Gamma^{(0)}(p_1, p_2; p_3, p_4)$ depends on only one combination of fourth components

$$\omega_1 + \omega_2 = \omega_3 + \omega_4 = \varpi,$$

where $P = (\mathbf{P}, \varpi)$. Therefore, $\Gamma^{(0)}(k, k'; P)$ does not depend on the fourth component of its first two arguments, and this allows us to carry out the integration with respect to ω in the right-hand side of (25.10):

$$\int d\omega\, G^{(0)}(\tfrac{1}{2}P + p)G^{(0)}(\tfrac{1}{2}P - p) = -\frac{2\pi i}{\varpi - (\mathbf{P}^2/4m) + 2\mu - (\mathbf{p}^2/m) + i\delta}.$$

Substituting this expression into (25.10), we find that the equation for $\Gamma^{(0)}(k, k'; P)$ becomes

$$\Gamma^{(0)}(k, k'; P) = U(\mathbf{k} - \mathbf{k}') + \frac{1}{(2\pi)^3}\int \frac{U(\mathbf{k} - \mathbf{p})\Gamma^{(0)}(p, k'; P)}{(\varkappa^2/m) - (\mathbf{p}^2/m) + i\delta}\, d\mathbf{p}, \tag{25.11}$$

where

$$\frac{\varkappa^2}{m} = \varpi - \frac{\mathbf{P}^2}{4m} + 2\mu.$$

Equation (25.11) cannot be solved in general form for an arbitrary interaction law, but, as we now show, its solution can be expressed in terms of the scattering amplitude of two colliding particles in the vacuum.

First we remind the reader of how to set up the problem of the scattering of a particle by a potential $U(\mathbf{r})$. The Schrödinger equation of a particle in the field $U(\mathbf{r})$ can be written in the form

$$(\nabla^2 + \mathbf{k}^2)\psi_{\mathbf{k}}(\mathbf{r}) = 2mU(\mathbf{r})\psi_{\mathbf{k}}(\mathbf{r}),$$

where $\mathbf{k}^2/2m$ is the eigenvalue of the particle's energy, and $\psi_{\mathbf{k}}(\mathbf{r})$ is the corresponding wave function. It is convenient to represent this equation in terms of a solution of Poisson's equation

$$\psi_{\mathbf{k}}(\mathbf{r}) = -\frac{m}{2\pi}\int \frac{e^{i|\mathbf{k}|\,|\mathbf{r}-\mathbf{r}'|}}{|\mathbf{r} - \mathbf{r}'|}\, U(\mathbf{r}')\psi_{\mathbf{k}}(\mathbf{r}')\, d\mathbf{r}' + \psi_{0\mathbf{k}}(\mathbf{r}), \tag{25.12}$$

where $\psi_{0\mathbf{k}}(\mathbf{r})$ is the wave function of a free particle with the same energy. The scattering amplitude is determined by the condition that at large distances from the scattering center the wave function should be the sum of a plane wave (a free particle) and an outgoing wave,[6] i.e.,

$$\psi_{\mathbf{k}}(\mathbf{r}) = e^{i\mathbf{k}\cdot\mathbf{r}} - f(\theta)\,\frac{e^{i|\mathbf{k}|\,|\mathbf{r}|}}{|\mathbf{r}|},$$

where θ is the angle of scattering relative to the direction of the vector $\mathbf{k}$. Comparing the behavior of (25.12) for large $|\mathbf{r}|$ with this definition, we obtain

$$f(\theta) = \frac{m}{2\pi}\int e^{-i\mathbf{k}'\cdot\mathbf{r}'}U(\mathbf{r}')\psi_{\mathbf{k}}(\mathbf{r}')\, d\mathbf{r}',$$

where the vector $\mathbf{k}'$ is directed along $\mathbf{r}$. Going over to the momentum representation for the wave function

$$\psi_{\mathbf{k}}(\mathbf{r}) = (2\pi)^{-3}\int \psi_{\mathbf{k}}(\mathbf{p})e^{i\mathbf{p}\cdot\mathbf{r}}\, d\mathbf{p},$$

we obtain

$$f(\theta) \equiv f(\mathbf{k}, \mathbf{k}') = \frac{m}{(2\pi)^4}\int U(\mathbf{k}' - \mathbf{p})\psi_{\mathbf{k}}(\mathbf{p})\, d\mathbf{p} \tag{25.13}$$

(the direction of $\mathbf{k}$ is that of the incident particle). The scattering amplitude

[6] Our definition of the scattering amplitude has a different sign from the usual definition (see e.g., L7).

usually refers to the values of (25.13) for $|\mathbf{k}| = |\mathbf{k}'|$, but we shall use (25.13) to define the scattering amplitude $f(\mathbf{k}, \mathbf{k}')$ more generally for arbitrary $\mathbf{k}$ and $\mathbf{k}'$. In the momentum representation, equation (25.12) takes the form

$$\psi_{\mathbf{k}}(\mathbf{p}) = (2\pi)^3\delta(\mathbf{k} - \mathbf{p}) + \frac{4\pi f(\mathbf{k}, \mathbf{p})}{\mathbf{k}^2 - \mathbf{p}^2 + i\delta}. \tag{25.14}$$

Substituting (25.14) into (25.13), we obtain

$$\frac{2\pi}{m} f(\mathbf{k}, \mathbf{k}') = U(\mathbf{k}' - \mathbf{k}) + \frac{1}{(2\pi)^3}\int \frac{U(\mathbf{k}' - \mathbf{p})[(2\pi/m) f(\mathbf{k}, \mathbf{p})\, d\mathbf{p}}{(\mathbf{k}^2/2m) - (\mathbf{p}^2/2m) + i\delta}. \tag{25.15}$$

We now return to equation (25.11). As is well known, the scattering of two colliding particles of masses m_1 and m_2, which interact with potential energy $U(\mathbf{r} - \mathbf{r}')$, reduces to the scattering by the potential $U(\mathbf{r})$ of a single particle, with reduced mass

$$m^* = \frac{m_1 + m_2}{m_1 m_2}.$$

Making the change

$$m \to m^* = \frac{m}{2}$$

everywhere in (25.15), we obtain

$$U(\mathbf{k} - \mathbf{k}') = \frac{4\pi}{m} f(\mathbf{k}', \mathbf{k}) - \frac{1}{(2\pi)^3}\int \frac{U(\mathbf{k} - \mathbf{p})(4\pi/m) f(\mathbf{k}', \mathbf{p})\, d\mathbf{p}}{(\mathbf{k}'^2/m) - (\mathbf{p}^2/m) + i\delta} \equiv \hat{L}\left(\frac{4\pi}{m} f\right),$$

where $\hat{L}$ denotes the operator appearing in the right-hand side. By subtracting identical expressions from both sides of (25.11), we can reduce (25.11) to the form

$$\begin{aligned}\Gamma^{(0)}(k, k'; P) &- \frac{1}{(2\pi)^3}\int \frac{U(\mathbf{k} - \mathbf{p})\Gamma^{(0)}(p, k'; P)}{(\mathbf{k}'^2/m) - (\mathbf{p}^2/m) + i\delta}\, d\mathbf{p} \\ &= U(\mathbf{k} - \mathbf{k}') + \frac{1}{(2\pi)^3}\int U(\mathbf{k} - \mathbf{p}) \\ &\times \left\{\frac{1}{(\varkappa^2/m) - (\mathbf{p}^2/m) + i\delta} - \frac{1}{(\mathbf{k}'^2/m) - (\mathbf{p}^2/m) + i\delta}\right\}\Gamma^{(0)}(p, k'; P)\, d\mathbf{p},\end{aligned} \tag{25.16}$$

where the left-hand side equals $\hat{L}(\Gamma^{(0)})$. Applying the operator $\hat{L}^{-1}$ to both sides of (25.16), we finally obtain the following equation for $\Gamma^{(0)}$:

$$\begin{aligned}\Gamma^{(0)}(k, k'; P) &= \frac{4\pi}{m} f(\mathbf{k}', \mathbf{k}) + \frac{1}{(2\pi)^3}\int \left[\frac{4\pi}{m} f(\mathbf{p}, \mathbf{k})\right] \\ &\times \left\{\frac{1}{(\varkappa^2/m) - (\mathbf{p}^2/m) + i\delta} - \frac{1}{(\mathbf{k}'^2/m) - (\mathbf{p}^2/m) + i\delta}\right\}\Gamma^{(0)}(p, k'; P)\, d\mathbf{p}.\end{aligned} \tag{25.17}$$

It is clear from (25.17) that to a first approximation $\Gamma^{(0)}(k, k'; P)$ equals $(4\pi/m) f(\mathbf{k}', \mathbf{k})$. The integral in the right-hand side of (25.17) converges, even if we assume that f and $\Gamma^{(0)}$ are constants, and hence the integral is of order $|\mathbf{k}|f^2/m$. As will be apparent from what follows, the important region of momenta is

$$|\mathbf{k}| \sim \sqrt{m\mu} \sim \sqrt{n_0 f},$$

i.e., $|\mathbf{k}|f \ll 1$, and we can confine ourselves to the first term for $\Gamma^{(0)}(k, k'; P)$. Moreover, we note that in this case, the dependence on $\mathbf{k}$ and $\mathbf{k}'$ can be neglected in the expression $f(\mathbf{k}, \mathbf{k}')$. For small energies, this dependence has the form of an expansion in powers of the ratio of a (the dimensions of the particles) to the wavelength $\lambda \sim 1/|\mathbf{k}|$. Since a is of the order of the scattering amplitude f, and since $|\mathbf{k}|f \ll 1$, we can finally write

$$\Gamma^{(0)}(k, k'; P) \approx \frac{4\pi}{m} f(0, 0) \equiv \frac{4\pi}{m} f_0. \tag{25.18}$$

25.5. The Green's function of a Bose gas in the low-density approximation. The spectrum. According to the above considerations, we have

$$\Sigma_{11}(p) = \frac{8\pi}{m} n_0 f_0, \quad \Sigma_{20}(p) = \Sigma_{02}(p) = \frac{4\pi}{m} n_0 f_0, \quad \mu = \frac{4\pi}{m} n_0 f_0. \tag{25.19}$$

Substitution of these expressions into (24.4) and (24.5) gives

$$G(p) = \frac{\omega + (\mathbf{p}^2/2m) + (4\pi/m)n_0 f_0}{\omega^2 - \varepsilon^2(\mathbf{p}) + i\delta},$$

$$\hat{G}(p) = -\frac{4\pi n_0 f_0}{m} \frac{1}{\omega^2 - \varepsilon^2(\mathbf{p}) + i\delta},$$

where

$$\begin{aligned}\varepsilon(\mathbf{p}) &= \sqrt{\left(\frac{\mathbf{p}^2}{2m} + \frac{4\pi n_0 f_0}{m}\right)^2 - \frac{16\pi^2 n_0^2 f_0^2}{m^2}} \\ &= \sqrt{\frac{\mathbf{p}^4}{4m^2} + \frac{4\pi n_0 f_0}{m^2}\mathbf{p}^2}\end{aligned} \tag{25.20}$$

is the spectrum of the system for small momenta. This expression for the spectrum differs from the expression (4.11) derived in Sec. 4 by having the exact s-wave scattering amplitude in place of the amplitude given by the Born approximation. It follows from (25.20) that for $|\mathbf{p}| \ll \sqrt{n_0 f_0}$, the quasi-particles have "acoustic dispersion," i.e.,

$$\varepsilon(\mathbf{p}) \approx |\mathbf{p}|\sqrt{\frac{4\pi n_0 f_0}{m^2}},$$

whereas for $|\mathbf{p}| \gg \sqrt{n_0 f_0}$, the quasi-particles become "almost free," i.e.,

$$\varepsilon(\mathbf{p}) \approx \varepsilon_0(\mathbf{p}) + \frac{4\pi n_0 f_0}{m}$$

(a spectrum of this latter type corresponds to a particle moving in a continuous medium characterized by some refractive index). The transition from the phonon region to the "free particle" region in the dispersion formula occurs when

$$|\mathbf{p}| \sim \sqrt{n_0 f_0} \ll \frac{1}{f_0}.$$

Thus, in both regions it is legitimate to regard the amplitudes as being approximately constant.

In conclusion, we note that there is no correlation at all between the present model and the properties of real helium. Aside from the fact that the low-density approximation does not pertain to liquid He II, it is worth pointing out that the spectrum given by (25.20) is actually unstable for small $\mathbf{p}$. In fact, for $\mathbf{p} \neq 0$ the velocity of the excitations

$$\frac{\partial \varepsilon}{\partial |\mathbf{p}|}$$

is greater than the velocity of sound

$$\sqrt{\frac{4\pi n_0 f_0}{m^2}},$$

i.e., the excitations can create phonons (see the next section). This leads to the appearance of attenuation in the spectrum, where the lifetimes of the excitations are inversely proportional to $|\mathbf{p}|^5$ for small $\mathbf{p}$. The spectrum of helium does not exhibit this instability for small $\mathbf{p}$.

26. Properties of the Spectrum of One-Particle Excitations Near the Cutoff Point

26.1. Statement of the problem. Of course, the spectrum of one-particle excitations in a real Bose liquid, i.e., in helium, cannot be calculated theoretically. There is linear dependence of the energy on the momentum (the phonon part of the spectrum) only for the smallest momenta. For larger momenta, the spectrum is no longer linear and its further behavior depends on the particular properties of the interactions between particles of the liquid.

The characteristic feature of the excitation spectrum of a Bose liquid compared to that of a Fermi liquid is that undamped (nonattenuated) Bose excitations can exist. From a mathematical point of view, this means that the solutions of equation (24.16) are real. At finite temperatures, the attenuation of excitations is due to the fact that they can collide with each other. At absolute zero, there are no real excitations. Therefore, the only mechanism which can lead to finite lifetimes of the excitations is the decay of the excitations into other excitations of lower energy, provided such a process is permitted by the laws of energy and momentum conservation. In a Fermi liquid, there can always be decay accompanied by formation of a particle and a hole, and as a result, the quasi-particles have finite lifetimes inversely proportional to $(|\mathbf{p}| - p_0)^2$. In a Bose liquid, undamped excitations can exist for sufficiently small momenta, and it is only when the momentum is increased that the energy of an excitation ultimately reaches a certain threshold value, above which the excitation is unstable with respect to decay into two or more excitations with less energy. This threshold, which is a singular point of the curve representing the spectrum, will be called the (*spectrum*) *cutoff point*. In the present section, we shall try to explain the character of this singularity. It will emerge in the course of our discussion that a complete analysis can be carried out without making any assumption whatsoever about the weakness of the interaction [see Pitayevski (P4)]. Our only assumption (and one which is certainly general enough from a physical standpoint) is that the cutoff point of the spectrum corresponds to a threshold for decay into two (and no more than two) excitations.

Momentum and energy must be conserved when an excitation decays into two excitations, a fact which is expressed by the equation

$$\varepsilon(\mathbf{p}) = \varepsilon(\mathbf{q}) + \varepsilon(\mathbf{p} - \mathbf{q}). \tag{26.1}$$

Here, $\mathbf{p}$ and $\varepsilon(\mathbf{p})$ are the momentum and energy of the decaying excitation, $\mathbf{q}$ and $\varepsilon(\mathbf{q})$ are the momentum and energy of one of the excitations formed in the decay, and $\mathbf{p} - \mathbf{q}$ and $\varepsilon(\mathbf{p} - \mathbf{q})$ are the momentum and energy of the other excitation formed in the decay. No decay is possible if for a given $\mathbf{p}$, equation (26.1) has no solutions for $\mathbf{q}$. Let p_c and $\varepsilon_c = \varepsilon(p_c)$ denote the momentum and energy at the decay threshold. Then the decay threshold is characterized by the fact that (26.1) has a solution for $\mathbf{q}$ if $\varepsilon = \varepsilon_c$, but no solution if $\varepsilon < \varepsilon_c$. This requires that for $|\mathbf{p}| = p_c$, the right-hand side of (26.1), regarded as a function of $\mathbf{q}$, should have a minimum for certain values of $\mathbf{q}$. For $|\mathbf{p}| = p_c$, the right-hand side of (26.1) depends on the two variables $|\mathbf{q}|$ and $\cos\theta$, where θ is the angle between the vectors $\mathbf{p}$ and $\mathbf{q}$, and can have a minimum both for $\theta = 0$ and for finite θ.

Suppose the right-hand side of (26.1) has a minimum for a certain momentum $\mathbf{q}$. Making a power-series expansion of $\varepsilon(\mathbf{q}) + \varepsilon(\mathbf{p} - \mathbf{q})$ up to terms of the second order in the increment $\Delta\mathbf{q}$, we have[7]

$$\begin{aligned}\varepsilon(\mathbf{q} + \Delta\mathbf{q}) + \varepsilon(\mathbf{p} - \mathbf{q} - \Delta\mathbf{q}) \approx \varepsilon(\mathbf{q}) + \varepsilon(\mathbf{p} - \mathbf{q}) + \frac{\partial\varepsilon(\mathbf{q})}{\partial\mathbf{q}_i}\Delta\mathbf{q}_i \\ - \frac{\partial\varepsilon(\mathbf{p} - \mathbf{q})}{\partial\mathbf{p}_i}\Delta\mathbf{q}_i + \frac{1}{2}\frac{\partial^2\varepsilon(\mathbf{q})}{\partial\mathbf{q}_i\partial\mathbf{q}_k}\Delta\mathbf{q}_i\,\Delta\mathbf{q}_k + \frac{1}{2}\frac{\partial^2\varepsilon(\mathbf{p} - \mathbf{q})}{\partial\mathbf{p}_i\partial\mathbf{p}_k}\Delta\mathbf{q}_i\,\Delta\mathbf{q}_k.\end{aligned}$$

At the point where the minimum is achieved, the linear terms must vanish. Obviously, there are two basic possibilities:

1. $\nabla_{\mathbf{q}}\varepsilon(\mathbf{q}) = \nabla_{\mathbf{p}}\varepsilon(\mathbf{p} - \mathbf{q}) \neq 0$. This corresponds to decay into two excitations which propagate in the direction of the vector $\mathbf{p}$ with the same velocity $\mathbf{v} = \nabla_{\mathbf{q}}\varepsilon$. Moreover, two subcases are possible: In the first,

[7] Summation over pairs of repeated indices ($i, k = 1, 2, 3$) is implied and $\partial/\partial\mathbf{p}_i \equiv \nabla_{\mathbf{p}_i}$, $\partial^2/\partial\mathbf{p}_i\partial\mathbf{p}_k \equiv \nabla_{\mathbf{p}_i}\nabla_{\mathbf{p}_k}$, etc.

which will be called *Case I*, one of the excitations has momentum arbitrarily close to zero. Then the velocity of the excitation at the point p_c is comparable to the velocity of sound c, and the excitation can create a phonon. In *Case II*, both of the excitations formed in the decay can have finite momenta.

2. $\nabla_{\mathbf{q}}\varepsilon(\mathbf{q}) = 0$, $\nabla_{\mathbf{p}}\varepsilon(\mathbf{p} - \mathbf{q}) = 0$. This requires that each excitation be formed with the momentum p_0 for which the excitation energy $\varepsilon(\mathbf{p})$ is a minimum. For liquid helium, this point of the spectrum corresponds to the value

$$p_0 = 1.92 \times 10^8 \text{ cm}^{-1}.$$

In the neighborhood of this point, the spectrum $\varepsilon(\mathbf{p})$ has the so-called "roton form"

$$\varepsilon(\mathbf{p}) = \Delta + \frac{(|\mathbf{p}| - p_0)^2}{2m^*} \qquad (||\mathbf{p}| - p_0| \ll p_0). \tag{26.2}$$

If $\varepsilon_c = 2\Delta$, the excitation decays into two rotons with momenta $\mathbf{q}$ and $\mathbf{q}_1$, where $|\mathbf{q}| = |\mathbf{q}_1| = p_0$ and $\varepsilon(\mathbf{q}) = \varepsilon(\mathbf{q}_1) = \Delta$. The angle θ between the two emitted rotons is determined by the condition that the sum of the momenta of the rotons should equal p_c. (This will be called *Case III*.)

The three cases just enumerated exhaust all the possible types of thresholds for decay into two excitations.

26.2. The basic system of equations. To investigate the form of the spectrum near the cutoff point, we use the quantum field theory methods developed earlier. Thus, we look for the form of the Green's function near the cutoff point, since the spectrum is determined by the poles of the Green's function. It is physically obvious that the singularities of the Green's function are related to diagrams in which one line divides into two lines, thereby representing graphically the process of decay of one excitation into two others.

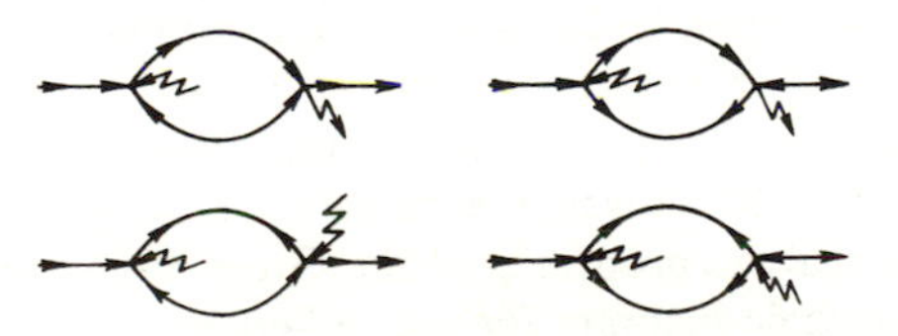

FIGURE 78

For example, consider the diagrams in Fig. 78, involving the different Green's functions G', $\hat{G}$ and $\check{G}$. Each of the loops shown in the figure represents a self-energy part, characterized by the fact that it consists of two triple vertices (where only lines corresponding to particles not in the condensate are considered) joined by two solid lines. The integral associated with a loop of this kind is

$$\int G(q)G(p-q)\Gamma_1\Gamma_2\,d\mathbf{q}\,d\omega', \tag{26.3}$$

where G can denote any of the three Green's functions G', $\hat{G}$ and $\check{G}$, and the vertex parts appearing to the right and to the left of these diagrams are denoted by Γ_1 and Γ_2. Suppose the external points have values of $\mathbf{p}$ and ω lying close to a pole of the function $\omega = \varepsilon(\mathbf{p})$. (It was shown in Sec. 24 that all three Green's functions have the same poles.) The singularity of the integral (26.3), if it has one, is associated with the region of integration (with respect to $\mathbf{q}$ and ω') for which the functions $G(q)$ and $G(p - q)$ are near their poles. According to (24.10), (24.12) and (24.14), near their poles these functions have the form

$$G(q) = \frac{A_1}{\omega' - \varepsilon(\mathbf{q}) + i\delta} \quad \text{or} \quad \frac{A_2}{\omega' + \varepsilon(\mathbf{q}) - i\delta}, \tag{26.4a}$$

and

$$G(p - q) = \frac{B_1}{\omega - \omega' - \varepsilon(\mathbf{p} - \mathbf{q}) + i\delta} \quad \text{or} \quad \frac{B_2}{\omega - \omega' + \varepsilon(\mathbf{p} - \mathbf{q}) - i\delta}, \tag{26.4b}$$

where in each case the first expression corresponds to a positive pole and the second corresponds to a negative pole. Substituting these expressions into (26.3), we see that the terms which interest us are terms of the form A_1B_1. In these terms we can carry out the integration with respect to ω' from $-\infty$ to ∞. Then the remaining integral with respect to $\mathbf{q}$ has the form

$$\int \frac{\Gamma_1\Gamma_2 AB\,d\mathbf{q}}{\varepsilon(\mathbf{q}) + \varepsilon(\mathbf{p} - \mathbf{q}) - \omega} \tag{26.5}$$

in some region of values of $\mathbf{q}$.

Whether or not (26.5) is singular depends on whether or not the denominator of the integrand vanishes for some value of $\mathbf{q}$. According to the analysis given in Sec. 26.1, the denominator is always greater than zero for $\omega < \varepsilon(p_c)$. The integrand of (26.5) first becomes infinite for $\omega = \varepsilon(p_c)$, i.e., the point $\omega = \varepsilon(p_c)$ is a singular point (in the usual mathematical sense of the word) of the integral (26.5). Moreover, the character of the singularity is determined only by the analytic properties of the Green's function, and does not depend on which specific diagram for the self-energy part was chosen from the diagrams shown in Fig. 78. This fact allows us to appreciably simplify our subsequent analysis. In fact, as just shown, to determine the character of a singularity, we only need the expressions for the Green's functions near their poles, and near their poles all three Green's functions have the same form. Therefore, if we are not interested in the magnitude of the regular terms or in various unimportant coefficients, there is no need to make a distinction between the functions G', $\hat{G}$ and $\check{G}$ near their poles, since the diagrams for all three Green's functions have the same structure.

For example, suppose we add the two equations (24.2) and introduce a new function

$$G_1(p) = G'(p) + \hat{G}(p).$$

Then $G_1(p)$ satisfies the equation

$$G_1(p) = G^{(0)}(p) + G^{(0)}(p)[\Sigma_{11}(p) + \Sigma_{20}(p)]G_1(p).$$

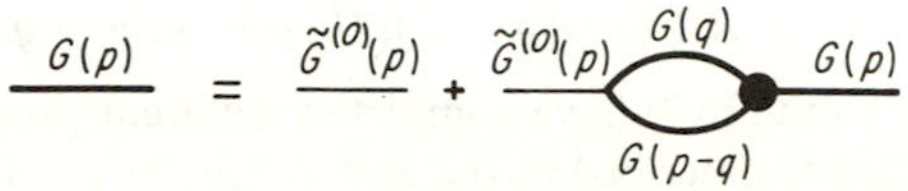

FIGURE 79

We divide the whole set of self-energy parts $\Sigma = \Sigma_{11} + \Sigma_{20}$ into diagrams Σ_0 which have no singularities at the point $\omega = \varepsilon_c$ and diagrams Σ_1 (like those shown in Fig. 78) which have singularities at $\omega = \varepsilon_c$. If we introduce another new function

$$\tilde{G}^{(0)}(p) = \frac{1}{(G^{(0)})^{-1}(p) - \Sigma_0(p)},$$

the resulting equation for G_1 can be written in the form of a Dyson equation, as shown schematically in Fig. 79. Since the character of the singularity is determined by the form of the Green's functions at the pole, where they are all the same except for coefficients, we can replace all internal lines for G', $\hat{G}$ or $\check{G}$ by G_1-lines. To the left of the loop appears a "bare" vertex $\Gamma^{(0)}$, which, from the standpoint of the general technique, represents the result of interaction of the particles in the condensate with three particles not in the condensate (as shown in Fig. 78, for example). To the right of the loop appears a triple vertex Γ (indicated by a heavy point), obtained from $\Gamma^{(0)}$ as a result of the interaction of the lines leaving $\Gamma^{(0)}$.

The diagrams expressing $\Gamma(p, p - q, q)$ in terms of $\Gamma^{(0)}(p, p - q, q)$ are shown in Fig. 80(a), where a square represents $\Gamma(p_1, p_2; p_3, p_4)$, the exact irreducible vertex part for scattering of particles not in the condensate by each other; Γ is the sum of all four-particle diagrams which cannot be divided between the external points p_1, p_2 and p_3, p_4 into two parts connected by only one or two lines. Summation of the diagrams shown in Fig. 80(a) leads to the simple equation shown schematically in Fig. 80(b). Thus,

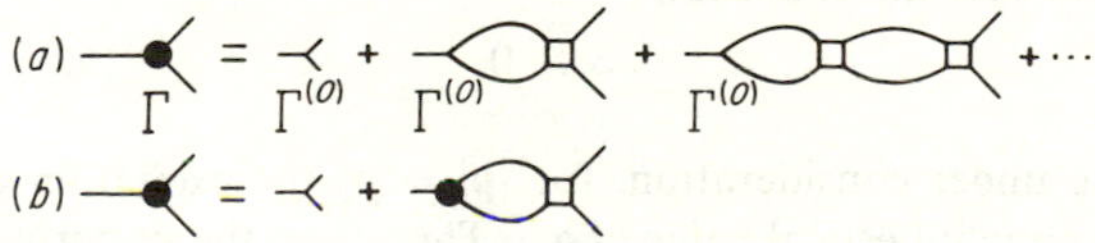

FIGURE 80

omitting the subscript on G_1, we can write the following analytical expressions for our two basic equations:

$$G^{-1}(p) - (\tilde{G}^{(0)})^{-1}(p) = \frac{i}{(2\pi)^4}\int \Gamma^{(0)}(p, p-q, q)G(q) \times G(p-q)\Gamma(p, p-q, q)\, d^4q, \tag{26.6}$$

$$\Gamma(p, p-q, q) = \Gamma^{(0)}(p, p-q, q) + \frac{i}{(2\pi)^4}\int \Gamma(p, p-k, k) \times G(k)G(p-k)\Gamma(k, p-k; p-q, q)\, d^4k. \tag{26.7}$$

The equations (26.6) and (26.7) have completely different properties near the three kinds of thresholds described at the end of Sec. 26.1. Therefore, these three cases must be discussed separately, as we now proceed to do.

26.3. Properties of the spectrum near the threshold for phonon creation. First we consider the properties of the excitation spectrum near a point where the velocity of the excitations becomes equal to the velocity of sound. Starting from this point, the excitation can create a phonon. In this case, the conservation laws (26.1) take the form

$$\varepsilon(\mathbf{p}) = \varepsilon(\mathbf{p} - \mathbf{q}) + \omega(\mathbf{q}), \tag{26.1'}$$

where $\omega(\mathbf{q})$ is the frequency of the phonon and $\mathbf{q}$ is its wave vector. For small $\mathbf{q}$, $\omega(\mathbf{q})$ has the form

$$\omega(\mathbf{q}) = c|\mathbf{q}| - \alpha|\mathbf{q}|^3. \tag{26.8}$$

Although the third-order terms in $\omega(\mathbf{q})$ will not be needed later, we assume that $\alpha > 0$, i.e., that the phonon spectrum is stable. The function $\varepsilon(\mathbf{p})$ has a singularity for $|\mathbf{p}| = p_c$. We make the assumption (to be justified by the final result) that this singularity appears in the terms of degree higher than two in the quantity $\Delta p = |\mathbf{p}| - p_c$, so that

$$\varepsilon(\mathbf{p}) \approx \varepsilon_c + c\,\Delta p + \beta\,(\Delta p)^2 \tag{26.9}$$

near p_c (by hypothesis, for $|\mathbf{p}| = p_c$ the velocity of the excitation $v = \partial\varepsilon/\partial|\mathbf{p}|$ equals the velocity of sound).

For $|\mathbf{p}| = p_c$ and $\cos\theta = 1$ (where θ is the angle between $\mathbf{p}$ and $\mathbf{q}$), the right-hand side of (26.1′) becomes

$$\varepsilon_c + \beta|\mathbf{q}|^2, \tag{26.10}$$

where we have taken account of (26.8) and (26.9). The point $|\mathbf{p}| = p_c$ is actually a threshold only if the expression (26.10) has a minimum for $\mathbf{q} = 0$, which requires that the condition

$$\beta > 0$$

be satisfied.

In the case under consideration, for $|\mathbf{p}| = p_c$ the excitation can create a phonon with an infinitesimal value of $\mathbf{q}$. Therefore, the important region for finding the singularity of the integral (26.6) is the region where the argument of one of the Green's functions, say $G(q)$, is small. For small $\mathbf{q}$ and ω, the Green's function is given by (24.19), i.e.,

$$G(q) = \frac{a}{\omega^2 - \omega^2(\mathbf{q}) + i\delta}. \tag{26.11}$$

[We cannot use the representation (26.4) for the function $G(q)$, since, for small q, the two poles almost coincide.] The Green's function has a singularity near $|\mathbf{p}| = p_c$, $\varepsilon = \varepsilon_c$. In keeping with (26.9), we assume that near its zero [i.e., near the pole of $G(p)$], the function $G^{-1}(p)$ has the form

$$\begin{gathered} G^{-1}(p) = A^{-1}[\Delta\varepsilon - c\,\Delta p - \beta(\Delta p)^2 + i\delta] \\ (\Delta p = |\mathbf{p}| - p_c, \quad \Delta\varepsilon = \omega - \varepsilon_c), \end{gathered} \tag{26.12}$$

plus higher-order terms, containing the singularity, which must now be determined.

Thus, we now examine the properties of the vertex part $\Gamma(p, p - q, q)$. For small $\mathbf{q}$, this vertex part represents a process in which a particle with momentum $\mathbf{p}$ emits a long-wavelength excitation, i.e., a phonon. It follows that Γ must be proportional to $|\mathbf{q}|$, the magnitude of the momentum of the emitted phonon, since, from a macroscopic point of view, this process consists of scattering of an excitation by density oscillations (sound). The interaction must vanish in the limit where the acoustic oscillations have infinitely long wavelength, since the excitation is not scattered in a homogeneous medium. Therefore, we have

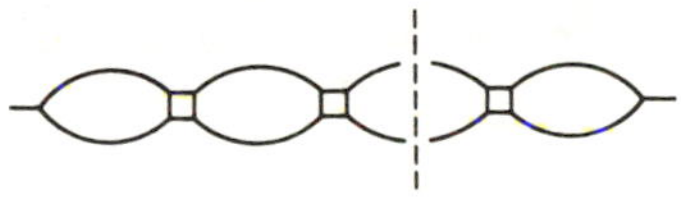

FIGURE 81

$$\Gamma(p, p - q, q) = g|\mathbf{q}| \tag{26.13}$$

in the region of small $|\mathbf{q}|$.

Next, consider the integral in the right-hand side of equation (26.6). According to the definition of Γ and $\Gamma^{(0)}$ (see Fig. 80), to every order in Γ this integral is a chain consisting of loops joined together by four-particle vertex functions (see Fig. 81). Each of these loops contributes to the singularity of the Green's function, and the contribution from each loop only has to be taken into account once, since we have assumed that the nonregular terms are small. Fixing our attention on one loop, we see that the sets of diagrams to the right and to the left of this loop can be summed independently and give rise to three-particle functions $\Gamma(p, p - q, q)$ at each vertex [recall the definition of the exact three-vertex function $\Gamma(p, p - q, q)$]. Thus, the small nonregular contribution to the reciprocal Green's function $G^{-1}(p)$ can be found by considering the nonregular part of the integral

$$\int \Gamma^2(p, p - q, q) G(q) G(p - q)\, d^4q. \tag{26.14}$$

Examining (26.14) in the region of small $\mathbf{q}$, and substituting into (26.14) the expressions (26.11)–(26.13) for $\Gamma(p, p - q, q)$, $G(q)$ and $G(p - q)$, we find

$$Aag^2 \int \frac{|\mathbf{q}|^4 \, d|\mathbf{q}| \, d\Omega \, d\omega'}{(\omega'^2 - c^2\mathbf{q}^2 + i\delta)[\omega' - \omega + \varepsilon(\mathbf{p} - \mathbf{q}) - i\delta]}. \tag{26.15}$$

The integration with respect to ω' from $-\infty$ to ∞ can be carried out, and reduces to evaluating the residue at the point $\omega' = c|\mathbf{q}|$. Omitting coefficients that are unimportant for our purposes, we can write (26.15) as

$$\int \frac{|\mathbf{q}|^3 \, d|\mathbf{q}| \, d\cos\theta}{c|\mathbf{q}| + \varepsilon(\mathbf{p} - \mathbf{q}) - \omega}, \tag{26.16}$$

after integrating with respect to ω'.

Although the integral (26.16) itself converges at its upper limit, it exhibits a singularity due to the behavior of its integrand in the region of small $|\mathbf{q}|$. Suppose we use (26.10) to expand the denominator of (26.16). Then we can set $\cos\theta = 1$ in the quadratic terms, since it is the small angles $\theta \ll 1$ that are important for finding the singularity. As a result, we obtain

$$\int \frac{|\mathbf{q}|^3 \, d|\mathbf{q}| \, d\cos\theta}{x + c|\mathbf{q}|(1 - \cos\theta) - 2\beta\,\Delta p|\mathbf{q}| + \beta|\mathbf{q}|^2} \propto \int |\mathbf{q}|^2 \ln(x - 2\beta|\mathbf{q}|\,\Delta p + \beta|\mathbf{q}|^2) \, d|\mathbf{q}|,$$

where we have written

$$x = c\,\Delta p - \Delta\varepsilon + \beta(\Delta p)^2.$$

Factoring the argument of the logarithm and integrating, we obtain

$$a_1\left(\frac{k_1}{2}\right)^3 \ln k_1 + a_2\left(\frac{k_2}{2}\right)^3 \ln k_2, \tag{26.17}$$

where

$$k_{1,2} = \beta\,\Delta p \pm \sqrt{(\beta\,\Delta p)^2 - \beta x}.$$

It is clear from (26.17) that $G^{-1}(p)$ actually has a singularity in terms of higher order than those used to derive the last expression. This fact justifies all the assumptions made above concerning the smallness of the nonregular terms.

We now determine the nonregular terms in the immediate neighborhood of the pole of $G(p)$, i.e., for

$$|x| \ll \beta\,(\Delta p)^2.$$

In this case, we can neglect the term involving k_2, and then we obtain

$$(\Delta p)^3 \ln(-\Delta p) \tag{26.18}$$

from (26.17). According to (26.12) and (26.18), in the neighborhood of a pole the Green's function has the form

$$G(p) = \frac{A}{\omega - \varepsilon_{\dot{c}} - c\,\Delta p - \beta\,(\Delta p)^2 - a\,(\Delta p)^3 \ln(-\Delta p)}.$$

This function determines the energy of an elementary excitation near the threshold. Below the threshold, there is no attenuation:

$$\varepsilon(\mathbf{p}) = \varepsilon_c + c(|\mathbf{p}| - p_c) + \beta(|\mathbf{p}| - p_c)^2 + a(|\mathbf{p}| - p_c)^3 \ln|p_c - |\mathbf{p}||.$$

Above the threshold, i.e., for $|\mathbf{p}| > p_c$, the energy of the excitation has a negative imaginary part, equal to $-a\pi(\Delta p)^3$:

$$\varepsilon(\mathbf{p}) = \varepsilon_c + c(|\mathbf{p}| - p_c) + \beta(|\mathbf{p}| - p_c)^2 + a(|\mathbf{p}| - p_c)^3 \ln|p_c - |\mathbf{p}|| - a\pi i(|\mathbf{p}| - p_c)^3.$$

In particular, this implies that a must be positive. Thus, for $|\mathbf{p}| > p_c$, there are no undamped excitations, and the lifetime of the excitations is inversely proportional to $(|\mathbf{p}| - p_c)^3$. The smallness of the attenuation near the threshold is related to the fact that the interaction with long-wavelength phonons is always weak, due to the presence of the factor $|\mathbf{q}|$ in Γ.

26.4. Properties of the spectrum near the threshold for decay into two excitations with parallel nonzero momenta. It follows from physical considerations that in the integration with respect to q in (26.6), the important values of the momentum $\mathbf{q}$ and the frequency ω' are those with which the excitations are formed near the decay threshold; however, these values do not correspond to singularities of the Green's functions of the excitations which are created. The only singularity associated with these values of the momentum and energy stems from the fact that in their neighborhood the given excitation may "coalesce" with another excitation, a process which is impossible at absolute zero, due to the absence of any real excitations. Therefore, the Green's functions appearing in the integrand of (26.6) have the simple form

$$G(q) = \frac{A}{\omega - \varepsilon(\mathbf{q}) + i\delta}$$

near their poles [see (26.4)], where $\varepsilon(\mathbf{q})$ is real and has no singularities in the given neighborhood of values of the vector $\mathbf{q}$. This fact greatly simplifies the subsequent analysis.

Now consider one of the loops in the set of chains which, according to Fig. 81 or equation (26.7), correspond to the right-hand side of equation (26.6). Obviously the quantities $\Gamma(p_1, p_2; p_3, p_4)$ and $\Gamma^{(0)}(p, p - q, q)$ associated with the vertices of the loop have no singularities. Hence, from now on, we shall always omit these terms in making calculations, on the grounds that they only lead to unimportant coefficients or to regular contributions to the Green's function. In studying a given loop, we restrict ourselves to a region of integration with respect to $\mathbf{q}$ which is close to the momentum $\mathbf{q}_0$ and energy ε_0 with which excitations are created. Using the expressions (26.4) for the Green's functions and integrating with respect to

ω', we find that the part of the integral (associated with the loop) which contains the singularity can be represented in the form

$$\int \frac{d\mathbf{q}}{\varepsilon(\mathbf{q}) + \varepsilon(\mathbf{p} - \mathbf{q}) - \omega}.$$

Since the expression $\varepsilon(\mathbf{q}) + \varepsilon(\mathbf{p} - \mathbf{q})$ must have a minimum for $|\mathbf{p}| = p_c$, it follows that it can be written as

$$\varepsilon(\mathbf{q}) + \varepsilon(\mathbf{p} - \mathbf{q}) \approx \varepsilon_c + v_c \Delta p + \alpha(\mathbf{q} - \mathbf{q}_0)^2 + \frac{\beta[(\mathbf{q} - \mathbf{q}_0)\cdot\mathbf{p}_c]^2}{p_c^2}$$

for values of $|\mathbf{p}|$ near p_c. Here, v_c is the velocity of each of the two excitations formed at the decay threshold, and $\mathbf{q}_0$ is the momentum of one of the excitations (recall that the excitations emitted in the decay have momenta directed along the vector $\mathbf{p}_c$). The coefficients α and β in this expansion are determined by the form of the functions $\varepsilon(\mathbf{p} - \mathbf{q})$ and $\varepsilon(\mathbf{q})$:

$$\alpha = \frac{v_c p_c}{2q_0(p_c - q_0)},$$

$$\beta = \frac{1}{2}\left[(\nabla_\mathbf{q})^2\varepsilon|_{\mathbf{q}=\mathbf{q}_0} + (\nabla_\mathbf{q})^2\varepsilon|_{\mathbf{q}=\mathbf{p}_c-\mathbf{q}_0} - \frac{v_c p_c}{q_0(p_c - q_0)}\right].$$

Introducing new variables

$$\mathbf{u} = \mathbf{q} - \mathbf{q}_0, \qquad \mathbf{u}\cdot\mathbf{p}_c = u p_c \cos\vartheta,$$

we obtain

$$\int \frac{u^2\, du\, d\cos\vartheta}{v_c \Delta p - \Delta\varepsilon + \alpha u^2 + \beta u^2 \cos^2\vartheta} \propto \sqrt{v_c \Delta p - \Delta\varepsilon}.$$

Summation over all the loops does not change the character of the singularity, since, unlike the phonon case, for the momentum values of interest ($|\mathbf{q}| \sim q_0$), the exact three-particle function $\Gamma(p, p - q, q)$ can neither vanish nor become infinite. Thus, the nonregular part of the reciprocal Green's function near p_c and ε_c has the form

$$a\sqrt{v_c \Delta p - \Delta\varepsilon}.$$

Since by hypothesis, the point $|\mathbf{p}| = p_c$, $\varepsilon = \varepsilon_c$ is a point of the spectrum, the function $G^{-1}(p)$ must vanish for $\Delta p = 0$, $\Delta\varepsilon = 0$, and hence for small Δp and $\Delta\varepsilon$, the regular part of $G^{-1}(p)$ must have the form $a_1 \Delta p + b_1 \Delta\varepsilon$. Thus, we finally have

$$G^{-1}(p) = A^{-1}[a\Delta p + \Delta\varepsilon + b\sqrt{v_c \Delta p - \Delta\varepsilon}].$$

The excitation energy is determined by the equation

$$G^{-1}(p) = 0. \tag{26.19}$$

Solving this equation, we obtain two roots

$$\Delta\varepsilon_{1,2} = -a\Delta p - \frac{b^2}{2} \pm \sqrt{ab^2 \Delta p + \frac{b^4}{4} + b^2 v_c \Delta p},$$

from which we must choose the value $\Delta\varepsilon_1$, with the plus sign in front of the radical, in order that $\Delta\varepsilon$ should approach 0 as $\Delta p \to 0$. Expanding the square root near the threshold, we find that

$$\varepsilon \approx \varepsilon_c + v_c(|\mathbf{p}| - p_c) - \left(\frac{a + v_c}{b}\right)^2 (\Delta p)^2$$

for small Δp. Substituting this expression into (26.19), we see that the condition

$$\frac{a + v_c}{b} > 0$$

must hold if (26.19) is to have a solution for small Δp and for negative Δp (beneath the threshold). For $|\mathbf{p}| > p_c$, when $a\Delta p + \Delta\varepsilon$ and $b\sqrt{v_c\Delta p - \Delta\varepsilon}$ are both positive, equation (26.19) has no solutions (either real or complex). Thus, in the present case, the curve representing the energy spectrum cannot be extended past the threshold point, and in fact terminates with slope v_c at this point.

26.5. Decay into two excitations emitted at an angle with each other. In this case also, the region of importance in the integration corresponds to the values of $|\mathbf{q}|$ with which the excitations are formed near the decay threshold. In this region, the Green's functions have the usual form (26.4), for the same reasons as in the preceding case. However, now we can no longer assert that the vertex part Γ is finite for $\varepsilon = \varepsilon_c$.

We begin, as before, by considering one of the loops in Fig. 81. The quantities associated with the vertices of the loop do not contain "dangerous" integrations, according to their definitions. Therefore, it is natural to assume that these quantities remain finite at the threshold point. As usual, we examine the integral over the region of values of $|\mathbf{q}|$ and ω' close to the values with which the excitations are formed near the threshold. The singularities of the given loop reduce to the singularities of the integral

$$\int \frac{d\mathbf{q}}{\varepsilon(\mathbf{q}) + \varepsilon(\mathbf{p} - \mathbf{q}) - \omega}, \tag{26.20}$$

where for $\varepsilon(\mathbf{q})$ and $\varepsilon(\mathbf{p} - \mathbf{q})$ we can use the expansion (26.2) of $\varepsilon(\mathbf{q})$ near the roton part of the spectrum. [We recall that Case III (p. 237) corresponds to decay into two rotons with momenta which equal p_0 in magnitude and make a finite angle θ_0 with each other, where $\cos\frac{1}{2}\theta_0 = p_c/2p_0$.] Substituting (26.2) into (26.20), we obtain

$$\int \frac{d\mathbf{q}}{2\Delta - \omega + (|\mathbf{q}| - p_0)^2/2m^* + (|\mathbf{q} - \mathbf{p}| - p_0)^2/2m^*}. \tag{26.21}$$

We now go over to cylindrical coordinates q_z', q_ρ' and ϕ, defined by the formulas

$$\begin{aligned} q_x &= (p_0 \sin\tfrac{1}{2}\theta_0 + q_\rho')\cos\phi, \\ q_y &= (p_0 \sin\tfrac{1}{2}\theta_0 + q_\rho')\sin\phi, \\ q_z &= p_0\cos\tfrac{1}{2}\theta_0 + q_z', \end{aligned} \tag{26.22}$$

where the z-axis is directed along the vector $\mathbf{p}$. Substituting (26.22) into (26.21) and neglecting higher powers of q'_z and q'_ρ, we obtain

$$\int \frac{dq'_\rho\, dq'_z}{2\Delta - \omega + (1/m^*)(\sin^2 \tfrac{1}{2}\theta_0\, q'^2_\rho + \cos^2 \tfrac{1}{2}\theta_0\, q'^2_z)}. \tag{26.23}$$

In the last expression, it is convenient to introduce polar coordinates r and ϑ, defined by

$$\frac{1}{\sqrt{m^*}}\, q'_\rho \sin \tfrac{1}{2}\theta_0 = r \cos \vartheta, \qquad \frac{1}{\sqrt{m^*}}\, q'_z \cos \tfrac{1}{2}\theta_0 = r \sin \vartheta.$$

As a result, (26.23) becomes

$$\int \frac{r\, dr}{2\Delta - \omega + r^2} \propto \ln (2\Delta - \omega).$$

Thus, each loop leads to a large term $\ln (2\Delta - \omega)$, depending only on the frequency ω of the external point. Suppose we fix our attention on one loop. Then, according to Fig. 80(a) and equation (26.7), the sums of all loops to the left and to the right of the given loop gives rise to exact three-particle vertex parts Γ, and hence, when $2\Delta - \omega$ is small, the principal term in the right-hand side of (26.6) has the form

$$\Gamma^2(p, p - q_0, q_0) \ln (2\Delta - \omega), \tag{26.24}$$

where $\mathbf{q}_0$ is some critical value of $\mathbf{q}$ (the vector $\mathbf{q}$ is the momentum of one of the rotons formed in the decay at the threshold point).

To determine $\Gamma(p, p - q, q)$ in the neighborhood $\mathbf{q} \sim \mathbf{q}_0$, we can solve equation (26.7). However, it is simpler to directly sum the terms of the series corresponding to this equation, by using the fact (mentioned above) that the principal term in each loop depends only on the frequency ω of the external point and is the same for each loop. Formally, the series in question is a geometric series, whose sum can be written as

$$\Gamma(p, p - q_0, q_0) \propto \frac{P}{1 + Q \ln [(2\Delta - \omega)/2\Delta]}.$$

Substituting this expression for the vertex part $\Gamma(p, p - q_0, q_0)$ into (26.24), we see that according to (26.6), the principal nonregular term in the function $G^{-1}(p)$ near the threshold is

$$\frac{a}{\ln [(2\Delta - \omega)/\alpha]}.$$

Finally, bearing in mind that by hypothesis $G^{-1}(p_c) = 0$, we find that

$$G^{-1}(p) = A^{-1}\left\{|\mathbf{p}| - p_c - \frac{a}{\ln [(2\Delta - \omega)/\alpha]}\right\}. \tag{26.25}$$

In this case, the equation

$$G^{-1}(p) = 0$$

gives the following expression for the energy spectrum for $|\mathbf{p}| < p_c$:

$$\varepsilon(\mathbf{p}) = 2\Delta - \alpha e^{-a/(p_c - |\mathbf{p}|)}.$$

[The exponential smallness of the quantity $\varepsilon(\mathbf{p}) - 2\Delta$ allows us to neglect powers of $\Delta\varepsilon$ in the expansion (26.25) of the regular part of $G^{-1}(p)$.] Thus, in the present case, as in the preceding one, the curve $\varepsilon(\mathbf{p})$ terminates at the point $|\mathbf{p}| = p_c$, but this time it has a horizontal tangent (of infinite order) at $|\mathbf{p}| = p_c$. It should be noted that in all the cases considered, the Green's function has a branch point for $|\mathbf{p}| = p_c$, $\omega = \varepsilon_c$.

We emphasize once again that the above analysis is not based on any concrete form of the interaction between particles nor on the assumption that this interaction is weak. We only made use of certain general basic relations between exact quantities defined by the diagram technique.[8]

27. Application of Field Theory Methods to a System of Interacting Bosons for $T \neq 0$

We conclude this chapter by considering the possibility of generalizing the theory presented above to the case of a system of interacting bosons at a finite temperature. In so doing, we naturally attempt to develop a general scheme which allows us to extend our previous methods. To this end, we begin with a thermodynamic description of the system, in which the chemical potential μ acts as the independent variable (instead of N, the total number of particles in the system). It has already been pointed out that such a description is impossible for an ideal Bose gas at temperatures below the point of Bose condensation, since the chemical potential of the gas, determined from the condition that the average number of particles in the system be constant, turns out to be identically zero in this temperature range.

As is well known, in an ideal Bose gas the number of bosons in the state with momentum $\mathbf{p}$ is

$$n_{\mathbf{p}} = \frac{1}{e^{[\varepsilon_0(\mathbf{p}) - \mu]/T} - 1}.$$

This distribution is meaningful only for negative μ. For sufficiently high temperatures, we have $\mu < 0$, and the point where μ vanishes is just the temperature at which Bose condensation occurs. For lower temperatures, we have to set μ identically equal to zero. The condition that the average number of particles in the system be constant is satisfied because of the particles "condensed on the lowest level." The number of such particles, as repeatedly noted above, is comparable to the total number of particles in the system, i.e., is proportional to the volume of the system.

---

[8] Recent experimental data [see H1, P7] indicates that decay into two rotons occurs in real He^4.

The formalism of the thermodynamic technique makes use of the temperature Green's function for noninteracting particles, which has Fourier components

$$\mathscr{G}^{(0)}(p) = \frac{1}{i\omega - \varepsilon_0(\mathbf{p}) + \mu}.$$

The temperature at which the system undergoes Bose condensation is the temperature at which the quantity

$$\frac{1}{\mu - \varepsilon_0(\mathbf{p})}$$

first exhibits a pole for zero momentum ($\mathbf{p} = 0$). An attempt to extend this expression into the region $\mu > 0$ involves us with a quantity which changes sign and becomes infinite for completely arbitrary values of $\mathbf{p}$. On the other hand, it is clear from our previous considerations that for $\mu < 0$, the temperature of a system of noninteracting particles exceeds the temperature where it undergoes Bose condensation.

Actually, these difficulties are of an artificial nature, and stem from the fact that there is no region in which perturbation theory can be applied to a gas of interacting bosons below the temperature of Bose condensation. For example, as we saw in Sec. 25, in the region of small momenta the separate terms of the perturbation series give rise to divergent expressions, and to obtain physical results, we have to sum a whole series of dominant perturbation-theory terms. In all similar situations, the general equations and the relations between various theoretical quantities have a wider meaning and can still be applied outside the region where perturbation theory is applicable. Thus, if interactions are taken into account, exact quantities like Green's functions have meaningful properties even at temperatures below the point of Bose condensation. As for the chemical potential μ, it is determined by the condition that the average number of particles in the system should equal the given number of particles. In the presence of interaction between the particles, μ is nowhere identically zero, and hence can be chosen from the outset as the independent variable, whose value for a given system is then determined from the condition that the number of particles be constant. There is no general argument which allows us to determine the sign of the chemical potential of a system of interacting bosons below the condensation point.[9]

In Sec. 23 we saw that the formulation of perturbation theory for $T = 0$

[9] In the model considered in Sec. 25, the quantity μ is positive, according to (25.19). (Note that f_0 is positive; negative f_0 would correspond to attraction between the particles, and such a system would be unstable to this approximation.) For real helium, $\mu < 0$ at temperatures below the λ-point, since otherwise helium could not be in equilibrium with its vapor at low temperatures. (As is well known, a necessary condition for this equilibrium is that the chemical potential of both phases be equal, and helium vapor is a dilute Boltzmann gas whose chemical potential is negative.)

involves the exact number of particles in the condensate. However, the number of particles in the condensate of an ideal gas, which is the starting point for subsequent derivations, does not appear anywhere in the theory. As we shall soon see, the situation remains the same at finite temperatures, and is a very important feature of the problem. In particular, this fact allows us to make precise just what is actually meant by the Bose-condensation temperature of a system of interacting particles. It is entirely clear that this temperature, which might be called the *λ-transition temperature*, need not coincide with the Bose-condensation temperature of an ideal gas. Physically, the λ-transition temperature is defined by the condition that the number of particles in the condensate should vanish. There is no general rule which allows us to determine the direction in which the transition temperature is shifted when the interaction is turned on. Therefore, in principle, the situation where the condensate can exist at temperatures above the condensation temperature of the ideal gas is physically possible, and so is the situation where turning on the interaction leads to the disappearance of the condensate present in the ideal gas.

At this point, it should be emphasized that in the formulation of the diagram technique for $T = 0$ (see Sec. 23), no use was made of the fact that the number of particles with zero momentum is infinite for an ideal gas. The characteristic feature of the perturbation theory developed in Sec. 23 is that we tried to set up our formalism in such a way as to take account of the operators ξ_0 and ξ_0^+ exactly, because of the special role of the particles with zero momentum. In other words, in dealing with the particles in the condensate, we did not make the usual statistical assumption to the effect that their contribution is relatively small. However, the remaining particles were treated in the usual way.

The same sort of approach is possible for finite temperatures. To set up our perturbation theory, we start from the representation of the Green's function in the form

$$\frac{\langle T(\psi_1 \psi_2 \cdots \psi_1^+ \psi_2^+ \cdots \mathscr{S})\rangle}{\langle \mathscr{S} \rangle}.$$

Here, the sign $\langle \cdots \rangle$, applied to a given expression, denotes the operation of taking the trace with respect to the states of the Hamiltonian $H_0 - \mu N$ for noninteracting particles, i.e.,

$$\langle \cdots \rangle = \frac{\operatorname{Sp}\{e^{(\mu N - H_0)/T} \cdots\}}{\operatorname{Sp} e^{(\mu N - H_0)/T}}. \tag{27.1}$$

Since the total number of particles is not conserved with μ as the independent variable, the thermodynamic average in (27.1) can be carried out for all the particles independently, including those with zero momentum. Therefore, just as in Sec. 23, each of the operations T and $\langle \cdots \rangle$ can be decomposed into two operations, i.e.,

$$T = T^0 T', \qquad \langle \cdots \rangle = \langle \langle \cdots \rangle' \rangle^0,$$

where the operations $T^0, \langle \cdots \rangle^0$ act on the particles in the condensate, and the operations $T', \langle \cdots \rangle'$ act on the other particles. The possibility of making this decomposition was an essential feature of our analysis for the case $T = 0$, and as before, it means that the particles with zero momentum play a special role. Therefore, we shall take these particles into account exactly.

Repeating almost word for word the argument leading to formula (23.14), we find that in order to use Matsubara's technique to calculate any Green's function of the particles not in the condensate, we have to know all the "exact" m-particle Green's functions of the particles in the state with zero momentum. Just as in (23.14), the Green's functions of the particles in the condensate are defined by the relations

$$\mathscr{G}_{0m}(\tau_1, \ldots, \tau_m; \tau_1', \ldots, \tau_m') = \frac{\langle T(\xi_0(\tau_1) \cdots \xi_0(\tau_m)\xi_0^+(\tau_1') \cdots \xi_0^+(\tau_m')\mathscr{S})\rangle}{\langle \mathscr{S} \rangle},$$

where the various τ are the "time" parameters of the technique for $T \neq 0$. This last expression can be written in a form analogous to (23.16), i.e., in terms of "Heisenberg" operators:

$$\mathscr{G}_{0m}(\tau_1, \ldots, \tau_m; \tau_1', \ldots, \tau_m') = \frac{\mathrm{Sp}\,\{e^{(\mu N - H)/T}T(\xi_0(\tau_1) \cdots \xi_0(\tau_m)\xi_0^+(\tau_1') \cdots \xi_0^+(\tau_m'))\}}{\mathrm{Sp}\,\{e^{(\mu N - H)/T}\}}. \quad (27.2)$$

Here, the "Heisenberg" operators $\xi_0(\tau)$ and $\xi_0^+(\tau)$ satisfy the equations

$$\frac{\partial}{\partial \tau}\xi_0(\tau) = [H - \mu N, \xi_0(\tau)],$$

$$\frac{\partial}{\partial \tau}\xi_0^+(\tau) = [H - \mu N, \xi_0^+(\tau)],$$

and are related to the ordinary Schrödinger operators by the formulas

$$\xi_0(\tau) = e^{(H - \mu N)\tau}\,\xi_0 e^{-(H - \mu N)\tau},$$

$$\xi_0^+(\tau) = e^{(H - \mu N)\tau}\,\xi_0^+ e^{-(H - \mu N)\tau}.$$

Thus, the Green's functions $\mathscr{G}_{0m}$ are Gibbs averages of chronological products of the operators ξ_0 and ξ_0^+.

In this connection, we recall that in quantum statistics the average of a quantity can be carried out in two equivalent ways. On the one hand, the average can be regarded as a quantum-mechanical average over the actual state of the system, where the state is characterized by the values of the energy and the particle number. On the other hand, the average can be carried out by using the Gibbs distribution, where the system is not regarded as closed; then, at a given temperature, there is a definite probability of the system occupying different quantum-mechanical states, with different values of the energy and particle number. The equivalence of these two methods of averaging is based on the fact that the Gibbs distribution has an extremely sharp maximum near the average values of the energy and particle number, so that, for example, the relative fluctuation of the energy, given by

$$\frac{\sqrt{\overline{(E - \bar{E})^2}}}{\bar{E}} \propto \frac{1}{\sqrt{\bar{N}}},$$

approaches zero as the dimensions of the system approach infinity. Thus, the energy and particle number of a closed system subjected to quantum-mechanical averaging are the same as the corresponding average values when the Gibbs distribution is used. From a thermodynamical point of view, the difference between the two methods of averaging is simply that in the first case the value of the averaged quantity is expressed in terms of the energy as a thermodynamical variable, while in the case of the Gibbs average the same value is expressed as a function of the temperature. The introduction of the chemical potential has the same statistical meaning. In view of all this, we shall regard (27.2) as an average over the quantum-mechanical state of the system:

$$\mathscr{G}_{0m}(\tau_1, \ldots, \tau_m; \tau_1', \ldots, \tau_m') = \langle \Phi^* | T(\xi_0(\tau_1) \cdots \xi_0(\tau_m) \xi_0^+(\tau_1') \cdots \xi_0^+(\tau_m')) | \Phi \rangle. \tag{27.3}$$

Thus, according to (27.2), the perturbation series for any Green's function of the particles not in the condensate involves exact Green's functions of the particles in the state with zero momentum. When written in the form (27.3), these Green's functions depend only on the properties of the operators ξ_0 and ξ_0^+ in the state Φ corresponding to a system of interacting particles. Therefore, if there is a condensate in this state, the operators ξ_0 and ξ_0^+ can be regarded as numbers, and then all the Green's functions of the form (27.2) should be replaced by products of the factors $\sqrt{n_0(T)}$:

$$\mathscr{G}_{0m}(\tau_1, \ldots, \tau_m; \tau_1', \ldots, \tau_m') = [n_0(T)]^m. \tag{27.4}$$

The validity of (27.4) can be verified by the same argument as used to derive the analogous formula for $T = 0$ (see Sec. 23).

In (27.3), $n_0(T)$ is the density of the particles in the condensate at a given temperature T, and hence the condition

$$n_0(T_\lambda) = 0$$

defines the transition temperature T_λ. As we have already mentioned, this temperature can be either higher or lower than the "Bose condensation" temperature T_0 of an ideal gas. In the latter case, the perturbation series for $T_0 > T > T_\lambda$ has a perfectly ordinary form, despite the fact that a condensate is present and the operators ξ_0, ξ_0^+ (in the interaction representation) are very large. We emphasize once more that this is related to the fact that the perturbation series involves the exact Green's functions of the particles in the condensate.

6

ELECTROMAGNETIC RADIATION IN AN ABSORBING MEDIUM

28. The Green's Functions of Radiation in an Absorbing Medium

The electromagnetic field plays a special role in the phenomena studied by statistical physics. In fact, all forces acting between particles of condensed matter (solids and liquids) are essentially of an electromagnetic nature. The distinctive feature of these forces is their short-range character, i.e., the fact that they fall off at distances of the order of interatomic distances. Moreover, the cohesive forces between the particles are chiefly due to these forces. In this chapter, however, we shall not study short-range forces. Instead, we confine ourselves to problems involving electromagnetic radiation whose wavelength is much greater than the distances between atoms. This category includes phenomena occurring when electromagnetic waves traverse matter, as well as various effects (known as *van der Waals forces*) related to long-range electromagnetic forces.

As is well known, the interaction of long-wavelength electromagnetic radiation with matter can be described in a purely macroscopic fashion by introducing a *complex dielectric constant*

$$\varepsilon(\omega) = \varepsilon'(\omega) + i\varepsilon''(\omega)$$

depending on the frequency ω of the radiation (see e.g., L9).[1] In this

[1] We assume that the magnetic permeability $\mu(\omega)$ equals unity, since it is appreciably different from unity only in narrow frequency ranges, which, as a rule, will not be important for our purposes.

section, we shall find the expressions for the Green's functions of electromagnetic radiation in an absorbing medium in terms of the dielectric constant of the medium.

In quantum mechanics, the electromagnetic field is usually described by the Schrödinger operators of the vector potential $\mathbf{A}(\mathbf{r})$ and the scalar potential $\varphi(\mathbf{r})$. We shall use "four-dimensional" notation to denote these operators,[2] i.e.,

$$\{A_\alpha\} = (\mathbf{A}, \varphi) \qquad (\alpha = 1, 2, 3, 0).$$

Together with the operators in the Schrödinger representation, we shall use the Heisenberg operators $A_\alpha(\mathbf{r}, t)$, defined by

$$A_\alpha(\mathbf{r}, t) = e^{i\hat{H}t} A_\alpha(\mathbf{r}) e^{-i\hat{H}t}, \tag{28.1}$$

in the ordinary way. The operators $A_\alpha(\mathbf{r}, t)$ are related to the operators of the electric and magnetic fields $\mathbf{E}(\mathbf{r}, t)$ and $\mathbf{H}(\mathbf{r}, t)$ by the usual relations

$$\begin{aligned} \mathbf{E}(\mathbf{r}, t) &= -\frac{\partial}{\partial t}\mathbf{A}(\mathbf{r}, t) - \operatorname{grad}\varphi(\mathbf{r}, t), \\ \mathbf{H}(\mathbf{r}, t) &= \operatorname{curl}\mathbf{A}(\mathbf{r}, t). \end{aligned} \tag{28.2}$$

The vector potential, as well as the operators representing it in the second-quantized representation, are not uniquely defined. There remains a certain arbitrariness related to the *gauge invariance* of the theory, which means that $A_\alpha(\mathbf{r}, t)$ can always be subjected to a *gauge transformation*

$$\begin{aligned} \mathbf{A}(\mathbf{r}, t) &\to \mathbf{A}(\mathbf{r}, t) + \operatorname{grad}\chi(\mathbf{r}, t), \\ \varphi(\mathbf{r}, t) &\to \varphi(\mathbf{r}, t) - \frac{\partial}{\partial t}\chi(\mathbf{r}, t), \end{aligned}$$

where χ is an arbitrary operator, since it is easily verified that $\mathbf{E}$ and $\mathbf{H}$, the quantities with direct physical meaning, are invariant under such a transformation.

The fact that the wavelength of the electromagnetic radiation is large means that we have a closed system of Maxwell equations for $\langle\mathbf{E}\rangle$ and $\langle\mathbf{H}\rangle$, the average values[3] of the electric and magnetic fields:

$$\begin{aligned} \operatorname{curl}\langle\mathbf{H}(\mathbf{r}, t)\rangle &= \frac{\partial}{\partial t}[\hat{\varepsilon}\langle\mathbf{E}(\mathbf{r}, t)\rangle], \\ \operatorname{curl}\langle\mathbf{E}(\mathbf{r}, t)\rangle &= -\frac{\partial}{\partial t}\langle\mathbf{H}(\mathbf{r}, t)\rangle. \end{aligned} \tag{28.3}$$

---

[2] In this chapter, we use Greek indices to label components of the four-dimensional vector potential, but Latin indices $i, k, \ldots$ are sometimes used to denote its spatial components. Moreover, summation is implied over all pairs of repeated indices, both Latin and Greek.

[3] Here the average is defined as the statistical average, e.g.,

$$\langle\mathbf{E}(\mathbf{r}, t)\rangle = \operatorname{Sp}\{e^{(F-\hat{H})/T}\,\mathbf{E}(\mathbf{r}, t)\},$$

where F is the free energy.

The dielectric constant $\hat{\varepsilon}$ appearing in these equations depends only on the properties of the medium. For absorbing media, $\hat{\varepsilon}$ represents an operator whose action on functions of time is described by

$$\hat{\varepsilon}\langle \mathbf{E}(\mathbf{r}, t)\rangle = \langle \mathbf{E}(\mathbf{r}, t)\rangle + \int_{-\infty}^{t} f(\mathbf{r}, t - t')\langle \mathbf{E}(\mathbf{r}, t')\rangle \, dt', \tag{28.4}$$

while, in Fourier components, the action of $\hat{\varepsilon}$ reduces simply to multiplying $\langle \mathbf{E}(\mathbf{r}, \omega)\rangle$ by

$$\varepsilon(\mathbf{r}, \omega) = 1 + \int_{0}^{\infty} e^{i\omega t} f(\mathbf{r}, t) \, dt, \tag{28.5}$$

the dielectric constant of the medium. In this case, the system of equations (28.3) becomes

$$\begin{aligned} \operatorname{curl} \langle \mathbf{H}(\mathbf{r}, \omega)\rangle &= -i\omega\varepsilon(\mathbf{r}, \omega)\langle \mathbf{E}(\mathbf{r}, \omega)\rangle, \\ \operatorname{curl} \langle \mathbf{E}(\mathbf{r}, \omega)\rangle &= i\omega\langle \mathbf{H}(\mathbf{r}, \omega)\rangle. \end{aligned} \tag{28.6}$$

It can be shown (see L9, Sec. 62) that the quantity $\varepsilon(\mathbf{r}, \omega)$ defined by (28.5) is an analytic function of the complex variable ω in the upper half-plane, and moreover has no zeros in the upper half-plane.

The properties of electromagnetic radiation at finite temperatures are determined by the temperature Green's function[4]

$$\begin{aligned} &\mathscr{D}_{\alpha\beta}(\mathbf{r}_1, \mathbf{r}_2; \tau_1 - \tau_2) \\ &= \begin{cases} -\operatorname{Sp} \{e^{(F - \hat{H})/T} e^{\hat{H}(\tau_1 - \tau_2)} A_\alpha(\mathbf{r}_1) e^{-\hat{H}(\tau_1 - \tau_2)} A_\beta(\mathbf{r}_2)\} & \text{for} \quad \tau_1 > \tau_2, \\ -\operatorname{Sp} \{e^{(F - \hat{H})/T} e^{-\hat{H}(\tau_1 - \tau_2)} A_\beta(\mathbf{r}_2) e^{\hat{H}(\tau_1 - \tau_2)} A_\alpha(\mathbf{r}_1)\} & \text{for} \quad \tau_1 < \tau_2. \end{cases} \end{aligned} \tag{28.7}$$

In order to express $\mathscr{D}_{\alpha\beta}$ in terms of $\varepsilon(\omega)$, the dielectric constant of the medium, we use the relation (established in Sec. 17) between the temperature Green's function and the retarded Green's function, defined in our case by

$$\begin{aligned} &D^R_{\alpha\beta}(\mathbf{r}_1, \mathbf{r}_2; t_1 - t_2) \\ &= \begin{cases} -i \operatorname{Sp} \{e^{(F - \hat{H})/T} [A_\alpha(\mathbf{r}_1, t_1) A_\beta(\mathbf{r}_2, t_2) - A_\beta(\mathbf{r}_2, t_2) A_\alpha(\mathbf{r}_1, t_1)]\} \\ \qquad\qquad\qquad\qquad\qquad\qquad \text{for} \quad t_1 > t_2, \\ 0 \qquad\qquad\qquad\qquad\qquad\qquad\quad \text{for} \quad t_1 < t_2. \end{cases} \end{aligned} \tag{28.8}$$

Since these results will be applied later to inhomogeneous media, we shall not regard $\mathscr{D}$ and D^R as functions of the differences of the space coordinates. In keeping with this, we shall also assume that the dielectric constant varies inside the medium, so that $\varepsilon(\omega) = \varepsilon(\mathbf{r}, \omega)$, as anticipated in (28.5).

We now repeat all the calculations of Sec. 17, omitting only the transformation to Fourier components with respect to the space coordinates.

---

[4] In formula (28.7) we introduce the free energy F instead of the thermodynamic potential Ω [cf. formulas (11.1) and (11.2)]. This change can be made since the chemical potential of the electromagnetic field is identically zero.

Then we easily obtain representations of the Lehmann type for $\mathscr{D}$ and D^R, i.e.,

$$\mathscr{D}_{\alpha\beta}(\mathbf{r}_1, \mathbf{r}_2; \omega_n) = \int_{-\infty}^{\infty} \frac{\rho_{\alpha\beta}(\mathbf{r}_1, \mathbf{r}_2; x)}{x - i\omega_n} dx, \tag{28.9}$$

$$D^R_{\alpha\beta}(\mathbf{r}_1, \mathbf{r}_2; \omega) = \int_{-\infty}^{\infty} \frac{\rho_{\alpha\beta}(\mathbf{r}_1, \mathbf{r}_2; x)}{x - \omega - i\delta} dx, \tag{28.10}$$

where

$$\rho_{\alpha\beta}(\mathbf{r}_1, \mathbf{r}_2; \omega) = -(2\pi)^3 \sum_{n,m} e^{(F-E_n)/T}(A_\alpha(\mathbf{r}_1))_{nm}(A_\beta(\mathbf{r}_2))_{mn}(1 - e^{-\omega_{mn}/T})\delta(\omega - \omega_{mn}).$$

It follows from (28.10) that $D^R(\omega)$ is an analytic function of ω in the upper half-plane. Comparing (28.10) and (28.9), we see that

$$\mathscr{D}_{\alpha\beta}(\mathbf{r}_1, \mathbf{r}_2; \omega_n) = D^R_{\alpha\beta}(\mathbf{r}_1, \mathbf{r}_2; i\omega_n) \tag{28.11}$$

for $\omega_n > 0$. To find $\mathscr{D}$ for $\omega_n < 0$, we note that $\mathscr{D}(\tau)$ is an even function of τ, since the operators of the electromagnetic field are real, i.e., $A^+_\alpha = A_\alpha$ (cf. Sec. 11.1). Therefore, the Fourier components $\mathscr{D}(\omega_n)$ are even in ω_n, which implies that[5]

$$\mathscr{D}_{\alpha\beta}(\mathbf{r}_1, \mathbf{r}_2; \omega_n) = D^R_{\alpha\beta}(\mathbf{r}_1, \mathbf{r}_2; i|\omega_n|) \tag{28.12}$$

for all ω_n.

Next, we calculate the retarded function, making important use of the gauge invariance of the theory. The tensor $D^R_{\alpha\beta}$ has a total of ten independent components (like any symmetric tensor of the second rank). However, we have at our disposal the arbitrariness associated with the gauge invariance. In fact, it is not the quantities $D^R_{\alpha\beta}$ themselves (formed from the components of the vector potential) which are physically significant, but only the six quantities which are formed from the operators $E_i(\mathbf{r}, t)$ by the same rules as used to form $D^R_{\alpha\beta}$ from $A_\alpha(\mathbf{r}, t)$ [see (28.8)]. Thus, there are only six physical conditions imposed on the ten quantities $D^R_{\alpha\beta}$, i.e., there are four arbitrary functions at our disposal. This arbitrariness can be exploited to make the components D^R_{00} and D^R_{i0} vanish. Obviously, this choice corresponds to choosing the gauge so that the scalar potential is zero. In this case, $\mathbf{E}$ and $\mathbf{H}$ are related to $\mathbf{A}$ by the formulas

$$\mathbf{E} = -\frac{\partial \mathbf{A}}{\partial t}, \qquad \mathbf{H} = \text{curl}\, \mathbf{A}. \tag{28.2'}$$

To express D^R_{ik} in terms of $\varepsilon(\omega)$, we proceed as follows: Suppose our system consisting of a body and equilibrium electromagnetic radiation, is in an

[5] We recall that like every Green's function for bosons, $\mathscr{D}(\omega_n)$ is nonzero only for "even" frequencies $\omega_n = 2n\pi T$.

external field created by external currents $\mathbf{j}^{\text{ext}}(\mathbf{r}, t)$. Then Maxwell's equations for the average fields take the form

$$\begin{aligned}\text{curl}\,\langle\mathbf{H}(\mathbf{r}, \omega)\rangle &= 4\pi\mathbf{j}^{\text{ext}}(\mathbf{r}, \omega) - i\omega\varepsilon(\mathbf{r}, \omega)\langle\mathbf{E}(\mathbf{r}, \omega)\rangle, \\ \text{curl}\,\langle\mathbf{E}(\mathbf{r}, \omega)\rangle &= i\omega\langle\mathbf{H}(\mathbf{r}, \omega)\rangle.\end{aligned} \tag{28.6'}$$

With the choice of gauge (28.2′), the average potential $\langle\mathbf{A}^{\text{ext}}(\mathbf{r}, t)\rangle$ satisfies the equation

$$[\varepsilon(\mathbf{r}, \omega)\omega^2\delta_{il} - \text{curl}_{im}\,\text{curl}_{ml}]\langle A_l(\mathbf{r}, \omega)\rangle = -4\pi j_i^{\text{ext}}(\mathbf{r}, \omega). \tag{28.13}$$

The solution of (28.13) has the form

$$\langle A_i^{\text{ext}}(\mathbf{r}, \omega)\rangle = -\int \bar{D}_{il}(\mathbf{r}, \mathbf{r}'; \omega) j_l^{\text{ext}}(\mathbf{r}', \omega)\, d\mathbf{r}', \tag{28.14}$$

where $\bar{D}$ is the Green's function (in the usual mathematical sense) of equation (28.13), equal to the solution of the equation

$$[\varepsilon(\mathbf{r}, \omega)\omega^2\delta_{il} - \text{curl}_{im}\,\text{curl}_{ml}]\bar{D}_{lk}(\mathbf{r}, \mathbf{r}'; \omega) = 4\pi\delta_{ik}\delta(\mathbf{r} - \mathbf{r}'). \tag{28.15}$$

Since $\varepsilon(\omega)$ is analytic in the upper half-plane, so is the function $\bar{D}$.

On the other hand, in the presence of external currents, $\langle\mathbf{A}^{\text{ext}}\rangle$ can be calculated directly from the definition (28.1). In this case, the Hamiltonian of the system has the form $\hat{H} + \hat{H}^{\text{ext}}$, where $\hat{H}$ is the Hamiltonian of the body and the radiation, while

$$\hat{H}^{\text{ext}} = -\int \mathbf{j}^{\text{ext}}(\mathbf{r}, t)\cdot\mathbf{A}(\mathbf{r})\, d\mathbf{r}.$$

Using the label "ext" to denote operators in the presence of an external field (and omitting the label in the absence of such a field), we have

$$\mathbf{A}^{\text{ext}}(\mathbf{r}, t) = e^{i(\hat{H} + \hat{H}^{\text{ext}})t}\mathbf{A}(\mathbf{r})e^{-i(\hat{H} + \hat{H}^{\text{ext}})t}.$$

Moreover, as in Sec. 6, we write

$$e^{-i(\hat{H} + \hat{H}^{\text{ext}})t} = e^{-i\hat{H}t}S_{\text{ext}}(t)$$

(cf. footnote 8, p. 57). If the external currents satisfy the condition $\mathbf{j}^{\text{ext}} \to 0$ as $t \to -\infty$, then $S_{\text{ext}}(t)$ has the form

$$S_{\text{ext}}(t) = T_t \exp\left\{-i\int_{-\infty}^{t} \hat{H}^{\text{ext}}(t')\, dt'\right\}$$

(cf. Sec. 6). In the presence of the external currents, the average value of the vector potential now becomes

$$\langle\mathbf{A}^{\text{ext}}(\mathbf{r}, t)\rangle = \langle S_{\text{ext}}^{-1}(t)\mathbf{A}(\mathbf{r}, t)S_{\text{ext}}(t)\rangle.$$

Expanding S_{ext} in a power series in H^{ext}, and retaining first-order terms in $\mathbf{j}^{\text{ext}}$, we find that

$$\begin{aligned}\langle A_i^{\text{ext}}(\mathbf{r}, t)\rangle = -i\int_{-\infty}^{t} dt' \int d\mathbf{r}'\, j_k^{\text{ext}}(\mathbf{r}', t')\langle[A_k(\mathbf{r}', t')A_i(\mathbf{r}, t) \\ - A_i(\mathbf{r}, t)A_k(\mathbf{r}', t')]\rangle.\end{aligned} \tag{28.16}$$

The expression (28.16) can be expressed in terms of the retarded function D^R_{ik} of the electromagnetic field. In fact, according to the definition (28.8), we have

$$\langle A_i^{\text{ext}}(\mathbf{r}, t)\rangle = -\int_{-\infty}^{\infty} dt' \int d\mathbf{r}'\, D^R_{ik}(\mathbf{r}, \mathbf{r}'; t - t') j_k^{\text{ext}}(\mathbf{r}', t').$$

Taking the Fourier components of this expression with respect to time, we finally obtain

$$\langle A_i^{\text{ext}}(\mathbf{r}, \omega)\rangle = -\int d\mathbf{r}'\, D^R_{ik}(\mathbf{r}, \mathbf{r}'; \omega) j_k^{\text{ext}}(\mathbf{r}', \omega). \tag{28.17}$$

Comparing (28.14) and (28.17), we see that since $\mathbf{j}^{\text{ext}}$ is arbitrary, D^R_{ik} coincides with $\bar{D}_{ik}$, the Green's function of equation (28.13). Thus, we arrive at the conclusion that D^R_{ik} also satisfies equation (28.15) [see Dzyaloshinski and Pitayevski (D3)]. Of course, the analytic properties of D^R_{ik} are exactly the same as those of $\varepsilon(\omega)$. The equation for $\mathscr{D}_{ik}(\omega_n)$ is obtained from (28.15), if we replace ω by $i|\omega_n|$:

$$[\varepsilon(\mathbf{r}, i|\omega_n|)\omega_n^2\delta_{il} + \operatorname{curl}_{im}\operatorname{curl}_{ml}]\mathscr{D}_{lk}(\mathbf{r}, \mathbf{r}'; \omega_n) = -4\pi\delta(\mathbf{r} - \mathbf{r}')\delta_{ik}. \tag{28.18}$$

For imaginary frequencies, the dielectric constant appearing in (28.18) is connected by a simple formula with $\varepsilon''(\omega)$, the imaginary part of $\varepsilon(\omega)$ for real frequencies (see L9, Sec. 62):

$$\varepsilon(i|\omega_n|) = 1 + \frac{2}{\pi}\int_0^{\infty} \frac{\omega\varepsilon''(\omega)}{\omega^2 + \omega_n^2}\, d\omega. \tag{28.19}$$

Since $\varepsilon'' > 0$ everywhere, it follows from (28.19) that $\varepsilon(i|\omega_n|)$ is a real, positive, monotonically decreasing function. Thus, by solving (28.14) or (28.18) we can express the Green's function of the electromagnetic field in terms of the imaginary part of the dielectric constant. In the case of inhomogeneous media, this is in general a very complicated problem. In subsequent sections, we shall consider the special case of layered media, for which a complete solution of the problem can be given.

Next, we turn to the case of a homogeneous medium, where ε does not depend on the coordinates. Since D^R and $\mathscr{D}$ now depend only on the difference $\mathbf{r} - \mathbf{r}'$, going over to Fourier components in (28.15), we obtain

$$\{[\omega^2\varepsilon(\omega) - k^2]\delta_{il} + k_i k_l\} D^R_{lj}(\mathbf{k}, \omega) = 4\pi\delta_{ij}. \tag{28.20}$$

This equation determines D^R in a gauge for which the scalar potential is zero. To find D^R in an arbitrary gauge, we use (28.20) to calculate the function $D^E_{ij} = \omega^2 D^R_{ij}$, where D^R_{ij} is the retarded function in the gauge $\varphi = 0$. The quantity defined in this way is already gauge-invariant, since it differs only by a constant term from the retarded function formed from the components of the operators of the electric field. The function D^E_{ij} satisfies the equation

$$\{[\omega^2\varepsilon(\omega) - k^2]\delta_{il} + k_i k_l\} D^E_{lj}(\mathbf{k}, \omega) = 4\pi\omega^2\delta_{ij}, \tag{28.21}$$

and is related to the function $D^R_{\alpha\beta}$ in an arbitrary gauge by the obvious formula [cf. (28.2)]

$$D^E_{ij} = \omega^2 D^R_{ij} - \omega k_i D^R_{0j} - \omega k_j D^R_{i0} + k_i k_j D^R_{00}, \tag{28.22}$$

where D^R_{00}, D^R_{i0} and D^R_{ij} denote the time components, the mixed components and the space components of $D^R_{\alpha\beta}$, respectively, in an arbitrary gauge. It follows from symmetry considerations that the vector D^R_{i0} must be directed along $\mathbf{k}$, the unit vector appearing in (28.21) and (28.22):

$$D^R_{i0} = D^R_{0i} = k_i d. \tag{28.23}$$

For the same reason, we have

$$D^R_{ij} = a\delta_{ij} + bk_ik_j. \tag{28.24}$$

Substituting (28.22)–(28.24) into (28.21), we obtain two equations determining D^R_{00}, a, b and d:

$$\begin{aligned} a[\varepsilon(\omega)\omega^2 - k^2] &= 4\pi, \\ a + \varepsilon(\omega)(\omega^2 b + D^R_{00} - 2\omega d) &= 0. \end{aligned} \tag{28.25}$$

Thus, we see that in the homogeneous case, $D^R_{\alpha\beta}$ is undetermined only to within two arbitrary functions, and not four functions, as in the case of an arbitrary inhomogeneous medium.

Some special cases of these formulas should be noted. If $b = d = 0$, we have

$$D^R_{ij} = \frac{4\pi\delta_{ij}}{\varepsilon(\omega)\omega^2 - k^2}, \qquad D^R_{00} = -\frac{4\pi}{\varepsilon(\omega)[\varepsilon(\omega)\omega^2 - k^2]}, \qquad D^R_{i0} = 0. \tag{28.26a}$$

Moreover, the case $\varphi = 0$ corresponds to

$$D^R_{ij} = \frac{4\pi}{\varepsilon(\omega)\omega^2 - k^2}\left[\delta_{ij} - \frac{k_ik_j}{\varepsilon(\omega)\omega^2}\right], \qquad D^R_{00} = D^R_{i0} = 0. \tag{28.26b}$$

Finally, in the "transverse gauge" (div $\mathbf{A} = 0$), we have

$$D^R_{ij} = \frac{4\pi}{\varepsilon(\omega)\omega^2 - k^2}\left(\delta_{ij} - \frac{k_ik_j}{k^2}\right), \quad D^R_{00} = \frac{4\pi}{\varepsilon(\omega)k^2}, \quad D^R_{i0} = 0. \tag{28.26c}$$

The corresponding formulas for the temperature Green's function $\mathscr{D}$ are obtained by changing ω to $i|\omega_n|$.

$$\mathscr{D}_{ij} = -\frac{4\pi\delta_{ij}}{\varepsilon(i|\omega_n|)\omega_n^2 + k^2}, \quad \mathscr{D}_{00} = \frac{4\pi}{\varepsilon(i|\omega_n|)[\varepsilon(i|\omega_n|)\omega_n^2 + k^2]}, \quad \mathscr{D}_{i0} = 0, \tag{28.27a}$$

$$\mathscr{D}_{ij} = -\frac{4\pi\delta_{ij}}{\varepsilon(i|\omega_n|)\omega_n^2 + k^2}\left[\delta_{ij} + \frac{k_ik_j}{\varepsilon(i|\omega_n|)\omega_n^2}\right], \qquad \mathscr{D}_{00} = \mathscr{D}_{i0} = 0, \tag{28.27b}$$

$$\mathscr{D}_{ij} = -\frac{4\pi\delta_{ij}}{\varepsilon(i|\omega_n|)\omega_n^2 + k^2}\left[\delta_{ij} - \frac{k_ik_j}{k^2}\right], \quad \mathscr{D}_{00} = \frac{4\pi}{\varepsilon(i|\omega_n|)k^2}, \quad \mathscr{D}_{i0} = 0. \tag{28.27c}$$

In many problems, the ordinary time-dependent Green's function of the electromagnetic field, defined by

$$D_{\alpha\beta}(\mathbf{r}_1 - \mathbf{r}_2, t_1 - t_2) = -i\,\mathrm{Sp}\,\{e^{(F-\hat{H})/T}T_t(A_\alpha(\mathbf{r}_1, t_1)A_\beta(\mathbf{r}_2, t_2))\} \tag{28.28}$$

turns out to be useful. As we saw in Sec. 17, the Fourier components of (28.28) are related to $D^R(\mathbf{k}, \omega)$ by the formulas

$$\begin{aligned} \operatorname{Re} D(\mathbf{k}, \omega) &= \operatorname{Re} D^R(\mathbf{k}, \omega), \\ \operatorname{Im} D(\mathbf{k}, \omega) &= \operatorname{Im} D^R(\mathbf{k}, \omega) \coth \frac{\omega}{2T}. \end{aligned} \tag{28.29}$$

We shall not give the complicated expressions for D obtained from (28.29).

The function D is of particular interest at the absolute zero of temperature, when it can be calculated by using the ordinary technique of quantum field theory (Chap. 2). Taking the limit $T \to 0$ in (28.29), we obtain

$$\begin{aligned} \operatorname{Re} D(\mathbf{k}, \omega) &= \operatorname{Re} D^R(\mathbf{k}, \omega), \\ \operatorname{Im} D(\mathbf{k}, \omega) &= \operatorname{Im} D^R(\mathbf{k}, \omega) \operatorname{sgn} \omega. \end{aligned} \tag{28.30}$$

Bearing in mind that the real part of $\varepsilon(\omega)$ is an even function, while the imaginary part is odd (see e.g., L9, Sec. 58), we easily see that to make the transformation from D^R to D for $T = 0$, we must replace ω by $|\omega|$ everywhere in the formulas (28.26). In particular, for the gauge (28.26a), we obtain

$$\begin{aligned} D_{ij} &= \frac{4\pi\delta_{ij}}{\varepsilon(|\omega|)\omega^2 - k^2}, \\ D_{00} &= -\frac{4\pi}{\varepsilon(|\omega|)[\varepsilon(|\omega|)\omega^2 - k^2]}, \qquad D_{0i} = 0. \end{aligned} \tag{28.31}$$

The formulas (28.31) are a generalization of the usual formulas for the photon Green's function in quantum electrodynamics (see e.g., A9). For the special case of transparent media $[\varepsilon''(\omega) = 0]$, they were obtained by Ryazanov (R2), who used another method.

29. Calculation of the Dielectric Constant

The problem of calculating the temperature Green's function $\mathscr{D}$ of the electromagnetic field in an absorbing medium can be approached differently, by using the diagram technique developed in Chap. 3. Since we are only interested in electromagnetic fields with wavelengths much greater than interatomic distances, we represent the interaction Hamiltonian of the particles and the field as a sum of two terms

$$\hat{H}_{\text{int}} = \hat{H}^{(1)}_{\text{int}} + \hat{H}^{(2)}_{\text{int}},$$

assigning the energies of the noninteracting particles and free photons to the zeroth-order Hamiltonian $\hat{H}_0$. Here, $\hat{H}^{(1)}_{\text{int}}$ contains the part of the interaction leading to the short-range forces mentioned at the beginning of the preceding section, and $\hat{H}^{(2)}_{\text{int}}$ describes the interaction between the particles and the long-wavelength electromagnetic field.

In the case of the gauge with zero scalar potential, we have

$$\hat{H}_{\text{int}}^{(2)} = -\int \mathbf{A}(\mathbf{r})\cdot\mathbf{j}(\mathbf{r})\,d\mathbf{r}, \tag{29.1}$$

where $\mathbf{j}(\mathbf{r})$ is the operator corresponding to the particle current density. The fact that the wavelengths appearing in (29.1) are long means that the Fourier-series expansion of $\mathbf{A}(\mathbf{r})$ only involves momenta $\mathbf{k}$ with magnitudes less than some cutoff momentum k_0, which is itself much less than $1/a$, the reciprocal interatomic distance. Therefore, all the integrals with respect to $\mathbf{k}$ which appear in the diagram technique must be cut off at the upper limit $k_0 \ll 1/a$. When the particles have nonrelativistic velocities (a condition satisfied in any macroscopic system), the current-density operator has the form

$$\mathbf{j}(\mathbf{r}) = \sum_a \left\{ -i\frac{e_a}{2m_a}\left[\psi_a^+(\mathbf{r})\,\nabla\psi_a(\mathbf{r}) - \nabla\psi_a^+(\mathbf{r})\psi_a(\mathbf{r})\right] - \frac{e_a^2}{2m_a}\,\mathbf{A}(\mathbf{r})\psi_a^+(\mathbf{r})_a\psi(\mathbf{r})\right\},$$

where the sum is over the particles of various kinds (see e.g., L7, Sec. 128).

We now use the number of long-wavelength photon lines to classify the diagrams for the corrections to the Green's function of the long-wavelength radiation. For simplicity, parts of diagrams which contain no such lines will be denoted by shaded polygons.[6] Clearly, these polygons can be regarded

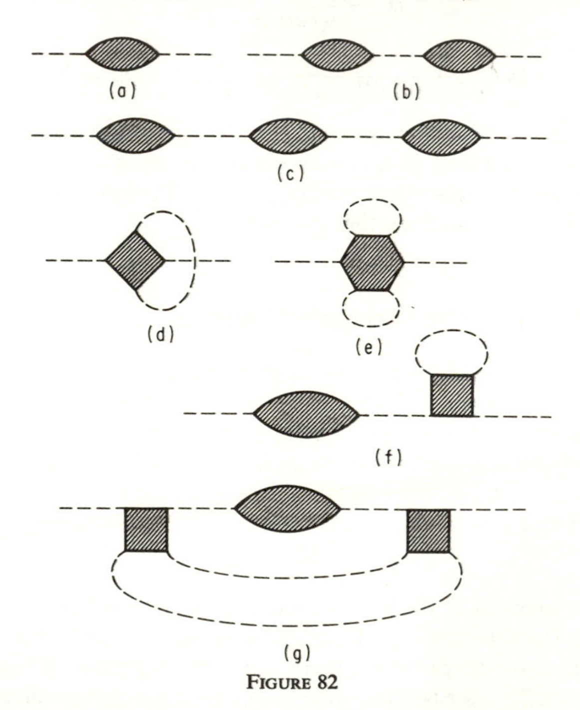

FIGURE 82

[6] This category includes the oval-shaped loop shown in Fig. 82.

as representing sums of all possible diagrams of the indicated type, where it is assumed that the perturbation series in terms of the charge e has first been reduced to a perturbation series in terms of the number of long-wavelength photon lines. The various types of diagrams corresponding to this series are shown in Fig. 82. The quantities associated with the shaded polygons are completely determined by the properties of the condensed body which is formed as a result of the action of the short-range forces.

It is clear at once from physical considerations that the contributions of diagrams (d), (e), (f) and (g) of Fig. 82 are negligibly small, since they correspond to various nonlinear processes, like scattering of light by light. This statement can be proved as follows: As already mentioned, all the integrals with respect to the momenta of the long-wavelength photon lines have to be cut off at some $k_0 \ll 1/a$. Therefore, a dimensionality argument shows that if an integration is carried out with respect to the momentum of any photon line, a small factor $k_0 a$ must be associated with the line. The only diagrams which involve no integrals at all with respect to the momenta of the long-wavelength photon lines are diagrams like (a), (b) and (c) in Fig. 82. Such a sequence of diagrams has already been summed in Sec. 10, in connection with the derivation of Dyson's equation. Thus, we can immediately write down the following equation[7] for $\mathscr{D}_{ij}$:

$$\begin{aligned}\mathscr{D}_{ij}(\mathbf{r}_1, \mathbf{r}_2; \omega_n) &= \mathscr{D}^{(0)}_{ij}(\mathbf{r}_1 - \mathbf{r}_2; \omega_n) \\ &\quad + \int d\mathbf{r}_3\, d\mathbf{r}_4\, \mathscr{D}^{(0)}_{il}(\mathbf{r}_1 - \mathbf{r}_3; \omega_n)\mathscr{T}_{lm}(\mathbf{r}_3, \mathbf{r}_4; \omega_n)\mathscr{D}_{mj}(\mathbf{r}_4, \mathbf{r}_2; \omega_n).\end{aligned} \tag{29.2}$$

The quantity $\mathscr{T}$ representing the contribution of the shaded loop shown in Fig. 82 is called the *polarization operator*. As the above considerations show, $\mathscr{T}$ is completely determined by the properties of the medium.

We now express the polarization operator $\mathscr{T}$ in terms of the dielectric constant of the system. To achieve this, we note that the Green's function $\mathscr{D}_{ij}$ of the long-wavelength radiation, in the gauge such that $\varphi = 0$, satisfies equation (28.18). Multiplying this equation from the left by the operator

$$\omega_n^2\delta_{ij} + \operatorname{curl}_{il}\operatorname{curl}_{lj},$$

and bearing in mind that $\mathscr{D}^{(0)}$ satisfies equation (28.18) with $\varepsilon = 1$, we obtain

$$\int d\mathbf{r}'\, \mathscr{T}_{il}(\mathbf{r}_1, \mathbf{r}'; \omega_n)\mathscr{D}_{lj}(\mathbf{r}', \mathbf{r}_2; \omega_n) = \frac{1}{4\pi}\left[\varepsilon(\mathbf{r}_1, i|\omega_n|) - 1\right]\omega_n^2\mathscr{D}_{ij}(\mathbf{r}_1, \mathbf{r}_2; \omega_n), \tag{29.3}$$

after a simple calculation. The desired formula for $\mathscr{T}$ then follows from (29.3):

$$\mathscr{T}_{ij}(\mathbf{r}_1, \mathbf{r}_2; \omega_n) = \frac{1}{4\pi}\left[\varepsilon(\mathbf{r}_1, i|\omega_n|) - 1\right]\omega_n^2\delta_{ij}\delta(\mathbf{r}_1 - \mathbf{r}_2). \tag{29.4}$$

---

[7] Anticipating the case of inhomogeneous media, we write (29.2) in the coordinate representation, carrying out a Fourier transformation only with respect to τ.

The polarization operator $\mathscr{T}$ turns out to be proportional to $\delta(\mathbf{r}_1 - \mathbf{r}_2)$. This is associated with the fact that the effects of space correlation are neglected in the macroscopic theory. Space correlations become important only in the frequency range where the *anomalous skin effect* occurs. The properties of $\mathscr{T}$ in this frequency range will be studied in Chap. 7. In the rest of this chapter, we shall be concerned with much larger frequencies, for which there is no anomalous skin effect.

In the case of a homogeneous body, $\mathscr{T}$ is a function of the difference variable $\mathbf{r}_1 - \mathbf{r}_2$, and ε does not depend on $\mathbf{r}$. Taking the Fourier transform of (29.4) with respect to $\mathbf{r}_1 - \mathbf{r}_2$, we obtain a simple formula relating the polarization operator $\mathscr{T}(\mathbf{k}, \omega_n)$ of the system to its dielectric constant:

$$\mathscr{T}_{ij}(\mathbf{k}, \omega_n) = \frac{1}{4\pi}[\varepsilon(i|\omega_n|) - 1]\,\omega_n^2\delta_{ij}.$$

This formula allows us to calculate the dielectric constant of a body at $T \neq 0$ by using the methods of quantum field theory. In fact, calculating the polarization operator of the system, we thereby find the values of $\varepsilon(\omega)$ for the discrete set of points $\omega_n = 2n\pi Ti$ along the imaginary axis. Then, recalling that $\varepsilon(\omega)$ is an analytic function of ω with no singularities in the upper half-plane, and repeating word for word the argument given in Sec. 17 involving $\mathscr{G}$ and G^R, but this time applied to ε, we arrive at the conclusion that to obtain $\varepsilon(\omega)$ it is sufficient to continue the quantity

$$\frac{4\pi}{3\omega_n^2}\mathscr{T}_{ii}(\omega_n)$$

analytically from the discrete set of points $\omega_n = 2n\pi Ti$ into the whole upper half-plane. Although this problem does not have a solution in general form, it can be solved in various special cases.

The problem of calculating $\varepsilon(\omega)$ becomes much simpler for $T = 0$. In this case, to calculate the polarization operator, we can use the time-dependent field theory technique described in Chap. 2. Repeating step by step all the calculations made in this section, but this time for the case of the time-dependent diagram technique, and bearing in mind what was said about the function $D_{ij}(\mathbf{k}, \omega)$ in Sec. 28, we obtain the formula

$$\Pi_{ij}(\omega) = \frac{1}{4\pi}[\varepsilon(|\omega|) - 1]\,\omega^2\delta_{ij} \tag{29.5}$$

(in the gauge with $\varphi = 0$). Moreover, recalling that

$$\varepsilon'(\omega) = \varepsilon'(-\omega), \qquad \varepsilon''(\omega) = -\varepsilon''(-\omega),$$

we can express ε in terms of Π:

$$\begin{aligned}\varepsilon'(\omega) &= 1 + \frac{4\pi}{3\omega^2}\,\mathrm{Re}\,\Pi_{ii}(\omega),\\ \varepsilon''(\omega) &= \frac{4\pi}{3\omega|\omega|}\,\mathrm{Im}\,\Pi_{ii}(\omega).\end{aligned} \tag{29.6}$$

Thus, calculating $\varepsilon(\omega)$ at $T = 0$ reduces to determining the polarization operator of the system.

30. Van der Waals Forces in an Inhomogeneous Dielectric

A long-wavelength electromagnetic field acts as the source of specific long-range forces, which can be called *van der Waals forces*, since they are of the same nature as the attractive forces between molecules at large distances. Although the contribution of these forces to the free energy of a body is very small compared to the contribution of the short-range cohesive forces, they lead to a qualitatively new effect, i.e., the nonadditivity of the free energy. It is just this nonadditivity, connected with the long-range character of the van der Waals forces, which enables us to separate their contribution to thermodynamic quantities.

The reason for the nonadditivity can be understood if we examine the relation between the van der Waals forces and the long-wavelength electromagnetic field. In fact, because of Maxwell's equations, any change in the density (with the concomitant change in electrical properties) of the medium inside a given region also leads to a change in the field outside the region. Therefore, the part of the free energy connected with the long-wavelength radiation is not determined exclusively by the properties of the medium at a given point, and hence is nonadditive. This explains such phenomena as the fact that the chemical potential of a thin liquid film on the surface of a solid body depends on the thickness of the film (see Sec. 31.3), or the fact that van der Waals forces can act as the source of interaction forces between solid bodies, in which case, the free energy depends on the distance between the bodies (see Sec. 31.1). It is obvious that in all these phenomena an important role is played by electromagnetic fields with wavelengths comparable to the relevant macroscopic distances (e.g., the thickness of a liquid film, or the distance between two bodies). This allows us to express the quantities of interest in terms of the dielectric constants $\varepsilon(\omega)$ of the various media involved.

To calculate the correction to the free energy due to the long-wavelength electromagnetic field, we use the diagram technique developed in Sec. 15 for calculating the thermodynamic potential Ω (in the case of photons, Ω coincides with the free energy F). Repeating the considerations of the preceding section, we see that only the

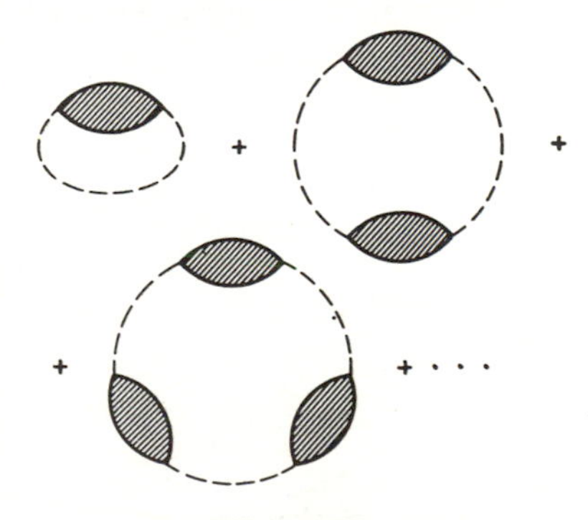

FIGURE 83

sequence of diagrams shown in Fig. 83 contributes to the free energy F. In this regard, we note that according to what was said in Sec. 15 about the coefficients associated with the diagrams for F, it might well turn out that we can no longer interpret the shaded loops in Fig. 83 as sums of all possible diagrams which do not contain long-wavelength photon lines. However, a closer look shows that this is not the case, since the coefficient associated with a given diagram is found to be $1/m$, where m is the number of photon lines.[8] This fact allows us to sum the shaded loops in Fig. 83 and associate with each of them the polarization operator $\mathscr{T}$ calculated in Sec. 29.

The sequence of diagrams shown in Fig. 83 corresponds to the following series for the free energy F:

$$\begin{aligned} F = F_0 - \frac{T}{2}\sum_{n=-\infty}^{\infty} \Big[&\int \mathscr{T}_{ik}(\mathbf{r}_1, \mathbf{r}_2; \omega_n)\mathscr{D}^{(0)}_{ki}(\mathbf{r}_2 - \mathbf{r}_1; \omega_n)\, d\mathbf{r}_1\, d\mathbf{r}_2 \\ &+ \frac{1}{2}\int \mathscr{T}_{ik}(\mathbf{r}_1, \mathbf{r}_2; \omega_n)\mathscr{D}^{(0)}_{kl}(\mathbf{r}_2 - \mathbf{r}_3; \omega_n) \\ &\times \mathscr{T}_{lp}(\mathbf{r}_3, \mathbf{r}_4; \omega_n)\mathscr{D}^{(0)}_{pi}(\mathbf{r}_4 - \mathbf{r}_1)\, d\mathbf{r}_1\, d\mathbf{r}_2\, d\mathbf{r}_3\, d\mathbf{r}_4 \\ &+ \cdots + \frac{1}{m}\int \mathscr{T}_{ik}(\mathbf{r}_1, \mathbf{r}_2; \omega_n)\mathscr{D}^{(0)}_{kl}(\mathbf{r}_2 - \mathbf{r}_3; \omega_n)\cdots \\ &\times \mathscr{T}_{qs}(\mathbf{r}_{2m-1}, \mathbf{r}_{2m}; \omega_n)\mathscr{D}^{(0)}_{si}(\mathbf{r}_{2m} - \mathbf{r}_1)\, d\mathbf{r}_1\cdots d\mathbf{r}_{2m} + \cdots \Big]. \end{aligned} \tag{30.1}$$

Here F_0 is the free energy of the body, and includes all corrections connected with short-range forces.

The series (30.1) cannot be summed directly. Instead of the free energy, we determine the correction to the pressure (more precisely, the correction to the stress tensor) due to interaction between the medium and the long-wavelength electromagnetic field. Suppose the body is subjected to a small deformation described by the displacement vector $\mathbf{u}(\mathbf{r})$. Then the change δF in the free energy is

$$-\int \mathbf{f}\cdot\mathbf{u}\, dV,$$

where $\mathbf{f}$ is the force acting on a unit volume of the deformed body. The corresponding change in F_0 is

$$\delta F_0 = \int \mathbf{u}\cdot \operatorname{grad} p_0\, dV,$$

[8] In Sec. 15, the number of topologically equivalent nth-order diagrams was $(n-1)!$, which led to the appearance of the factor

$$\frac{1}{n} = \frac{(n-1)!}{n!}.$$

In our case, where

$$\hat{H}_{\text{int}} = \hat{H}^{(1)}_{\text{int}} + \hat{H}^{(2)}_{\text{int}}$$

(see Sec. 29), the coefficient which appears when the exponential is expanded is $1/l!(2m)!$, where m is the number of photon lines, and l is the number of vertices associated with $\hat{H}^{(1)}_{\text{int}}$. A calculation like that made in Sec. 15 shows that the number of topologically equivalent diagrams is $l!(2m-1)!$, which implies the assertion made in the text.

where $p_0(\rho, T)$ is the pressure corresponding to the given density ρ and temperature T, without taking any corrections into account. As for the series (30.1), only the polarization operator $\mathcal{H}$ is changed by the displacement, since only $\mathcal{H}$ depends on the properties of the medium. In fact, we have

$$\delta\mathcal{H}_{ik}(\mathbf{r}_1, \mathbf{r}_2; \omega_n) = \frac{1}{4\pi}\omega_n^2\delta_{ik}\delta(\mathbf{r}_1 - \mathbf{r}_2)\delta\varepsilon(\mathbf{r}_1, i|\omega_n|)$$

[see (29.4)].

If we calculate the variation of (30.1), the coefficients $1/m$ are cancelled, and the result is

$$\delta F = \delta F_0 - \frac{T}{8\pi}\sum_{n=-\infty}^{\infty}\omega_n^2\int d\mathbf{r}\,\delta\varepsilon(\mathbf{r}, i|\omega_n|)$$

$$\times\Big[\mathcal{D}_{ii}^{(0)}(\mathbf{r}-\mathbf{r};\omega_n) + \int\mathcal{D}_{ik}^{(0)}(\mathbf{r}-\mathbf{r}_1;\omega_n)\mathcal{H}_{kl}(\mathbf{r}_1, \mathbf{r}_2;\omega_n)$$

$$\times\,\mathcal{D}_{li}^{(0)}(\mathbf{r}_2-\mathbf{r};\omega_n)\,d\mathbf{r}_1\,d\mathbf{r}_2 + \int\mathcal{D}_{ik}^{(0)}(\mathbf{r}-\mathbf{r}_1;\omega_n)\mathcal{H}_{kl}(\mathbf{r}_1, \mathbf{r}_2;\omega_n)$$

$$\times\,\mathcal{D}_{lp}^{(0)}(\mathbf{r}_2-\mathbf{r}_3;\omega_n)\mathcal{H}_{pq}(\mathbf{r}_3, \mathbf{r}_4;\omega_n)\mathcal{D}_{qi}^{(0)}(\mathbf{r}_4-\mathbf{r};\omega_n)\,d\mathbf{r}_1\,d\mathbf{r}_2\,d\mathbf{r}_3\,d\mathbf{r}_4 + \cdots\Big].$$

The series in brackets is just the series for the Green's function $\mathcal{D}$ of the long-wavelength photons corresponding to the sequence of diagrams (a), (b), (c), . . . of Fig. 82, and hence

$$\delta F = \delta F_0 - \frac{T}{8\pi}\sum_{n=-\infty}^{\infty}\omega_n^2\int\mathcal{D}_{ii}(\mathbf{r}, \mathbf{r};\omega_n)\,\delta\varepsilon(\mathbf{r}, i|\omega_n|)\,d\mathbf{r}.$$

Recalling that $\mathcal{D}$ is an even function of ω_n, we finally obtain

$$\delta F = \delta F_0 - \frac{T}{4\pi}{\sum_{n=0}^{\infty}}'\omega_n^2\int\mathcal{D}_{ii}(\mathbf{r}, \mathbf{r};\omega_n)\,\delta\varepsilon(\mathbf{r}, i\omega_n)\,d\mathbf{r}, \tag{30.2}$$

where $\omega_n = 2n\pi T$, and the prime on the summation sign means that the term with $n = 0$ is taken with weight $\frac{1}{2}$.

The variation $\delta\varepsilon$ is connected with the displacement $\mathbf{u}$ by the formula[9]

$$\delta\varepsilon = -\mathbf{u}\cdot\operatorname{grad}\varepsilon - \rho\frac{\partial\varepsilon}{\partial\rho}\operatorname{div}\mathbf{u}.$$

---

[9] The change in ε at a given point consists of two terms: The first is connected with the transport of ε by the medium, i.e.,

$$\delta_1\varepsilon = \varepsilon(\mathbf{r}-\mathbf{u}) - \varepsilon(\mathbf{r}) = -\mathbf{u}\cdot\operatorname{grad}\varepsilon,$$

while the second is connected with the change in density due to the deformation:

$$\delta_2\varepsilon = \frac{\partial\varepsilon}{\partial\rho}\delta\rho = -\frac{\partial\varepsilon}{\partial\rho}\rho\operatorname{div}\mathbf{u}.$$

Substituting this expression for $\delta\varepsilon$ into (30.2) and integrating by parts, we obtain the following expression for **f**:

$$\mathbf{f} = -\operatorname{grad} p_0 - \frac{T}{4\pi}\sum_{n=0}^{\infty}{}' \omega_n^2 \mathscr{D}_{ii}(\mathbf{r}, \mathbf{r}; \omega_n) \operatorname{grad} \varepsilon(\mathbf{r}, i\omega_n) + \frac{T}{4\pi}\sum_{n=0}^{\infty}{}' \omega_n^2 \operatorname{grad}\left[\mathscr{D}_{ii}(\mathbf{r}, \mathbf{r}; \omega_n)\rho \frac{\partial \varepsilon(\mathbf{r}, i\omega_n)}{\partial \rho}\right]. \tag{30.3}$$

Using (30.3), we can easily calculate the correction to the chemical potential of the body. Noting that $\mathbf{f} = 0$ corresponds to mechanical equilibrium, we equate (30.3) to zero. Then, after a simple calculation, we obtain

$$-\rho \operatorname{grad}\left[\mu_0(\rho, T) - \frac{T}{4\pi}\sum_{n}{}' \omega_n^2 \mathscr{D}_{ii}(\mathbf{r}, \mathbf{r}; \omega_n)\frac{\partial \varepsilon}{\partial \rho}\right] = 0, \tag{30.4}$$

where we have taken account of the relations

$$\operatorname{grad} \varepsilon(\rho, T) = \frac{\partial \varepsilon}{\partial \rho}\operatorname{grad} \rho, \qquad dp_0(\rho, T) = \rho\, d\mu_0(\rho, T), \tag{30.5}$$

which hold for constant temperature (μ_0 is the unperturbed chemical potential per unit volume). As is well known, the condition for any inhomogeneous body to be in mechanical equilibrium is that the chemical potential be constant throughout the body. Therefore, it follows at once from (30.4) that

$$\mu(\rho, T) = \mu_0(\rho, T) - \frac{T}{4\pi}\sum_{n}{}' \omega_n^2 \mathscr{D}_{ii}(\mathbf{r}, \mathbf{r}; \omega_n)\frac{\partial \varepsilon}{\partial \rho}. \tag{30.6}$$

Next, we calculate the stress tensor. To do this, we must bring the expression (30.3) for the force **f** into the form

$$f_i = \frac{\partial \sigma_{ik}}{\partial x_k}. \tag{30.7}$$

We first introduce two extra functions

$$\begin{aligned} \mathscr{D}_{ik}^E(\mathbf{r}, \mathbf{r}'; \omega_n) &= -\omega_n^2 \mathscr{D}_{ik}(\mathbf{r}, \mathbf{r}'; \omega_n), \\ \mathscr{D}_{ik}^H(\mathbf{r}, \mathbf{r}'; \omega_n) &= \operatorname{curl}_{il} \operatorname{curl}'_{km} \mathscr{D}_{lm}(\mathbf{r}, \mathbf{r}'; \omega_n), \end{aligned} \tag{30.8}$$

besides the Green's function $\mathscr{D}_{ik}(\mathbf{r}, \mathbf{r}'; \omega_n)$. With this new notation, we can write the components of the force as

$$\begin{aligned} f_i = -\frac{\partial p_0}{\partial x_i} &+ \frac{T}{4\pi}\sum_{n=0}^{\infty}{}' \frac{\partial}{\partial x_i}\left[\varepsilon(\mathbf{r}, i\omega_n)\mathscr{D}_{kk}^E(\mathbf{r}, \mathbf{r}; \omega_n) - \rho(\mathbf{r})\frac{\partial \varepsilon(\mathbf{r}, i\omega_n)}{\partial \rho}\mathscr{D}_{kk}^E(\mathbf{r}, \mathbf{r}; i\omega_n)\right] \\ &- \frac{T}{4\pi}\sum_{n=0}^{\infty}{}' \varepsilon(\mathbf{r}, i\omega_n)\frac{\partial}{\partial x_i}\mathscr{D}_{kk}^E(\mathbf{r}, \mathbf{r}; \omega_n). \end{aligned} \tag{30.9}$$

Then we need only transform the last term in (30.9). Except for the factor $T/4\pi$ and the summation, this term can be written in the form

$$\varepsilon(\mathbf{r}')\frac{\partial}{\partial x_i}\mathscr{D}^E_{kk}(\mathbf{r},\mathbf{r}')+\varepsilon(\mathbf{r})\frac{\partial}{\partial x_i'}\mathscr{D}^E_{kk}(\mathbf{r},\mathbf{r}'), \tag{30.10}$$

where we intend to set $\mathbf{r}=\mathbf{r}'$ at the end of the calculation. After some obvious manipulations, (30.10) becomes

$$\begin{aligned}
&2\frac{\partial}{\partial x_k}\varepsilon(\mathbf{r})\mathscr{D}^E_{ik}(\mathbf{r},\mathbf{r})-\frac{\partial}{\partial x_k}\varepsilon(\mathbf{r})\mathscr{D}^E_{ki}(\mathbf{r},\mathbf{r}')\\
&\quad-\frac{\partial}{\partial x_k'}\varepsilon(\mathbf{r}')\mathscr{D}^E_{ik}(\mathbf{r},\mathbf{r}')+\varepsilon(\mathbf{r}')\left[\frac{\partial}{\partial x_i}\mathscr{D}^E_{kk}(\mathbf{r},\mathbf{r}')-\frac{\partial}{\partial x_k}\mathscr{D}^E_{ik}(\mathbf{r},\mathbf{r}')\right]\\
&\quad+\varepsilon(\mathbf{r})\left[\frac{\partial}{\partial x_i'}\mathscr{D}^E_{kk}(\mathbf{r},\mathbf{r}')-\frac{\partial}{\partial x_k'}\mathscr{D}^E_{ki}(\mathbf{r},\mathbf{r}')\right].
\end{aligned}\tag{30.11}$$

We now observe that equation (28.18) for the Green's function $\mathscr{D}$ leads to the following identities:

$$\begin{aligned}
&\frac{\partial}{\partial x_k}\varepsilon(\mathbf{r})\mathscr{D}^E_{ki}(\mathbf{r},\mathbf{r}')=-4\pi\frac{\partial}{\partial x_i}\delta(\mathbf{r}-\mathbf{r}'),\\
&\frac{\partial}{\partial x_k'}\varepsilon(\mathbf{r}')\mathscr{D}^E_{ik}(\mathbf{r},\mathbf{r}')=4\pi\frac{\partial}{\partial x_i}\delta(\mathbf{r}-\mathbf{r}'),\\
&\frac{\partial}{\partial x_k}\mathscr{D}^H_{ik}(\mathbf{r},\mathbf{r}')=-\frac{\partial}{\partial x_k'}\mathscr{D}^H_{ik}(\mathbf{r},\mathbf{r}')=-4\pi\frac{\partial}{\partial x_i}\delta(\mathbf{r}-\mathbf{r}'),\\
&\varepsilon(\mathbf{r}')\left[\frac{\partial}{\partial x_k}\mathscr{D}^E_{ik}(\mathbf{r},\mathbf{r}')-\frac{\partial}{\partial x_i}\mathscr{D}^E_{kk}(\mathbf{r},\mathbf{r}')\right]\\
&\qquad=-\frac{\partial}{\partial x_k'}\mathscr{D}^H_{ki}(\mathbf{r},\mathbf{r}')+\frac{\partial}{\partial x_i'}\mathscr{D}^H_{kk}(\mathbf{r},\mathbf{r}')+8\pi\frac{\partial\delta(\mathbf{r}-\mathbf{r}')}{\partial x_i},\\
&\varepsilon(\mathbf{r})\left[\frac{\partial}{\partial x_k'}\mathscr{D}^E_{ki}(\mathbf{r},\mathbf{r}')-\frac{\partial}{\partial x_i'}\mathscr{D}^E_{kk}(\mathbf{r},\mathbf{r}')\right]\\
&\qquad=-\frac{\partial}{\partial x_k}\mathscr{D}^H_{ik}(\mathbf{r},\mathbf{r}')+\frac{\partial}{\partial x_i}\mathscr{D}^H_{kk}(\mathbf{r},\mathbf{r}')-8\pi\frac{\partial\delta(\mathbf{r}-\mathbf{r}')}{\partial x_i}.
\end{aligned}\tag{30.12}$$

Substituting (30.12) into (30.11) and setting $\mathbf{r}=\mathbf{r}'$, we obtain

$$\varepsilon(\mathbf{r})\frac{\partial}{\partial x_i}\mathscr{D}^E_{kk}(\mathbf{r},\mathbf{r})=2\frac{\partial}{\partial x_k}\varepsilon(\mathbf{r})\mathscr{D}^E_{ik}(\mathbf{r},\mathbf{r})+2\frac{\partial}{\partial x_k}\mathscr{D}^H_{ik}(\mathbf{r},\mathbf{r})-\frac{\partial}{\partial x_i}\mathscr{D}^H_{kk}(\mathbf{r},\mathbf{r}). \tag{30.13}$$

Finally, substituting (30.13) into (30.9), we find that the force $\mathbf{f}$ can be represented in the form (30.7), with a stress tensor given by

$$\begin{aligned}
\sigma_{ik}=-\delta_{ik}p_0(\rho,T)-\frac{T}{2\pi}\sum_{n=0}^{\infty}{}'\Big\{&-\frac{1}{2}\delta_{ik}\left[\varepsilon(\mathbf{r},i\omega_n)-\rho\frac{\partial\varepsilon(\mathbf{r},i\omega_n)}{\partial\rho}\right]\mathscr{D}^E_{ll}(\mathbf{r},\mathbf{r};\omega_n)\\
&-\frac{1}{2}\delta_{ik}\mathscr{D}^H_{ll}(\mathbf{r},\mathbf{r};\omega_n)+\varepsilon(\mathbf{r},i\omega_n)\mathscr{D}^E_{ik}(\mathbf{r},\mathbf{r};\omega_n)+\mathscr{D}^H_{ik}(\mathbf{r},\mathbf{r};\omega_n)\Big\}.
\end{aligned}\tag{30.14}$$

The formula (30.14) still has no direct physical meaning, since it involves the functions $\mathscr{D}^E(\mathbf{r}, \mathbf{r}')$ and $\mathscr{D}^H(\mathbf{r}, \mathbf{r}')$ which become infinite for $\mathbf{r} = \mathbf{r}'$. This is due to the fact that the short-wavelength electromagnetic oscillations make an infinite contribution to $\mathscr{D}_{ik}$, unless we introduce an appropriate cutoff. However, the short-wavelength oscillations are not relevant as far as effects due to inhomogeneity of a body are concerned, since the contribution of such oscillations is the same for both homogeneous and inhomogeneous bodies with the same value of ε at the given point. The contribution we are interested in, which comes from the long-wavelength oscillations and is independent of the character of the cutoff, can be obtained from formula (30.14) by carrying out a suitable subtraction. In fact, in (30.14) the Green's function $\mathscr{D}^E_{ik}(\mathbf{r}, \mathbf{r})$ [and similarly for $\mathscr{D}^H_{ik}(\mathbf{r}, \mathbf{r})$] should be regarded as the limit

$$\lim_{\mathbf{r}\to\mathbf{r}'} [\mathscr{D}^E_{ik}(\mathbf{r}, \mathbf{r}') - \bar{\mathscr{D}}^E_{ik}(\mathbf{r}, \mathbf{r}')], \tag{30.15}$$

where $\bar{\mathscr{D}}^E$ is the Green's function of a homogeneous medium whose dielectric constant is the same as that of the inhomogeneous body at the point where the stress tensor is being calculated. To avoid superfluous complications, we shall continue to write the formula (30.14) in its present form, with the understanding that the limit (30.15) has already been taken. The same remark applies to formula (30.6) for the chemical potential, which can be written as

$$\mu(\rho, T) = \mu_0(\rho, T) + \frac{T}{4\pi} \sum_{n=0}^{\infty}{}' \frac{\partial \varepsilon(\mathbf{r}, i\omega_n)}{\partial \rho} \mathscr{D}^E_{ii}(\mathbf{r}, \mathbf{r}; \omega_n), \tag{30.16}$$

if we use (30.8).

It should be noted that we regard systems consisting of several bodies, each of which is homogeneous, as belonging to the category of inhomogeneous media. In this case, when (28.18) is solved, the components $\mathscr{D}_{ik}$ must satisfy certain conditions at the boundaries between the bodies. In equation (28.18), the components of $\mathbf{r}$ are the independent variables, and the components of $\mathbf{r}'$ act as parameters. Therefore, the boundary conditions we have in mind involve the variable $\mathbf{r}$ and are the conditions corresponding to continuity of the tangential components of the electric and magnetic fields. Since the point $\mathbf{r}$ corresponds to the index i in the tensor $\mathscr{D}_{ik}$, it is the tangential components of the tensors $\mathscr{D}^E_{ik}$ and $\mathscr{D}^H_{ik}$ relative to this index that must be continuous.

In principle, formulas (30.15) and (30.16), obtained by Dzyaloshinski and Pitayevski (D3), solve the problem of calculating the van der Waals part of the thermodynamic quantities describing a body. This problem reduces to solving equation (28.18) for the Green's function $\mathscr{D}_{ik}$.

31. Molecular Interaction Forces

31.1. Interaction forces between solid bodies. We now use the general theory developed above to calculate the van der Waals forces acting between two solid bodies whose surfaces are very close together, where it is assumed that the space between the bodies is filled with some liquid. We shall use the indices 1 and 2 to denote quantities pertaining to the two solid bodies, and the index 3 to denote quantities pertaining to the medium between the bodies. Although it is assumed that the region between the two bodies is bounded by two infinite parallel planes, it should be kept in mind that in a correct formulation of the problem, we must assume that at least one of the bodies has finite dimensions and is surrounded on all sides by the medium 3, and then we must determine the total force acting on this body. However, since the molecular forces fall off very rapidly with distance, the resulting force can actually be attributed entirely to forces acting across the narrow gap separating the two bodies.

The total force acting on body 2 can be calculated as the total momentum current flowing into body 2 from the surrounding medium 3, and equals the integral of the momentum current density evaluated over any surface surrounding the body. In this regard, we must take into consideration the fact that the medium 3 is in thermodynamic equilibrium, one condition for which is that its chemical potential μ should be constant, where μ is given by formula (30.16). Since the corrections to the density of the medium connected with the long-wavelength fluctuations of the field are small, we can regard ρ as constant throughout the medium 3. Then, according to (30.5), the change in the chemical potential $\mu_0(\rho, T)$ is the same as the change in the quantity $p_0(\rho, T)/\rho$. Therefore, the condition $\mu =$ const can be written in the form

$$p_0(\rho, T) + \frac{T}{4\pi}\sum_{n=0}^{\infty}{}' \rho \frac{\partial \varepsilon_3(i\omega_n)}{\partial \rho} \mathscr{D}^E_{ll}(\mathbf{r}, \mathbf{r}; \omega_n) = \text{const.} \tag{31.1}$$

Because of the condition (31.1), part of the total stress tensor (30.14) turns out to be a constant uniform pressure throughout the liquid, and makes no contribution at all to the total force acting on the body. Thus, to determine the force in question, it is actually sufficient to write the stress tensor in medium 3 in the form

$$\sigma'_{ik} = -\frac{T}{2\pi}\sum_{n=0}^{\infty}{}' \Big\{\varepsilon_3(i\omega_n)\Big[\mathscr{D}^E_{ik}(\mathbf{r}, \mathbf{r}; \omega_n) - \frac{1}{2}\,\delta_{ik}\mathscr{D}^E_{ll}(\mathbf{r}, \mathbf{r}; \omega_n)\Big] + \mathscr{D}^H_{ik}(\mathbf{r}, \mathbf{r}; \omega_n) - \frac{1}{2}\,\delta_{ik}\mathscr{D}^H_{ll}(\mathbf{r}, \mathbf{r}; \omega_n)\Big\}. \tag{31.2}$$

Let the x-axis be perpendicular to the planes bounding the region between the bodies, and let the width of this region be l (so that the surfaces of the bodies 1 and 2 are the planes $x = 0$ and $x = l$). Then the force F acting on a unit area of the surface of body 2 equals

$$\begin{aligned} F(l) &= \sigma'_{xx}(l) \\ &= \frac{T}{4\pi}\sum_{n=0}^{\infty}{}' \{\varepsilon_3(i\omega_n)[\mathscr{D}^E_{yy}(l, l; \omega_n) + \mathscr{D}^E_{zz}(l, l; \omega_n) - \mathscr{D}^E_{xx}(l, l; \omega_n)] \\ &\quad + \mathscr{D}^H_{yy}(l, l; \omega_n) + \mathscr{D}^H_{zz}(l, l; \omega_n) - \mathscr{D}^H_{xx}(l, l; \omega_n)\} \end{aligned} \tag{31.3}$$

where a positive force corresponds to attraction between the bodies, and a negative force corresponds to repulsion.

Because of the homogeneity of the problem relative to the direction of the y and z-axes, the Green's function $\mathscr{D}_{ik}(\mathbf{r}, \mathbf{r}'; \omega_n)$ involves y, y', z, z' only through the differences $y - y', z - z'$. Suppose we take the Fourier transform with respect to these variables, i.e.,

$$\mathscr{D}_{ik}(x, x'; \mathbf{q}; \omega_n) = \int e^{-iq_y(y-y')-iq_z(z-z')}\mathscr{D}_{ik}(\mathbf{r}, \mathbf{r}'; \omega_n)\, d(y - y')\, d(z - z')$$

and direct the y-axis along the vector $\mathbf{q}$. Then the equations (28.18) for the Green's function become

$$\left(w^2 - \frac{d^2}{dx^2}\right)\mathscr{D}_{zz}(x, x') = -4\pi\delta(x - x'),$$

$$\left(\varepsilon\omega_n^2 - \frac{d^2}{dx^2}\right)\mathscr{D}_{yy}(x, x') + iq\frac{d}{dx}\mathscr{D}_{xy}(x, x') = -4\pi\delta(x - x'),$$

$$w^2\mathscr{D}_{xy}(x, x') + iq\frac{d}{dx}\mathscr{D}_{yy}(x, x') = 0,$$

$$w^2\mathscr{D}_{xx}(x, x') + iq\frac{d}{dx}\mathscr{D}_{xy}(x, x') = -4\pi\delta(x - y),$$

$$\left(\varepsilon\omega_n^2 - \frac{d^2}{dx^2}\right)\mathscr{D}_{xy}(x, x') + iq\frac{d}{dx}\mathscr{D}_{xx}(x, x') = 0,$$

where

$$w = \sqrt{\varepsilon\omega_n^2 + q^2},$$

and x' acts as a parameter. (The components $\mathscr{D}_{xz}$ and $\mathscr{D}_{yz}$ of the Green's function vanish, since the equations for them are homogeneous). The solution of this system reduces to the solution of just two equations

$$\begin{aligned}\left(w^2 - \frac{d^2}{dx^2}\right)\mathscr{D}_{zz}(x, x') &= -4\pi\delta(x - x'),\\ \left(w^2 - \frac{d^2}{dx^2}\right)\mathscr{D}_{yy}(x, x') &= -\frac{4\pi w^2}{\varepsilon\omega_n^2}\delta(x - x'),\end{aligned} \tag{31.4}$$

since from a knowledge of $\mathscr{D}_{yy}$, we can immediately determine $\mathscr{D}_{xy}$ and $\mathscr{D}_{xx}$:

$$\begin{aligned}\mathscr{D}_{xy} &= -\frac{iq}{w^2}\frac{d}{dx}\mathscr{D}_{yy},\\ \mathscr{D}_{xx} &= -\frac{iq}{w^2}\frac{d}{dx}\mathscr{D}_{xy} - \frac{4\pi}{w^2}\delta(x - x').\end{aligned} \tag{31.5}$$

The boundary conditions, corresponding to continuous tangential components of the electric and magnetic fields, reduce to the requirement that the quantities $\mathscr{D}^E_{yk}$, $\mathscr{D}^H_{yk}$, $\mathscr{D}^E_{zk}$, $\mathscr{D}^H_{zk}$ (or equivalently, the quantities $\mathscr{D}_{yk}$, $\mathscr{D}_{zk}$,

$\text{curl}_{yl}\,\mathscr{D}_{lk}, \text{curl}_{zl}\,\mathscr{D}_{lk})$ be continuous. Using the first of the equations (31.5), we find that the quantities

$$\mathscr{D}_{zz}, \quad \frac{d\mathscr{D}_{zz}}{dx}, \quad \mathscr{D}_{yy}, \quad \frac{\varepsilon}{w^2}\frac{d\mathscr{D}_{yy}}{dx} \tag{31.6}$$

must be continuous on the boundaries of the gap. Since we are only interested in the Green's function in the gap, we can confine ourselves to the case $0 < x' < l$ from the outset. In the region 3 ($0 < x < l$), the functions $\mathscr{D}_{yy}$ and $\mathscr{D}_{zz}$ are given by the equations (31.4) with

$$\varepsilon = \varepsilon_3, \qquad w = w_3 = \sqrt{\varepsilon_3\omega_n^2 + q^2}.$$

In the regions 1 ($x < 0$) and 2 ($x > l$), these functions satisfy the same equations without right-hand sides (since here we always have $x \neq x'$), and with ε_1, w_1 or ε_2, w_2 instead of ε, w.

In the present case, the remark made at the end of Sec. 30 in connection with the limit (30.15) reduces to the requirement that in the gap we subtract from all the $\mathscr{D}$-functions their values for $\varepsilon_1 = \varepsilon_2 = \varepsilon_3$, $w_1 = w_2 = w_3$. In particular, it follows that we can drop the term involving the δ-function in the second of the equations (31.5), so that in the gap $\mathscr{D}_{xy}$ and $\mathscr{D}_{xx}$ are given by

$$\mathscr{D}_{xy} = -\frac{iq}{w_3^2}\frac{d}{dx}\mathscr{D}_{yy}, \quad \mathscr{D}_{xx} = -\frac{iq}{w_3^2}\frac{d}{dx}\mathscr{D}_{xy}. \tag{31.7}$$

Before solving these equations, we make one further remark. The general solution of the equations (31.4) has the form

$$f_1(x - x') + f_2(x + x').$$

Using (31.4), (31.7) and the definition of the functions $\mathscr{D}^E$ and $\mathscr{D}^H$, we can show that the parts of the Green's functions depending on the sum $x + x'$ make no contribution at all to the expression (31.3) for the force. We shall not pursue this point, since the result is already obvious from physical considerations. In fact, if $x = x'$ in a solution of the form $f_2(x + x')$, we obtain a momentum flow in the gap which depends on the coordinates and hence contradicts the law of conservation of momentum. Therefore, from now on we shall as a rule only derive expressions for parts of the Green's functions $\mathscr{D}^+$, which depend only on $x - x'$.

We now find $\mathscr{D}_{zz}$, which satisfies the equations

$$\left(w_3^2 - \frac{d^2}{dx^2}\right)\mathscr{D}_{zz} = -4\pi\delta(x - x') \quad \text{for} \quad 0 < x < l,$$

$$\left(w_1^2 - \frac{d^2}{dx^2}\right)\mathscr{D}_{zz} = 0 \quad \text{for} \quad x < 0,$$

$$\left(w_2^2 - \frac{d^2}{dx^2}\right)\mathscr{D}_{zz} = 0 \quad \text{for} \quad x > l.$$

Solving these equations, we have

$$\mathscr{D}_{zz} = Ae^{w_1 x} \quad \text{for} \quad x < 0,$$

$$\mathscr{D}_{zz} = Be^{-w_2 x} \quad \text{for} \quad x > l.$$

$$\mathscr{D}_{zz} = C_1 e^{w_3 x} + C_2 e^{-w_3 x} - \frac{2\pi}{w_3} e^{-w_3|x-x'|} \quad \text{for} \quad 0 < x < l.$$

Determining the constants A, B, C_1 and C_2 from the conditions that $\mathscr{D}_{zz}$ and $d\mathscr{D}_{zz}/dx$ be continuous on the boundary, we find that

$$\mathscr{D}^{+}_{zz} = \frac{4\pi}{w_3 \Delta} \cosh [w_3(x - x')] - \frac{2\pi}{w_3} e^{-w_3|x-x'|} \quad \text{for} \quad 0 < x < l,$$

where

$$\Delta = 1 - e^{2w_3 l} \frac{w_1 + w_3}{w_1 - w_3} \frac{w_2 + w_3}{w_2 - w_3}. \tag{31.8}$$

Subtracting the value of $\mathscr{D}^{+}_{zz}$ for $w_1 = w_2 = w_3$ (which makes Δ infinite), we finally obtain

$$\mathscr{D}^{+}_{zz} = \frac{4\pi}{w_3 \Delta} \cosh [w_3 (x - x')]. \tag{31.9}$$

Similarly, solving the equation for $\mathscr{D}_{yy}$, we have

$$\mathscr{D}^{+}_{yy} = \frac{4\pi w_3}{\omega_n^2 \varepsilon_3 \bar{\Delta}} \cosh [w_3(x - x')], \tag{31.10}$$

$$\bar{\Delta} = 1 - e^{2w_3 l} \frac{\varepsilon_1 w_3 + \varepsilon_3 w_1}{\varepsilon_1 w_3 - \varepsilon_3 w_1} \frac{\varepsilon_2 w_3 + \varepsilon_3 w_2}{\varepsilon_2 w_3 - \varepsilon_3 w_2}, \tag{31.11}$$

after an appropriate subtraction. Then, using the equations (31.7), we find that

$$\mathscr{D}^{+}_{xy} = - \frac{4\pi i q}{\omega_n^2 \varepsilon_3 \bar{\Delta}} \sinh [w_3(x - x')],$$

$$\mathscr{D}^{+}_{xx} = - \frac{4\pi q^2}{\omega_n^2 \varepsilon_3 w_3 \bar{\Delta}} \cosh [w_3(x - x')]. \tag{31.12}$$

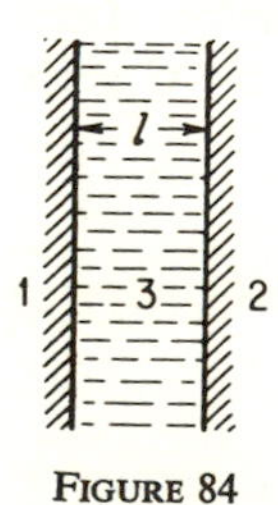

FIGURE 84

Next, calculating the quantities $\mathscr{D}^{E}_{ik}(x, x'; q, \omega_n)$, $\mathscr{D}^{H}_{ik}(x, x'; q, \omega_n)$ and substituting them in (31.3), we obtain

$$F(l) = - \frac{T}{2\pi} \sum_{n=0}^{\infty}{}' \int_0^\infty q \, w_3 \left(\frac{1}{\Delta} + \frac{1}{\bar{\Delta}} \right) dq.$$

Going over to a new variable of integration p defined by

$$q = \sqrt{\varepsilon_3}\, \omega_n \sqrt{p^2 - 1}$$

and returning to the usual system of units, we arrive at the final result, due to Dzyaloshinski, Lifshitz and Pitayevski (D1): The expression for the force $F(l)$ acting on a unit area of each of two bodies separated by a gap of width l filled with a liquid (see Fig. 84) is

$$F(l) = \frac{T}{\pi c^3} \sum_{n=0}^{\infty}{}' \varepsilon_3^{3/2}\omega_n^3 \int_1^{\infty} p^2 \left\{ \left[\frac{s_1 + p}{s_1 - p} \frac{s_2 + p}{s_2 - p} \exp\{2p\omega_n l\sqrt{\varepsilon_3}/c\} - 1 \right]^{-1} + \left[\frac{s_1 + (p\varepsilon_1/\varepsilon_3)}{s_1 - (p\varepsilon_1/\varepsilon_3)} \frac{s_2 + (p\varepsilon_2/\varepsilon_3)}{s_2 - (p\varepsilon_2/\varepsilon_3)} \exp\{2p\omega_n l\sqrt{\varepsilon_3}/c\} - 1 \right]^{-1} \right\} dp. \tag{31.13}$$

Here

$$s_1 = \sqrt{(\varepsilon_1/\varepsilon_3) - 1 + p^2}, \quad s_2 = \sqrt{(\varepsilon_2/\varepsilon_3) - 1 + p^2}, \quad \omega_n = \frac{2n\pi T}{\hbar},$$

ε_1, ε_2 and ε_3 are functions of the imaginary frequency $i\omega_n$ [$\varepsilon = \varepsilon(i\omega_n)$], and we explicitly write the velocity of light c and Planck's constant $\hbar$. For $\varepsilon_3 = 1$, i.e., for the case of bodies separated by a vacuum, this formula was first obtained by Lifshitz (L13) by another approach, without using the methods of quantum field theory.

The general formula (31.13) is very complicated. However, it can be considerably simplified by using the fact that the effect of temperature on the interaction forces between the bodies is usually very unimportant.[10] The point is that due to the presence of the exponentials in the integrand of (31.3), the main contribution to the sum is made by the terms for which $\omega_n \sim c/l$ or $n \sim \hbar c/lT$. Thus, in the case $lT/\hbar c \ll 1$, it is the large values of n that are important, and in (31.13) we can go from summation to integration with respect to $dn = (\hbar/2\pi T)\, d\omega$. Then the temperature does not appear explicitly in the formula, and we obtain the following result [where $\varepsilon = \varepsilon(i\omega)$]:

$$F(l) = \frac{\hbar}{2\pi^2 c^3} \int_0^{\infty} d\omega \int_1^{\infty} p^2\omega^3\varepsilon_3^{3/2} \left\{ \left[\frac{s_1 + p}{s_1 - p} \frac{s_2 + p}{s_2 - p} \exp\{2p\omega l\sqrt{\varepsilon_3}/c\} - 1 \right]^{-1} + \left[\frac{s_1 + (p\varepsilon_1/\varepsilon_3)}{s_1 - (p\varepsilon_1/\varepsilon_3)} \frac{s_2 + (p\varepsilon_2/\varepsilon_3)}{s_2 - (p\varepsilon_2/\varepsilon_3)} \exp\{2p\omega l\sqrt{\varepsilon_3}/c\} - 1 \right]^{-1} \right\} dp. \tag{31.14}$$

Formula (31.14) is still quite complicated, but it can be greatly simplified in two important limiting cases.

First, we consider the limiting case of "small" separations, by which we mean separations small compared to the wavelengths λ_0 characteristic of the absorption spectra of the given bodies. The temperatures which are meaningful for condensed matter are in any event small compared to the relevant values of $\hbar\omega$ (e.g., in the visible region of the spectrum), and hence the inequality $lT/\hbar c \ll 1$ certainly holds. Due to the presence of the exponential factor $\exp\{2p\omega\sqrt{\varepsilon_3}/c\}$ in the denominators of the integrand of (31.14), when we integrate with respect to p, the main contribution comes from values of p such that $p\omega l/c \sim 1$. Thus $p \gg 1$, and hence the main terms can be found

[10] Here, in speaking of the effect of temperature, we do not consider the effect of the temperature dependence of the dielectric constant itself.

by setting $s_1 \approx s_2 \approx p$. To this approximation, the first term in braces in (31.14) vanishes. Then, if we introduce the new variable of integration $x = 2lp\omega\sqrt{\varepsilon_3}/c$, (31.14) becomes

$$F(l) = \frac{\hbar}{16\pi^2 l^3} \int_0^\infty dx \int_0^\infty x^2 \left(\frac{\varepsilon_1 + \varepsilon_3}{\varepsilon_1 - \varepsilon_3} \frac{\varepsilon_2 + \varepsilon_3}{\varepsilon_2 - \varepsilon_3} e^x - 1 \right)^{-1} d\omega \quad (31.15)$$

(to this approximation, the lower limit of the integration with respect to x can be changed to 0). In this case, the force turns out to be inversely proportional to the cube of the separation, which might have been expected, in view of the ordinary laws for the van der Waals forces between two molecules. The function $\varepsilon(i\omega) - 1$ falls off monotonically to zero as ω increases. Therefore, starting from some $\omega \sim \omega_0$ the values of ω cease to make an important contribution to the integral. The condition that l be small means that we must have $l \ll c/\omega_0$.

Next, we consider the opposite limiting case of "large" separations $l \gg \lambda_0$, which, however, are not so large that the inequality $lT/\hbar c \ll 1$ is violated. Introducing the new variable of integration $x = 2pl\omega/c$ in the general formula (31.14), and keeping p rather than ω as the second variable [unlike (31.15)], we find that

$$F(l) = \frac{\hbar c}{32\pi^2 l^4} \int_0^\infty dx \int_1^\infty \frac{x^3}{p^2} \varepsilon_3^{3/2} \left\{ \left[\frac{s_1 + p}{s_1 - p} \frac{s_2 + p}{s_2 - p} \exp\{x\sqrt{\varepsilon_3}\} - 1 \right]^{-1} + \left[\frac{s_1 + (p\varepsilon_1/\varepsilon_3)}{s_1 - (p\varepsilon_1/\varepsilon_3)} \frac{s_2 + (p\varepsilon_2/\varepsilon_3)}{s_2 - (p\varepsilon_2/\varepsilon_3)} \exp\{x\sqrt{\varepsilon_3}\} - 1 \right]^{-1} \right\} dp,$$

where $\varepsilon = \varepsilon(ixc/2pl)$. Due to the presence of the factor $\exp\{x\sqrt{\varepsilon_3}\}$ in the denominator, the values $x \approx 1/\sqrt{\varepsilon_3}$ make the main contribution to the integral with respect to x, and since $p \geqslant 1$, the argument of the function ε is close to zero for large l, in the important region of the values of the variables. Accordingly, we can simply replace $\varepsilon_1, \varepsilon_2, \varepsilon_3$ by their values at $\omega = 0$, i.e., by the electrostatic values of the dielectric constant. Thus, making the substitution $x \to x/\sqrt{\varepsilon_{30}}$, we finally obtain

$$F(l) = \frac{hc}{32\pi^2 l^4 \sqrt{\varepsilon_{30}}} \int_0^\infty dx \int_1^\infty \frac{x^3}{p^2} \left\{ \left[\frac{s_{10} + p}{s_{10} - p} \frac{s_{20} + p}{s_{20} - p} e^x - 1 \right]^{-1} + \left[\frac{s_{10} + (p\varepsilon_{10}/\varepsilon_{30})}{s_{10} - (p\varepsilon_{10}/\varepsilon_{30})} \frac{s_{20} + (p\varepsilon_{20}/\varepsilon_{30})}{s_{20} - (p\varepsilon_{20}/\varepsilon_{30})} e^x - 1 \right]^{-1} \right\} dp, \quad (31.16)$$

where

$$s_{10} = \sqrt{(\varepsilon_{10}/\varepsilon_{30}) - 1 + p^2}, \qquad s_{20} = \sqrt{(\varepsilon_{20}/\varepsilon_{30}) - 1 + p^2},$$

and $\varepsilon_{10}, \varepsilon_{20}, \varepsilon_{30}$ are the electrostatic values of the dielectric constant.

Finally, we consider the case of high temperatures. If $lT/\hbar c \gg 1$, then, from all the terms in the sum (31.13) we need only keep the first term. However, we cannot set $n = 0$ directly in (31.13), since this leads to an indeterminate expression (the factor ω_n^3 vanishes, but the integral with respect to p diverges). This difficulty can be avoided by first introducing a new variable of integration $x = 2p\omega_n l\sqrt{\varepsilon_{30}}/c$ instead of p (this causes the factor ω_n^3 to disappear). Then, setting $\omega_n = 0$, we obtain

$$F(l) = \frac{T}{16\pi l^3}\int_0^\infty x^2\left[\frac{\varepsilon_{10}+\varepsilon_{30}}{\varepsilon_{10}-\varepsilon_{30}}\frac{\varepsilon_{20}+\varepsilon_{30}}{\varepsilon_{20}-\varepsilon_{30}}e^x - 1\right]^{-1}dx. \tag{31.17}$$

Thus, for sufficiently large separations l, the interaction forces fall off more slowly and again behave like l^{-3}, with a coefficient which depends on the temperature and the electrostatic values of the dielectric constant.

31.2. Interaction forces between molecules in a solution. We now show how to go from the macroscopic formula (31.14) to the interaction of individual molecules in a vacuum. To achieve this, we make the formal assumption that both bodies are sufficiently "dilute." From the standpoint of macroscopic electrodynamics, this means that their dielectric constants are close to unity, i.e., that the differences $\varepsilon_1 - 1$ and $\varepsilon_2 - 1$ are small.

We begin by considering the case of "small" distances. From formula (31.15) with $\varepsilon_3 = 1$, we obtain

$$\begin{aligned} F(l) &= \frac{\hbar}{64\pi^2 l^3}\int_0^\infty\int_0^\infty x^2 e^{-x}(\varepsilon_1 - 1)(\varepsilon_2 - 1)\,dx\,d\omega \\ &= \frac{\hbar}{32\pi^2 l^3}\int_0^\infty [\varepsilon_1(i\omega) - 1][\varepsilon_2(i\omega) - 1]\,d\omega, \end{aligned} \tag{31.18}$$

to the required accuracy. Using (28.19) to express $\varepsilon(i\omega)$ in terms of $\varepsilon''(\omega)$ where ω is real, we find that

$$\begin{aligned} &\int_0^\infty [\varepsilon_1(i\omega) - 1][\varepsilon_2(i\omega) - 1]\,d\omega \\ &\quad = \frac{4}{\pi^2}\int_0^\infty\int_0^\infty\int_0^\infty \frac{\omega_1\omega_2\varepsilon_1''(\omega_1)\,\varepsilon_2''(\omega_2)}{(\omega_1^2+\omega^2)(\omega_2^2+\omega^2)}\,d\omega_1\,d\omega_2\,d\omega \\ &\quad = \frac{2}{\pi}\int_0^\infty\int_0^\infty \frac{\varepsilon_1''(\omega_1)\varepsilon_2''(\omega_2)}{\omega_1+\omega_2}\,d\omega_1\,d\omega_2, \end{aligned}$$

which gives the expression

$$F(l) = \frac{\hbar}{16\pi^3 l^3}\int_0^\infty\int_0^\infty \frac{\varepsilon_1''(\omega_1)\varepsilon_2''(\omega_2)}{\omega_1+\omega_2}\,d\omega_1\,d\omega_2 \tag{31.19}$$

for the force $F(l)$. This force corresponds to an interaction between the particles with potential energy[11]

[11] If the potential energy of the interaction between the molecules 1 and 2 is $U = -a/R^6$, then the total energy of the interaction between all pairs of molecules in two half-spaces separated by a gap of width l is

$$U = -\frac{a\pi N_1 N_2}{12 l^2},$$

and the force is

$$F = -\frac{dU}{dl} = \frac{a\pi N_1 N_2}{6 l^3}.$$

This explains the correspondence between (31.19) and (31.20).

$$U(R) = -\frac{3\hbar}{8\pi^4 R^6 N_1 N_2} \int_0^\infty \int_0^\infty \frac{\varepsilon_1''(\omega_1)\varepsilon_2''(\omega_2)}{\omega_1 + \omega_2} \, d\omega_1 \, d\omega_2, \tag{31.20}$$

where R is the distance between the molecules, and N_1, N_2 are the numbers of molecules per unit volume in the first and second bodies, respectively. The imaginary part of the dielectric constant is related to the spectral density of the "oscillator strengths" $f(\omega)$ [familiar from spectroscopy] by the formula

$$\omega\varepsilon''(\omega) = \frac{2\pi^2 e^2}{m} N f(\omega)$$

(see e.g., L9, Sec. 62). Substitution of this expression into (31.20) gives

$$U(R) = -\frac{3\hbar e^4}{2m^2 R^6} \int_0^\infty \int_0^\infty \frac{f_1(\omega_1) f_2(\omega_2)}{\omega_1 + \omega_2} \, d\omega_1 \, d\omega_2 \tag{31.21}$$

which is exactly the same as the well-known London formula (see E2), obtained by applying ordinary perturbation theory to the case of a dipole interaction between two molecules. For example, consider the interaction of two hydrogen atoms. Using the familiar expression

$$f_{0n} = \frac{2m}{\hbar^2} (E_n - E_0) |x_{0n}|^2$$

for the oscillator strengths associated with the transition between the states E_0 and E_n (where x_{0n} is the corresponding matrix element of the coordinates of the electron in the atom), and transforming (31.21) from an integral over the frequency to a sum over the energy levels of the atom, we obtain the expression

$$U(R) = -\frac{6e^4}{R^6} \sum_{n,m} \frac{|x_{0n}|^2 |x_{0m}|^2}{E_n - E_0 + E_m - E_0}$$

for the potential energy of two hydrogen atoms.

For "large" distances, the formula for the attractive force between two "dilute" bodies has the form

$$\begin{aligned} F(l) &= \frac{\hbar c}{32\pi^2 l^4} (\varepsilon_{10} - 1)(\varepsilon_{20} - 1) \int_0^\infty x^3 e^{-x} \, dx \int_1^\infty \frac{1 - 2p^2 + 2p^4}{8p^6} \, dp \\ &= \frac{\hbar c}{l^4} \frac{23}{640\pi^2} (\varepsilon_{10} - 1)(\varepsilon_{20} - 1). \end{aligned} \tag{31.22}$$

This force corresponds to an interaction between two molecules with potential energy

$$U(R) = -\frac{23\hbar c}{4\pi R^7} \alpha_1 \alpha_2, \tag{31.23}$$

where α_1 and α_2 are the static polarizabilities of the two molecules ($\varepsilon_0 = 1 - 4\pi N\alpha$). Formula (31.23) coincides with the result of a quantum-mechanical calculation by Casimir and Polder (C1) of the attraction between two molecules at sufficiently large distances, when the effect of retardation becomes important.

We now consider the interaction of two molecules in a liquid [see Pitayevski (P5)]. To do this, let the first "body" be a weak solution of molecules of one kind, with concentration N_1 (the number of particles per cm^3), and let the second "body" be a weak solution of molecules of another kind, with concentration N_2, where the solvent is the same in both cases. Moreover, let the gap be filled with pure solvent. For small concentrations of the solutes, the dielectric constants ε_1 and ε_2 of the solutions differ only slightly from the dielectric constant of the pure solvent, which we denote by $\varepsilon_3 = \varepsilon$. Thus we have

$$\varepsilon_1 = \varepsilon + N_1\left(\frac{\partial \varepsilon_1}{\partial N_1}\right)_{N_1=0},$$

$$\varepsilon_2 = \varepsilon + N_2\left(\frac{\partial \varepsilon_2}{\partial N_2}\right)_{N_2=0},$$

to terms of the first order in the concentrations. If we retain only terms of the same order in formula (31.15) for the force at "small" distances, we obtain

$$F(l) = \frac{\hbar}{32\pi^2 l^3} N_1 N_2 \int_0^\infty \left(\frac{\partial \varepsilon_1(i\omega)}{\partial N_1}\right)_{N_1=0}\left(\frac{\partial \varepsilon_2(i\omega)}{\partial N_2}\right)_{N_2=0}\frac{d\omega}{\varepsilon^2(i\omega)}$$

[cf. the transition to formula (31.18)]. This force corresponds to an interaction between the dissolved molecules with potential energy

$$U(R) = -\frac{3\hbar}{16\pi^3 R^6}\int_0^\infty \left(\frac{\partial \varepsilon_1(i\omega)}{\partial N_1}\right)_{N_1=0}\left(\frac{\partial \varepsilon_2(i\omega)}{\partial N_2}\right)_{N_2=0}\frac{d\omega}{\varepsilon^2(i\omega)}. \tag{31.24}$$

Similarly, we obtain

$$U(R) = -\frac{23\hbar c}{64\pi^3 \varepsilon_0^{5/2} R^7}\left(\frac{\partial \varepsilon_{10}}{\partial N_1}\right)_{N_1=0}\left(\frac{\partial \varepsilon_{20}}{\partial N_2}\right)_{N_2=0} \tag{31.25}$$

for the potential energy at "large" distances. We see that the interaction forces between the molecules are no longer given in terms of their polarizabilities in the case where the molecules of the solute interact strongly with the solvent.

31.3. A thin liquid film on the surface of a solid body. As our last application of the general theory of van der Waals forces, we calculate the thermodynamic properties of a thin liquid film on the surface of a solid body, where it is assumed, of course, that the thickness l of the film is large compared to interatomic distances. Earlier, we derived a formula [see (30.16)] expressing the chemical potential of a liquid per unit mass in terms of the Green's function of the long-wavelength electromagnetic field in the liquid. However, in the present problem, this formula is not suitable for two reasons. In the first place, it involves a quantity about which nothing is known experimentally, i.e., the quantity $\partial\varepsilon/\partial\rho$ over the whole frequency range. In the

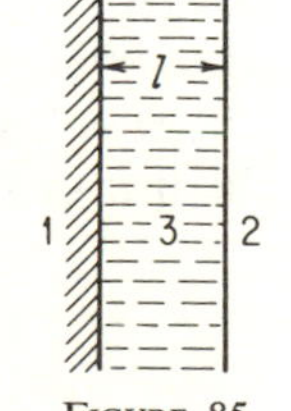

FIGURE 85

second place, it gives the chemical potential μ as a function of ρ, whereas we usually need to know μ as a function of the pressure p.

Consider a liquid film 3 lying on the surface of a solid body 1, and let the film be in equilibrium with its vapor 2 (see Fig. 85). As far as its electromagnetic properties are concerned, we shall regard the vapor as a vacuum, we shall set its dielectric constant equal to zero everywhere ($\varepsilon_2 = 1$). According to the condition for mechanical equilibrium, the normal component σ_{xx} of the stress tensor must be continuous at the surface of the film. From this we obtain the equation

$$p = p_0(\rho, T) - \bar{\sigma}_{xx},$$

where p is the pressure of the vapor, $p_0(\rho, T)$ is the pressure of the "bulk" liquid at the given density and pressure, and $\bar{\sigma}_{xx}$ denotes the set of all terms except the first in the expression (30.14) for the stress tensor in the film. Solving this equation for ρ, we find the density as a function of the pressure[12]

$$\rho = \rho_0(p + \bar{\sigma}_{xx}, T). \tag{31.26}$$

Substituting (31.26) into formula (30.16) for the chemical potential, we obtain

$$\mu = \mu_0(p + \bar{\sigma}_{xx}, T) + \frac{T}{4\pi} \sum_{n=0}^{\infty}{}' \frac{\partial \varepsilon(i\omega_n)}{\partial \rho} \mathscr{D}_{ii}^E(\mathbf{r}, \mathbf{r}; \omega_n), \tag{31.27}$$

where $\mu_0(p, T)$ is the chemical potential of the bulk liquid. Expanding μ_0 in powers of the small quantity $\bar{\sigma}_{xx}$, and using the thermodynamic relation

$$\left(\frac{\partial \mu}{\partial p}\right)_T = \frac{1}{\rho},$$

we can reduce (31.27) to the form

$$\mu(p, T) = \mu_0(p, T) + \frac{1}{\rho} \bar{\sigma}_{xx} + \frac{T}{4\pi} \sum_{n=0}^{\infty}{}' \frac{\partial \varepsilon(i\omega_n)}{\partial \rho} \mathscr{D}_{ii}^E(\mathbf{r}, \mathbf{r}; \omega_n). \tag{31.28}$$

Finally, substituting the expression (30.14) for $\bar{\sigma}_{xx}$ into (31.28), we find that the term involving $\partial\varepsilon/\partial\rho$ drops out, and all that remains is

$$\mu(p, T) = \mu_0(p, T) + \frac{1}{\rho} \sigma'_{xx}.$$

Here, σ'_{xx} is the xx-component of the "reduced" stress tensor (31.2), and is constant throughout the film (since the momentum flow is constant). According to (31.3), it is just σ'_{xx} that gives the force $F(l)$.

Now let ζ denote the "van der Waals part" of the chemical potential of the film (per unit volume of the liquid), i.e., let

$$\mu = \mu_0 + \frac{\zeta}{\rho}. \tag{31.29}$$

---

[12] The quantity $\bar{\sigma}_{xx}$ is also a function of ρ, but since it represents a small correction to the pressure, we can set $\rho = \rho_0(p, T)$ in (31.26).

Then, according to what was just said,

$$\zeta = \sigma'_{xx} = F(l). \tag{31.30}$$

In the limit $l \to \infty$ (i.e., for the bulk liquid), ζ goes to zero. Thus, the quantity ζ in which we are interested can be determined without making any special calculations. In fact, ζ is given by the formulas for $F(l)$ derived above [the general formula (31.13) and the subsequent limiting formulas], provided only that we set $\varepsilon_2 = 1$.

The reader who is particularly interested in the problems treated in this section will find further details in the papers D1, D2 and L13.

7

THEORY OF SUPERCONDUCTIVITY

32. Background Information. Choice of a Model

32.1. The phenomenon of superconductivity. One of the most important and difficult problems of quantum statistics is the problem of superconductivity. As is well known, at sufficiently low temperatures, many metals undergo a phase transition to a new, "superconducting" state (or phase). In this state, the thermodynamic and electric properties of the metal are radically different from its properties in the normal state. Perhaps the most striking manifestation of this is the fact that when a metal is cooled down to a certain critical temperature, it suddenly ceases to exhibit any resistance to electric current, i.e., no energy dissipation takes place when current flows in a superconductor. The transition from the normal state to the superconducting state is a phase transition of the second kind, characterized by the fact that the heat capacity of the metal has a discontinuity at the transition temperature.

Experiments have also shown that the magnetic properties of a superconductor are radically different from the comparatively simple properties of a normal metal. In fact, a magnetic current cannot effectively penetrate a bulk superconductor (the *Meissner-Ochsenfeld effect*). Thus, if a superconductor is placed in a constant magnetic field, the *penetration depth*, i.e., the effective depth (measured from the surface of the superconductor) of the region where the magnetic field differs appreciably from zero, is quite small ($\sim 10^{-5}$ to 10^{-6} cm).

In recent years, remarkable success has been achieved in explaining the phenomenon of superconductivity. It turns out that to develop a theory of superconductivity, we have to make extensive use of the methods of quantum field theory. These methods will be discussed in detail, as soon as we have said a bit more about the physical aspects of the problem.

For quite some time, it has been clear that there is a close resemblance between the phenomena of superconductivity and superfluidity. This is immediately apparent from the fact that no external potential difference is required to maintain an electric current in a superconductor, i.e., the external sources are not required to do any work. Thus, in the sense that the electrons are the carriers of electric current in a metal, superconductivity is a kind of superfluidity of the electron liquid. In treating the superfluidity of liquid helium (see Chap. 1), we discussed in detail the properties of the energy spectrum of the excitations that are necessary for superfluidity to occur. However, it should be pointed out at once that for small momenta, the spectrum of a superconductor cannot have the form which is appropriate for liquid helium. In fact, the spectrum of liquid helium starts with an acoustic (i.e., phonon) branch, and, as is well known, the propagation of sound involves long-wavelength density oscillations. However, density changes in the electron liquid in a metal demand a rather large amount of energy, since such changes are hindered by the Coulomb forces acting between the electrons and the lattice, and between the electrons themselves. Since changing the density of the electron liquid violates the condition of electrical neutrality, the corresponding spectrum of long-wavelength oscillations begins at some finite frequency (as in the case of a plasma), which in a metal is actually quite large (~ 1 ev $\sim 10^4$ deg K). Of course, this argument does not apply to the short-wavelength excitations, whose wave vectors have magnitudes of the order of the reciprocal of the interatomic distance, and as we know, it is just these electronic excitations that play the main role in a normal metal. According to the results of Chap. 1, a sufficient condition for superfluidity to occur is that these short-wavelength excitations be separated from the ground state by a *gap*, i.e., that the spectrum have the form shown in Fig. 86. We note that apart from these considerations, the fact that superconductors have spectra of the indicated type is implied by experimental data on the electronic part of the heat capacity at low temperatures, which show that the dependence of the heat capacity on the temperature has the form $e^{-\Delta/T}$.

We have no intention of discussing the various phenomenological theories of superconductivity, which in some cases have been rather successful in explaining the experimental data. Instead, we merely note that none of these theories has been able to clarify the microscopic mechanism of the phenomenon.

The discovery in 1950 of the *isotope effect* (see M3, M4, R1) was the key to an understanding of the relative importance of the various interactions in a metal in contributing to the phenomenon of superconductivity. It was found that the *critical temperature* T_c (i.e., the temperature at which the

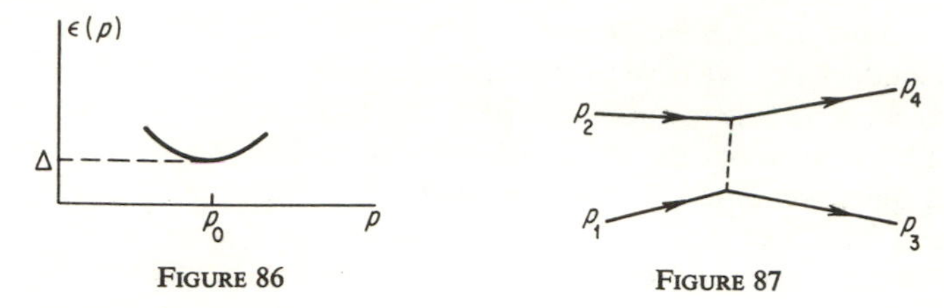

FIGURE 86 FIGURE 87

metal undergoes the transition from the normal to the superconducting state) of the various isotopes of a metal obeys the law $T_c \propto M^{-1/2}$, where M is the mass of the given isotope. Independently, Fröhlich (F3) conjectured that the interaction between electrons and phonons is the basic mechanism responsible for superconductivity (a mechanism which clearly depends on the mass of the ions in an essential way).

32.2. The model. The interaction Hamiltonian. As we have already seen on p. 78, in the present case the interaction Hamiltonian has the form

$$H_{\text{int}} = g \int \psi_\alpha^+(\mathbf{r})\psi_\alpha(\mathbf{r})\varphi(\mathbf{r})\, d\mathbf{r}. \tag{32.1}$$

We want to calculate the matrix element for electron-electron scattering in which two electrons exchange one phonon. This process is shown schematically in Fig. 87, where the dashed line represents the exchange of one phonon. In the matrix element, this line corresponds to the phonon D-function

$$g^2 D(\mathbf{p}_3 - \mathbf{p}_1, \varepsilon_3 - \varepsilon_1) = g^2 \frac{u^2(\mathbf{p}_3 - \mathbf{p}_1)^2}{(\varepsilon_3 - \varepsilon_1)^2 - u^2(\mathbf{p}_3 - \mathbf{p}_1)^2}, \tag{32.2}$$

where $\mathbf{p}_3 - \mathbf{p}_1$ and $\varepsilon_3 - \varepsilon_1$ are the momentum change and energy change, respectively, of one of the colliding electrons. Near the Fermi surface, the momentum change due to the collision is generally of order p_0 (i.e., $u|\mathbf{p}_3 - \mathbf{p}_1|$ is of the order of the Debye frequency ω_D, since $p_0 \sim a^{-1}$). However, the energy change can be small, i.e., $|\varepsilon_3 - \varepsilon_1| \ll \omega_D$, and then the effective interaction, determined by the expression (32.2), reduces simply to the constant $-g^2$ and hence is attractive.

It was shown in 1956 by Cooper (C3) that the effective attraction between electrons near the Fermi surface, due to electron-phonon interaction, must lead to bound pairs of electrons, regardless of how weak the attraction may be. Since formation of pairs is "energetically favorable," turning on the interaction causes the ground state of the system to be "rebuilt." The amount of energy needed to excite the system then equals the energy of formation of a bound pair, which acts as a gap in the excitation spectrum. With this idea as the starting point, it has been possible to construct a complete theory of superconductivity, which explains the abundant data accumulated over several decades of intensive study of the phenomenon. We have chosen a formulation of the theory which differs from the original versions (see B1, B7, B8), and is based on the use of the methods of quantum field theory. In our opinion, this approach has many advantages; not only is it simple and elegant, but it also makes it possible to obtain a number of important new results.

Before going on, we note that the electron-phonon interaction is not the only interaction between electrons in a metal, and in fact, the colliding electrons experience Coulomb forces. Therefore, the effective interaction between the electrons will be either attractive or repulsive, depending on the ratio between the sizes of the electron-phonon attraction and the Coulomb repulsion. In its general form, the problem of taking both interactions into account for actual metals is very complicated, especially since real superconductors are anisotropic. In this regard, it should be pointed out that the contemporary theory of superconductivity deals primarily with a simple model where the electrons have a quadratic dispersion law and where it is postulated in advance that the interaction between the electrons, whose energies lie in a narrow region near the Fermi surface,[1] is attractive in nature. Moreover, in the treatment given below, it will be assumed (for simplicity) that the interaction between the electrons is constant and sufficiently weak in this energy region.

So far, no theory of superconductivity has yet been constructed which is based on concepts of the Fermi liquid and also takes anisotropy into account. However, it is interesting to note that despite the crudeness of the present model, the theory not only explains the phenomenon qualitatively, but also leads to good quantitative agreement with the available experimental data.

Having made these remarks, we now write the effective Hamiltonian for electron-electron interaction in the second-quantized representation:

$$H_{\text{int}} = \frac{\lambda}{2(2\pi)^3} \sum_{\mathbf{p}_1+\mathbf{p}_2=\mathbf{p}_3+\mathbf{p}_4} a^+_{\mathbf{p}_1\sigma_1} a^+_{\mathbf{p}_2\sigma_2} a_{\mathbf{p}_3\sigma_2} a_{\mathbf{p}_4\sigma_1} \theta_{\mathbf{p}_1}\theta_{\mathbf{p}_2}\theta_{\mathbf{p}_3}\theta_{\mathbf{p}_4}. \tag{32.3}$$

Here $\lambda < 0$, and the $\theta_{\mathbf{p}}$ are cutoff factors, i.e.,

$$\theta_{\mathbf{p}} = \begin{cases} 1 & \text{for} \quad |\varepsilon(\mathbf{p}) - \varepsilon_F| < \omega_D, \\ 0 & \text{for} \quad |\varepsilon(\mathbf{p}) - \varepsilon_F| > \omega_D, \end{cases}$$

whose presence means that only electrons with energies in a narrow range of width $2\omega_D$ near the Fermi surface ($\omega_D \ll \varepsilon_F$) participate in the interaction. Below, we shall often write this Hamiltonian in terms of the operators $\psi_\alpha(\mathbf{r})$ and $\psi^+_\beta(\mathbf{r})$ in the coordinate representation:

$$H_{\text{int}} = \frac{\lambda}{2} \int \psi^+_\alpha(\mathbf{r})\psi^+_\beta(\mathbf{r})\psi_\beta(\mathbf{r})\psi_\alpha(\mathbf{r})\, d\mathbf{r}. \tag{32.4}$$

Of course, in (32.4) it is understood that the values of the four arguments of the ψ-operators are actually distinct, in keeping with the presence of the

[1] Obviously, the width of this region is of the order of the maximum energy of the emitted phonons, i.e., of order ω_D, where ω_D is the Debye frequency.

factors $\theta_{\mathbf{p}}$ in the expression (32.3) for the Hamiltonian. Thus, a more exact way of writing (32.4) would be

$$H_{\text{int}} = \frac{\lambda}{2}\int \theta(\mathbf{r}-\boldsymbol{\xi}_1)\theta(\mathbf{r}-\boldsymbol{\xi}_2)\theta(\mathbf{r}-\boldsymbol{\xi}_3)\theta(\mathbf{r}-\boldsymbol{\xi}_4) \times \psi_\alpha^+(\boldsymbol{\xi}_1)\psi_\beta^+(\boldsymbol{\xi}_2)\psi_\beta(\boldsymbol{\xi}_3)\psi_\alpha(\boldsymbol{\xi}_4)\,d\mathbf{r}\,d\boldsymbol{\xi}_1\,d\boldsymbol{\xi}_2\,d\boldsymbol{\xi}_3\,d\boldsymbol{\xi}_4, \tag{32.5}$$

where $\theta(\mathbf{x})$ is the Fourier transform of $\theta_{\mathbf{p}}$:

$$\theta(\mathbf{x}) = \frac{1}{(2\pi)^3}\int e^{i\mathbf{p}\cdot\mathbf{x}}\,\theta_{\mathbf{p}}\,d\mathbf{p}. \tag{32.6}$$

By going over to the Fourier representation, it is easily verified that the function $\theta(\mathbf{x})$ behaves like a delta function, i.e.,

$$\int \theta(\mathbf{x}-\mathbf{y})f(\mathbf{y})\,d\mathbf{y} = f(\mathbf{x})$$

if the function $f(\mathbf{x})$ has nonzero Fourier components $f_{\mathbf{p}}$ only for momenta $\mathbf{p}$ near the Fermi surface, and in the theory given here we shall deal with just such functions. All this explains what is meant by (32.4).

33. The Cooper Phenomenon. Instability of the Ground State of a System of Noninteracting Fermions with Respect to Arbitrarily Weak Attraction between the Particles

33.1. The equation for the vertex part. We now consider the properties of a system with the interaction (32.4), by studying the vertex part $\Gamma_{\alpha\beta,\gamma\delta}(p_1, p_2; p_3, p_4)$ at absolute zero. Writing the perturbation series for Γ, we find that to a first approximation it equals

$$\lambda(\delta_{\alpha\gamma}\delta_{\beta\delta} - \delta_{\alpha\delta}\delta_{\gamma\delta}). \tag{33.1}$$

The diagrams corresponding to the second term in the perturbation series are shown in Fig. 88. As we know, diagrams (a) and (c) are associated with singularities in the vertex part of the "zero-sound" type, i.e., these singularities are important for small momentum transfer. On the other hand, diagrams of the form (b) are associated with singularities in $\Gamma_{\alpha\beta,\gamma\delta}(p_1, p_2; p_3, p_4)$ for small values of the total four-momentum $q = p_1 + p_2$.

Let us examine this last case further. Using the specific properties of our model, we can obtain more detailed information about the vertex part in the

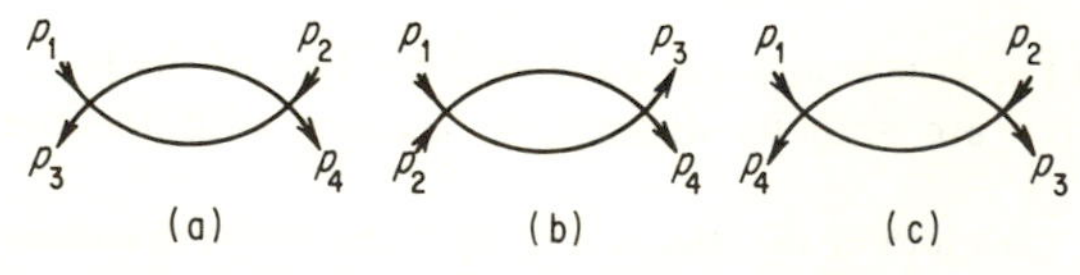

FIGURE 88

region of small q than is given by the general result (20.8). The matrix element for the diagram shown in Fig. 88(b) equals

$$\lambda^2 \frac{i}{(2\pi)^4} (\delta_{\alpha\gamma}\delta_{\beta\delta} - \delta_{\alpha\delta}\delta_{\gamma\beta}) \int G(k)G(q-k)\, d^4k,$$

where

$$q = (\mathbf{q}, \omega_0) = (\mathbf{p}_1 + \mathbf{p}_2, \omega_1 + \omega_2).$$

Replacing the Green's functions by their explicit expressions, and integrating by parts, we obtain

$$\begin{aligned} &\frac{\lambda^2}{(2\pi)^3} (\delta_{\alpha\gamma}\delta_{\beta\delta} - \delta_{\alpha\delta}\delta_{\beta\gamma}) \int \frac{d\mathbf{k}}{\omega_0 - \varepsilon_0(\mathbf{k}) - \varepsilon_0(\mathbf{q}-\mathbf{k}) + 2\mu + i\delta} \\ &\qquad \text{for} \quad \varepsilon_0(\mathbf{k}) > \mu, \quad \varepsilon_0(\mathbf{q}-\mathbf{k}) > \mu, \\ &-\frac{\lambda^2}{(2\pi)^3} (\delta_{\alpha\gamma}\delta_{\beta\delta} - \delta_{\alpha\delta}\delta_{\beta\gamma}) \int \frac{d\mathbf{k}}{\omega_0 - \varepsilon_0(\mathbf{k}) - \varepsilon_0(\mathbf{q}-\mathbf{k}) + 2\mu - i\delta} \\ &\qquad \text{for} \quad \varepsilon_0(\mathbf{k}) < \mu, \quad \varepsilon_0(\mathbf{q}-\mathbf{k}) < \mu. \end{aligned} \tag{33.2}$$

In the model under discussion, only electrons in a narrow range near the Fermi energy $\varepsilon_F \approx \mu$ interact, and hence the integration with respect to $\mathbf{k}$ in the integrals (33.2) is restricted by the conditions

$$|\varepsilon_0(\mathbf{k}) - \mu| < \omega_D, \qquad |\varepsilon_0(\mathbf{q}-\mathbf{k}) - \mu| < \omega_D.$$

Assuming that $v|\mathbf{q}|$, $\omega_0 \ll \omega_D$, we go over to an integration with respect to $\xi = v(|\mathbf{k}| - p_0)$, in the usual way. Then, neglecting changes of order $v|\mathbf{q}|$, ω_0 in the upper limits of integration, we transform (33.2) into

$$\begin{aligned} &-\frac{\lambda^2 m p_0}{2\pi^2} (\delta_{\alpha\gamma}\delta_{\beta\delta} - \delta_{\alpha\delta}\delta_{\beta\gamma}) \\ &\qquad \times \int_0^{\omega_D} d\xi \int_0^1 \left[\frac{1}{\omega_0 + 2\xi + v|\mathbf{q}|x - i\delta} + \frac{1}{2\xi + v|\mathbf{q}|x - \omega_0 - i\delta}\right] dx, \end{aligned}$$

where $x = \cos\theta$ (θ is the angle between the vectors $\mathbf{q}$ and $\mathbf{k}$) and the remaining integrations are completely elementary. In this way, we find that diagram (b) of Fig. 88 corresponds to the expression

$$\begin{aligned} -\frac{\lambda^2 m p_0}{2\pi^2} (\delta_{\alpha\gamma}\delta_{\beta\delta} - \delta_{\alpha\delta}\delta_{\beta\gamma}) \Bigg[1 &+ \frac{1}{2} \ln \frac{2\omega_D - i\delta}{\omega_0 + v|\mathbf{q}| - i\delta} \\ &+ \frac{1}{2} \ln \frac{2\omega_D - i\delta}{-\omega_0 + v|\mathbf{q}| - i\delta} \\ &+ \frac{\omega_0}{2v|\mathbf{q}|} \left(\ln \frac{\omega_0 - i\delta}{\omega_0 + v|\mathbf{q}| - i\delta} + \ln \frac{-\omega_0 + v|\mathbf{q}| - i\delta}{-\omega_0 - i\delta} \right) \Bigg], \end{aligned} \tag{33.3}$$

where in choosing the branches of the logarithm, we have used the condition that the integral of the first term in brackets is positive for $\omega_0 > 0$, while the integral of the second term is positive for $\omega_0 < 0$.

For small $v|\mathbf{q}|$ and ω_0, the principal term in (33.3) has the form

$$-\lambda^2 \frac{mp_0}{2\pi^2}(\delta_{\alpha\gamma}\delta_{\beta\delta} - \delta_{\alpha\delta}\delta_{\beta\gamma}) \ln \frac{\omega_D}{\max\{\omega_0, v|\mathbf{q}|\}}.$$

Therefore, for $\omega_D \gg \omega_0$, $\omega_D \gg v|\mathbf{q}|$, the fact that the interaction parameter λ is small can be compensated by the fact that the logarithm is large, and as a result, this term becomes of the same order of magnitude as the first term (33.1) of the perturbation series. Thus, to obtain the vertex part in the region of small $v|\mathbf{q}|$ and ω_0, where

$$\lambda \ln \frac{\omega_D}{\max\{\omega_0, v|\mathbf{q}|\}} \sim 1,$$

we have to sum all the principal terms of the perturbation series, as in Chap. 4. With this in mind, we write the equation for the vertex part in a form such that the terms leading to the singularities of $\Gamma_{\alpha\beta,\gamma\delta}(p_1, p_2; p_3, p_4)$ for small $q = p_1 + p_2$ are isolated:

$$\begin{aligned}\Gamma_{\alpha\beta,\gamma\delta}(p_1, p_2; p_3, p_4) = \tilde{\Gamma}_{\alpha\beta,\gamma\delta}(p_1, p_2; p_3, p_4) \\ + \frac{i}{2(2\pi)^4}\int \tilde{\Gamma}_{\alpha\beta,\xi\eta}(p_1, p_2; k, q - k) \\ \times G(k)G(q-k)\Gamma_{\xi\eta,\gamma\delta}(k, q-k; p_3, p_4)\, d^4k.\end{aligned} \tag{33.4}$$

Here, $\tilde{\Gamma}_{\alpha\beta,\gamma\delta}(p_1, p_2; p_3, p_4)$ is the sum of all matrix elements whose diagrams are irreducible in the sense that they cannot be divided into two parts which are joined by two electron lines, and are such that one part contains only the two ingoing external lines while the other part contains only the two outgoing external lines.

The kernel of the integral equation (33.4) contains a large logarithmic term which comes from integrating the two Green's functions. Because of the smallness of the interaction parameter, we can replace $\tilde{\Gamma}$ by the first nonvanishing terms in its perturbation series, since they do not contain large quantities. However, it is our aim to calculate the kernel of equation (33.4) without restricting ourselves to terms of order $\lambda \ln(\omega_D/\omega_0) \sim 1$, and in fact we shall try to find an expression for the kernel up to terms of order λ. At first glance, it seems that to do this we need to know the quantity $\tilde{\Gamma}$ up to terms of order λ^2 in the perturbation series, since the logarithmic interaction in (33.4) can compensate one power of λ. Thus, we examine the second-order terms in the perturbation series for $\tilde{\Gamma}$, where the corresponding diagrams are shown in Fig. 89. Considering diagram (a), for example, we find that its matrix element is

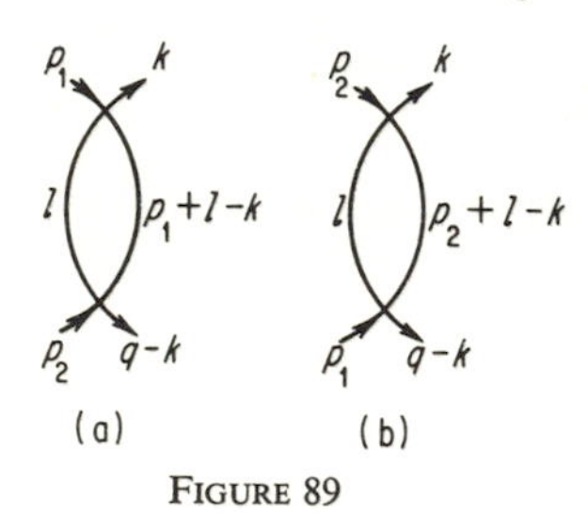

FIGURE 89

$$\lambda^2 \int G(l)G(l - k + p_1)\, d^4l,$$

except for numerical factors. Replacing the Green's functions by their explicit expressions, and integrating by parts, we obtain

$$\lambda^2 \int \frac{dl}{\omega_k - \omega_1 + \varepsilon_0(\boldsymbol{l} - \mathbf{k} + \mathbf{p}_1) - \varepsilon_0(\boldsymbol{l})} \quad \text{for} \quad \varepsilon_0(\boldsymbol{l}) > \mu, \quad \varepsilon_0(\boldsymbol{l} - \mathbf{k} + \mathbf{p}_1) < \mu,$$

$$-\lambda^2 \int \frac{dl}{\omega_k - \omega_1 + \varepsilon_0(\boldsymbol{l} - \mathbf{k} + \mathbf{p}_1) - \varepsilon_0(\boldsymbol{l})} \quad \text{for} \quad \varepsilon_0(\boldsymbol{l}) < \mu, \quad \varepsilon_0(\boldsymbol{l} - \mathbf{k} + \mathbf{p}_1) > \mu. \tag{33.5}$$

However, the regions of integration with respect to $\boldsymbol{l}$ are actually much narrower than if they were determined only by the indicated inequalities. This is connected with the properties of our model, where the only electrons which can interact are those with momenta near the Fermi momentum $|v(|\mathbf{p}| - p_0)| < \omega_D$. The region of integration associated with the first expression in (33.5) is represented by the shaded areas in Fig. 90, and the region of integration associated with the second expression is represented by the dark areas.[2] In both cases, the value of the difference $\varepsilon(\boldsymbol{l} - \mathbf{k} + \mathbf{p}_1) - \varepsilon(\boldsymbol{l})$ in the denominator of the integrand equals ω_D in the region of integration, which has a volume of order $m^2\omega_D^2/p_0$. Therefore, the matrix elements for the diagrams shown in Fig. 89 are of order $\lambda^2(m^2\omega_D/p_0)$, i.e., the ratio of their contribution to that from a simple vertex is of order $\lambda m p_0(\omega_D/\varepsilon_F)$. [In the present model, $\lambda m p_0$ ($\ll 1$) is a small dimensionless parameter, as can be seen from (33.3)]. Since the quantity ω_D is $\ll \varepsilon_F$, because of its physical meaning, this extra smallness cannot be compensated in the given regions by the largeness of the logarithm. It follows that in equation (33.4) for the quantity Γ, we can confine ourselves to the simple first-order vertex (33.1).

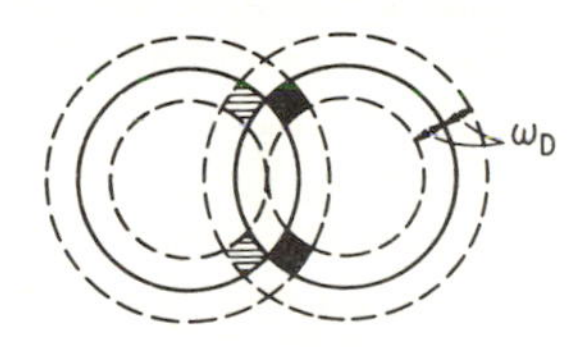

FIGURE 90

The resulting equation for the vertex part can now be solved easily. In fact, we note that according to (33.3), $\Gamma_{\alpha\beta,\gamma\delta}(p_1, p_2; p_3, p_4)$ depends only on the sum of variables $q = p_1 + p_2$. Therefore, the integral in the right-hand side of (33.4) reduces to the integral in the expression for the second-order matrix elements. Since this integral equals (33.3), we finally have

$$\Gamma_{\alpha\beta,\gamma\delta}(p_1, p_2; p_3, p_4) \equiv \Gamma(q)(\delta_{\alpha\gamma}\delta_{\beta\delta} - \delta_{\alpha\delta}\delta_{\beta\gamma}),$$

$$\Gamma(q) = \lambda\left\{1 + \left(\frac{\lambda m p_0}{2\pi^2}\right)\left[1 + \ln\left|\frac{2\omega_D}{\omega_0}\right| + \frac{\pi i}{2} + \frac{1}{2}\ln\left|\frac{\omega_0^2}{\omega_0^2 - v^2|\mathbf{q}|^2}\right| + \frac{\omega_0}{2v|\mathbf{q}|}\ln\left|\frac{\omega_0 - v|\mathbf{q}|}{\omega_0 + v|\mathbf{q}|}\right|\right]\right\}^{-1}. \tag{33.6}$$

33.2. Properties of the vertex part. For simplicity, we first examine the expression (33.6) when $\mathbf{q} = 0$. If ω_0 is real and positive, we have

$$\Gamma(\omega_0) = \frac{\lambda}{1 + \left(\dfrac{\lambda m p_0}{2\pi^2}\right)\left[\ln\left|\dfrac{2\omega_D}{\omega_0}\right| + \dfrac{\pi i}{2}\right]}. \tag{33.7}$$

---

[2] Note that $|\mathbf{k} - \mathbf{p}_1| < 2p_0$, since otherwise there is no region of integration at all.

Next, we consider $\Gamma(\omega_0)$ as a function of the complex variable ω_0, defining it as the analytic continuation of (33.7) into the upper half-plane $\mathrm{Im}\,\omega_0 > 0$. We then have

$$\Gamma(\omega_0) = \frac{\lambda}{1 + \left(\dfrac{\lambda m p_0}{2\pi^2}\right)\left[\ln\left|\dfrac{2\omega_D}{\omega_0}\right| + \dfrac{\pi i}{2} - i\varphi\right]}.$$

Thus, if the interaction is attractive ($\lambda < 0$), the quantity $\Gamma(\omega_0)$ has a pole at the point $\omega_0 = i\varpi$, where

$$\varpi = 2\omega_D e^{-2\pi^2/|\lambda| m p_0}. \tag{33.8}$$

In the neighborhood of the pole, $\Gamma(\omega_0)$ has the form

$$\Gamma(\omega_0) = -\frac{2\pi^2}{m p_0}\frac{i\varpi}{\omega_0 - i\varpi}.$$

The connection between this result and Cooper's ideas concerning the formation of bound electron pairs (see Sec. 32.2) should be noted. The vertex part $\Gamma_{\alpha\beta,\gamma\delta}(p_1, p_2; p_3, p_4)$ is related to the Fourier components of the two-particle Green's function by formula (10.17). Therefore, the presence of a pole in the quantity Γ implies that the same pole is present in the two-particle Green's function. The fact that pairs are formed means that the ground state of the gas of noninteracting fermions (which is our starting point) is unstable, and application of arbitrarily weak attractive forces between the particles leads to "rebuilding" of the whole system. The existence of this instability is reflected by the fact that the vertex part, regarded as a function of the complex variable $\omega_0 = \omega_1 + \omega_2$, has a pole in the upper half-plane. This pole, being purely imaginary, determines the relaxation time of the unstable ground state. Moreover, because of the uncertainty principle, this time corresponds to the binding energy of the actual pair. The pairs in the rebuilt ground state behave like "Bose structures", and any number of them can accumulate on the level of least energy, just as in the case of ordinary bosons. In the superconducting state, these pairs lie on a level for which the momentum of the pair as a whole is zero, in complete analogy with the phenomenon of "Bose condensation" for ordinary bosons.

When $v|\mathbf{q}|$ is nonzero, the expression (33.6) can be written in the following form ($\omega_0 > v|\mathbf{q}|$):

$$\Gamma(\mathbf{q}, \omega_0) = \lambda\left\{1 + \left(\frac{\lambda m p_0}{2\pi^2}\right)\left[1 + \ln\left|\frac{2\omega_D}{\omega_0}\right| + \frac{\pi i}{2} - \frac{1}{2}\ln\left(1 - \frac{v^2|\mathbf{q}|^2}{\omega_0^2}\right) + \frac{\omega_0}{2v|\mathbf{q}|}\ln\left(\frac{\omega_0 - v|\mathbf{q}|}{\omega_0 + v|\mathbf{q}|}\right)\right]\right\}^{-1}.$$

After making the analytic continuation into the half-plane $\mathrm{Im}\,\omega_0 > 0$ and using the definition (33.8) of ϖ, we find that

$$\Gamma(\mathbf{q}, \omega_0) = -\frac{2\pi^2}{m p_0}\left\{\ln\frac{\omega_0}{i\varpi} - 1 + \frac{1}{2}\ln\left(1 - \frac{v^2|\mathbf{q}|^2}{\omega_0^2}\right) - \frac{\omega_0}{2v|\mathbf{q}|}\ln\left(\frac{\omega_0 - v|\mathbf{q}|}{\omega_0 + v|\mathbf{q}|}\right)\right\}^{-1}. \tag{33.9}$$

For small $v|\mathbf{q}| \ll \varpi$, we have

$$\Gamma(\mathbf{q}, \omega_0) = -\frac{2\pi^2}{mp_0} \frac{i\varpi}{\omega_0 - i\varpi + i(v^2|\mathbf{q}|^2/6\varpi)}.$$

The pole of $\Gamma(\mathbf{q}, \omega_0)$ as a function of $|\mathbf{q}|$ occurs when the denominator of (33.9) vanishes. For small $v|\mathbf{q}|$, this takes place for

$$\omega_0 = i\varpi\left(1 - \frac{v^2|\mathbf{q}|^2}{6\varpi^2}\right),$$

i.e., the absolute value of ω_0 is decreased. For a certain value $v|\mathbf{q}|_{\max}$, which is easily found to be

$$v|\mathbf{q}|_{\max} = e\,\varpi, \tag{33.10}$$

the pole occurs for $\omega_0 = 0$, and then Γ has no pole at all for larger values of $v|\mathbf{q}|$. Since $\mathbf{q}$ is the total momentum of a system of two particles, this means that only those electrons which are moving with almost opposite momenta exhibit a tendency to form pairs.

33.3. Determination of the critical temperature. We note once again that the above considerations testify to the instability of the ground state of a system of particles subject to attractive forces at low temperatures. This instability stems from the fact that two particles whose center of mass is almost at rest can form a bound pair. The bound pairs behave like bosons which "condense" on the level of lowest energy. The temperature at which this instability first appears is just the temperature at which the metal goes from the normal to the superconducting state. To determine this critical temperature, we can use our analogy with the Bose gas. In the weak-interaction model, i.e., to the approximation in which scattering of the particles by each other is neglected, the bound pairs form an ideal gas. As we know, the temperature Green's function of an ideal Bose gas has the form

$$\mathscr{G}(\mathbf{q}, i\omega_n) = \left(i\omega_n - \frac{\mathbf{q}^2}{2m} + \mu\right)^{-1} \tag{33.11}$$

representing the values at the points $\omega = i\omega_n = i2n\pi T$ of a function $G^R(\mathbf{q}, \omega)$ which is analytic in the upper half-plane of the complex variable ω. For $\omega_n = 0$, this quantity equals

$$\left(\mu - \frac{\mathbf{q}^2}{2m}\right)^{-1}. \tag{33.12}$$

For some temperature $T = T_0$, called the temperature of "Bose condensation" and determined by the condition $\mu = 0$, (33.12) first becomes infinite at the point $\mathbf{q} = 0$.

For a bound pair, the analog of this boson Green's function is the two-particle fermion Green's function (16.5), which, regarded as a function of the variables

$$\mathbf{q} = \mathbf{p}_1 + \mathbf{p}_2, \qquad \omega_{0n} = (\omega_1 + \omega_2)_n \tag{33.13}$$

corresponding to the center of mass of the pair, must have the same analytic properties as (33.11) at the transition point. The one-particle fermion Green's functions in the expression (16.5) have no singularities at all in the variables (33.13). Therefore, we consider the vertex part $\Gamma_T(\mathbf{q}, \omega_0)$,[3] defining it as the analytic function (in the upper half-plane of the complex variable ω_0) which has the same values at the points $\omega_0 = i(\omega_1 + \omega_2)_n$ as the vertex part in the temperature technique, i.e., $\Gamma_T(\mathbf{q}, \omega_0)$ is the analytic continuation of the temperature vertex part

$$\mathcal{T}_{\alpha\beta,\gamma\delta}(\mathbf{p}_1, \omega_1, \mathbf{p}_2, \omega_2; \mathbf{p}_3, \omega_3, \mathbf{p}_4, \omega_4) \equiv \Gamma_T(\mathbf{q}, \omega_0)(\delta_{\alpha\gamma}\delta_{\beta\delta} - \delta_{\alpha\delta}\delta_{\beta\gamma}).$$

(We shall see later that to the approximation of interest, this quantity, like (33.9), actually depends only on the variables $\mathbf{q}$ and ω_0.) Then, on the basis of the considerations just given, we assume that $\Gamma_T(\mathbf{q}, \omega_0)$ has a pole in the half-plane $\operatorname{Im} \omega_0 > 0$ at temperatures below the critical temperature, while at the critical temperature itself, $\Gamma_T(\mathbf{q}, \omega_0)$ first has a pole for $\omega_0 = 0$.

The required equation for the temperature vertex part has the same structure as (33.4), i.e.,

$$\begin{aligned} \mathcal{T}_{\alpha\beta,\gamma\delta}&(\mathbf{p}_1, \omega_1, \mathbf{p}_2, \omega_2; \mathbf{p}_3, \omega_3, \mathbf{p}_4, \omega_4) \\ &= \tilde{\mathcal{T}}_{\alpha\beta,\gamma\delta}(\mathbf{p}_1, \omega_1, \mathbf{p}_2, \omega_2; \mathbf{p}_3, \omega_3, \mathbf{p}_4, \omega_4) \qquad (33.14) \\ &\quad - \frac{T}{2(2\pi)^3} \sum_{\omega'} \int \tilde{\mathcal{T}}_{\alpha\beta,\xi\eta}(\mathbf{p}_1, \omega_1, \mathbf{p}_2, \omega_2; \mathbf{k}, \omega'; \mathbf{q} - \mathbf{k}, \omega_0 - \omega') \\ &\quad \times \mathcal{G}(k)\mathcal{G}(q - k)\mathcal{T}_{\xi\eta,\gamma\delta}(\mathbf{k}, \omega', \mathbf{q} - \mathbf{k}, \omega_0 - \omega'; \mathbf{p}_3, \omega_3, \mathbf{p}_4, \omega_4)\, d\mathbf{k}, \end{aligned}$$

where $\tilde{\mathcal{T}}$ is again the sum of the matrix elements of all diagrams which cannot be separated (by making vertical cuts) into two parts joined by two identically directed lines. For the same reasons as before, we can restrict ourselves to the first-order terms in the perturbation series for $\tilde{\mathcal{T}}_{\alpha\beta,\gamma\delta}(\mathbf{p}_1, \omega_1, \mathbf{p}_2, \omega_2; \mathbf{p}_3, \omega_3, \mathbf{p}_4, \omega_4)$. The problem of finding the vertex part then reduces to evaluating the sum and integral in the matrix element

$$-\frac{\lambda^2}{(2\pi)^3} T \sum_{\omega'} \int \mathcal{G}(k)\mathcal{G}(q - k)\, d\mathbf{k}. \qquad (33.15)$$

Substituting the Green's functions (14.6) into (33.15), we obtain an elementary sum with respect to ω'_n, which is easily calculated.

We shall not evaluate (33.15) for arbitrary values of $|\mathbf{q}|$, ω_0, since it is clear from homogeneity considerations that a pole first appears in $\Gamma_T(\mathbf{q}, \omega_0)$ when $|\mathbf{q}| = \omega_0 = 0$, just as in the Bose gas. Therefore, we need only find the solution of (33.14) for $|\mathbf{q}| = \omega_0 = 0$, and the point at which this solution becomes infinite determines the temperature of the transition from the normal to the superconducting state. For $|\mathbf{q}| = \omega_0 = 0$, we can write the integral in (33.15) as

$$-\frac{\lambda^2}{2\pi^2} mp_0 \int_0^{\omega_D} \tanh\left(\frac{\xi}{2T}\right)\frac{d\xi}{\xi} = -\lambda^2 \frac{mp_0}{2\pi^2}\left(\ln\frac{\omega_D}{2T} - \int_0^\infty \frac{\ln x\, dx}{\cosh^2 x}\right). \qquad (33.16)$$

---

[3] Spin variables will be omitted everywhere.

(After integrating by parts, the remaining integral converges, and hence the limit $x = \omega_D/2T$ can be replaced by infinity.) The integral in the right-hand side of (33.16) equals $-\ln(2\gamma/\pi)$, where $\ln\gamma = C \approx 0.577$ is Euler's constant, and hence

$$\Gamma_T(0, 0) = \mathscr{T}(0, 0) = \frac{\lambda}{1 + \dfrac{\lambda m p_0}{2\pi^2}\ln\left(\dfrac{2\gamma\omega_D}{\pi T}\right)}.$$

Near the critical temperature, this expression can be written as

$$\mathscr{T}(0, 0) = -\frac{2\pi^2}{m p_0}\frac{T_c}{T - T_c}, \tag{33.17}$$

where the critical temperature T_c equals

$$T_c = \frac{\gamma}{\pi}2\omega_D e^{-2\pi^2/|\lambda| m p_0}. \tag{33.18}$$

Moreover,

$$\varpi = \frac{\pi}{\gamma}T_c, \tag{33.19}$$

where ϖ is the quantity defined by (33.8), characterizing the instability of the system at absolute zero.

34. The Basic System of Equations for a Superconductor

34.1. A superconductor at absolute zero. We now derive a system of equations for the Green's functions describing the properties of a metal in the superconducting state [see Gorkov (G6)], restricting ourselves at first to the case of zero temperature. In the present model, the total Hamiltonian of the system of electrons in the second-quantized representation has the form

$$\hat{H} = \int\left[-\left(\psi^+\frac{\nabla^2}{2m}\psi\right) + \frac{\lambda}{2}(\psi^+(\psi^+\psi)\psi)\right]d\mathbf{r},$$

where $(\psi^+\psi) \equiv \psi_\alpha^+\psi_\alpha$, and the operators $\psi(\mathbf{r})$, $\psi^+(\mathbf{r})$ in the Schrödinger representation satisfy the usual commutation relations

$$\begin{aligned} \{\psi_\alpha(\mathbf{r}), \psi_\beta^+(\mathbf{r}')\} &= \delta_{\alpha\beta}\delta(\mathbf{r} - \mathbf{r}'), \\ \{\psi_\alpha(\mathbf{r}), \psi_\beta(\mathbf{r}')\} &= \{\psi_\alpha^+(\mathbf{r}), \psi_\beta^+(\mathbf{r}')\} = 0. \end{aligned} \tag{34.1}$$

Next, we go over to the Heisenberg representation, in which the operators $\tilde{\psi}$ and $\tilde{\psi}^+$ depend on time and obey the following operator equations:

$$\begin{aligned} \left(i\frac{\partial}{\partial t} + \frac{\nabla^2}{2m}\right)\tilde{\psi}_\alpha(x) - \lambda(\tilde{\psi}^+(x)\tilde{\psi}(x))\tilde{\psi}_\alpha(x) &= 0, \\ \left(i\frac{\partial}{\partial t} - \frac{\nabla^2}{2m}\right)\tilde{\psi}_\alpha^+(x) + \lambda\tilde{\psi}_\alpha^+(x)(\tilde{\psi}^+(x)\tilde{\psi}(x)) &= 0. \end{aligned} \tag{34.2}$$

Then an equation for the Green's function

$$G_{\alpha\beta}(x, x') = -i\langle T(\tilde{\psi}_{\alpha}(x)\tilde{\psi}^{+}_{\beta}(x'))\rangle$$

of the system can be obtained from (34.2), in the obvious way:

$$\left(i\frac{\partial}{\partial t} + \frac{\nabla^2}{2m}\right)G_{\alpha\beta}(x, x') + i\lambda\langle T((\tilde{\psi}^{+}(x)\tilde{\psi}(x))\tilde{\psi}_{\alpha}(x)\tilde{\psi}^{+}_{\beta}(x'))\rangle = \delta(x - x'). \tag{34.3}$$

Equation (34.3) involves an average of a product of four $\tilde{\psi}$-operators for a system of noninteracting electrons. Using Wick's theorem, we can decompose this average into averages of pairs of the operators $\tilde{\psi}$ and $\tilde{\psi}^{+}$. For interacting particles, the product of four $\tilde{\psi}$-operators can be expressed in terms of the vertex part, i.e., it already includes the contributions from various scattering processes. In the weak-interaction model under consideration, these scattering processes involving collisions of particles can be neglected, but at the same time, it must be borne in mind that the ground state of the system differs from the usual state with a filled Fermi sphere, because of the presence of bound pairs of electrons. As already noted (see p. 288), such pairs behave like "Bose structures" and hence any number of them can accumulate on the level with lowest energy. If there is no external field, and if we neglect scattering processes, the pairs obviously "condense" in the state where each pair is at rest.

Now consider the operator products $\tilde{\psi}\tilde{\psi}$ and $\tilde{\psi}^{+}\tilde{\psi}^{+}$. The first of these products annihilates two electrons, and the second creates two electrons. In particular, the two electrons may be in a bound state, i.e., the operators $\tilde{\psi}\tilde{\psi}$ and $\tilde{\psi}^{+}\tilde{\psi}^{+}$ contain terms corresponding to the annihilation and creation of bound pairs, including pairs lying on the lowest level. Since there are many pairs of the latter type (the number of such pairs is proportional to the number of particles), the corresponding contribution to the operators $\tilde{\psi}\tilde{\psi}$ and $\tilde{\psi}^{+}\tilde{\psi}^{+}$ can be regarded as a number, just as was done for a system of bosons. We observe that in a metal, there are special reasons for not considering pairs which do not lie on the lowest level. In fact, a bound pair of electrons with finite momentum (associated with its motion as a whole) represents a Bose excitation with spin zero, and as we have already noted (see p. 281), it follows from the charge-neutrality condition in a metal that such excitations require a considerable amount of energy (~ 1 ev), which is much larger than the characteristic energies encountered in constructing a theory of superconductivity.

Returning to equation (34.3) for the Green's function, and taking the above remarks into account, we can write explicit expressions for averages of four $\tilde{\psi}$-operators. For example, we have

$$\begin{aligned}\langle T(\tilde{\psi}_{\alpha}(x_1)\tilde{\psi}_{\beta}(x_2)\tilde{\psi}^{+}_{\gamma}(x_3)\tilde{\psi}^{+}_{\delta}(x_4))\rangle &= -\langle T(\tilde{\psi}_{\alpha}(x_1)\tilde{\psi}^{+}_{\gamma}(x_3))\rangle\langle T(\tilde{\psi}_{\beta}(x_2)\tilde{\psi}^{+}_{\delta}(x_4))\rangle \\ &+\langle T(\tilde{\psi}_{\alpha}(x_1)\tilde{\psi}^{+}_{\delta}(x_4))\rangle\langle T(\tilde{\psi}_{\beta}(x_2)\tilde{\psi}^{+}_{\gamma}(x_3))\rangle \\ &+\langle N|T(\tilde{\psi}'_{\alpha}(x_1)\tilde{\psi}_{\beta}(x_2))|N + 2\rangle\langle N + 2|T(\tilde{\psi}^{+}_{\gamma}(x_3)\tilde{\psi}^{+}_{\delta}(x_4))|N\rangle,\end{aligned} \tag{34.4}$$

where $|N\rangle$ and $|N+2\rangle$ are ground states of systems with N and $N+2$ particles, respectively. This way of writing things means that we have neglected all effects of scattering of particles by each other, and the presence of the interaction has been taken into account only to the extent that it leads to the formation of bound pairs. The third term in the right-hand side of (34.4) has been written in complete analogy with the case of a Bose gas, in keeping with the fact that a large number of bound pairs are "condensed on the lowest level." The quantity

$$\langle N|T(\tilde{\psi}\tilde{\psi})|N+2\rangle\langle N+2|T(\tilde{\psi}^{+}\tilde{\psi}^{+})|N\rangle,$$

obviously has the same order of magnitude as the density of pairs, and it is easily seen that we can write

$$\begin{aligned} \langle N|T(\tilde{\psi}_{\alpha}(x)\tilde{\psi}_{\beta}(x'))|N+2\rangle &= e^{-2i\mu t}F_{\alpha\beta}(x-x'), \\ \langle N+2|T(\tilde{\psi}_{\alpha}^{+}(x)\tilde{\psi}_{\beta}^{+}(x'))|N\rangle &= e^{2i\mu t}F_{\alpha\beta}^{+}(x-x'). \end{aligned} \tag{34.5}$$

In the homogeneous problem (i.e., in the absence of an external field), the Green's function $G(x, x')$ depends only on the coordinate difference $x - x'$. The extra dependence on t in (34.5) comes from the general quantum-mechanical formula for the time derivative of an arbitrary operator $\tilde{A}(t)$:

$$\frac{\partial}{\partial t}\langle N|\tilde{A}(t)|N+2\rangle = i(E_N - E_{N+2})\langle N|\tilde{A}(t)|N+2\rangle.$$

By definition, the chemical potential μ equals $\partial E/\partial N$, and hence the energy difference $E_{N+2} - E_N$ equals 2μ.

We now substitute (34.4) into the equation (34.3) for the Green's function. In so doing, we shall everywhere omit the first two terms in the right-hand side of (34.4), since, as is easily verified, they lead to an additive correction to the chemical potential in the equations for the functions G, F, F^{+}, and hence are of no interest. As a result, we obtain the following equation[4] connecting G and F^{+}:

$$\left(i\frac{\partial}{\partial t} + \frac{\nabla^2}{2m}\right)\hat{G}(x-x') - i\lambda\hat{F}(0+)\hat{F}^{+}(x-x') = \delta(x-x'). \tag{34.6}$$

The quantity $\hat{F}(0+)$ is defined as

$$F_{\alpha\beta}(0+) = e^{2i\mu t}\langle N|\tilde{\psi}_{\alpha}(x)\tilde{\psi}_{\beta}(x)|N+2\rangle \equiv \lim_{\substack{\mathbf{r}\to\mathbf{r}' \\ t\to t'+0}} F_{\alpha\beta}(x-x'). \tag{34.7}$$

An equation for $\hat{F}^{+}(x-x')$ can be obtained in a similar way, by using the second of the equations (34.2):

$$\left(i\frac{\partial}{\partial t} - \frac{\nabla^2}{2m} - 2\mu\right)\hat{F}^{+}(x-x') + i\lambda\hat{F}^{+}(0+)\hat{G}(x-x') = 0. \tag{34.8}$$

---

[4] Here $\hat{G}$, $\hat{F}$, $\hat{F}^{+}$ denote the matrices whose elements are $G_{\alpha\beta}$, $F_{\alpha\beta}$, $F_{\alpha\beta}^{+}$, respectively, and products of $\hat{G}$, $\hat{F}$, $\hat{F}^{+}$ are matrix products.

According to (34.7), we have

$$F^{+}_{\alpha\beta}(0+) = e^{-2i\mu t}\langle N + 2|\tilde{\psi}^{+}_{\alpha}(x)\tilde{\psi}^{+}_{\beta}(x)|N\rangle. \tag{34.9}$$

In the absence of interactions depending on the spins of the particles, the Green's function $G(x - x')$ is proportional to the unit matrix in the spin variables, i.e.,

$$\hat{G}_{\alpha\beta}(x - x') = \delta_{\alpha\beta}G(x - x').$$

The functions $\hat{F}$ and $\hat{F}^{+}$ are proportional to a matrix $\hat{I}$ which is antisymmetric in its indices. In fact, since the operators $\tilde{\psi}_{\alpha}(x)$ and $\tilde{\psi}_{\beta}(x')$ anticommute when evaluated at the same instant of time, we have

$$F_{\alpha\beta}(\mathbf{r} - \mathbf{r}', 0) = -F_{\beta\alpha}(\mathbf{r}' - \mathbf{r}, 0).$$

It follows that

$$[F^{+}_{\alpha\beta}(\mathbf{r} - \mathbf{r}', 0)]^{*} = -F_{\alpha\beta}(\mathbf{r} - \mathbf{r}', 0), \tag{34.10}$$

and in particular,

$$[F^{+}_{\alpha\beta}(0+)]^{*} = -F_{\alpha\beta}(0+). \tag{34.11}$$

It is convenient to write F and F^{+} in the form

$$\begin{aligned} \hat{F}^{+}(x - x') &= \hat{I}F^{+}(x - x'), \\ \hat{F}(x - x') &= \hat{I}F(x - x'), \end{aligned} \tag{34.12}$$

where $(\hat{I}^{2})_{\alpha\beta} = -\delta_{\alpha\beta}$. We see from (34.10) that $F^{+}(x - x')$ and $F(x - x')$ are connected by the relation

$$[F^{+}(\mathbf{r} - \mathbf{r}', 0)]^{*} = F(\mathbf{r} - \mathbf{r}', 0).$$

The antisymmetry of $\hat{F}$ and $\hat{F}^{+}$ in their spin indices corresponds to the fact that the bound pairs are in a singlet state. To within a constant factor, the function $F_{\alpha\beta}(\mathbf{r} - \mathbf{r}', 0)$ can obviously be regarded as the wave function of a bound pair of particles (the center of mass of the pair is at rest).

Eliminating all dependence on the spin variables, we can now write (34.6) and (34.8) in the form

$$\begin{aligned} \left(i\frac{\partial}{\partial t} + \frac{\nabla^{2}}{2m}\right)G(x - x') - i\lambda F(0+)F^{+}(x - x') &= \delta(x - x'), \\ \left(i\frac{\partial}{\partial t} - \frac{\nabla^{2}}{2m} - 2\mu\right)F^{+}(x - x') + i\lambda F^{+}(0+)G(x - x') &= 0, \end{aligned} \tag{34.13}$$

where $F^{*}(0+) = F^{+}(0+)$. Finally, taking the Fourier components of these equations, we obtain[5]

$$\begin{aligned} \left(\omega - \frac{\mathbf{p}^{2}}{2m}\right)G(p) - i\lambda F(0+)F^{+}(p) &= 1, \\ \left(\omega + \frac{\mathbf{p}^{2}}{2m} - 2\mu\right)F^{+}(p) + i\lambda F^{+}(0+)G(p) &= 0. \end{aligned} \tag{34.14}$$

[5] This system of equations closely resembles the system of equations for the functions G' and $\hat{G}$ for a Bose system, but it should be kept in mind that then the operators ξ_{0} and ξ_{0}^{+} (corresponding to particles in the condensate) are the analogs of the functions F and F^{+}. Therefore, we use the notation G and F^{+}, instead of G' and $\hat{G}$ for bosons.

So far, we have used thermodynamic variables such that the number of particles is specified. It is much more convenient to use the chemical potential μ as the independent variable. As usual, the transformation to these new variables can be achieved by setting $\omega = \omega' + \mu$. In this way, we rewrite the system (34.14) as

$$\begin{aligned}(\omega - \xi)G(p) - i\lambda F(0+)F^+(p) &= 1,\\ (\omega + \xi)F^+(p) + i\lambda F^+(0+)G(p) &= 0,\end{aligned} \tag{34.15}$$

where $\xi = v(|\mathbf{p}| - p_0)$, $p_0 \approx \sqrt{2m\mu}$ is the Fermi momentum and $v = p_0/m$. (The prime on ω' has been dropped.) Solving (34.15), we obtain

$$G(p) = \frac{\omega + \xi}{\omega^2 - \xi^2 - \Delta_0^2},$$

$$F^+(p) = -i\lambda \frac{F^+(0+)}{\omega^2 - \xi^2 - \Delta_0^2}$$

where

$$\Delta_0^2 = \lambda^2 |F^+(0+)|^2. \tag{34.16}$$

The determinant of the system (34.15) vanishes at the points $\omega = \pm\varepsilon(\mathbf{p})$, where $\varepsilon(\mathbf{p}) = \sqrt{\xi^2 + \Delta_0^2}$. Therefore, our solution of (34.15) has been determined only to within arbitrary terms of the form

$$A_1(\mathbf{p})\delta[\omega - \varepsilon(\mathbf{p})] + A_2(\mathbf{p})\delta[\omega + \varepsilon(\mathbf{p})].$$

As a boundary condition determining the choice of the functions A_1 and A_2 appearing in the functions G and F^+, we can use Landau's result, which states that the sign of the imaginary part of the Green's function G is the opposite of the sign of ω, and that the function

$$G^R(p, \omega) = \operatorname{Re} G(p, \omega) + i \operatorname{Im} G(p, \omega) \operatorname{sgn} \omega$$

must be analytic with no singularities in the upper half-plane (see Sec. 7.2). It is easily verified that the solution of (34.15) satisfying these requirements is[6]

$$G(p) = \frac{u_\mathbf{p}^2}{\omega - \varepsilon(\mathbf{p}) + i\delta} + \frac{v_\mathbf{p}^2}{\omega + \varepsilon(\mathbf{p}) - i\delta}, \tag{34.17}$$

$$F^+(p) = -i\lambda \frac{F^+(0+)}{[\omega - \varepsilon(\mathbf{p}) + i\delta][\omega + \varepsilon(\mathbf{p}) - i\delta]}, \tag{34.18}$$

where the functions $u_\mathbf{p}^2$ and $v_\mathbf{p}^2$ are given by

$$\mu_\mathbf{p}^2 = \frac{1}{2}\left[1 + \frac{\xi}{\varepsilon(\mathbf{p})}\right], \qquad v_\mathbf{p}^2 = \frac{1}{2}\left[1 - \frac{\xi}{\varepsilon(\mathbf{p})}\right]. \tag{34.19}$$

---

[6] We have chosen $F^+(0+)$ to be real. This is always possible in the absence of an external field, since the equations (34.13) are invariant under the transformation

$$F(x - x') \to F(x - x')\exp\{2i\chi\}, \qquad F(0+) \to F(0+)\exp\{2i\chi\},$$
$$F^+(x - x') \to F^+(x - x')\exp\{-2i\chi\}, \qquad F^+(0+) \to F^+(0+)\exp\{-2i\chi\},$$

with constant phase χ. For further details, see Sec. 34.2.

The positive pole $\omega = \varepsilon(\mathbf{p})$ in the Green's function (34.17) determines the excitation spectrum $\varepsilon(\mathbf{p}) = \sqrt{\xi^2 + \Delta_0^2}$. This spectrum has a gap Δ_0 which we can determine as follows, starting from the relation

$$F^+(0) = \frac{1}{(2\pi)^4} \int F^+(p)\, d\mathbf{p}\, d\omega. \tag{34.20}$$

Substituting (34.18) into (34.20), we obtain the equation

$$1 = -\frac{\lambda}{2(2\pi)^3} \int \frac{d\mathbf{p}}{\sqrt{\xi^2 + \Delta_0^2}}. \tag{34.21}$$

To prevent this integral from diverging, we introduce a cutoff, using the condition that in the present model, only electrons with energies in a layer of thickness $2\omega_D$ about the Fermi surface participate in the interaction. Carrying out the integration, we find that

$$1 = -\frac{\lambda}{2\pi^2} mp_0 \ln \frac{2\omega_D}{\Delta_0},$$

and hence

$$\Delta_0 = 2\omega_D e^{-1/\eta}, \tag{34.22}$$

where

$$\eta = \frac{|\lambda| mp_0}{2\pi^2}.$$

Comparing (34.22) and (33.18), we find that the size of the gap in the energy spectrum at absolute zero is connected with the initial temperature by the relation

$$\Delta_0 = \frac{\pi}{\gamma} T_c. \tag{34.23}$$

34.2. The equations in the presence of an external electromagnetic field. Gauge invariance. If the superconductor is in an external field (e.g., an electromagnetic field), the system of equations (34.13) becomes a bit more complicated. First of all, we note that in an external field, G, F and F^+ are functions of both coordinates x and x', and not just the difference $x - x'$.

The electromagnetic field can be introduced in the usual way, by making the change

$$\nabla \to \nabla - ie\mathbf{A} \quad \text{or} \quad \nabla \to \nabla + ie\mathbf{A}, \tag{34.24}$$

depending on whether the differentiation is applied to the operator $\tilde{\psi}$ or the operator $\tilde{\psi}^+$. (It is usually convenient to choose the gauge in which the scalar potential is zero.) Then the equations for G and F^+ become

$$\begin{aligned} &\left(i\frac{\partial}{\partial t} + \frac{1}{2m}[\nabla_\mathbf{r} - ie\mathbf{A}(\mathbf{r})]^2 + \mu\right) G(x, x') \\ &\qquad\qquad -i\lambda F(x, x) F^+(x, x') = \delta(x - x'), \\ &\left(i\frac{\partial}{\partial t} - \frac{1}{2m}[\nabla_\mathbf{r} + ie\mathbf{A}(\mathbf{r})]^2 - \mu\right) F^+(x, x') \\ &\qquad\qquad + i\lambda F^+(x, x) G(x, x') = 0, \end{aligned} \tag{34.25}$$

and are obviously gauge invariant. If the vector potential undergoes the gauge transformation

$$\mathbf{A} \to \mathbf{A} + \nabla_{\mathbf{r}}\chi, \tag{34.26}$$

then

$$\begin{aligned} G(x, x') &\to G(x, x') \exp \{ie[\chi(\mathbf{r}) - \chi(\mathbf{r}')]\}, \\ F(x, x') &\to F(x, x') \exp \{ie[\chi(\mathbf{r}) + \chi(\mathbf{r}')]\}, \\ F^+(x, x') &\to F^+(x, x') \exp \{-ie[\chi(\mathbf{r}) + \chi(\mathbf{r}')]\}, \end{aligned} \tag{34.27}$$

while the "gap" $|\lambda|F(x, x)$ or $|\lambda|F^+(x, x)$, which in an external field is generally a function of x, transforms according to

$$\begin{aligned} F(x, x) &\to F(x, x) \exp \{2ie\chi(\mathbf{r})\}, \\ F^+(x, x) &\to F^+(x, x) \exp \{-2ie\chi(\mathbf{r})\}. \end{aligned} \tag{34.28}$$

The gauge invariance of the equations (34.25) will allow us to give a consistent treatment of the properties of a superconductor in a magnetic field (see Sec. 37.1). It should be emphasized that this gauge invariance is connected with the expression (32.4) for the interaction Hamiltonian. Strictly speaking, the Hamiltonian (32.3) is not gauge invariant, which is, of course, a feature of the particular model chosen. It is easily verified that in this model, the quantity

$$\overline{F(x, x)} = \int \theta(\mathbf{r} - \mathbf{y})\theta(\mathbf{r} - \mathbf{z})F(y, z)\, d\mathbf{z}\, d\mathbf{y}$$

and a similar quantity $\overline{F^+(x, x)}$, appear in the equations (34.25), instead of $F(x, x)$ and $F^+(x, x)$ [the values of the functions F and F^+ for identical arguments]. The function $F(\mathbf{y}, \mathbf{z}; t = t')$, being the wave function of a pair, is correlated over a distance $\xi_0 \sim v/T_c$, called the *coherence distance*, of the order of the dimensions of a pair, and falls off rapidly for $|\mathbf{y} - \mathbf{z}| \gg \xi_0$. Furthermore, as already mentioned in Sec. 32.2, the function θ behaves like a δ-function, with a maximum whose width is of order v/ω_D. Thus, changing $\overline{F(x, x)}$ to $F(x, x)$ leads to an error of order T_c/ω_D, which is always small for real superconductors.

34.3. A superconductor at finite temperatures. We conclude this section by considering the problem of extending the above considerations to the case of nonzero temperature. Clearly, such a generalization can be made by using the technique for $T \neq 0$, presented in Chap. 3. In the superconducting state, the system is characterized by the nonzero averages

$$\mathscr{F}(x, x') = \frac{\langle T_\tau(\psi(x)\psi(x')\mathscr{S})\rangle}{\langle\mathscr{S}\rangle},$$

$$\mathscr{F}^+(x, x') = \frac{\langle T_\tau(\bar{\psi}(x)\bar{\psi}(x')\mathscr{S})\rangle}{\langle\mathscr{S}\rangle},$$

where the averaging operation and the operators $\psi(x)$, $\bar{\psi}(x)$ have the same meaning as in Chap. 3. (Recall that the chemical potential has been chosen as the independent thermodynamic variable.) Suppose we regard the Gibbs averages in the definitions of $\mathscr{F}(x, x')$ and $\mathscr{F}^+(x, x')$ as quantum-mechanical averages over a state with energy equal to the average energy $\bar{E}$ and particle number equal to the average number of particles. Then the annihilation or creation of a bound pair of electrons belonging to the set of bound pairs in the "Bose condensate" has practically no effect on the overall state of the system, since the number of such pairs is very large (in fact, proportional to the total number of particles in the system). In other words, just as at $T = 0$, for a system in the superconducting state at $T \neq 0$ there are terms involving the products $\psi\psi$ and $\bar{\psi}\bar{\psi}$ which can be regarded as numbers.

We shall assume that (thermodynamic) averages of products of four ψ-operators can be written in terms of the Green's function

$$\mathscr{G}(x, x') = -\frac{\langle T_\tau(\psi(x)\bar{\psi}(x')\mathscr{S})\rangle}{\langle\mathscr{S}\rangle}$$

and the functions $\mathscr{F}(x, x')$, $\mathscr{F}^+(x, x')$, exactly as was done in the expression (34.4) for the case $T = 0$. Just as before, this means that the effects of scattering of particles by one another are neglected. Thus we have

$$\begin{aligned}&\frac{\langle T_\tau(\psi_\alpha(x_1)\psi_\beta(x_2)\bar{\psi}_\gamma(x_3)\bar{\psi}_\delta(x_4)\mathscr{S})\rangle}{\langle\mathscr{S}\rangle} \\ &= -\mathscr{G}_{\alpha\gamma}(x_1, x_3)\mathscr{G}_{\beta\delta}(x_2, x_4) + \mathscr{G}_{\alpha\delta}(x_1, x_4)\mathscr{G}_{\beta\gamma}(x_2, x_3) + \mathscr{F}_{\alpha\beta}(x_1, x_2)\mathscr{F}^+_{\gamma\delta}(x_3, x_4).\end{aligned} \tag{34.29}$$

The corresponding derivation of equations for the functions $\mathscr{G}$ and $\mathscr{F}^+$ will not be given here, since it is completely analogous to the derivation of (34.13). We merely write down the final result

$$\begin{aligned}&\left(-\frac{\partial}{\partial\tau} + \frac{\nabla^2}{2m} + \mu\right)\mathscr{G}(x - x') + \Delta\mathscr{F}^+(x - x') = \delta(x - x'), \\ &\left(\frac{\partial}{\partial\tau} + \frac{\nabla^2}{2m} + \mu\right)\mathscr{F}^+(x - x') - \Delta^*\mathscr{G}(x - x') = 0,\end{aligned} \tag{34.30}$$

where

$$\Delta = |\lambda|\mathscr{F}(0+), \qquad \Delta^* = |\lambda|\mathscr{F}^+(0+). \tag{34.31}$$

It is sometimes necessary to determine the function $\mathscr{F}(x - x')$. As is easily verified, $\mathscr{F}(x - x')$ satisfies the equation

$$\left(-\frac{\partial}{\partial\tau} + \frac{\nabla^2}{2m} + \mu\right)\mathscr{F}(x - x') - \Delta\mathscr{G}(x' - x) = 0,$$

where the function $\mathscr{G}$ appears with the arguments x and x' reversed.

It is not hard to see that the set of four equations for the functions $\mathscr{G}(x - x')$, $\mathscr{F}^+(x - x')$, $\mathscr{F}(x - x')$ and $\mathscr{G}(x' - x)$ can be written as a single matrix equation

$$\begin{pmatrix} \left(-\frac{\partial}{\partial\tau} + \frac{\nabla^2}{2m} + \mu\right) & \Delta \\ -\Delta^* & \left(\frac{\partial}{\partial\tau} + \frac{\nabla^2}{2m} + \mu\right)\end{pmatrix}\begin{pmatrix}\mathscr{G}(x - x') & \mathscr{F}(x - x') \\ \mathscr{F}^+(x - x') & -\mathscr{G}(x' - x)\end{pmatrix} = \hat{1}. \tag{34.32}$$

In other words, the four functions in question form a single matrix Green's function for the operator in the left-hand side of (34.32).

As we know, in the temperature technique all quantities are expanded in Fourier series with respect to the frequency, and not in Fourier integrals. Just as in the case of the Green's function in Chap. 3, suppose we introduce Fourier components for the functions $\mathscr{F}$ and $\mathscr{F}^+$, i.e., suppose we write

$$\begin{aligned} \mathscr{F}^+(x - x') &= \frac{T}{(2\pi)^3} \sum_n e^{-i\omega_n \tau} \int e^{i\mathbf{p}\cdot\mathbf{r}} \mathscr{F}^+_{\omega_n}(\mathbf{p})\, d\mathbf{p}, \\ \mathscr{F}(x - x') &= \frac{T}{(2\pi)^3} \sum_n e^{-i\omega_n \tau} \int e^{i\mathbf{p}\cdot\mathbf{r}} \mathscr{F}_{\omega_n}(\mathbf{p})\, d\mathbf{p}, \end{aligned} \tag{34.33}$$

where $\omega_n = (2n + 1)\pi T$. Then the system (34.30) corresponds to the system

$$\begin{aligned} (i\omega - \xi)\mathscr{G}_\omega(\mathbf{p}) + \Delta \mathscr{F}^+_\omega(\mathbf{p}) &= 1, \\ (i\omega + \xi)\mathscr{F}^+_\omega(\mathbf{p}) + \Delta^* \mathscr{G}_\omega(\mathbf{p}) &= 0, \end{aligned} \tag{34.34}$$

with the solution

$$\mathscr{G}_\omega(\mathbf{p}) = -\frac{i\omega + \xi}{\omega^2 + \xi^2 + \Delta^2}, \qquad \mathscr{F}^+_\omega(\mathbf{p}) = \frac{\Delta^*}{\omega^2 + \xi^2 + \Delta^2}. \tag{34.35}$$

Moreover, we note that in the absence of an external field, the functions $\mathscr{F}$ and $\mathscr{F}^+$ are equal, and the quantity Δ is real. Unlike the situation encountered in the case of the system (34.15), the solution (34.35) is unique. This is connected with the fact that the analytic properties of functions in the temperature technique are defined uniquely. The size of the gap can be found from the condition

$$1 = \frac{|\lambda| T}{(2\pi)^3} \sum_n \int \frac{d\mathbf{p}}{\omega_n^2 + \xi^2 + \Delta^2}. \tag{34.36}$$

The sum over frequencies in (34.36) is easily calculated, and as a result, instead of the condition (34.21) for $T = 0$, we obtain a new relation determining the size of the gap for $T \neq 0$:

$$1 = \frac{|\lambda| m p_0}{2\pi^2} \int_0^{\omega_D} \frac{\tanh [\sqrt{\xi^2 + \Delta^2(T)}/2T]}{\sqrt{\xi^2 + \Delta^2(T)}}\, d\xi. \tag{34.37}$$

At the point where the phase transition occurs, i.e., at the temperature $T = T_c$, the gap $\Delta(T)$ vanishes, and, as must be the case, the condition (34.37) goes into the condition (33.15) determining the critical temperature T_c.

35. Derivation of the Equations of the Theory of Superconductivity in the Phonon Model

We now derive the equations of the theory of superconductivity, using a "phonon model" in which the electrons interact with one another via the electron-phonon interaction (only the case $T = 0$ will be considered).

Although this model suffers from the same defect as the model considered above (since it does not take account of the Coulomb forces acting in the metal), it does, of course, have a more direct physical meaning than the model involving four-fermion interactions. However, the latter model is somewhat more convenient for obtaining practical results. The basic advantage of the phonon model is the fact that the Hamiltonian (32.1) for electron-phonon interactions is gauge invariant from the very outset, unlike the Hamiltonian (32.3) for four-fermion interactions, which is only approximately gauge invariant, because of the inequality $T_c \ll \omega_D$. In general, the inequality $T_c \ll \omega_D$ is only satisfied in the weak-coupling approximation.[7] It will be shown later that the weak-coupling approximation is not essential in the theory of superconductivity, and that it is actually only the ratio $\omega_D/\varepsilon_F \ll 1$ which plays the role of the small parameter in this theory ($\omega_D/\varepsilon_F \sim u/v \sim 10^{-2}$ to 10^{-3}, where u is the velocity of sound in the metal, and v is the velocity of the electrons at the Fermi surface).[8]

Thus, let the interaction Hamiltonian of the system of electrons and phonons have the form

$$H_{\text{int}} = g \int (\psi^+(\mathbf{r})\psi(\mathbf{r}))\varphi(\mathbf{r})\, d\mathbf{r}.$$

If the system is in the superconducting state, then, in addition to the Green's function G, its properties are characterized by two more functions F and F^+. Therefore, instead of the usual Dyson equation (see Sec. 21), we must in general study three equations, connecting the three functions

$$\begin{aligned} \hat{G}_{\alpha\beta}(x, x') &= -i\langle T(\tilde{\psi}_\alpha(x)\tilde{\psi}_\beta^+(x'))\rangle = \delta_{\alpha\beta}G(x - x'), \\ \hat{F}_{\alpha\beta}^+(x, x') &= \langle T(\tilde{\psi}_\alpha^+(x)\tilde{\psi}_\beta^+(x'))\rangle = I_{\alpha\beta}F^+(x - x'), \\ \hat{F}_{\alpha\beta}(x, x') &= \langle T(\tilde{\psi}_\alpha(x)\tilde{\psi}_\beta(x'))\rangle = -I_{\alpha\beta}F(x - x'). \end{aligned} \tag{35.1}$$

As we shall see, the equation for the phonon Green's function

$$D(x_1 - x_2) = -i\langle T(\varphi(x_1)\varphi(x_2))\rangle,$$

remains practically unchanged.

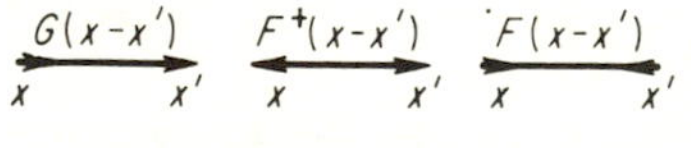

$G(x-x')$ $F^+(x-x')$ $F(x-x')$

x x' x x' x x'

FIGURE 91

The equation for the Green's function can be obtained by using the diagram technique of perturbation theory. Just as in the case of a system of bosons below the point of "Bose condensation," the set of all possible perturbation-theory diagrams is enlarged by the appearance of lines corresponding to the functions F and F^+. Suppose we represent the functions G, F^+ and F by thick lines equipped with two arrows (actually arrowheads), as shown in Fig. 91, whose directions at the points x and x' are chosen in

---

[7] However, the condition $T_c \ll \omega_D$ is always satisfied in real metals.

[8] For an analysis of the electron-phonon interaction in the theory of superconductivity, see B6, B7.

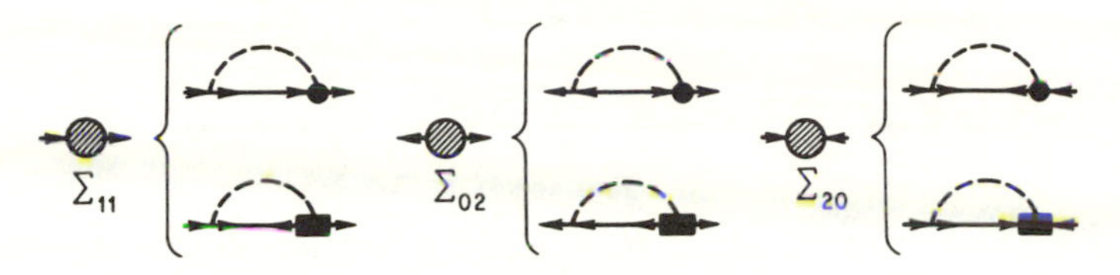

FIGURE 92

keeping with (35.1), i.e., an arrow directed away from a given point corresponds to the operator $\tilde{\psi}$, while an arrow directed towards a given point corresponds to the operator $\tilde{\psi}^+$. Then it is easily seen, in complete analogy with the Bose gas, that there are three types of irreducible self-energy parts, shown in Fig. 92 and denoted by $\Sigma_{11}(x, x')$, $\Sigma_{20}(x, x')$, $\Sigma_{02}(x, x')$. In Fig. 92, a dashed line represents a phonon D-function, a point represents a simple vertex or a factor g, and a heavy point or rectangle represents the modification of a simple vertex due to the various electron-phonon interactions.

Taking Fourier components of all quantities, we consider one of the self-energy diagrams, say $\Sigma_{11}(p)$. It is not hard to see that to within terms of order ω_D/ε_F, we can neglect all phonon corrections to the three-vertex part in the simplest diagram for $\Sigma_{11}(p)$, indicated in Fig. 93. In fact, as shown in Sec. 21, the values of the D-function (and of the phonon vertex) for phonons whose momenta are of the order of the Fermi momentum of the electrons are the important values in the integrand of the corresponding analytical expression. For this reason, the estimate made in Sec. 21 of the corrections to the phonon vertex coming from electron-phonon interactions is still valid in the present case, since the size of these corrections is determined by the values of the Green's functions in energy and momentum regions far from the Fermi surface. Moreover, it is clear that the Fourier components of the electron Green's function for a metal in the superconducting state are different from their values for a normal metal only in a narrow region near the Fermi surface, with excitation energies no larger than the order of magnitude of the maximum energy of the phonons (of order ω_D), and accordingly, the functions $F^+(p)$ and $F(p)$, which are peculiar to the superconducting state, are nonzero only in this region. In the diagrams of Fig. 92, for the self-energy parts Σ_{11}, Σ_{20} and Σ_{02}, we show in each case two different types of diagram, depending on which modification of the phonon vertex is chosen. According to what has just been said, we can immediately omit the diagrams of the second type, where the phonon vertices are indicated by heavy rectangles, since diagrams of this type can be constructed only by using the Green's functions F and F^+ characteristic of superconductors. Thus, for the phonon vertices in the irreducible self-energy parts Σ_{11}, Σ_{20} and Σ_{02}, we can restrict ourselves to the zeroth-order approximation of perturbation theory.

FIGURE 93

FIGURE 94

For the same reasons, the phonon Green's function $D(x_1 - x_2)$, whose Fourier components can be found directly by using (21.21), remains practically unchanged. The structure of the equations for the functions G and F^+ is shown in Fig. 94, and is clear without further explanation. In the coordinate representation, the analytical versions of these equations are

$$\begin{aligned}\left(i\frac{\partial}{\partial t} + \frac{\nabla^2}{2m} + \mu\right)G(x - x')\\ &= \delta(x - x') + g^2 i \int G(x - z)D(x - z)G(z - x')\, d^4z\\ &\quad + g^2 i \int F(x - z)D(x - z)F^+(z - x')\, d^4z.\\ \left(-i\frac{\partial}{\partial t} + \frac{\nabla^2}{2m} + \mu\right)F^+(x - x')\\ &= g^2 i \int G(z - x)D(z - x)F^+(z - x')\, d^4z\\ &\quad + g^2 i \int F^+(x - z)D(x - z)G(z - x')\, d^4z.\end{aligned} \tag{35.2}$$

The effect of an electromagnetic field can be included in these equations in the usual way, just as was done in Sec. 34.2. We emphasize that the resulting system is completely gauge invariant, unlike the system (34.13) in which the gauge invariance is only approximate, to within terms of order T_c/ω_D. Unfortunately, the system (35.2) is much more complicated than (34.13) and contains nonlinear integral terms which make it unsuitable for solution in the coordinate representation, as required in a variety of problems involving inhomogeneous magnetic fields. However, as a rule, equivalent practical results are obtained by using either model.

For the homogeneous problem (i.e., in the absence of a magnetic field), transforming to the momentum representation in (35.2), we easily obtain the following equations involving the Fourier components of all quantities:

$$\begin{aligned}(\omega - \xi - g^2 i\overline{G}_\omega)G(p) - g^2 i\overline{F}_\omega F^+(p) &= 1,\\ (-\omega - \xi - g^2 i\overline{G}_{-\omega})F^+(p) - g^2 i\overline{F^+_\omega}\, G(p) &= 0.\end{aligned} \tag{35.3}$$

Here we have introduced the notation

$$\begin{aligned}\overline{G}_\omega &= \frac{1}{(2\pi)^4}\int G(p - k)D(k)\, d^4k,\\ \overline{F}_\omega &= \frac{1}{(2\pi)^4}\int F(p - k)D(k)\, d^4k,\\ \overline{F^+_\omega} &= \frac{1}{(2\pi)^4}\int F^+(p - k)D(k)\, d^4k,\end{aligned} \tag{35.4}$$

where $\bar{G}_{-\omega}(\mathbf{p}) = \bar{G}_\omega(-\mathbf{p})$. The system (35.3) is completely analogous to the system (34.15), derived on p. 295. The only difference is that whereas in (34.15) the quantities $\bar{G}_\omega$, $\bar{F}_\omega$ and $\overline{F_\omega^+}$ are constant in the region $|v(|\mathbf{p}| - p_0)| < \omega_D$ near the Fermi surface and vanish outside this region, these quantities are in general functions of $\mathbf{p}$ and ω, which fall off smoothly to zero for $|v(|\mathbf{p}| - p_0)| \gg \omega_D$, $\omega \gg \omega_D$.

Solving for the functions G and F^+ in terms of $\bar{G}_\omega$, $\bar{F}_\omega$ and $\overline{F_\omega^+}$, we find that

$$G(p) = \frac{\omega - g^2 i\bar{G}_\omega + \xi}{(\omega - \xi - g^2 i\bar{G}_\omega)(\omega + \xi + g^2 i\bar{G}_{-\omega}) - g^4|\overline{F_\omega^+}|^2},$$

$$F^+(p) = \frac{-ig^2\overline{F_\omega^+}}{(\omega - \xi - g^2 i\bar{G}_\omega)(\omega + \xi + g^2 i\bar{G}_{-\omega}) - g^4|\overline{F_\omega^+}|^2}.$$

Substituting these expressions into the equations (35.4) which define $\bar{G}_\omega$ and $\overline{F_\omega^+}$, we obtain two integral equations for $\bar{G}_\omega$ and $\overline{F_\omega^+}$. These equations have been solved by Eliashberg (E3). Omitting computational details, we merely give the final result. It turns out that for small energies, the excitation spectrum has the form

$$\varepsilon(\mathbf{p}) = \sqrt{\xi^2 + \Delta_0^2},$$

where, however, in keeping with (21.45), the quantity $\xi = v_1(|\mathbf{p}| - p_0)$ involves v_1, the renormalized velocity at the Fermi surface. At absolute zero, the energy gap is related to $|\overline{F_{\omega=0}^+}|$ by the formula

$$\Delta_0 = g^2 \frac{v_1}{v_0} |\overline{F_{\omega=0}^+}|.$$

In the weak-interaction limit (i.e., for $g^2 \ll 1$), these formulas agree with the results of the preceding section.

36. The Thermodynamics of Superconductors

36.1. Temperature dependence of the energy gap. We now examine in more detail how the size of the gap in the energy spectrum depends on the temperature. We first consider the case of low temperatures ($T \ll T_c$), carrying out a suitable series expansion of the condition (34.37). We have the identity

$$\frac{1}{\eta} = \int_0^{\omega_D} \frac{d\xi}{\sqrt{\xi^2 + \Delta^2}} - 2\int_0^\infty \frac{d\xi}{\sqrt{\xi^2 + \Delta^2}} \frac{1}{\exp\{\sqrt{\xi^2 + \Delta^2}/T\} + 1}, \tag{36.1}$$

where $\eta = |\lambda| m p_0/2\pi^2$. (Since the second integral converges, we can set its upper limit equal to ∞.) Expanding the exponential in the integrand of the second integral in (36.1), transforming to an integral with respect to ε, and using the definitions of the appropriate Bessel functions, we can represent equation (36.1) as a series of Bessel functions of order zero, i.e.,

$$\ln\frac{\Delta_0}{\Delta} = 2\sum_{n=1}^{\infty}(-1)^{n+1}K_0\left(\frac{n\Delta}{T}\right) \tag{36.2}$$

where $\Delta_0 \equiv \Delta(T = 0)$. For low temperatures, $\Delta \gg T$ and hence, using the asymptotic expansion of the Bessel functions, we obtain

$$\Delta = \Delta_0 - \sqrt{2\pi T\Delta_0}\left(1 - \frac{T}{8\Delta_0}\right)e^{-\Delta_0/T}. \tag{36.3}$$

To determine the behavior of the gap for temperatures near the transition temperature T_c, it is most convenient to start from the relation (34.36). Near T_c the size of the gap is small, and hence in (34.36), we can carry out an expansion in powers of Δ^2/T^2:

$$\frac{1}{\eta} = T\sum_n \int_{-\omega_D}^{\omega_D} d\xi\left[\frac{1}{\omega_n^2 + \xi^2} - \frac{\Delta^2}{(\omega_n^2 + \xi^2)^2} + \frac{\Delta^4}{(\omega_n^2 + \xi^2)^3} + \cdots\right].$$

Interchanging the order of summation over the frequencies and integration over ξ in the (convergent) terms of the right-hand side, we obtain

$$\frac{1}{\eta} = \int_0^{\omega_D}\frac{d\xi}{\xi}\tanh\frac{\xi}{2T} - \frac{\Delta^2}{(\pi T)^2}\sum_{n=0}^{\infty}\frac{1}{(2n + 1)^3} + \frac{3}{4}\frac{\Delta^4}{(\pi T)^4}\sum_{n=0}^{\infty}\frac{1}{(2n + 1)^5} + \cdots \tag{36.4}$$

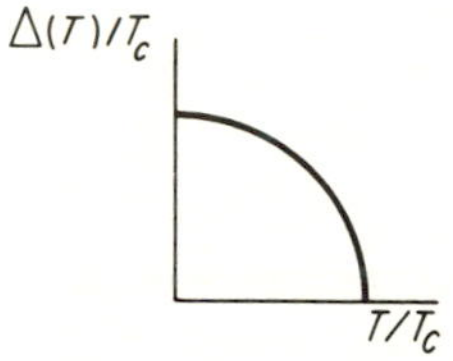

FIGURE 95

Expressing the series appearing in (36.4) in terms of the Riemann zeta function, i.e., writing

$$\sum_{n=0}^{\infty}\frac{1}{(2n + 1)^z} = \frac{2^z - 1}{2^z}\zeta(z), \tag{36.5}$$

and substituting (36.5) into (36.4), we find that

$$\ln\frac{T}{T_c} = -\frac{7\zeta(3)}{8}\frac{\Delta^2}{(\pi T)^2} + \frac{93\zeta(5)}{128}\frac{\Delta^4}{(\pi T)^4} + \cdots$$

Then, to a first approximation, we find that the size of the gap near T_c is

$$\Delta = \pi T_c\sqrt{\frac{8}{7\zeta(3)}}\sqrt{1 - \frac{T}{T_c}} \approx 3.06T_c\sqrt{1 - \frac{T}{T_c}}. \tag{36.6}$$

The curve representing the behavior of the gap as a function of the temperature over the whole temperature range is shown in Fig. 95.

36.2. Heat capacity. To find the various thermodynamic quantities characterizing a superconductor, we use the relation derived earlier [cf. pp. 95, 140] for the derivative of the thermodynamic potential with respect to the interaction parameter, i.e.,

$$\frac{\partial\Omega}{\partial\lambda} = \frac{1}{\lambda}\langle H_{\text{int}}\rangle,$$

where in our case, H_{int} is the expression (32.4). Retaining terms in the average which are nonzero only in the superconducting phase, we find that

$$\frac{\partial\Omega}{\partial|\lambda|} = -\frac{1}{\lambda^2}|\Delta|^2,$$

since $\lambda < 0\,(V = 1)$. The relation between $1/|\lambda|$ and Δ at a given temperature is given by (36.1). Therefore, the difference between the values of the thermodynamic potential for the superconducting and normal states of the metal is

$$\Omega_s - \Omega_n = \int_0^\Delta \frac{d(1/|\lambda|)}{d\Delta}\Delta^2\,d\Delta.$$

According to the general principles of statistical physics, this correction is the same for all the thermodynamic potentials, since it is expressed in appropriate variables. Using (36.2) which implies

$$\frac{1}{|\lambda|} = \frac{mp_0}{2\pi^2}\left[\ln\frac{2\omega_D}{\Delta} - 2\sum_{n=1}^{\infty}(-1)^{n+1}K_0\left(\frac{n\Delta}{T}\right)\right],$$

and also using the formula

$$K_0'(x) = -K_1(x)$$

(familiar from the theory of Bessel functions), we obtain

$$F_s - F_n = -\left(\frac{mp_0}{2\pi^2}\right)\left[\frac{\Delta^2}{2} - 2\sum_{n=1}^{\infty}(-1)^{n+1}\frac{T^2}{n^2}\int_0^{n\Delta/T}K_1(x)x^2\,dx\right].$$

To evaluate this expression, we write

$$\int_0^{n\Delta/T}K_1(x)x^2\,dx = 2 - \int_{n\Delta/T}^{\infty}K_1(x)x^2\,dx,$$

and then note that since $\Delta/T \gg 1$ at low temperatures, we need only calculate the integral on the right for $n = 1$ [this can be done by using the asymptotic expansion of the function $K_1(x)$]. The remaining sum with respect to n is easily evaluated, and the final result is found to be

$$F_s - F_n = \frac{mp_0T^2}{6} - \frac{mp_0}{2\pi^2}\left[\frac{\Delta^2}{2} + \sqrt{2\pi\Delta_0^3T}\left(1 + \frac{15}{8}\frac{T}{\Delta_0}\right)e^{-\Delta_0/T}\right]. \quad (36.7)$$

The first term in the right-hand side of (36.7) is the negative of the principal term in the expansion of the free energy of the normal metal in powers of T. As is well known, this term leads to the formula

$$C_n = \frac{mp_0T}{3},$$

i.e., the electronic part of the heat capacity of the normal metal is a linear function of T. Substituting (36.3) into (36.7), we find that in the superconducting phase at low temperatures, the entropy is

$$S_s = \frac{mp_0}{\pi^2}\sqrt{\frac{2\pi\Delta_0^3}{T}}\,e^{-\Delta_0/T},$$

while the heat capacity is

$$C_s = \frac{mp_0}{\pi^2}\sqrt{\frac{2\pi\Delta_0^5}{T^3}}\, e^{-\Delta_0/T}.$$

Then, using (34.23), we obtain the following expression for the ratio of the heat capacity of the superconducting metal in the region $T \ll T_c$ to the heat capacity of the normal metal for $T = T_c$:

$$\frac{C_s(T)}{C_n(T_c)} = \frac{3}{\gamma}\sqrt{\frac{2}{\pi}}\left(\frac{\Delta_c}{T}\right)^{3/2} e^{-\Delta_0/T}. \tag{36.8}$$

Next, to derive asymptotic formulas for the temperature region near T_c, we start from the expansion (36.4):

$$\delta\frac{1}{|\lambda|} = -\left(\frac{mp_0}{2\pi^2}\right)\frac{7\zeta(3)}{(2\pi T)^2}\Delta\delta\Delta.$$

Using (36.6), we find that the difference in free energies is

$$F_s - F_n = -\left(\frac{mp_0}{2\pi^2}\right)\frac{7\zeta(3)}{16(\pi T)^2}\Delta^4 = -\frac{2mp_0T_c^2}{7\zeta(3)}\left(1 - \frac{T}{T_c}\right)^2, \tag{36.9}$$

and hence the entropy in the superconducting phase equals

$$S_s = -\frac{4mp_0T_c}{7\zeta(3)}\left(1 - \frac{T}{T_c}\right) + S_n.$$

Carrying out another differentiation and retaining principal terms, we find that the heat capacity of the superconductor at the transition point is given by

$$C_s(T_c) = C_n(T_c) + \frac{4mp_0}{7\zeta(3)}T_c.$$

Thus, when the metal makes the phase transition to the superconducting state, its heat capacity undergoes a jump equal to

$$\frac{4mp_0T_c}{7\zeta(3)}.$$

Taking account of higher-order terms in $T_c - T$ in the expansion (36.9), we find the following expression for the ratio of the heat capacity of the superconducting metal in the region near T_c to the heat capacity of the normal metal for $T = T_c$

$$\frac{C_s(T)}{C_n(T_c)} = 2.43 + 3.77\left(\frac{T}{T_c} - 1\right).$$

The behavior of the electronic part of the heat capacity of a metal is summarized in Fig. 96.

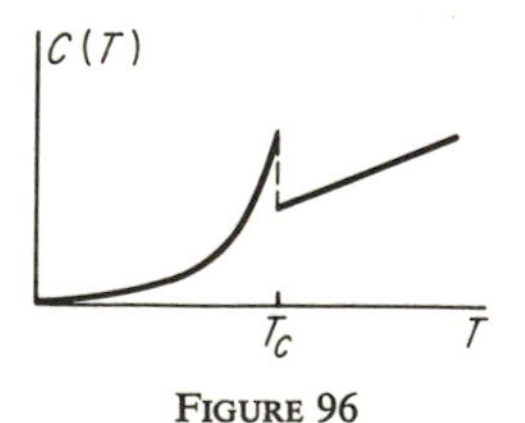

FIGURE 96

36.3. The critical magnetic field. An important thermodynamic quantity in the theory of superconductivity is the *critical* (*magnetic*) *field* H_c. In fact, at a given temperature $T < T_c$, a metal in a magnetic field can be either in the superconducting state or in the normal state, depending on whether or not the magnetic field is less than H_c. Suppose a superconductor is placed in a magnetic field $\mathbf{H}$. Then the surface current screening the field $\mathbf{H}$ produces a magnetic moment $\mathbf{M}$ which interacts with $\mathbf{H}$. The extra energy (per unit volume) associated with this interaction is $-\frac{1}{2}\mathbf{H}\cdot\mathbf{M}$. Let the superconductor be of cylindrical shape, and let $\mathbf{H}$ be parallel to the axis of the cylinder. Calculating the surface current from the condition that the magnetic field be zero in the body of the superconductor, and determining the magnetic moment due to this current, we find that the extra magnetic energy is $H^2/8\pi$, i.e., the free energy of the superconductor in the magnetic field is

$$F_{sH} = F_s + \frac{H^2}{8\pi}.$$

Therefore, as the magnetic field is increased at a given temperature, the metal makes a transition from the superconducting phase to the normal phase. This phase transition of the first kind takes place at a critical field H_c determined by

$$\frac{H_c^2}{8\pi} = F_n - F_s.$$

Once again, we consider only limiting cases. For low temperatures ($T \ll T_c$), neglecting exponentially small terms, we find from (36.3) and (36.7) that

$$H_c(0) = \sqrt{\frac{2mp_0}{\pi}}\,\Delta_0 = T_c\frac{\pi}{\gamma}\sqrt{\frac{2mp_0}{\pi}} \tag{36.10}$$

and

$$H_c(T) = H_c(0)\left(1 - \frac{\gamma^2}{3}\frac{T^2}{T_c^2}\right). \tag{36.11}$$

In the temperature region near T_c, we use formula (36.9) and express $H_c(T)$ near T_c in terms of $H_c(0)$, as given by (36.10). This leads to the following expression for $H_c(T)$ as a function of T:

$$H_c(T) = H_c(0)\gamma\sqrt{\frac{8}{7\zeta(3)}}\left(1 - \frac{T}{T_c}\right) \approx 1.73H_c(0)\left(1 - \frac{T}{T_c}\right). \tag{36.12}$$

It should be noted that the experimental data is usually described by the curve

$$H_c(T) = H_c(0)\left(1 - \frac{T^2}{T_c^2}\right). \tag{36.13}$$

In both limiting cases, the theoretical formulas (36.11) and (36.12) are in satisfactory agreement with the experimental formula (36.13) [see B1, K5].

37. A Superconductor in a Weak Electromagnetic Field

37.1. A weak constant magnetic field. Next, we study the electromagnetic properties of superconductors. For the time being, we shall only consider the behavior of superconductors in "weak" fields, i.e., in fields which are small compared to the critical magnetic field. We assume that a superconductor with a plane surface occupies the half-space $z < 0$, and is situated in a constant magnetic field, directed parallel to its surface, as shown in Fig. 97. In terms of the vector potential $\mathbf{A} = (A_x, A_y, A_z)$, we have

$$\mathbf{H} = \operatorname{curl} \mathbf{A}.$$

In vacuum, $\mathbf{H}$ is constant, and we can choose

$$A_y = -Hz, \qquad A_x = A_z = 0 \tag{37.1}$$

(say) as the components of the vector potential. The action of the magnetic field produces a current in the superconductor, and the Maxwell equation

$$\nabla^2 \mathbf{A} = -4\pi \mathbf{j} \tag{37.2}$$

relates the field distribution in the superconductor to the current density $\mathbf{j}$.

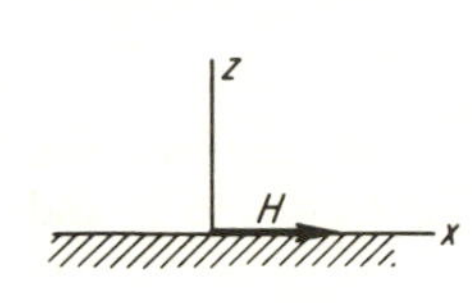

FIGURE 97

Since the current density is itself due to the presence of the field, $\mathbf{j}$ is proportional to $\mathbf{A}$, to the approximation which is linear in the field. Then it follows by a homogeneity argument that the relation between the current density and the field in an infinite superconductor must have the general form

$$\mathbf{j}(x) = -\int Q(x - y)\mathbf{A}(y)\, d^4y, \tag{37.3}$$

or in Fourier components,

$$\mathbf{j}(k) = -Q(k)\mathbf{A}(k). \tag{37.3'}$$

In what follows, we shall not concern ourselves with the detailed solution of the electromagnetic problem defined by (37.2) and (37.3) in the half-space $z < 0$. Instead, we shall just derive an expression for the kernel $Q(x - y)$, with a view to demonstrating how the methods of quantum field theory are applied to this case.

The current density $\mathbf{j}(x)$ at a given point is (as usual) the thermodynamic average of the familiar quantum-mechanical expression for the current-density operator in the second-quantized representation:

$$\hat{\mathbf{j}}(x) = \frac{ie}{2m}(\nabla_{\mathbf{r}'} - \nabla_{\mathbf{r}})_{\mathbf{r}' \to \mathbf{r}}(\tilde{\bar{\psi}}(x')\tilde{\psi}(x)) - \frac{e^2\mathbf{A}(x)}{m}(\tilde{\bar{\psi}}(x)\tilde{\psi}(x)). \tag{37.4}$$

Therefore, we can immediately write the current density $\mathbf{j}(x)$ in terms of the Green's function of the system:

$$\mathbf{j}(x) = 2\left[\frac{ie}{2m}(\nabla_{\mathbf{r}'} - \nabla_{\mathbf{r}})\mathscr{G}(x, x') - \frac{e^2\mathbf{A}(x)}{m}\mathscr{G}(x, x')\right]_{\mathbf{r}' \to \mathbf{r},\, \tau' \to \tau + 0}. \tag{37.5}$$

We now proceed to find the Green's function, or more exactly, the corrections to the Green's function which are of the first order in the magnetic field. In a constant magnetic field, the Green's functions $\mathscr{G}$, $\mathscr{F}$, $\mathscr{F}^+$ do not depend on the variables τ_1 and τ_2 separately, but only on the "time" difference $\tau_1 - \tau_2$. Thus, going over to Fourier components $\mathscr{G}_\omega$ and $\mathscr{F}_\omega^+$, we can write the following system of equations for the values of these quantities in a constant magnetic field [cf. (34.25)]:

$$\left(i\omega + \frac{1}{2m}[\nabla_{\mathbf{r}} - ie\mathbf{A}(\mathbf{r})]^2 + \mu\right)\mathscr{G}_\omega(\mathbf{r}, \mathbf{r}') + \Delta(\mathbf{r})\mathscr{F}_\omega^+(\mathbf{r}, \mathbf{r}') = \delta(\mathbf{r} - \mathbf{r}'),$$
$$\left(-i\omega + \frac{1}{2m}[\nabla_{\mathbf{r}} + ie\mathbf{A}(\mathbf{r})]^2 + \mu\right)\mathscr{F}_\omega^+(\mathbf{r}, \mathbf{r}') - \Delta^*(\mathbf{r})\mathscr{G}_\omega(\mathbf{r}, \mathbf{r}') = 0. \tag{37.6}$$

Now suppose we write the Green's functions $\mathscr{G}$ and $\mathscr{F}^+$ in the form

$$\mathscr{G} = \mathscr{G}_0 + \mathscr{G}^{(1)}, \quad \mathscr{F} = \mathscr{F}_0 + \mathscr{F}^{(1)}, \quad \mathscr{F}^+ = \mathscr{F}_0^+ + \mathscr{F}^{+(1)},$$

where $\mathscr{G}_0$, $\mathscr{F}_0$, $\mathscr{F}_0^+$ are the Green's functions in the absence of the field, and $\mathscr{G}^{(1)}$, $\mathscr{F}^{(1)}$, $\mathscr{F}^{+(1)}$ are the corrections which are linear in the field. Linearizing the equations (37.6), we obtain

$$\left(i\omega + \frac{\nabla^2}{2m} + \mu\right)\mathscr{G}_\omega^{(1)}(\mathbf{r}, \mathbf{r}') + \Delta^{(0)}\mathscr{F}_\omega^{+(1)}(\mathbf{r}, \mathbf{r}')$$
$$= -\Delta^{(1)}(\mathbf{r})\mathscr{F}_{0\omega}^+(\mathbf{r} - \mathbf{r}') + \frac{ie}{2m}(\nabla\cdot\mathbf{A} + \mathbf{A}\cdot\nabla)\mathscr{G}_{0\omega}(\mathbf{r} - \mathbf{r}'),$$
$$\left(-i\omega + \frac{\nabla^2}{2m} + \mu\right)\mathscr{F}_\omega^{+(1)}(\mathbf{r}, \mathbf{r}') - \Delta^{(0)}\mathscr{G}_\omega^{(1)}(\mathbf{r}, \mathbf{r}')$$
$$= \Delta^{*(1)}(\mathbf{r})\mathscr{G}_{0\omega}(\mathbf{r} - \mathbf{r}') - \frac{ie}{2m}(\nabla\cdot\mathbf{A} + \mathbf{A}\cdot\nabla)\mathscr{F}_{0\omega}^+(\mathbf{r} - \mathbf{r}'). \tag{37.7}$$

Using these equations, we can quite easily express $\mathscr{G}_\omega^{(1)}(\mathbf{r}, \mathbf{r}')$ and $\mathscr{F}_\omega^{+(1)}(\mathbf{r}, \mathbf{r}')$ in terms of the quantities appearing in the right-hand side of (37.7). To do this, it is convenient to use the expression (34.32) for the inverse of the operator appearing in the left-hand side of (37.7). However, before writing the corresponding result for $\mathscr{G}_\omega^{(1)}(\mathbf{r}, \mathbf{r}')$, we first examine the structure of the equations (37.7) in more detail.

The system (37.6), and consequently the system (37.7), is gauge invariant, i.e., invariant under the transformations (34.26) and (34.27). Therefore, in calculating the current density $\mathbf{j}(\mathbf{r})$, which in the linear approximation equals

$$\mathbf{j}(\mathbf{r}) = \frac{ie}{m}T\sum_\omega(\nabla_{\mathbf{r}'} - \nabla_{\mathbf{r}})_{\mathbf{r}'\to\mathbf{r}}\,\mathscr{G}_\omega^{(1)}(\mathbf{r}, \mathbf{r}') - \frac{e^2\mathbf{A}(\mathbf{r})N}{m}, \tag{37.8}$$

the final result can only depend on the transverse part of the vector potential $\mathbf{A}$. In other words, adding the gradient of any scalar to $\mathbf{A}$, i.e., changing $\mathbf{A}$ to $\mathbf{A} + \nabla_{\mathbf{r}}\chi$, cannot change the current density $\mathbf{j}(\mathbf{r})$. However, the functions $\mathscr{G}_\omega^{(1)}(\mathbf{r}, \mathbf{r}')$, $\mathscr{F}_\omega^{+(1)}(\mathbf{r}, \mathbf{r}')$ are by no means invariant under such a transformation of the vector potential, and the same applies to the quantities $\Delta^{(1)}(\mathbf{r})$, $\Delta^{*(1)}(\mathbf{r})$ appearing in the right-hand side of (37.7), which must satisfy the equation

$$\Delta^{*(1)}(\mathbf{r}) = |\lambda| T \sum_{\omega} \mathscr{F}_{\omega}^{+(1)}(\mathbf{r}, \mathbf{r}).$$

When the vector potential $\mathbf{A}(\mathbf{r})$ is arbitrary, $\Delta^*(\mathbf{r})$ is generally an unknown function of $\mathbf{A}$. Nevertheless, because of the homogeneity of the problem, it can be asserted that the function $\Delta^{*(1)}(\mathbf{r})$, being a scalar, depends only on the quantity div $\mathbf{A}$, to the approximation which is linear in the field. Therefore, if we choose the gauge so that

$$\operatorname{div} \mathbf{A} = 0,$$

the problem can be simplified considerably, since with this choice of $\mathbf{A}(\mathbf{r})$, the function $\Delta^{*(1)}(\mathbf{r})$ vanishes identically. The results derived below apply only to such a "purely transverse" vector potential.

The method just given (in which $\Delta^{*(1)}$, $\Delta^{(1)}$ vanish), can be generalized in such a way as to be applicable to problems that do not have spatial homogeneity in the absence of the field (e.g., the problem of a superconductor of finite dimensions in a magnetic field). In such cases, we can form a scalar from the field $\mathbf{A}(\mathbf{r})$ and the vector $\mathbf{r}$ (or any other vector characterizing the problem). Thus, in the general case of an inhomogeneous problem, the functions $\Delta^{(1)}(\mathbf{r})$, $\Delta^{*(1)}(\mathbf{r})$ in equations (37.7)[9] now depend on both the longitudinal and the transverse components of the vector $\mathbf{A}(\mathbf{r})$. However, it is always possible to choose the longitudinal part $\mathbf{A}_{\text{long}} = \operatorname{grad} \chi$ in such a way that $\Delta^{(1)}, \Delta^{*(1)}$ vanish. The function χ which achieves this can be determined from the condition div $\mathbf{j} = 0$, i.e., from the law of charge conservation.

We now return to the equations (37.7) for an infinite superconductor. Setting $\Delta^{(1)}(\mathbf{r}) = 0$ and using (34.32), we find the following expression for $\mathscr{G}_{\omega}^{(1)}(\mathbf{r}, \mathbf{r}')$, the correction to the Green's function which is linear in the field:

$$\begin{aligned} \mathscr{G}_{\omega}^{(1)}(\mathbf{r}, \mathbf{r}') = \frac{ie}{m} \int \{ & \mathscr{G}_{0\omega}(\mathbf{r} - \boldsymbol{l})[\mathbf{A}(\boldsymbol{l}) \cdot \nabla_l] \mathscr{G}_{0\omega}(\boldsymbol{l} - \mathbf{r}') \\ & + \mathscr{F}_{0\omega}(\boldsymbol{l} - \mathbf{r})[\mathbf{A}(\boldsymbol{l}) \cdot \nabla_l] \mathscr{F}_{0\omega}^{+}(\mathbf{r}' - \boldsymbol{l})\} \, d\boldsymbol{l}. \end{aligned} \tag{37.9}$$

In this formula, we have already used the condition div $\mathbf{A} = 0$, and as a result, we can assume that the differentiation in the operator $\mathbf{A}(\boldsymbol{l}) \cdot \nabla_l$ applies only to the functions $\mathscr{G}_{0\omega}(\boldsymbol{l} - \mathbf{r}')$ and $\mathscr{F}_{0\omega}^{+}(\mathbf{r}' - \boldsymbol{l})$. According to (37.8), the current density equals

$$\begin{aligned} \mathbf{j}(\mathbf{r}) = \frac{e^2}{m^2} T \sum_{\omega} (\nabla_{\mathbf{r}} - \nabla_{\mathbf{r}'}) \int \{ & \mathscr{G}_{0\omega}(\mathbf{r} - \boldsymbol{l})[\mathbf{A}(\boldsymbol{l}) \cdot \nabla_l] \mathscr{G}_{0\omega}(\boldsymbol{l} - \mathbf{r}') \\ & + \mathscr{F}_{0\omega}(\boldsymbol{l} - \mathbf{r})[\mathbf{A}(\boldsymbol{l}) \cdot \nabla_l] \mathscr{F}_{0\omega}^{+}(\mathbf{r}' - \boldsymbol{l})\}_{\mathbf{r}' \to \mathbf{r}} \, d\boldsymbol{l} - \frac{Ne^2}{m} \mathbf{A}(\mathbf{r}). \end{aligned} \tag{37.10}$$

---

[9] The functions $\mathscr{G}_0$ and $\mathscr{F}_0^{+}$ in (37.7) must now be understood to be the Green's functions of the given body, subject to the appropriate boundary conditions.

At this point, it is convenient to go over to the Fourier representation, by introducing the Fourier components of the current density $\mathbf{j(r)}$ and the vector potential $\mathbf{A(r)}$ in the usual way, i.e.,

$$\mathbf{j}(\mathbf{r}) = \frac{1}{(2\pi)^3}\int \mathbf{j}(\mathbf{k})e^{i\mathbf{k}\cdot\mathbf{r}}\,d\mathbf{k}, \qquad \mathbf{A}(\mathbf{r}) = \frac{1}{(2\pi)^3}\int \mathbf{A}(\mathbf{k})e^{i\mathbf{k}\cdot\mathbf{r}}\,d\mathbf{k}.$$

The equation relating the components $\mathbf{j(k)}$ and $\mathbf{A(k)}$ is then

$$\mathbf{j}(\mathbf{k}) = -\frac{2e^2T}{(2\pi)^3m^2}\sum_{\omega}\int \mathbf{p}[\mathbf{p}\cdot\mathbf{A}(\mathbf{k})][\mathscr{G}_{\omega}(\mathbf{p}_+)\mathscr{G}_{\omega}(\mathbf{p}_-) + \mathscr{F}_{\omega}(\mathbf{p}_+)\mathscr{F}^+_{\omega}(\mathbf{p}_-)]\,d\mathbf{p} - \frac{Ne^2}{m}\mathbf{A}(\mathbf{k}), \tag{37.11}$$

where $\mathbf{p}_{\pm} = \mathbf{p} \pm \frac{1}{2}\mathbf{k}$, and $\mathscr{G}_{\omega}$, $\mathscr{F}_{\omega}$, $\mathscr{F}^+_{\omega}$ are given by (34.35). We emphasize once again that our result pertains only to the case where the gauge is such that the vector potential is purely transverse. However, (37.11) remains valid for any gauge, provided we replace $\mathbf{A(k)}$ by its transverse part.

In a superconductor, the field $\mathbf{A(r)}$ and the current $\mathbf{j(r)}$ vary in distances of the order of the penetration depth δ (usually $\sim 10^{-5}$), i.e., in distances much larger than atomic distances. Therefore, in (37.11) the components $\mathbf{j(k)}$ and $\mathbf{A(k)}$ are only important in the region

$$k \sim \frac{1}{\delta} \ll p_0.$$

We shall see below that the contribution to the integral in (37.11) comes mainly from a narrow region of values of $|\mathbf{p}|$ near the Fermi surface, of order $||\mathbf{p}| - p_0| \sim |\mathbf{k}|$. Moreover, (37.11) involves only the two vectors $\mathbf{k}$ and $\mathbf{A(k)}$, with $\mathbf{k}\cdot\mathbf{A} = 0$. Expressing the variable $\mathbf{p}$ in a system of spherical coordinates with $\mathbf{k}$ as the polar axis and averaging over angles in the azimuthal plane, we immediately find that the vector $\mathbf{j(k)}$ has the same direction as the vector $\mathbf{A(k)}$. In fact, these considerations show that the result of substituting the expressions (34.35) for the functions $\mathscr{G}_{\omega}$ and $\mathscr{F}^+_{\omega}$ into (37.11) is

$$\mathbf{j}(\mathbf{k}) = -\frac{Ne^2}{m}\bar{Q}(\mathbf{k})\mathbf{A}(\mathbf{k}),$$

where

$$\bar{Q}(\mathbf{k}) = 1 + \frac{3T}{4}\sum_{\omega}\int_0^{\pi}\sin^3\theta\,d\theta\int_{-\infty}^{+\infty}d\xi\,\frac{(i\omega + \xi_+)(i\omega + \xi_-) + \Delta^2}{(\omega^2 + \xi_+^2 + \Delta^2)(\omega^2 + \xi_-^2 + \Delta^2)}. \tag{37.12}$$

(Here, we use the fact that $p_0^3/3\pi^2 = N$.) For small $\mathbf{k}$, we have

$$\xi_{\pm} = \xi \pm \frac{\mathbf{v}\cdot\mathbf{k}}{2}.$$

To make further calculations, we must bear in mind that for large ω and

ξ, the integrand in the right-hand side of (37.12) behaves like ω^{-2} for $\omega \gg \xi$, and like ξ^{-2} for $\xi \gg \omega$. Therefore, strictly speaking, the integral over ξ and the sum over the frequencies ω diverge. To understand what is at issue here, we consider the singularities of this expression for a normal metal (i.e., for $\Delta = 0$):

$$\frac{3}{4}T\sum_{\omega}\int d\xi \int_0^{\pi} \sin^3\theta\, d\theta \frac{1}{(i\omega - \xi_+)(i\omega - \xi_-)}. \tag{37.13}$$

First we note that in (37.13), it is of vital importance in which order the summation over the frequencies and the integration over ξ are performed. In fact, if we first integrate over ξ, the poles of the integrand lie on one side of the real axis for any sign of ω, and hence the resulting expression turns out to be zero. On the other hand, suppose we first sum over the frequencies $\omega = (2n + 1)\pi T$. Then it is easily seen that the result of summing this simple series is

$$\frac{3}{8}\int_0^{\pi} \sin^3\theta\, d\theta \int \frac{d\xi}{\mathbf{v}\cdot\mathbf{k}}\left(\tanh\frac{\xi_-}{2T} - \tanh\frac{\xi_+}{2T}\right) = -1. \tag{37.14}$$

The reason for the dependence of (37.13) on the relative order of integration and summation is contained in the fact that (37.13) is formally divergent. However, it can be seen that the essential point here is the following: If we first sum over the frequencies, it turns out that the resulting sum is nonzero only in a very narrow energy region near the Fermi surface [it follows from (37.14) that the width of this region is $\sim\mathbf{v}\cdot\mathbf{k}$]. In this region, the integral with respect to the momentum is rapidly convergent, and it is only for this reason that we can write the expression for the energy of the excitations, measured from the Fermi surface, in the form

$$\xi = \frac{\mathbf{p}^2 - p_0^2}{2m} \approx v(|\mathbf{p}| - p_0).$$

Therefore, in integrals of the given type, we must always calculate the sum first, and only afterwards integrate with respect to ξ. Otherwise, the integral with respect to ξ will include the region $||\mathbf{p}| - p_0| \sim p_0$, and then it is no longer appropriate to write all quantities as expansions near the Fermi surface, as we have done.

Having made these remarks, we now describe a method which allows us to avoid the necessity of evaluating the rather complicated sum over the frequencies in (37.12). To the integrand in (37.12) we add and subtract the similar expression for the normal metal [cf. (37.13)]. Then the integral and the sum over frequencies of the difference between the integrands converge rapidly, and hence we can interchange the order of summation and integration. The corresponding expression for the normal metal, given by (37.14), cancels the first term on the right (equal to 1) in (37.12). Integrating with respect to ξ, we obtain

$$\bar{Q}(\mathbf{k}) = \frac{3\pi T}{4}\sum_{\omega}\int_{-1}^{+1} \frac{(1 - \beta^2)\, d\beta}{\sqrt{\omega^2 + \Delta^2}} \frac{\Delta^2}{\omega^2 + \Delta^2 + \frac{1}{4}v^2|\mathbf{k}|^2\beta^2}. \tag{37.15}$$

It is difficult to transform the kernel $\bar{Q}(\mathbf{k})$ any further without making some assumptions about the size of $|\mathbf{k}|$. It is clear from the structure of the integrand in (37.15) that only the ratio of the quantity $v|\mathbf{k}|$ to the transition temperature T_c is important. In fact, for $T \ll T_c$, the gap Δ_0 is of order T_c, while near T_c, i.e., for $|T - T_c| \ll T_c$, the gap is small, but then $\omega = (2n + 1)\pi T \sim T_c$. The quantity $\xi_0 \sim v/T_c$, with the dimensions of length, is the "coherence distance" of the bound electrons (see p. 297), and acts as a characteristic parameter in the modern theory of superconductivity. The penetration depth δ can be either larger or smaller than ξ_0. In the first limiting case, where $\delta \gg \xi_0$, the important values of $|\mathbf{k}|$, which are of order $1/\delta$, satisfy the inequality $v|\mathbf{k}| \ll T_c$, while in the second limiting case, where $\delta \ll \xi_0$, they satisfy the inequality $v|\mathbf{k}| \gg T_c$.

We begin by examining the first limiting case, where $v|\mathbf{k}| \ll T_c$. In the expression (37.15), we retain only the first term in an expansion with respect to $v|\mathbf{k}|$:

$$\bar{Q}(\mathbf{k}) = \frac{3\pi}{4}\Delta^2 T \sum_{\omega} \int_{-1}^{+1} \frac{(1 - \beta^2)\, d\beta}{(\omega^2 + \Delta^2)^{3/2}} = \pi T \Delta^2 \sum_{\omega} \frac{1}{(\omega^2 + \Delta^2)^{3/2}}. \tag{37.16}$$

Thus, for $\delta \gg \xi_0$, the kernel $\bar{Q}(\mathbf{k})$ does not depend on $\mathbf{k}$, and the relation between the current and the field is of a local character, in the sense that the current at a given point $\mathbf{r}$ is determined only by the field $\mathbf{A}(\mathbf{r})$ at that point:

$$\mathbf{j}(\mathbf{r}) = -\frac{e^2 N_s}{m}\mathbf{A}(\mathbf{r}). \tag{37.17}$$

An equation of this form was first suggested by London and London (L14, L15), and therefore superconductors for which $\delta \gg \xi_0$ will henceforth be called *superconductors of the London type*. The function $N_s(T)$ can be thought of as the number of "superconducting" electrons. Using formula (37.16), we can express the ratio $N_s(T)/N$ as a function of the temperature. It should be emphasized that the gap Δ appearing in (37.16) is the equilibrium gap at the given temperature in the absence of the field, and is determined by the condition (34.37). At $T = 0$ the sum over frequencies can be replaced by an integral, by setting $2\pi T\delta n = d\omega$. Evaluating this integral, we find that at $T = 0$ the number of "superconducting electrons" equals N, the total number of electrons.

Near T_c the quantity $\Delta(T)$ is small compared to T_c and ω. Thus, neglecting Δ^2 in the denominator of (37.16), we obtain the series

$$\frac{N_s(T)}{N} = \frac{2\Delta^2}{\pi^2 T^2} \sum_{n>0} \frac{1}{(2n + 1)^3},$$

which has already been evaluated on p. 304. Using the expression (36.6) for the size of the gap near T_c, we find that in this case

$$\frac{N_s(T)}{N} = 2\left(1 - \frac{T}{T_c}\right).$$

Next, we consider the second limiting case, where $v|\mathbf{k}| \gg T_c$. The integrand in (37.15) has poles at the points

$$\frac{v|\mathbf{k}|\beta}{2} = \pm i\sqrt{\omega^2 + \Delta^2}.$$

Since $v|\mathbf{k}|$ is large, this means that the integrand has a sharp maximum in the range of angles $\beta \approx T_c/v|\mathbf{k}| \ll 1$. Therefore, in the numerator we can neglect β^2 compared to unity, and the remaining expression is a rapidly converging integral with respect to β, whose integrand falls off like β^{-2} in the region $T_c/v|\mathbf{k}| \ll \beta \ll 1$. Substituting $v|\mathbf{k}|\beta = x$ and going over to infinite limits of integration, we can evaluate this integral by using residues:

$$\bar{Q}(\mathbf{k}) = \frac{3T\pi^2}{v|\mathbf{k}|} \sum_{\omega>0} \frac{\Delta^2}{\omega^2 + \Delta^2} = \frac{3\pi^2}{4v|\mathbf{k}|} \Delta \tanh \frac{\Delta}{2T}. \tag{37.18}$$

We see that in this case the kernel $\bar{Q}(\mathbf{k})$ depends on $\mathbf{k}$ in an essential way. Therefore, if the penetration depth of the field is such that $\delta \ll \xi_0$, the relation (37.3) is nonlocal, i.e., the value of the current density $\mathbf{j}(\mathbf{r})$ at a given point $\mathbf{r}$ is determined by the values of the vector potential in a whole neighborhood of $\mathbf{r}$, with linear dimensions of order ξ_0. The fact that there is a nonlocal relation between the field and current in certain superconductors was first predicted by Pippard (P2), on the basis of an analysis of experimental data. Accordingly, superconductors for which $\delta \ll \xi_0$ will henceforth be called *superconductors of the Pippard type.*

The following important fact should be pointed out at once: As already mentioned, in deciding which of these two cases actually occurs, all that matters is the ratio of the penetration depth δ to the parameter $\xi_0 \sim v/T_c$. Therefore, if the condition $\delta \ll \xi_0$ holds at sufficiently low temperatures, then, since δ increases as we approach T_c, it follows that at temperatures sufficiently close to T_c the opposite case prevails, i.e., δ becomes much larger than ξ_0. In other words, the superconductor is always of the London type for temperatures in the immediate neighborhood of T_c. A considerable number of known superconductors are of the Pippard type over almost the whole temperature region, and only go over into superconductors of the London type in a very narrow neighborhood of T_c, i.e., for $T_c - T \ll T_c$. The remaining pure superconductors are of an intermediate type at low temperatures, and then have a rather marked "London region" of temperatures near T_c. The case of superconducting alloys will be considered in Sec. 39.3.

Now that the expression (37.15) for the kernel $\bar{Q}(\mathbf{k})$ is at our disposal, we can use Maxwell's equations to solve the problem of the penetration of a magnetic field into a superconductor with a plane surface. In the London case, this problem has a particularly simple solution. Substituting (37.17) into (37.2), and assuming that all quantities are functions of the coordinate z

only, we find that the distribution of the vector potential in the superconductor is given by the expression

$$A_y(z) = -H(0)\,\delta e^{z/\delta},$$

where

$$\delta = \sqrt{\frac{m}{4\pi N_s e^2}} \tag{37.19}$$

is the *London penetration depth.*[10] The solution of the analogous problem in the Pippard case is very complicated and requires the use of special mathematical methods. We shall not discuss this case here, but instead we refer the reader who is particularly interested in the theory of superconductivity (and the results to which it leads) to the original literature (see e.g., B1, K5).

37.2. A superconductor in an alternating field. So far, we have only considered the properties of superconductors in a constant magnetic field. However, the behavior of superconductors in an alternating electromagnetic field is also a problem of great physical interest (in particular, the nature of absorption and reflection of electromagnetic radiation incident upon the surface of a superconductor). The thermodynamic or equilibrium approach to this problem, which was used in our previous treatment, is not directly applicable to the case of a variable field. In this case, the analytic relations derived in Sec. 17, connecting the various time-dependent functions with the corresponding functions in the temperature technique, turn out to be extremely useful.

Thus, consider an infinite superconductor, in which there is an alternating field $\mathbf{A}$ of frequency ω (as before, we assume that the scalar potential φ is zero). Clearly, formula (37.3) still describes the relation between the field and the current which appears in the superconductor because of the action of the field. The difference between the case of an alternating field and the case of a constant field is that we must now know the Fourier components $Q(\mathbf{k}, \omega)$ for nonzero ω [the kernel $\bar{Q}(\mathbf{k})$ defined previously obviously satisfies the relation $\bar{Q}(\mathbf{k}) \equiv Q(\mathbf{k}, 0)$]. We again start from the quantum-mechanical expression for the current-density operator

$$\hat{\mathbf{j}}(x) = \frac{ie}{2m}(\nabla_{\mathbf{r}'} - \nabla_{\mathbf{r}})_{\mathbf{r}'\to\mathbf{r}}\tilde{\psi}^+(x')\tilde{\psi}(x) - \frac{e^2}{m}\mathbf{A}(x)\tilde{\psi}^+(x)\tilde{\psi}(x)$$

$$\equiv \hat{\mathbf{j}}_1(x) - \frac{e^2}{m}\mathbf{A}(x)\tilde{\psi}^+(x)\tilde{\psi}(x),$$

[10] In ordinary units,

$$\delta = \sqrt{\frac{mc^2}{4\pi N_s e^2}}$$

where the operators are written in the Heisenberg representation and depend on the field. According to (6.28), the relation between these operators and the corresponding operators in the interaction representation is given by the formula

$$\mathring{\mathbf{j}} = S^{-1}(t)\mathbf{j}S(t),$$

where

$$S(t) = T\exp\left\{i\int_{-\infty}^{t} \mathbf{j}(x)\cdot\mathbf{A}(x)\,d^4x\right\}.$$

To the approximation which is linear in the field, we have

$$\mathring{\mathbf{j}}_\alpha(x) = \mathbf{j}_{1\alpha}(x) - \frac{e^2}{m}\mathbf{A}_\alpha(x)\psi^+(x)\psi(x) + i\int_{-\infty}^{t}[\mathbf{j}_{1\alpha}(x), \mathbf{j}_{1\beta}(y)]\mathbf{A}_\beta(y)\,d^4y.$$

The current density in the superconductor at a given point and at a given time is given by the average

$$\mathbf{j}(x) = \langle\mathring{\mathbf{j}}(x)\rangle = \sum_m e^{(\Omega+\mu N_m - E_m)/T}\langle m|\mathring{\mathbf{j}}(x)|m\rangle.$$

Since $\langle\mathbf{j}_1\rangle \equiv 0$, it follows that

$$\mathbf{j}_\alpha(x) = -\frac{e^2N}{m}\mathbf{A}_\alpha(x) + \int P^R_{\alpha\beta}(x-y)\mathbf{A}_\beta(y)\,d^4y, \tag{37.20}$$

where we have introduced the function

$$P^R_{\alpha\beta}(x-y) = \begin{cases} i\langle[\mathbf{j}_\alpha(x), \mathbf{j}_\beta(y)]\rangle & \text{for} \quad t_x > t_y, \\ 0 & \text{for} \quad t_x < t_y. \end{cases} \tag{37.21}$$

Then, going over to Fourier components, we find that the kernel of the integral relation (37.3) has the form

$$Q_{\alpha\beta}(\mathbf{k}, \omega) = \frac{e^2N}{m}\delta_{\alpha\beta} - P^R_{\alpha\beta}(\mathbf{k}, \omega). \tag{37.22a}$$

Next, we consider the same problem, using the finite-temperature technique. Formally regarding the field $\mathbf{A}(\mathbf{r}, \tau)$ and the current density $\mathbf{j}(\mathbf{r}, \tau)$ as functions of the "time" parameter τ, we have the relation

$$\mathbf{j}_\alpha(\mathbf{r}, \tau) = -\frac{e^2N}{m}\mathbf{A}_\alpha(\mathbf{r}, \tau) + \int d\mathbf{r}\int_0^\beta \mathscr{P}_{\alpha\beta}(x-y)\mathbf{A}_\beta(y)\,d\tau_y,$$

instead of (37.20), where

$$\mathscr{P}_{\alpha\beta}(\mathbf{r}-\mathbf{r}', \tau-\tau') = \langle T_\tau(\mathbf{j}_{\alpha 1}(\mathbf{r}, \tau)\mathbf{j}_{1\beta}(\mathbf{r}', \tau'))\rangle. \tag{37.23}$$

Moreover, taking Fourier components of the temperature quantities, we now have

$$Q_{\alpha\beta}(\mathbf{k}, \omega_0) = \frac{e^2N}{m}\delta_{\alpha\beta} - \mathscr{P}_{\alpha\beta}(\mathbf{k}, \omega_0), \tag{37.22b}$$

instead of (37.22a), where ω_0 ranges over the discrete values $\omega_0 = 2n\pi T$.

We now prove quite generally that the Fourier components $P^R_{\alpha\beta}(\mathbf{k}, \omega)$ and $\mathscr{P}_{\alpha\beta}(\mathbf{k}, \omega_0)$ are the values of one and the same function of the complex variable ω, which is analytic in the upper half-plane, provided the values of ω are taken along the real axis in the first case and at the points $\omega = i\omega_0$ in the second case. The method of proof is identical with that used in the previous chapters. Writing (37.21) and (37.23) as sums over intermediate states, we find that the appropriate expressions for the Fourier components of the functions (37.21) and (37.23) are

$$\begin{aligned} P^R_{\alpha\beta}(\mathbf{k}, \omega) &= -\sum_{m,p} \rho_{pm}(\mathbf{k}) \frac{1}{\omega - \omega_{pm} + i\delta}, \\ \mathscr{P}_{\alpha\beta}(\mathbf{k}, \omega_0) &= -\sum_{m,p} \rho_{pm}(\mathbf{k}) \frac{1}{i\omega_0 - \omega_{pm}}, \end{aligned} \tag{37.24}$$

where

$$\rho_{pm}(\mathbf{k}) = e^{(\Omega + \mu N_m - E_m)/T}(1 - e^{-\omega_{pm}/T})(\mathbf{j}_{\alpha 1})_{mp}(\mathbf{j}_{\beta 1})_{pm}(2\pi)^3\delta(\mathbf{k} - \mathbf{k}_{pm}).$$

It is clear from (37.24) that $P^R_{\alpha\beta}(\mathbf{k}, \omega)$ is obtained from $\mathscr{P}_{\alpha\beta}(\mathbf{k}, \omega_0)$ by replacing ω_0 by $-i\omega$, where the values of $P^R_{\alpha\beta}(\mathbf{k}, \omega)$ along the real axis must be defined as the limit as ω approaches the real axis from above. Thus, using the Matsubara technique to calculate $\mathscr{P}_{\alpha\beta}(\mathbf{k}, \omega_0)$, and then continuing this function to real frequencies by setting $P^R_{\alpha\beta}(\mathbf{k}, \omega) = \mathscr{P}_{\alpha\beta}(\mathbf{k}, -i\omega)$ in such a way that the resulting function has no singularities in the upper half-plane of the complex variable ω, we can in principle find the kernel $Q_{\alpha\beta}(\mathbf{k}, \omega)$ which determines the relation between $\mathbf{j}$ and $\mathbf{A}$ in the case of an alternating electromagnetic field.

With this as our objective, we proceed to examine formally the equations for the temperature quantities $\mathscr{G}$ and $\mathscr{F}^+$ in a field of the form

$$\mathbf{A}(\mathbf{r}, \tau) = \mathbf{A}(\mathbf{k}, \omega_0)e^{i\mathbf{k}\cdot\mathbf{r} - i\omega_0\tau},$$

which depends on τ. For the Fourier components of the current density, we now have

$$\begin{aligned} \mathbf{j}(\mathbf{k}, \omega_0) = -\frac{2e^2}{(2\pi)^3 m^2} T \sum_{\omega'} \int \mathbf{p}[\mathbf{p}\cdot\mathbf{A}(\mathbf{k}, \omega_0)][\mathscr{G}(p_+)\mathscr{G}(p_-) \\ + \mathscr{F}(p_+)\mathscr{F}^+(p_-)]\, d\mathbf{p} - \frac{Ne^2}{m}\mathbf{A}(\mathbf{k}, \omega_0) \end{aligned}$$

instead of (37.11), where

$$p_\pm = (\mathbf{p}_\pm, \omega_\pm) = (\mathbf{p} \pm \tfrac{1}{2}\mathbf{k}, \omega' \pm \tfrac{1}{2}\omega_0).$$

Repeating the whole argument which led to (37.15) in the case of a constant field, we obtain the following expression, after integrating with respect to ξ:

$$\begin{aligned} Q(\mathbf{k}, \omega_0) &= \frac{3\pi T}{4} \sum_{\omega'} \int_{-1}^{+1} d\beta\,(1 - \beta^2) \\ &\times \left\{ \frac{i(\omega_+ + \sqrt{\omega_+^2 + \Delta^2})[i(\omega_- + \sqrt{\omega_+^2 + \Delta^2}) - v|\mathbf{k}|\beta] + \Delta^2}{\sqrt{\omega_+^2 + \Delta^2}[\omega_-^2 + \Delta^2 + (v|\mathbf{k}|\beta - i\sqrt{\omega_+^2 + \Delta^2})^2]} \right. \\ &+ \left. \frac{i(\omega_- + \sqrt{\omega_-^2 + \Delta^2})[i(\omega_+ + \sqrt{\omega_-^2 + \Delta^2}) + v|\mathbf{k}|\beta] + \Delta^2}{\sqrt{\omega_-^2 + \Delta^2}[\omega_+^2 + \Delta^2 + (v|\mathbf{k}|\beta + i\sqrt{\omega_-^2 + \Delta^2})^2]} \right\}. \end{aligned} \tag{37.25}$$

To simplify (37.25) further, we have to make certain assumptions about the size of $v|\mathbf{k}|$, just as before. In what follows, we shall confine ourselves to the case of greatest practical interest, where $v|\mathbf{k}| \gg T_c$, $v|\mathbf{k}| \gg \omega_0$. In this case, by the previous argument, the main contribution to the expression for $Q(\mathbf{k}, \omega_0)$ comes from the range of angles $\beta \sim T_c/v|\mathbf{k}|$, $\beta \sim \omega'/v|\mathbf{k}|$, and therefore, in the numerator of the integrand we can neglect β^2 compared to unity. For $\beta \gg T_c/v|\mathbf{k}|$, $\beta \gg \omega'/v|\mathbf{k}|$, the remaining expression in braces falls off more slowly than in the case $T = 0$, i.e., like β^{-1}. For this reason, it is convenient to regroup terms in (37.25), by isolating the terms which fall off more slowly:

$$Q(\mathbf{k}, \omega_0) = \frac{3\pi T}{4} \sum_{\omega'} \int_{-1}^{1} d\beta$$

$$\times \left\{ \frac{\Delta^2 - (\omega_+ + \sqrt{\omega_+^2 + \Delta^2})(\omega_- + \sqrt{\omega_-^2 + \Delta^2})}{\sqrt{\omega_+^2 + \Delta^2}[\omega_-^2 + \Delta^2 + (v|\mathbf{k}|\beta - i\sqrt{\omega_+^2 + \Delta^2})^2]} \right.$$

$$+ \frac{\Delta^2 - (\omega_- + \sqrt{\omega_-^2 + \Delta^2})(\omega_+ + \sqrt{\omega_+^2 + \Delta^2})}{\sqrt{\omega_-^2 + \Delta^2}[\omega_+^2 + \Delta^2 + (v|\mathbf{k}|\beta + i\sqrt{\omega_-^2 + \Delta^2})^2]}$$

$$- i \frac{\omega_+ + \sqrt{\omega_+^2 + \Delta^2}}{\sqrt{\omega_+^2 + \Delta^2}[v|\mathbf{k}|\beta - i(\sqrt{\omega_+^2 + \Delta^2} + \sqrt{\omega_-^2 + \Delta^2})]}$$

$$\left. + i \frac{\omega_- + \sqrt{\omega_-^2 + \Delta^2}}{\sqrt{\omega_-^2 + \Delta^2}[v|\mathbf{k}|\beta + i(\sqrt{\omega_+^2 + \Delta^2} + \sqrt{\omega_-^2 + \Delta^2})]} \right\}.$$

Carrying out the integration and taking the limit $v|\mathbf{k}| \to \infty$, we obtain

$$Q(\mathbf{k}, \omega_0) = \frac{3\pi^2 T}{4v|\mathbf{k}|} \sum_{\omega'} \left(1 + \frac{\Delta^2 - \omega'(\omega' - \omega_0)}{\sqrt{\omega'^2 + \Delta^2}\sqrt{(\omega' - \omega_0)^2 + \Delta^2}}\right), \tag{37.26}$$

which goes into (37.18) when $\omega_0 = 0$.

Since in evaluating the sum in (37.25), ω' ranges over the values $\omega' = (2n + 1)\pi T$, we can represent $Q(\mathbf{k}, \omega_0)$ in the form of a contour integral

$$Q(\mathbf{k}, \omega_0) = \frac{3\pi i}{16v|\mathbf{k}|} \int_C \left(1 + \frac{\Delta^2 - \omega'(\omega' - \omega_0)}{\sqrt{\omega'^2 + \Delta^2}\sqrt{(\omega' - \omega_0)^2 + \Delta^2}}\right) \tan \frac{\omega'}{2T}\, d\omega', \tag{37.27}$$

where the contour C consists of the two parts C_+ and C_- shown in Fig. 98. The choice of the analytic branches of the functions $\sqrt{\omega'^2 + \Delta^2}$ and $\sqrt{(\omega' - \omega_0)^2 + \Delta^2}$ is clear from the figure. On the cuts, the values of these functions are purely imaginary, while the imaginary part is positive to the right of the upper cut and to the left of the lower cut. Next, we transform the integral along the contours C_+ and C_- into an integral along the four contours $C_+^{(1,2)}$ and $C_-^{(1,2)}$ shown in Fig. 99. It is easily seen that the integrals along $C_+^{(2)}$ and $C_-^{(2)}$, regarded

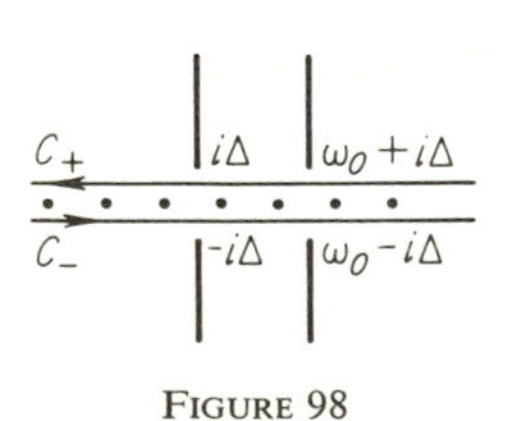

FIGURE 98

formally as functions of ω_0, have singularities for $\omega_0 = (2n + 1)\pi T \pm i\Delta$, since then the contour of integration goes through the points $\omega' = (2n + 1)\pi T$, where the function $\tan (\omega'/2T)$ becomes infinite. Therefore, to determine the branch of the function which is analytic in the upper half-plane of the variable $\omega = i\omega_0$, we have to transform the expression (37.27) for the special values $\omega = 2n\pi Ti$ in such a way that the contour of integration does not go through singularities of the integrand when we later extend (37.27) to

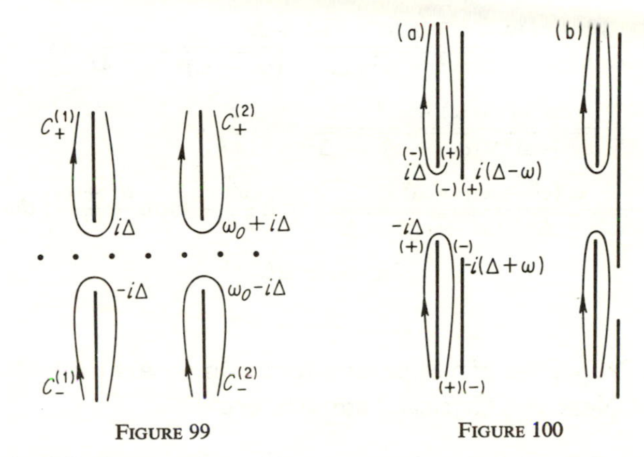

FIGURE 99

FIGURE 100

arbitrary values of ω. To accomplish this, we note that if $\omega_0 = 2n\pi T$, then, because of the periodicity of $\tan (\omega'/2T)$, the integral along the contour $C_-^{(2)}$ equals the integral along the contour $C_+^{(1)}$, and similarly for the contours $C_+^{(2)}$ and $C_-^{(1)}$. (This can be seen, for example, by introducing the new variable of integration $\omega' - \omega_0 = -u$ in the integral along the contour $C_-^{(2)}$.) Therefore, (37.27) can be written in the form

$$Q(\mathbf{k}, \omega_0) = \frac{3\pi i}{8v|\mathbf{k}|}\left(\int_{C_-^{(1)}} + \int_{C_+^{(1)}}\right)$$

$$\times \left(1 + \frac{\Delta^2 - \omega'(\omega' - \omega_0)}{\sqrt{\omega'^2 + \Delta^2}\sqrt{(\omega' - \omega_0)^2 + \Delta^2}}\right) \tan \frac{\omega'}{2T}\, d\omega'.$$

This expression, regarded formally as a function of the variable $\omega = i\omega_0$, represents an analytic function in the upper half-plane, since in the given case, the contour of intègration never goes through any singularities of the integrand, if $\operatorname{Im} \omega > 0$.

We are interested in the quantity $Q(\mathbf{k}, \omega)$ for $\omega > 0$, which can now be written down immediately. When this is done, two cases arise, i.e., $\omega < 2\Delta$ and $\omega > 2\Delta$. In both cases, the calculation can be carried out in an elementary fashion, in keeping with Fig. 100, where in parentheses we indicate the choice of the sign of the imaginary part of the functions on different sides of the cuts. The final result is given by the following expressions (A6, A7, M2):

$$Q(\mathbf{k}, \omega) = \frac{3\pi}{4v|\mathbf{k}|}\left\{\int_{\Delta}^{\Delta+\omega} \frac{\omega'(\omega' - \omega) + \Delta^2}{\sqrt{\omega'^2 - \Delta^2}\sqrt{\Delta^2 - (\omega' - \omega)^2}} \tanh \frac{\omega'}{2T} d\omega' \right.$$
$$\left. + i \int_{\Delta}^{\infty} \frac{\omega'(\omega' + \omega) + \Delta^2}{\sqrt{\omega'^2 - \Delta^2}\sqrt{(\omega' + \omega)^2 - \Delta^2}} \left(\tanh \frac{\omega'}{2T} - \tanh \frac{\omega' + \omega}{2T}\right) d\omega'\right\} \tag{a}$$

$$\text{for} \quad \omega < 2\Delta,$$

$$Q(\mathbf{k}, \omega) = \frac{3\pi}{4v|\mathbf{k}|}\left\{\int_{\omega-\Delta}^{\omega+\Delta} \frac{\omega'(\omega' - \omega) + \Delta^2}{\sqrt{\omega'^2 - \Delta^2}\sqrt{\Delta^2 - (\omega' - \omega)^2}} \tanh \frac{\omega'}{2T} d\omega' \right.$$
$$+ i \int_{\Delta}^{\omega-\Delta} \frac{\omega'(\omega' - \omega) + \Delta^2}{\sqrt{\omega'^2 - \Delta^2}\sqrt{(\omega' - \omega)^2 - \Delta^2}} \tanh \frac{\omega'}{2T} d\omega' \tag{b}$$
$$\left. + i \int_{\Delta}^{\infty} \frac{\omega'(\omega' + \omega) + \Delta^2}{\sqrt{\omega'^2 - \Delta^2}\sqrt{(\omega' + \omega)^2 - \Delta^2}} \left(\tanh \frac{\omega'}{2T} - \tanh \frac{\omega' + \omega}{2T}\right) d\omega'\right\}$$

$$\text{for} \quad \omega > 2\Delta.$$

38. The Properties of a Superconductor in an Arbitrary Magnetic Field Near the Critical Temperature

The properties of superconductors near the critical temperature constitute a special case, since then the size of the gap is small enough to cause all the equations to become much simpler. It is easily seen from the results of Sec. 36 [see (36.4)] that in this case the equations can be expanded in the quantity $1 - (T/T_c) \ll 1$. Moreover, as already noted in the preceding section, near T_c the penetration depth δ of a weak magnetic field is $\gg \xi_0$, i.e., all quantities in the field, including the field itself, vary over distances much larger than $\xi_0 \sim v/T_c$, the characteristic parameter of the theory. This fact allows us to construct a theory for this temperature region, which describes the behavior of superconductors in arbitrary magnetic fields, of the order of the critical field [see Gorkov (G7)].

With this aim, we again write down the equations (37.6), i.e.,

$$\left(i\omega + \frac{1}{2m}[\nabla_{\mathbf{r}} - ie\mathbf{A}(\mathbf{r})]^2 + \mu\right)\mathscr{G}_\omega(\mathbf{r}, \mathbf{r}') + \Delta(\mathbf{r})\mathscr{F}^+_\omega(\mathbf{r}, \mathbf{r}') = \delta(\mathbf{r} - \mathbf{r}'),$$

$$\left(-i\omega + \frac{1}{2m}[\nabla_{\mathbf{r}} + ie\mathbf{A}(\mathbf{r})]^2 + \mu\right)\mathscr{F}^+_\omega(\mathbf{r}, \mathbf{r}') - \Delta^*(\mathbf{r})\mathscr{G}_\omega(\mathbf{r}, \mathbf{r}') = 0,$$

together with the equation determining the size of the gap:

$$\Delta^*(\mathbf{r}) = |\lambda| T \sum_\omega \mathscr{F}^+_\omega(\mathbf{r}, \mathbf{r}). \tag{38.1}$$

Since $|\Delta|$ is small, we can expand the function $\mathscr{F}^+_\omega(\mathbf{r}, \mathbf{r}')$ in powers of $|\Delta|$, and then, substituting this expansion into (38.1), we find an equation for $\Delta^*(\mathbf{r})$. In doing this, it is convenient to introduce the Fourier components of the Green's functions $\tilde{\mathscr{G}}^{(0)}_\omega(\mathbf{r}, \mathbf{r}')$ for electrons in the normal metal [in the given field $\mathbf{A}(\mathbf{r})$]. The equation satisfied by $\tilde{\mathscr{G}}^{(0)}_\omega(\mathbf{r}, \mathbf{r}')$ can be written in two ways, either as

$$\left(i\omega + \frac{1}{2m}[\nabla_{\mathbf{r}} - ie\mathbf{A}(\mathbf{r})]^2 + \mu\right)\tilde{\mathscr{G}}^{(0)}_\omega(\mathbf{r}, \mathbf{r}') = \delta(\mathbf{r} - \mathbf{r}'), \tag{38.2a}$$

or as

$$\left(i\omega + \frac{1}{2m}[\nabla_{\mathbf{r}'} + ie\mathbf{A}(\mathbf{r}')]^2 + \mu\right)\tilde{\mathscr{G}}^{(0)}_\omega(\mathbf{r}, \mathbf{r}') = \delta(\mathbf{r} - \mathbf{r}'). \tag{38.2b}$$

Using the function $\tilde{\mathscr{G}}^{(0)}_\omega(\mathbf{r}, \mathbf{r}')$ and equation (38.2b), we can transform the system of equations for $\mathscr{G}_\omega$ and $\mathscr{F}^+_\omega$ into integral form:

$$\begin{aligned} \mathscr{G}_\omega(\mathbf{r}, \mathbf{r}') &= \tilde{\mathscr{G}}^{(0)}_\omega(\mathbf{r}, \mathbf{r}') - \int \tilde{\mathscr{G}}^{(0)}_\omega(\mathbf{r}, l)\Delta(l)\mathscr{F}^+_\omega(l, \mathbf{r}')\,dl, \\ \mathscr{F}^+_\omega(\mathbf{r}, \mathbf{r}') &= \int \tilde{\mathscr{G}}^{(0)}_{-\omega}(l, \mathbf{r})\Delta^*(l)\mathscr{G}_\omega(l, \mathbf{r}')\,dl. \end{aligned} \tag{38.3}$$

Before going further, we derive an expression for the function $\tilde{\mathscr{G}}^{(0)}_\omega(\mathbf{r}, \mathbf{r}')$. In the absence of the magnetic field, $\mathscr{G}^{(0)}_\omega(\mathbf{r} - \mathbf{r}')$ equals

$$\mathscr{G}^{(0)}_\omega(R) = \begin{cases} -\dfrac{m}{2\pi R}\, e^{ip_0R - (|\omega|R/v)} & \text{for} \quad \omega > 0, \\ -\dfrac{m}{2\pi R}\, e^{-ip_0R - (|\omega|R/v)} & \text{for} \quad \omega < 0, \end{cases} \tag{38.4}$$

where $R = |\mathbf{R}| = |\mathbf{r} - \mathbf{r}'|$. This can be verified by direct substitution of (38.4) into (38.2a) in the absence of the field, or by using our familiar expansion for the Fourier components $\mathscr{G}^{(0)}_\omega(\mathbf{p}) = (i\omega - \xi)^{-1}$:

$$\begin{aligned} \mathscr{G}^{(0)}_\omega(R) &= \frac{1}{(2\pi)^3}\int \mathscr{G}^{(0)}_\omega(\mathbf{p})\, e^{i\mathbf{p}\cdot\mathbf{R}}\, d\mathbf{p} \\ &= \frac{m}{(2\pi)^2 iR}\int \frac{e^{ip_0R + i(\xi R/v)} - e^{-ip_0R - i(\xi R/v)}}{i\omega - \xi}\, d\xi. \end{aligned}$$

[Of course, we are interested in the form of the function $\mathscr{G}^{(0)}_\omega(R)$ at distances large compared to atomic distances, i.e., such that $Rp_0 \gg 1$.] Carrying out the integration with respect to ξ, we obtain (38.4).

The function $\mathscr{G}^{(0)}_\omega(R)$ is rapidly oscillating; since $p_0R \gg 1$, this fact allows us to use a kind of quasi-classical approach to determine the function $\tilde{\mathscr{G}}^{(0)}_\omega(\mathbf{r}, \mathbf{r}')$ in the presence of the magnetic field. In fact, suppose we look for $\tilde{\mathscr{G}}^{(0)}_\omega(\mathbf{r}, \mathbf{r}')$ in the form

$$\tilde{\mathscr{G}}^{(0)}_\omega(\mathbf{r}, \mathbf{r}') = e^{i\varphi(\mathbf{r}, \mathbf{r}')}\mathscr{G}^{(0)}_\omega(\mathbf{r} - \mathbf{r}'), \tag{38.5}$$

where $\varphi(\mathbf{r}, \mathbf{r}) = 0$. Then, substituting (38.5) into (38.2) and differentiating only principal terms, we find the following equation for $\varphi(\mathbf{r}, \mathbf{r}')$, the correction to the action:

$$\mathbf{n}\cdot\nabla_{\mathbf{r}}\varphi(\mathbf{r}, \mathbf{r}') = e\mathbf{n}\cdot\mathbf{A}(\mathbf{r}) \qquad \left(\mathbf{n} = \frac{\mathbf{R}}{|\mathbf{R}|}\right). \tag{38.6}$$

In (38.6) we have dropped all terms which are quadratic in **A**, since for the magnetic fields H in which we are interested, the quantity ep_0/H (the *Larmor radius*) is very large compared to the penetration depth δ, i.e., $p_0 \gg eA \sim eH\delta$.

Returning to (38.3), we now expand these equations in powers of $|\Delta(\mathbf{r})|$. It follows from (38.4) and (38.6) that the expansion for $\mathscr{F}^{+}_{\omega}(\mathbf{r}, \mathbf{r}')$ need only be carried out to terms of the third order in $|\Delta|$ inclusive, while the expansion for the Green's function $\mathscr{G}_{\omega}(\mathbf{r}, \mathbf{r}')$ need only be carried out to terms of the second order in $|\Delta|$:

$$\mathscr{G}_{\omega}(\mathbf{r}, \mathbf{r}') = \tilde{\mathscr{G}}^{(0)}_{\omega}(\mathbf{r}, \mathbf{r}') - \int \tilde{\mathscr{G}}^{(0)}_{\omega}(\mathbf{r}, \boldsymbol{l})\Delta(\boldsymbol{l})\tilde{\mathscr{G}}^{(0)}_{\omega}(\mathbf{m}, \mathbf{r}')\Delta^*(\mathbf{m})\tilde{\mathscr{G}}^{(0)}_{-\omega}(\mathbf{m}, \boldsymbol{l})\, d\mathbf{m}\, d\boldsymbol{l}. \tag{38.7}$$

Substituting (38.7) into the second of the equations (38.3), we find an expansion for $\mathscr{F}^{+}_{\omega}(\mathbf{r}, \mathbf{r}')$. Then, using this expansion and (38.1), we obtain the following equation for $\Delta^*(\mathbf{r})$:

$$\begin{aligned}\Delta^*(\mathbf{r}) = {} & |\lambda| T \sum_{\omega} \int \tilde{\mathscr{G}}^{(0)}_{\omega}(\boldsymbol{l}, \mathbf{r})\Delta^*(\boldsymbol{l})\tilde{\mathscr{G}}^{(0)}_{-\omega}(\boldsymbol{l}, \mathbf{r})\, d\boldsymbol{l} \\ & - |\lambda| T \sum_{\omega} \int \tilde{\mathscr{G}}^{(0)}_{\omega}(\boldsymbol{l}, \mathbf{m})\Delta(\mathbf{m})\tilde{\mathscr{G}}^{(0)}_{\omega}(\mathbf{s}, \mathbf{r})\Delta^*(\mathbf{s})\tilde{\mathscr{G}}^{(0)}_{-\omega}(\mathbf{s}, \mathbf{m})\Delta^*(\boldsymbol{l})\tilde{\mathscr{G}}^{(0)}_{-\omega}(\boldsymbol{l}, \mathbf{r})\, d\boldsymbol{l}\, d\mathbf{m}\, d\mathbf{s}.\end{aligned} \tag{38.8}$$

In (38.8) the important distances in the integrals are of order ξ_0, since, as can be seen from (38.4) and (38.5), the function $\tilde{\mathscr{G}}^{(0)}_{\omega}(\mathbf{r}, \mathbf{r}')$ falls off exponentially for $|\mathbf{r} - \mathbf{r}'| > \xi_0$. Moreover, the gap $\Delta(\mathbf{r})$ and the field $\mathbf{A}(\mathbf{r})$ change over distances of the order of the penetration depth, which is much larger than ξ_0 near the critical temperature. For the same reasons, the phase $\varphi(\mathbf{r}, \mathbf{r}')$ in (38.5) can be written in the form

$$\varphi(\mathbf{r}, \mathbf{r}') \approx e\mathbf{A}(\mathbf{r})\cdot(\mathbf{r} - \mathbf{r}').$$

Near T_c,

$$A(\mathbf{r}) \sim H\delta \propto \sqrt{1 - \frac{T}{T_c}},$$

and hence the phase $\varphi(\mathbf{r}, \mathbf{r}')$ is small and the exponential can be expanded with respect to φ.

We begin by examining the first term in the right-hand side of (38.8). Suppose that

$$K(\boldsymbol{l}, \mathbf{r}) = T \sum_{\omega} \tilde{\mathscr{G}}^{(0)}_{\omega}(\boldsymbol{l}, \mathbf{r})\tilde{\mathscr{G}}^{(0)}_{-\omega}(\boldsymbol{l}, \mathbf{r}) = K_0(\boldsymbol{l} - \mathbf{r}) \exp\{2ie\mathbf{A}(\mathbf{r})\cdot(\boldsymbol{l} - \mathbf{r})\}.$$

Then, using the representation (38.4) for the Green's function in coordinate space, and carrying out the sum over frequencies, we find the following expression for $K_0(R)$:

$$K_0(R) = \frac{m^2 T}{(2\pi R)^2} \frac{1}{\sinh(2\pi RT/v)}. \tag{38.9}$$

As we have already noted, all quantities change only slightly in distances of order ξ_0. Therefore, in the integral

$$\int K_0(\boldsymbol{l} - \mathbf{r}) \exp\{2ie\mathbf{A}(\mathbf{r})\cdot(\boldsymbol{l} - \mathbf{r})\}\Delta^*(\boldsymbol{l})\, d\boldsymbol{l},$$

we can make series expansions of all quantities in powers of $\boldsymbol{l} - \mathbf{r}$ near the point $\mathbf{r}$. Retaining only terms up to the second order in $\boldsymbol{l} - \mathbf{r}$, we obtain

$$\Delta^*(\mathbf{r}) \int K_0(R)\, d\mathbf{R} + \frac{1}{6}[\nabla_{\mathbf{r}} + 2ie\mathbf{A}(\mathbf{r})]^2\Delta^*(\mathbf{r}) \int K_0(R)R^2\, d\mathbf{R}. \tag{38.10}$$

According to (38.9), the function $K_0(R)$ goes to infinity like R^{-3} at $R = 0$, and hence the first of the two integrals on the right is formally divergent. A cutoff for this divergent expression is best introduced in momentum space. As a result, we find the familiar expression

$$\int K_0(R)\, d\mathbf{R} = \frac{mp_0}{2\pi^2} \int_0^{\omega_D} \tanh\left(\frac{\xi}{2T}\right)\frac{d\xi}{\xi}.$$

The integrand of the second integral in (38.10) has no singularity at $R = 0$. Evaluating this integral directly in the coordinate representation, we obtain

$$\int R^2 K_0(R)\, d\mathbf{R} = \frac{7\zeta(3)v^2}{8(\pi T)^2} \frac{mp_0}{2\pi^2}.$$

Next, we examine the second term in the right-hand side of (38.10), which is of the third order in $|\Delta|$. The dependence of $\Delta(\mathbf{r})$ on the coordinates can be neglected in this term, which can then be written as

$$\left(\frac{mp_0}{2\pi^2}\right)\Delta^*(\mathbf{r})|\Delta(\mathbf{r})|^2 T \sum_{\omega} \int \frac{d\xi}{(\omega^2 + \xi^2)^2} = \left(\frac{mp_0}{2\pi^2}\right)\frac{7\zeta(3)}{8(\pi T)^2}\Delta^*(\mathbf{r})|\Delta(\mathbf{r})|^2.$$

Collecting all the results just obtained, we finally find that near the critical temperature $\Delta^*(r)$ satisfies the equation

$$\left\{\frac{1}{4m}[\nabla_{\mathbf{r}} + 2ie\mathbf{A}(\mathbf{r})]^2 + \frac{1}{\beta}\left[\frac{T_c - T}{T_c} - \frac{7\zeta(3)}{8(\pi T_c)^2}|\Delta(\mathbf{r})|^2\right]\right\}\Delta^*(\mathbf{r}) = 0, \tag{38.11}$$

where

$$\beta = \frac{7\zeta(3)}{6(\pi T_c)^2}\varepsilon_F.$$

In the absence of the field, $\Delta(\mathbf{r})$ does not depend on the coordinates and equation (38.11) agrees with the initial terms of the expansion (36.4).

We now calculate the current density $\mathbf{j}(\mathbf{r})$. Of course, formula (37.5) of the preceding section is still valid in the present case. However, using the fact that the quantity Δ is small compared to T_c, we shall only write the first terms of the expansions in powers of **A** and Δ. We note that the second term in the right-hand side of (37.5) can be written in the form

$$-\lim_{t'\to t+0} \frac{2e^2\mathbf{A}(\mathbf{r})}{m}\mathscr{G}(\mathbf{r}, \mathbf{r}) \equiv -\frac{e^2}{m}\mathbf{A}(\mathbf{r})N \tag{38.12}$$

where N is the total electron density, equal to the electron density in the normal metal. A change in N would violate the condition of electrical neutrality.[11] Substituting (38.7) into (38.5) and using (38.12), we find that

$$T\sum_{\omega}\frac{ie}{m}(\nabla_{\mathbf{r}'} - \nabla_{\mathbf{r}})_{\mathbf{r}'\to\mathbf{r}}\tilde{\mathscr{G}}_{\omega}^{(0)}(\mathbf{r}, \mathbf{r}') - \frac{e^2}{m}N\mathbf{A}(\mathbf{r}) \equiv 0,$$

since the current in a normal metal is zero in a constant magnetic field. Thus we have

$$\mathbf{j}(\mathbf{r}) = \frac{ie}{m}(\nabla_{\mathbf{r}'} - \nabla_{\mathbf{r}})_{\mathbf{r}'\to\mathbf{r}}T\sum_{\omega}\delta\mathscr{G}_{\omega}(\mathbf{r}, \mathbf{r}'),$$

where

$$\delta\mathscr{G}_{\omega}(\mathbf{r}, \mathbf{r}') = -\int\tilde{\mathscr{G}}_{\omega}^{(0)}(\mathbf{r}, \mathbf{l})\Delta(\mathbf{l})\tilde{\mathscr{G}}_{\omega}^{(0)}(\mathbf{m}, \mathbf{r}')\Delta^*(\mathbf{m})\tilde{\mathscr{G}}_{-\omega}^{(0)}(\mathbf{m}, \mathbf{l})\,d\mathbf{m}\,d\mathbf{l}. \tag{38.13}$$

Then, substituting (38.5) into (38.13), expanding all quantities near the point **r** up to first-order terms, and dropping terms which vanish when integrated with respect to the angular variables, we obtain

$$\mathbf{j}(\mathbf{r}) = \left[\frac{ie}{m}\left(\Delta\frac{\partial\Delta^*}{\partial\mathbf{r}} - \Delta^*\frac{\partial\Delta}{\partial\mathbf{r}}\right) - \frac{4e^2|\Delta|^2}{m}\mathbf{A}(\mathbf{r})\right]C, \tag{38.14}$$

where $\partial/\partial\mathbf{r} \equiv \nabla_{\mathbf{r}}$, and

$$C = \frac{1}{3}T\sum_{\omega}\int\left\{\left[\mathscr{G}_{\omega}^{(0)}(\mathbf{m} - \mathbf{r})\frac{\partial}{\partial\mathbf{r}}\mathscr{G}_{\omega}^{(0)}(\mathbf{r} - \mathbf{l}) - \mathscr{G}_{\omega}^{(0)}(\mathbf{r} - \mathbf{l})\frac{\partial}{\partial\mathbf{r}}\mathscr{G}_{\omega}^{(0)}(\mathbf{m} - \mathbf{r})\right]\cdot\mathbf{m}\right\}\mathscr{G}_{-\omega}^{(0)}(\mathbf{m} - \mathbf{l})\,d\mathbf{m}\,d\mathbf{l}.$$

It is convenient to calculate C in Fourier components, by replacing **r** by $i\nabla_{\mathbf{p}}$ in the usual way. We omit the details of this rather simple calculation, which leads to the final result

$$C = \frac{7\zeta(3)N}{16(\pi T_c)^2}.$$

The system of equations (38.11) and (38.14) describes the properties of a superconductor in a constant magnetic field near the critical temperature T_c. Suppose we introduce the wave function

$$\psi(\mathbf{r}) = \sqrt{\frac{7\zeta(3)N}{8(\pi T_c)^2}}\,\Delta(\mathbf{r}), \tag{38.15}$$

[11] The solution (38.7) satisfies this condition, i.e., if

$$\mathscr{G}_{\omega}(\mathbf{r}, \mathbf{r}') = \mathscr{G}_{\omega}^{(0)}(\mathbf{r}, \mathbf{r}') + \delta\mathscr{G}_{\omega}(\mathbf{r}, \mathbf{r}'),$$

then

$$T\sum_{\omega}\delta\mathscr{G}_{\omega}(\mathbf{r}, \mathbf{r}) \equiv 0.$$

proportional to $\Delta(\mathbf{r})$. Then, taking the complex conjugate of (38.11) and using the expression (38.15) to replace $\Delta(\mathbf{r})$ by $\psi(\mathbf{r})$ everywhere, we can write the system (38.11) and (38.14) in the following form:

$$\begin{gathered}\left\{\frac{1}{4m}[\nabla_{\mathbf{r}} - 2ie\mathbf{A}(\mathbf{r})]^2 + \frac{1}{\beta}\left[\frac{T_c - T}{T_c} - \frac{1}{N}|\psi|^2\right]\right\}\psi(\mathbf{r}) = 0,\\ \mathbf{j}(\mathbf{r}) = -\frac{2ie}{4m}(\psi^* \nabla_{\mathbf{r}}\psi - \psi \nabla_{\mathbf{r}}\psi^*) - \frac{(2e)^2}{2m}\mathbf{A}(\mathbf{r})|\psi|^2.\end{gathered} \tag{38.16}$$

The reason for introducing the wave function $\psi(\mathbf{r})$ is now apparent, i.e., the equations (38.16) resemble the quantum-mechanical equations for a particle of mass $2m$ and charge $2e$. Physically, this result is entirely understandable, since $\Delta(\mathbf{r})$ is a quantity proportional to the wave function of a bound pair (more exactly, the wave function of the pair relative to its center of mass). It is interesting to note that equations of this form were proposed in the phenomenological theory of Ginzburg and Landau (G5), where, however, e appeared instead of $2e$. Apart from this important change, the new theory of superconductivity confirms the correctness of the Ginzburg-Landau theory near T_c, and moreover enables us to calculate the constants appearing in the Ginzburg-Landau theory.

We conclude this section by noting once again that our derivation makes use of the fact that all quantities vary only slightly over distances of order ξ_0. It is easily seen from (38.13) and (37.19) that in the general case all quantities vary over distances of the order of the London penetration depth near the critical temperature T_c. However, for superconductors of the Pippard type this penetration depth becomes larger than ξ_0 only in the immediate neighborhood of T_c. Therefore, for superconductors of the Pippard type, the region of applicability of our equations is a very narrow region of temperatures near T_c. For superconductors of the London type (or of the intermediate type), these equations are applicable in a rather wide region of temperatures near T_c. From the experimental standpoint, this is the interesting region, and it should be pointed out that in this region the equations (38.16) lead to a very good agreement between theory and experimental data.

39. Theory of Superconducting Alloys

39.1. Statement of the problem. We conclude our discussion of the theory of superconductivity by studying the interesting problem of the properties of *superconducting alloys*, by which we mean superconductors containing *impurities*, i.e., atoms of foreign elements or lattice defects of other kinds [see Abrikosov and Gorkov (A3, A4)]. In the normal state, these lattice defects cause the *residual resistance* of the metal. In the superconducting state, the impurities play a different role. As already indicated, the interaction between the electrons in a superconductor causes a definite spatial coherence (or correlation) to be established between them. In particular, the dependence of the various Green's functions in the coordinate representation on their spatial arguments at distances of order ξ_0 (the effective size of a pair) undergoes an essential change when the metal makes the transition from the normal to the superconducting state. In an alloy, the electrons are scattered by the impurities, and since this scattering takes place randomly at arbitrary angles and the scattered electrons have very small wavelengths, the correlation between the electrons is very sensitive to the scattering processes. This means that impurity scattering must decrease the spatial coherence between the electrons.

For very low concentrations, the role of the impurities is small, but an increase in the concentration leads to a decrease in the coherence distance of the electrons in the superconductor. For sufficiently high concentrations, the role of the coherence distance ξ_0 is taken over by the mean free path of the electrons, and for such concentrations we have a right to expect new characteristic properties of the superconductor to appear. It is not our aim to give an exhaustive treatment of the whole subject here. Instead, we shall confine ourselves to an analysis of a single aspect of the theory, i.e., the properties of superconducting alloys in a weak constant magnetic field (see Sec. 39.3). However, this will allow us to demonstrate the full scope of the characteristic field theory techniques which are so useful in studying problems of this kind.

It has already been remarked in Sec. 37.1 that, as far as their electromagnetic properties in weak fields are concerned, the majority of real superconductors are of the "nonlocal" type (i.e., of the Pippard type or of the intermediate type). If a superconductor of this type is placed in an electromagnetic field, the current density at a given point is determined by the values of the field in a whole region around the point. In fact, this nonlocal effect lies at the very foundation of any theory based on Cooper's ideas on the formation of bound pairs of electrons. The finite dimensions of the pair give rise to coherence between electrons at distances of order $\xi_0 \sim 10^{-4}$ cm, an effect which also manifests itself in the nonlocal relation between the current and the field, if the field varies over distances appreciably less than ξ_0 (these distances are of the order of the penetration depth of the field). In the opposite case, i.e., for superconductors of the London type, the field varies only slightly over the distances of order ξ_0 which are important in the integral relation (37.3), and hence the field at the point $\mathbf{r}$ can be brought out in front of the integral. The considerations just given, concerning the role of impurities in superconductors, show that for sufficiently high impurity concentrations, a superconducting alloy must be of the London type. Since the mean free path begins to take over the role of the coherence distance as the impurity concentration is increased, a point arrives at which the mean free path becomes less than the penetration depth, and then the London case prevails.

Before going on, we have to clarify one further point. In real superconductors, the order of magnitude of ξ_0 is 10^{-4} cm, but according to the above remarks, the superconductor exhibits new properties for concentrations such that the mean free path becomes comparable to the penetration depth, whose order of magnitude is 10^{-5} cm. It is an extremely important fact that this new behavior takes place for concentrations that are still quite low ($\sim 1\%$). The point is that for large impurity concentrations, we are essentially dealing with a new substance, whose properties have nothing in common with the original superconductor. In particular, the properties of the electron-phonon interaction change, and so does the temperature at which the transition to the superconducting phase occurs. These changes in the basic properties of the lattice can be neglected for sufficiently low concentrations. However, even these low impurity concentrations cause essential changes in the behavior of the superconductor in a magnetic field. At the same time, it is interesting to note that the thermodynamic properties of the superconducting alloy are practically the same as those of the pure superconductor, as confirmed by experiment.

The ordinary methods, based on the transport equation and used (for example) to study the problem of the residual resistance of a normal metal, turn out to be unsuitable for solving the problem just posed. Accordingly, in what follows, we shall again resort to the methods of quantum field theory.

39.2. The residual resistance of a normal metal. In order to clarify our subsequent analysis, we first illustrate the technique to be used later by applying it to the problem of calculating the residual resistance of a normal metal at the absolute zero of temperature [see Abrikosov and Gorkov (A3, A4), Edwards (E1)]. Of course, the results obtained in this case are completely equivalent to those generally known, which were derived by using the transport equation.

It is well known that the presence of impurities in a normal metal leads to a finite conductivity σ. Therefore, for sufficiently small frequencies, the current density $\mathbf{j}$ in the presence of an applied homogeneous electric field $\mathbf{E}$ is given by the formula

$$\mathbf{j} = \sigma \mathbf{E}.$$

Introducing the vector potential $\mathbf{A}$ in the usual way, i.e., writing $\mathbf{E} = -\partial \mathbf{A}/\partial t$, we can represent this relation in the following form (for a monochromatic field component):

$$\mathbf{j}_\omega = i\omega\sigma \mathbf{A}_\omega.$$

Then our relation agrees with (37.3′), with a kernel $Q(\mathbf{k}, \omega)$ which is simply equal to

$$Q(\mathbf{k}, \omega) = -i\omega\sigma.$$

It is our intention to find $Q(\mathbf{k}, \omega)$ by using the methods of quantum field theory.

Bearing in mind the difference between the definition of the Green's function in the field theory technique for $T = 0$ and in the corresponding technique for $T \neq 0$, we have

$$\mathbf{j}(x) = \frac{e}{m}(\nabla_{\mathbf{r}'} - \nabla_{\mathbf{r}})_{\mathbf{r}' \to \mathbf{r}} \lim_{t' \to t+0} G(x, x') - \frac{Ne^2}{m}\mathbf{A}(x)$$

instead of (37.5). Expanding the Green's function in the usual way up to linear terms in the field, we obtain

$$\begin{aligned}\mathbf{j}(x) = -\frac{ie^2}{2m^2}(\nabla_{\mathbf{r}'} - \nabla_{\mathbf{r}})_{\mathbf{r}' \to \mathbf{r}} \int \mathbf{A}(y) \cdot (\nabla_{\mathbf{y}'} - \nabla_{\mathbf{y}})_{\mathbf{y}' \to \mathbf{y}} \\ \times G^{(0)}(x, y')G^{(0)}(y, x')\, d^4y - \frac{Ne^2}{m}\mathbf{A}(x).\end{aligned} \tag{39.1}$$

where the functions $G^{(0)}(x, y)$ are the Green's functions in the absence of the field. We note that these functions no longer depend just on the argument difference $x - y$ as has always been the case until now. In (39.1) we assume that the functions $G^{(0)}(x, y)$ incorporate the interaction between the electrons and the impurity atoms, but from now on, the Green's functions of the metal containing impurities will be written without the superscript, e.g., $G(x, y)$, and the superscript will be used to denote the Green's functions of the pure metal, e.g., $G^{(0)}(x - y)$. The interaction between the electrons and the impurity atoms corresponds to the Hamiltonian

$$H_{\text{int}} = \sum_a H_a,$$

where

$$H_a = \int u(\mathbf{r} - \mathbf{r}_a)\psi^+(x)\,\psi(x)\, d\mathbf{r}.$$

Before transforming (39.1) further, we first find the function $G(x, x')$.

When impurities are present, the Green's function is not given by the expression (7.7). Instead, we write the Green's function in the form

$$G(x, x') = \frac{1}{(2\pi)^4}\int G(\mathbf{p}, \mathbf{p}'; \omega)e^{i\mathbf{p}\cdot\mathbf{r} - i\mathbf{p}'\cdot\mathbf{r}' - i\omega(t - t')}\, d\mathbf{p}\, d\mathbf{p}'\, d\omega. \tag{39.2}$$

By the usual rules of field theory, the function $G(\mathbf{p}, \mathbf{p}'; \omega)$ corresponds to the sum of diagrams shown in Fig. 101, where $G^{(0)}(p)$ is associated with each line, and impurity vertices are indicated by crosses. A factor

$$u(\mathbf{q})e^{i\mathbf{q}\cdot\mathbf{r}_a}\delta(\omega - \omega')$$

is associated with each impurity vertex, where $u(\mathbf{q})$ is the Fourier component of the potential $u(\mathbf{r})$, and $\mathbf{q}$ is the momentum transfer. Summing the diagrams, we obtain the following integral equation for $G(\mathbf{p}, \mathbf{p}' : \omega)$:

$$\begin{aligned}G(\mathbf{p}, \mathbf{p}'; \omega) = \delta(\mathbf{p} - \mathbf{p}')G^{(0)}(p) \\ + \frac{1}{(2\pi)^3}\sum_a G^{(0)}(p)\int u(\mathbf{p} - \mathbf{p}'')e^{i(\mathbf{p} - \mathbf{p}'')\cdot\mathbf{r}_a}G(\mathbf{p}'', \mathbf{p}'; \omega)\, d\mathbf{p}''.\end{aligned} \tag{39.3}$$

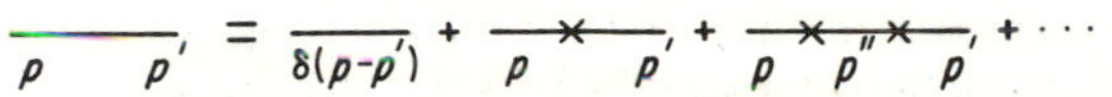

$p \quad p' = \delta(p-p') + p \quad p' + p \quad p'' \quad p' + \cdots$

FIGURE 101

The exact solution of (39.3) does not interest us. Since the impurity atoms are randomly distributed throughout the metal, we have to average all expressions over the position of each impurity atom. In doing this, we use the important fact that the average distance between impurity atoms is much larger than the lattice spacing, because of our assumption that the impurity concentration is low. As a result, the averaging can be carried out over volumes with dimensions that are large compared to the interatomic distances. After performing this averaging, the Green's function $G(\mathbf{p}, \mathbf{p}'; \omega)$ obviously becomes

$$\overline{G(\mathbf{p}, \mathbf{p}'; \omega)} = G(p)\delta(\mathbf{p} - \mathbf{p}'). \tag{39.4}$$

The momenta $\mathbf{p}$, $\mathbf{p}'$ of interest here have magnitudes of the order of the Fermi momentum p_0, which in turn is of the order of the reciprocal of the

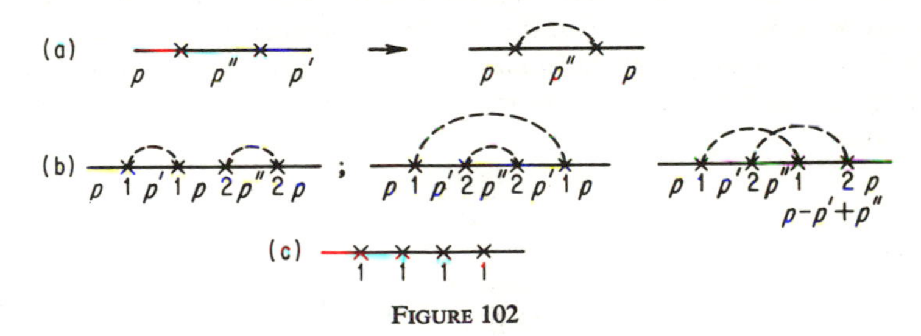

FIGURE 102

interatomic distance. This fact immediately makes it simpler to carry out the averaging. In making the calculations, we shall use the Born approximation, i.e., we shall assume that

$$p_0^3 \int u(\mathbf{r})\, d\mathbf{r} \ll \varepsilon_F.$$

It can be shown that the final result, expressed in terms of collision times, is still valid in the general case.

The simplest diagram for $G(\mathbf{p}, \mathbf{p}'; \omega)$ contains only one cross. Averaging over the position of the impurity atom, we obtain a constant, i.e.,

$$\overline{u(\mathbf{q})e^{i\mathbf{q}\cdot\mathbf{r}_a}} = u(0),$$

which can be included in the ground-state energy and will henceforth be regarded as zero. The diagram which is next in complexity contains two crosses [see Fig. 102(a)]. If these crosses refer to different atoms, the matrix element contains the factor

$$u(\mathbf{p}'' - \mathbf{p}')u(\mathbf{p} - \mathbf{p}'')e^{i(\mathbf{p}-\mathbf{p}'')\cdot\mathbf{r}_a + i(\mathbf{p}''-\mathbf{p}')\cdot\mathbf{r}_b},$$

whose average vanishes. However, if scattering by the same atom takes place at the two crosses and if $\mathbf{p} = \mathbf{p}'$, then the average value of the diagram in question [without the external $G^{(0)}(p)$] is nonzero and equals

$$\frac{1}{V} \int |u(\mathbf{p} - \mathbf{p}')|^2 G^{(0)}(p') \frac{d\mathbf{p}'}{(2\pi)^3}, \tag{39.5}$$

where V is the volume of the system. [To obtain this result, it is convenient to transform from integrals with respect to the momentum to discrete sums in formulas (39.2) and (39.3), and then go back to integrals, after evaluating the average.]

In what follows, we shall be interested in values of $\mathbf{p}$ whose magnitudes are close to p_0. As in Sec. 21.3, the integral in (39.5) can be divided into two parts, one with respect to values of $\mathbf{p}'$ far from the Fermi surface, and the other with respect to values of $\mathbf{p}'$ near the Fermi surface. (We can choose the limits of the second integral with respect to $|\mathbf{p}'|$ to be symmetric in $|\mathbf{p}'| - p_0$.) The integral over the distant region gives a real constant, which, together with $u(0)$ represents a renormalization of the chemical potential and need not be considered. In the second integral, we can regard $u(\mathbf{p} - \mathbf{p}')$ as a slowly varying function. Substituting the expression (7.7) for $G^{(0)}(p)$ into (39.5), and summing over the impurity atoms (which simply means multiplying by the number of atoms), we find that the main contribution to the self-energy is

$$\frac{i \operatorname{sgn} \omega}{2\tau},$$

where

$$\frac{1}{\tau} = \frac{nmp_0}{(2\pi)^2} \int |u(\theta)|^2 \, d\Omega, \tag{39.6}$$

and θ is the angle between the vectors $\mathbf{p}$ and $\mathbf{p}'$. According to (39.6), τ is the mean time between collisions in the Born approximation, where n is the number of impurity atoms per unit volume. Thus, it is clear that the region near the Fermi surface, where $v(|\mathbf{p}| - p_0) \sim 1/\tau$, plays the main role in our integrals.

From this point of view, not all of the diagrams are equivalent. For example, suppose we compare the three diagrams shown in Fig. 102(b), where the dashed lines join crosses referring to the same atom. It is not hard to see that in the expressions corresponding to the first two diagrams, the integration with respect to $\mathbf{p}'$ and $\mathbf{p}''$ can be carried out near the Fermi surface for arbitrary angles between the momenta. On the other hand, in the integral corresponding to the third diagram, the requirement that the arguments of all the Green's functions be near the Fermi surface leads to a restriction on the angles. As a result, the contribution of such a diagram turns out to be $(vp_0G)^{-1}$ times smaller than the contributions of the other diagrams. Since the values of ω and ξ needed below are of order τ^{-1}, where τ is the mean time between collisions, the factor $(vp_0G)^{-1}$ can be estimated as $(p_0l)^{-1}$, where $l = v\tau$ is the mean free path.

Moreover, it is easily shown that diagrams containing more than two crosses corresponding to the same atom make a small contribution. For example, suppose the total contribution (from all the impurity atoms) of diagrams like the first diagram shown in Fig. 102(b) is compared with the contribution of diagrams like that shown in Fig. 102(c). The first set of diagrams gives rise to a quantity of order

$$\frac{1}{\tau^2}\, G^{(0)}(p) \sim \frac{1}{\tau},$$

while the second set of diagrams gives rise to a quantity of order

$$\frac{1}{\tau}\,\frac{u^2(\mathbf{q})}{v^2}\, p_0^4 \sim \frac{1}{\tau\varepsilon_F^2}\left[p_0^3\int u\, d\mathbf{r}\right]^2 \ll \frac{1}{\tau}$$

(if the Born approximation is valid). It follows that we need only consider diagrams containing two crosses for each impurity atom.

Summing all the "important" diagrams [i.e., only the "paired" diagrams which contain no intersecting dashed lines as in the third diagram of Fig. 102 (b)], we obtain the following equation for the G-function:

$$G(p) = G^{(0)}(p) + \frac{n}{(2\pi)^3}\, G^{(0)}(p) \int |u(\mathbf{p}-\mathbf{p}')|^2 G(p')G(p)\, d\mathbf{p}'. \quad (39.7)$$

If the Born approximation is not used, we have to take account of diagrams with many crosses corresponding to a single impurity atom, but it can be shown that the resulting change is equivalent to replacing the Born amplitude $u(\theta)$ by the total scattering amplitude. The same applies to the rest of our calculations, and hence we can interpret $u(\theta)$ as the total scattering amplitude in any relevant formula.

The solution of equation (39.7) is

$$G(p) = \frac{1}{\omega - \xi - \bar{G}_\omega},$$

where $\bar{G}_\omega$ satisfies the equation

$$\bar{G}_\omega = \frac{n}{(2\pi)^3}\int |u(\mathbf{p}-\mathbf{p}')|^2 \frac{1}{\omega - \xi' - \bar{G}_\omega}\, d\mathbf{p}'.$$

Taking $\bar{G}_\omega$ to be purely imaginary, i.e., setting $\bar{G}_\omega = -i\beta$, and calculating the integral on the right by the method used in connection with (39.5), we find that

$$\beta = \frac{\operatorname{sgn}\beta}{2\tau},$$

where τ is defined by (39.6). Comparing the function obtained in this way with the result for the case of a small number of impurities ($G \to G^{(0)}$), we obtain

$$\beta = \frac{\operatorname{sgn}\omega}{2\tau}$$

or

$$G(p) = \frac{1}{\omega - \xi + (i\omega/2|\omega|\tau)}. \tag{39.8}$$

Going over to the x-representation, we easily see that the entire change in G as compared to $G^{(0)}$ reduces to multiplication by an exponentially damped factor:

$$G(x - x') = G^{(0)}(x - x')e^{-|\mathbf{r}-\mathbf{r}'|/2l}. \tag{39.9}$$

In fact, after carrying out the angular integrations, we have

$$G(x - x') \propto \int p\,dp\,d\omega\,\frac{e^{-i\omega(t-t')}\sin pR}{R[\omega - \xi + (i\omega/2|\omega|\tau)]}$$

$$\propto m\int d\xi\,d\omega\,e^{-i\omega(t-t')}\,\frac{e^{i[p_0+(\xi/v)]R} - e^{-i[p_0+(\xi/v)]R}}{R[\omega - \xi + (i\omega/2|\omega|\tau)]}.$$

Then, integrating with respect to ξ and taking the residue, we obtain (39.9).

We now turn to the calculation of the kernel $Q(\mathbf{k}, \omega)$. Taking Fourier components, we find it convenient to write the expression for Q obtained from (39.1) in the form

$$Q_{\alpha\beta}(\mathbf{k}, \omega) = \frac{Ne^2}{m}\delta_{\alpha\beta} - \frac{2ie^2}{m^2}\int \mathbf{p}'_\alpha \mathbf{\Pi}_\beta(p'_+, p'_-)\,d\mathbf{p}'\,d\omega', \tag{39.10}$$

where

$$p_\pm = (\mathbf{p}' \pm \tfrac{1}{2}\mathbf{k}, \omega' \pm \tfrac{1}{2}\omega).$$

In (39.10), one of the photon vertices is singled out and the second is contained in $\mathbf{\Pi}_\beta(p_+, p_-)$, which to within a constant factor can be regarded as the result of inserting the photon vertex $\mathbf{p}'_\beta$ into an electron line. Substituting this vertex into the electron Green's function $G(\mathbf{p}, \mathbf{p}'; \omega)$, we obtain

$$\mathbf{\Pi}(p'_+, p'_-) = \frac{1}{(2\pi)^3}\int G(\mathbf{p}'_+, \mathbf{p}''_+; \omega' + \tfrac{1}{2}\omega)G(\mathbf{p}''_-, \mathbf{p}'_-; \omega' - \tfrac{1}{2}\omega)\mathbf{p}''d\mathbf{p}''. \tag{39.11}$$

The functions $G(\mathbf{p}, \mathbf{p}'; \omega)$ appearing in (39.11) correspond to the sum of diagrams shown in Fig. 101, and they satisfy equation (39.3). In averaging over the positions of the impurity atoms, it must be borne in mind that the average of the product of two Green's functions does not equal the product of the separate averages.

For a pure metal, the expression (39.11) corresponds to the diagram shown in Fig. 103(a). After averaging over the positions of the impurity atoms, in addition to the simple diagrams corresponding to the transition from the

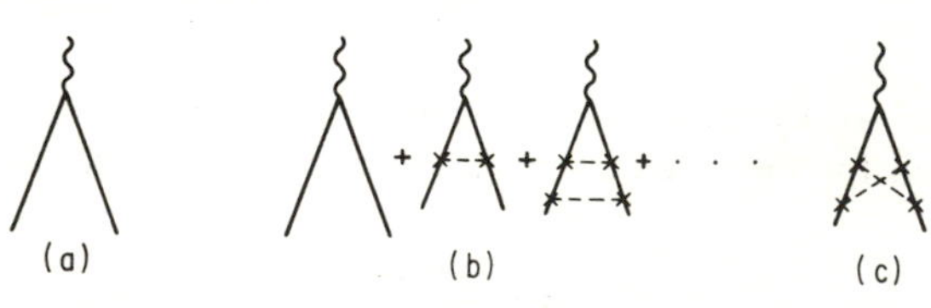

FIGURE 103

zeroth-order Green's functions $G^{(0)}(p)$ to the Green's functions $G(p)$ given by (39.8), the diagrams shown in Fig. 103(b) turn out to make important contributions to (39.11). The large contribution from these diagrams is due to the fact that at a vertex the photon momentum $k \ll p_0$, and hence the main contribution to the integral comes from the momentum region near the Fermi surface. A diagram of any other type, e.g., the diagram shown in Fig. 103(c), makes a much smaller contribution, since one of the integrations is over a momentum region far from the Fermi surface. Thus, averaging the quantity (39.11) reduces to summing the "ladder diagrams" shown in Fig. 103(b).

The integral equation for $\mathbf{\Pi}(p_+, p_-)$ has the form

$$\mathbf{\Pi}(p_+, p_-) = G(p_+)G(p_-)\left[\mathbf{p} + \frac{n}{(2\pi)^3}\int |u(\mathbf{p} - \mathbf{p}')|^2 \mathbf{\Pi}(p'_+, p'_-)\, d\mathbf{p}'\right], \quad (39.12)$$

and the following two limiting cases are possible:

1. *The anomalous skin effect* ($v|\mathbf{k}| \gg 1/\tau$). It is easy to see that in this case, the integral in the right-hand side of (39.12) is negligibly small, of order $(v|\mathbf{k}|\tau)^{-1} \ll 1$.
2. *The normal skin effect* ($v|\mathbf{k}| \ll 1/\tau$). In this case (the one in which we are interested), we can set $\mathbf{p}_+ = \mathbf{p}_-$.

The vector obtained by evaluating the integral in (39.12) is obviously directed along $\mathbf{p}$. Suppose we write

$$\mathbf{p}\Lambda(\omega', \omega) = \frac{n}{(2\pi)^3}\int |u(\mathbf{p} - \mathbf{p}')|^2 \mathbf{\Pi}(p'_+, p'_-)\, d\mathbf{p}'. \quad (39.13)$$

Then $\Lambda(\omega, \omega')$ can be regarded as independent of $|\mathbf{p}|$, since $|\mathbf{p}| \approx p_0$. Multiplying (39.12) by

$$\frac{n}{(2\pi)^3}\,|u(\boldsymbol{l} - \mathbf{p})|^2,$$

and integrating with respect to $\mathbf{p}$, we obtain

$$\boldsymbol{l}\Lambda(\omega', \omega) = \frac{n}{(2\pi)^3}\int |u(\boldsymbol{l} - \mathbf{p})|^2 \mathbf{p}G(p_+)G(p_-)[1 + \Lambda(\omega', \omega)]\, d\mathbf{p}. \quad (39.14)$$

Substituting the expression (39.8) for $G(p)$ into (39.14), we easily find that $\Lambda(\omega, \omega')$ is nonzero only for $|\omega'| < \frac{1}{2}\omega$, since otherwise, according to (7.7), both poles in (39.14) [for the integration with respect to ξ] lie in the same half-plane. In this interval $\Lambda(\omega', \omega)$ does not depend on ω'. Integrating with respect to ξ, and using the relation

$$\cos\theta = \cos\theta' \cos\theta'' + \sin\theta' \sin\theta'' \cos(\phi' - \phi''), \quad (39.15)$$

we obtain

$$\Lambda(\omega', \omega) = \begin{cases} \dfrac{i}{\tau_1}\dfrac{1}{\omega + (i/\tau_{tr})} & \text{for} \quad \omega'^2 < \dfrac{\omega^2}{4}, \\[2ex] 0 & \text{for} \quad \omega'^2 > \dfrac{\omega^2}{4}, \end{cases} \quad (39.16)$$

where

$$\frac{1}{\tau_1} = \frac{1}{\tau} - \frac{1}{\tau_{tr}}, \qquad \frac{1}{\tau_{tr}} = \frac{nmp_0}{(2\pi)^2}\int |u(\theta)|^2(1 - \cos\theta)\, d\Omega. \tag{39.17}$$

Substituting (39.16) into (39.12) and (39.10), and carrying out the integration (with the assumption that $\omega\tau \ll 1$), we find that

$$Q_{\alpha\beta}(\omega) = -i\omega\sigma\delta_{\alpha\beta}.$$

As was to be expected, the conductivity

$$\sigma = \frac{Ne^2\tau_{tr}}{m}$$

involves τ_{tr}, the "transport" time between collisions.

Thus, we see that the calculation of various characteristics of a metal, averaged over the positions of the impurity atoms, can be carried out by using a special kind of field theory technique. In our case, the averaging reduces to pair averages of scattering by the same atoms, each of which corresponds in a diagram to a dashed line connecting two crosses. Such a line carries momentum $\mathbf{q}$ and corresponds in a matrix element to a factor $n|u(\mathbf{q})|^2$, which plays the role of a D-function for the dashed line. The frequency of an electron line is not changed at any vertex to which a dashed line is attached. A very important fact is that a small contribution is made by any dashed line "bridging" a vertex at which an electron line undergoes a large momentum change ($q \sim p_0$). In particular, for this reason we can neglect diagrams with intersecting dashed lines [the relative order of magnitude of the contributions made by such lines is $(p_0 l)^{-1} \ll 1$].

39.3. Electromagnetic properties of superconducting alloys. We now use the above method to study superconductors containing impurities, where the case of arbitrary temperatures will be considered from the very beginning. First we write the equations of the superconductor in the field due to the presence of the impurities:

$$\left[i\omega + \frac{\nabla^2}{2m} + \mu - \sum_{\mathbf{r}_a} u(\mathbf{r} - \mathbf{r}_a)\right]\mathscr{G}_\omega(\mathbf{r}, \mathbf{r}') + \Delta(\mathbf{r})\mathscr{F}^+_\omega(\mathbf{r}, \mathbf{r}') = \delta(\mathbf{r} - \mathbf{r}'),$$

$$\left[-i\omega + \frac{\nabla^2}{2m} + \mu - \sum_{\mathbf{r}_a} u(\mathbf{r} - \mathbf{r}_a)\right]\mathscr{F}^+_\omega(\mathbf{r}, \mathbf{r}') - \Delta^*(\mathbf{r})\mathscr{G}_\omega(\mathbf{r}, \mathbf{r}') = 0.$$

Of course, just as before, we are only interested in the quantities $\mathscr{G}$ and $\mathscr{F}^+$ averaged over the impurities. To carry out this averaging, we have to expand all the Green's functions in series like (39.3) [see Fig. 101] in powers of the interaction potential in the presence of impurities. In so doing, it must be kept in mind that in general the introduction of impurities also changes the gap $\Delta(\mathbf{r})$, $\Delta^*(\mathbf{r})$. This might greatly complicate the diagram technique since the corrections to Δ must themselves be determined from an integral equation, because of the condition $\Delta(\mathbf{r}) = |\lambda|\mathscr{F}(x, x)$. Nevertheless, it turns out that after performing this averaging, $\overline{\Delta(\mathbf{r})} = \Delta^{(0)}$and all the corrections vanish. We could convince ourselves of this fact directly by examining the structure of the corrections to $\Delta(\mathbf{r})$. However, we shall assume this fact from the outset, writing $\Delta(\mathbf{r}) = \Delta^{(0)}$ This assumption will be confirmed later by the final result, according to which all quantities of the type $\mathscr{F}(x, x)$ do not change when nonmagnetic impurities are added. Therefore, the only difference between this problem and the one discussed above is that the superconductor is now described by three Green's functions $\mathscr{G}$, $\mathscr{F}$ and $\mathscr{F}^+$. This compels us to modify our diagram technique somewhat. It is easily seen that the resulting modification is entirely in keeping with the considerations of Sec. 35, and leads to the appearance of $\mathscr{F}$ and $\mathscr{F}^+$-lines in the diagrams for $\mathscr{G}$, as well as $\mathscr{G}$-lines in the diagrams for $\mathscr{F}$ and $\mathscr{F}^+$.

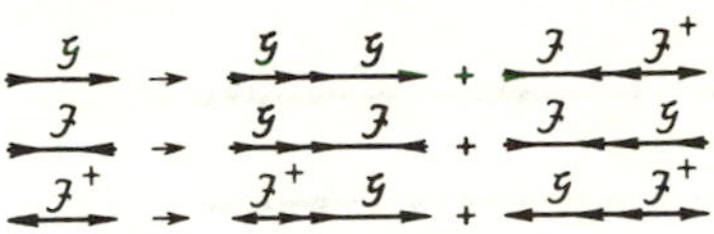

FIGURE 104

The Hamiltonian for interaction with the impurities contains operator products $\psi\bar{\psi}$. Therefore, when an impurity vertex is inserted into an electron line, two possibilities arise for each of the lines $\mathscr{G}$, $\mathscr{F}$ and $\mathscr{F}^+$. These possibilities are shown in Fig. 104. The result can be written in the form

$$\mathscr{G}(x, x') \to \mathscr{G}(x, y)\mathscr{G}(y, x') - \mathscr{F}(x, y)\mathscr{F}^+(y, x'),$$
$$\mathscr{F}^+(x, x') \to \mathscr{F}^+(x, y)\mathscr{G}(y, x') + \mathscr{G}(y, x)\mathscr{F}^+(y, x'),$$
$$\mathscr{F}(x, x') \to \mathscr{G}(x, y)\mathscr{F}(y, x') + \mathscr{F}(x, y)\mathscr{G}(x', y).$$

Instead of (39.3), we have the following equations for the functions $\mathscr{G}$ and $\mathscr{F}^+$:

$$\begin{aligned}\mathscr{G}(\mathbf{p}, \mathbf{p}'; \omega) = {}& \mathscr{G}^{(0)}(p)\delta(\mathbf{p} - \mathbf{p}') \\ & + \frac{1}{(2\pi)^3}\Big[\mathscr{G}^{(0)}(p) \int u(\mathbf{p} - \mathbf{p}'') \sum_a e^{i(\mathbf{p}-\mathbf{p}'')\cdot \mathbf{r}_a}\, \mathscr{G}(\mathbf{p}'', \mathbf{p}'; \omega)\, d\mathbf{p}'' \\ & - \mathscr{F}^{(0)}(p) \int u(\mathbf{p} - \mathbf{p}'') \sum_a e^{i(\mathbf{p}-\mathbf{p}'')\cdot \mathbf{r}_a}\, \mathscr{F}^+(\mathbf{p}'', \mathbf{p}'; \omega)\, d\mathbf{p}''\Big],\\ \mathscr{F}^+(\mathbf{p}, \mathbf{p}'; \omega) = {}& \mathscr{F}^{+(0)}(p)\delta(\mathbf{p} - \mathbf{p}') \\ & + \frac{1}{(2\pi)^3}\Big[\mathscr{F}^{+(0)}(p) \int u(\mathbf{p} - \mathbf{p}'') \sum_a e^{i(\mathbf{p}-\mathbf{p}'')\cdot \mathbf{r}_a}\, \mathscr{G}(\mathbf{p}'', \mathbf{p}'; \omega)\, d\mathbf{p}'' \\ & + \mathscr{G}^{(0)}(-p) \int u(\mathbf{p} - \mathbf{p}'') \sum_a e^{i(\mathbf{p}-\mathbf{p}'')\cdot \mathbf{r}_a}\, \mathscr{F}^+(\mathbf{p}'', \mathbf{p}'; \omega)\, d\mathbf{p}''\Big].\end{aligned} \tag{39.18}$$

In principle, we have to consider one more equation, for the function $\mathscr{F}(\mathbf{p}, \mathbf{p}'; \omega)$. However, for a pure superconductor,

$$\mathscr{F}^{(0)}(x, x') = \mathscr{F}^{+(0)}(x, x')$$

in the absence of a field, and without giving the proof, we note that the same is true for superconducting alloys after averaging the equations (39.18) over the positions of the impurity atoms.

The generalization of the averaging technique developed in Sec. 39.2 to the case of finite temperatures, and its application to superconductors, can be carried out quite simply. When an electron is scattered by a static impurity, only the three components of its momentum change. Therefore, as before, we associate the factor $n|u(\mathbf{q})|^2$ with each dashed line. Moreover, the frequency of an electron line is not changed at an impurity vertex. All the estimates allowing us to neglect diagrams with intersecting dashed lines, and also diagrams in which a dashed line "bridges" a vertex where the momentum transfer is of the order of the Fermi momentum, are still valid. In making these estimates, it is essential to use the properties of the Green's functions of the normal metal. The equations for the averaged functions $\mathscr{G}(p)$ and $\mathscr{F}^+(p)$ are shown schematically in Fig. 105, and their structure is

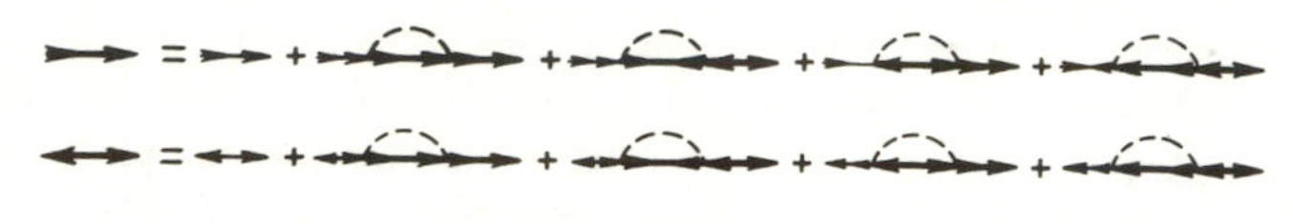

FIGURE 105

apparent without further explanations. We note that from a diagram point of view, the equations of Fig. 105 resemble the equations of Sec. 35 (see Fig. 94) for a system with electron-phonon interactions. The difference is that in Fig. 105 there are "zeroth-order" lines of all kinds, i.e., $\mathscr{G}^{(0)}(p)$, $\mathscr{F}^{+(0)}(p)$ and $\mathscr{F}^{(0)}(p)$.

Using the explicit expressions for the functions $\mathscr{G}^{(0)}$ and $\mathscr{F}^{+(0)}$, corresponding to a pure superconductor, we can bring the system of equations shown in Fig. 105 into the very simple form

$$\begin{aligned} (i\omega - \xi - \overline{\mathscr{G}}_\omega)\mathscr{G}(p) + (\Delta + \overline{\mathscr{F}_\omega^+})\mathscr{F}^+(p) &= 1, \\ (i\omega + \xi + \overline{\mathscr{G}}_{-\omega})\mathscr{F}^+(p) + (\Delta + \overline{\mathscr{F}_\omega^+})\mathscr{G}(p) &= 0, \end{aligned} \tag{39.19}$$

where

$$\begin{aligned} \overline{\mathscr{G}}_\omega &= \frac{n}{(2\pi)^3}\int |u(\mathbf{p} - \mathbf{p}')|^2 \mathscr{G}(p')\, d\mathbf{p}', \\ \overline{\mathscr{F}_\omega^+} &= \frac{n}{(2\pi)^3}\int |u(\mathbf{p} - \mathbf{p}')|^2 \mathscr{F}^+(p')\, d\mathbf{p}', \\ \overline{\mathscr{G}}_{-\omega}(\mathbf{p}) &= \overline{\mathscr{G}}_\omega(-\mathbf{p}), \end{aligned} \tag{39.20}$$

and we have written $p' = (\mathbf{p}', \omega)$. The solution of the system (39.19) is

$$\begin{aligned} \mathscr{G}(p) &= -\frac{i\omega - \overline{\mathscr{G}}_\omega + \xi}{(i\omega - \overline{\mathscr{G}}_\omega)^2 + \xi^2 + (\Delta + \overline{\mathscr{F}_\omega^+})^2}, \\ \mathscr{F}^+(p) &= \frac{\Delta + \overline{\mathscr{F}_\omega^+}}{(i\omega - \overline{\mathscr{G}}_\omega)^2 + \xi^2 + (\Delta + \overline{\mathscr{F}_\omega^+})^2} \end{aligned} \tag{39.21}$$

(it is shown below that $\overline{\mathscr{G}}_\omega = -\overline{\mathscr{G}}_{-\omega}$). Substituting (39.21) into (39.20), we obtain two equations determining $\overline{\mathscr{G}}_\omega$ and $\overline{\mathscr{F}^+_\omega}$. We see that $\overline{\mathscr{G}}_\omega$ contains a constant term (as before), signifying an additive correction to the chemical potential, which does not depend on the temperature and comes from integrating with respect to $\mathbf{p}'$ over a region far from the Fermi surface. Therefore, this term is the same as for the normal metal:

$$\delta\mu \approx \frac{n}{(2\pi)^3}\int |u(\mathbf{p}-\mathbf{p}')|^2 \frac{d\xi}{\xi}.$$

After subtracting $\delta\mu$, the remaining part of $\overline{\mathscr{G}}_\omega$, as can be seen from (39.20) and (39.21), is determined by exactly the same integral as $\overline{\mathscr{F}^+_\omega}$, and hence

$$-\frac{\overline{\mathscr{G}}_\omega}{i\omega} = \frac{\overline{\mathscr{F}^+_\omega}}{\Delta}.$$

If we introduce the notation

$$\tilde{\Delta} = \Delta + \overline{\mathscr{F}^+_\omega} = \eta_\omega \Delta, \qquad i\tilde{\omega} = i\omega - \overline{\mathscr{G}}_\omega = i\eta_\omega \omega,$$

the function η_ω satisfies the equation

$$\eta_\omega = 1 + \frac{\eta_\omega}{2\pi}\int \frac{d\xi}{\xi^2 + (\omega^2 + \Delta^2)\eta_\omega^2},$$

whose solution is

$$\eta_\omega = 1 + \frac{1}{2\tau\sqrt{\omega^2+\Delta^2}}. \tag{39.22}$$

Thus, the functions $\mathscr{G}(p)$ and $\mathscr{F}^+(p)$, which are averaged over the positions of the impurity atoms, can be obtained from the corresponding functions for the pure superconductor, by making the substitution

$$\omega \to \eta_\omega \omega, \qquad \Delta \to \eta_\omega \Delta. \tag{39.23}$$

It is not hard to see that just as in the case of the normal metal, in coordinate space these formulas imply that the zeroth-order functions are multiplied by $e^{-R/2l}$. In particular, it follows that the quantity

$$\Delta = |\lambda| \mathscr{F}^+(x, x)$$

is the same in the alloy as in the pure superconductor. Since, as we saw in Sec. 36, the thermodynamic properties of a superconductor depend only on Δ, this justifies our earlier statement that the thermodynamic properties of a superconductor remain the same in the presence of impurities with a sufficiently low concentration.

Strictly speaking, in the electron-electron interaction model under consideration, this conclusion is true only to within terms of order $1/\omega_D\tau \sim 10^{-6}$ cm$/l$. However, in the more realistic phonon model, there is a frequency cutoff, and consequently no such terms arise. In other respects, both models give the same results. It should also be noted that in an anisotropic superconductor, the thermodynamics depend on the impurity concentration. For example (see T1, and P. Hohenberg, private communication), the variation of T_c has the form

$$\frac{T_{c0} - T_c}{T_{c0}} = \frac{\pi}{8} \frac{\overline{\Delta^2} - (\bar{\Delta})^2}{(\bar{\Delta})^2} \frac{1}{T_{c0}\tau}$$

for $1/T_c\tau \ll 1$, where $\bar{\Delta}$ and $\overline{\Delta^2}$ denote the averages of $\Delta(\mathbf{p})$ and $\Delta^2(\mathbf{p})$ over all directions of the momentum. In the known cases, $[\overline{\Delta^2} - (\bar{\Delta})^2]/(\bar{\Delta})^2$ is of order 10^{-2}.

We now turn to the problem of determining the temperature dependence of the penetration depth of a weak static magnetic field in a superconducting alloy. According to (37.10), the expression for the current density $\mathbf{j}(\mathbf{r})$ has the form

$$\mathbf{j}(\mathbf{r}) = \frac{e^2}{m^2} T \sum_{\omega} (\nabla_{\mathbf{r}} - \nabla_{\mathbf{r}'})_{\mathbf{r}' \to \mathbf{r}} \int \{\mathscr{G}_\omega(\mathbf{r}, \boldsymbol{l})[\mathbf{A}(\boldsymbol{l}) \cdot \nabla_l] \mathscr{G}_\omega(\boldsymbol{l}, \mathbf{r}')$$

$$+ \mathscr{F}_\omega(\boldsymbol{l}, \mathbf{r})[\mathbf{A}(\boldsymbol{l}) \cdot \nabla_l] \mathscr{F}_\omega^+(\mathbf{r}', \boldsymbol{l})\} \, d\boldsymbol{l} - \frac{Ne^2}{m} \mathbf{A}(\mathbf{r}),$$

to the approximation which is linear in the field, where, however, the functions $\mathscr{G}_\omega(\mathbf{r}, \mathbf{r}')$ and $\mathscr{F}_\omega^+(\mathbf{r}, \mathbf{r}')$ include the interaction with the impurity atoms. Averaging this expression over the positions of the impurity atoms, and transforming to Fourier components, we can represent the kernel $Q_{\alpha\beta}(\mathbf{k})$ in the form

$$Q_{\alpha\beta}(\mathbf{k}) = \frac{Ne^2}{m} \delta_{\alpha\beta} + \frac{2e^2 T}{(2\pi)^3 m^2} \sum_{\omega'} \int \mathbf{p}'_\alpha \mathbf{\Pi}_\beta^{(1)}(p'_+, p'_-) \, d\mathbf{p}' \qquad (39.24)$$

$[p'_\pm = p' \pm \frac{1}{2}k,\ k = (\mathbf{k}, 0)]$, where $\mathbf{\Pi}^{(1)}(p_+, p_-)$ is the Fourier component of

$$\mathbf{\Pi}^{(1)}(x - y, y - x') = -\frac{i}{2}(\nabla_{\mathbf{y}} - \nabla_{\mathbf{y}'})_{\mathbf{y}' \to \mathbf{y}} \overline{[\mathscr{G}(x, y')\mathscr{G}(y, x') - \mathscr{F}^+(y', x')\mathscr{F}(x, y)]},$$

i.e.,

$$\mathbf{\Pi}^{(1)}(x - y, y - x') = \frac{T^2}{(2\pi)^6} \sum_{\omega_+, \omega_-} \int \mathbf{\Pi}^{(1)}(p_+, p_-)$$

$$\times \, e^{i\mathbf{p}_+ \cdot (\mathbf{x} - \mathbf{y}) - i\omega_+ (\tau_x - \tau_y)} e^{i\mathbf{p}_- \cdot (\mathbf{y} - \mathbf{x}') - i\omega_- (\tau_y - \tau_{x'})} d\mathbf{p}_+ \, d\mathbf{p}_-.$$

As in the case of the normal metal, the average of a product of two Green's functions does not equal the product of the two separate averages. To carry out the averaging over the positions of the impurity atoms, we again have to sum a whole set of diagrams. Since the superconductivity modifies the Green's functions only near the Fermi surface, the required diagrams are of the "ladder type" (see Fig. 106), as in Sec. 39.2. However, the fact that

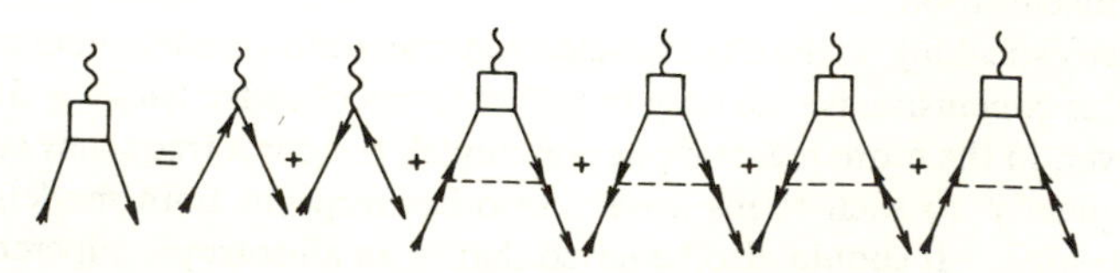

FIGURE 106

superconductors involve three different Green's functions leads to equations which are considerably more complicated than equation (39.12) [corresponding to the sum of diagrams in Fig. 103 for a normal metal].

It is clear from Fig. 106 that to determine $\mathbf{\Pi}^{(1)}(p_+, p_-)$, we need to know three more quantities, whose diagrams differ from $\mathbf{\Pi}^{(1)}(p_+, p_-)$ by having electron lines with arrows pointing in other directions. Each of these quantities is characterized by its own combination of the functions $\mathscr{G}$, $\mathscr{F}$ and $\mathscr{F}^+$:

$$\mathbf{\Pi}^{(2)}(x - y, y - x)$$
$$= -\frac{i}{2}(\nabla_{\mathbf{y}} - \nabla_{\mathbf{y}'})_{\mathbf{y}' \to \mathbf{y}}[\overline{\mathscr{F}^+(x, y')\mathscr{G}(y, x')} + \overline{\mathscr{G}(y, x)\mathscr{F}^+(y', x')}],$$

$$\mathbf{\Pi}^{(3)}(x - y, y - x')$$
$$= -\frac{i}{2}(\nabla_{\mathbf{y}} - \nabla_{\mathbf{y}'})_{\mathbf{y}' \to \mathbf{y}}[\overline{\mathscr{G}(y, x)\mathscr{G}(x', y')} - \overline{\mathscr{F}^+(x, y')\mathscr{F}(y, x')}],$$

$$\mathbf{\Pi}^{(4)}(x - y, y - x')$$
$$= -\frac{i}{2}(\nabla_{\mathbf{y}} - \nabla_{\mathbf{y}'})_{\mathbf{y}' \to \mathbf{y}}[\overline{\mathscr{G}(x, y')\mathscr{F}(y, x')} + \overline{\mathscr{F}(x, y)\mathscr{G}(x', y')}].$$

Thus, in the present case, we have to solve a system of four equations for the Fourier components $\mathbf{\Pi}^{(i)}(p_+, p_-)$, $i = 1, 2, 3, 4$, instead of the single equation (39.12).

The way in which the equations are constructed is clear from Fig. 106. For example, introducing the functions

$$\mathbf{\Lambda}^{(i)}(\omega') = \frac{n}{(2\pi)^3}\int |u(\mathbf{p} - \mathbf{p}')|^2 \mathbf{\Pi}^{(i)}(p'_+, p'_-)\, d\mathbf{p}' \qquad (i = 1, 2, 3, 4), \qquad (39.25)$$

we have

$$\mathbf{\Pi}^{(1)}(p_+, p_-) = \mathbf{p}[\mathscr{G}(p_+)\mathscr{G}(p_-) + \mathscr{F}(p_+)\mathscr{F}^+(p_-)]$$
$$+\mathscr{G}(p_+)\mathscr{G}(p_-)\mathbf{\Lambda}^{(1)}(\omega) - \mathscr{F}^+(p_+)\mathscr{G}(p_-)\mathbf{\Lambda}^{(2)}(\omega)$$
$$-\mathscr{F}(p_+)\mathscr{F}^+(p_-)\mathbf{\Lambda}^{(3)}(\omega) - \mathscr{G}(p_+)\mathscr{F}(p_-)\mathbf{\Lambda}^{(4)}(\omega),$$

and similarly for the other three equations. Substituting these equations into (39.25), we obtain a system of equations for the quantities $\mathbf{\Lambda}^{(i)}(\omega)$. We shall not write this system out completely, since in its general form it can only be solved when the scattering is spherically symmetric. However, we shall not be interested in the solution of the system in the general case. For very low impurity concentrations, the properties of a superconducting alloy are close to those of the pure superconductor, and, as we have already mentioned, most pure superconductors are of the Pippard type or of the intermediate type. When impurities are introduced, the coherence distance decreases, until, for a sufficiently high impurity concentration, the superconductor becomes a superconducting alloy of the London type, whose electrodynamics are now "local." Whether or not the electromagnetic behavior is local in a given case depends on the ratio between δ, the penetration depth, and l, the mean free path for scattering by impurity atoms (or equivalently, on the ratio between the sizes of the characteristic quantities $|\mathbf{k}| \sim \delta^{-1}$ and l^{-1}).

From now on, we shall assume that the impurity concentration is such that the pure superconductor has turned into a superconducting alloy of the London type ($|\mathbf{k}|l \ll 1$). In this case, the system of equations mentioned above becomes greatly simplified. In fact, we can neglect the quantity $\mathbf{k}$ in $\Pi^{(1)}(p_+, p_-)$, so that $\mathbf{p}_+ = \mathbf{p}_-$. Then it turns out that

$$\boldsymbol{\Lambda}^{(1)}(\omega) = -\boldsymbol{\Lambda}^{(3)}(\omega),$$
$$\boldsymbol{\Lambda}^{(2)}(\omega) = \boldsymbol{\Lambda}^{(4)}(\omega),$$

and

$$\boldsymbol{\Lambda}^{(1)}(\omega) = \mathbf{p}\Lambda^{(1)}(\omega),$$
$$\boldsymbol{\Lambda}^{(2)}(\omega) = \mathbf{p}\Lambda^{(2)}(\omega).$$

Therefore,

$$\boldsymbol{\Pi}^{(1)}(p, p) = \mathbf{p}\{\mathscr{G}^2(p) + [\mathscr{F}^+(p)]^2\}\{1 + \Lambda^{(1)}(\omega)\} - 2\mathbf{p}\mathscr{G}(p)\mathscr{F}^+(p)\Lambda^{(2)}(\omega), \tag{39.26}$$

and the system of equations for the functions $\Lambda^{(i)}(\omega)$ takes the form

$$\begin{aligned}\Lambda^{(1)}(\omega) &= [\overline{\mathscr{G}^2_\omega} + \overline{(\mathscr{F}^+_\omega)^2}][1 + \Lambda^{(1)}(\omega)] - 2\overline{\mathscr{G}_\omega\mathscr{F}^+_\omega}\Lambda^{(2)}(\omega),\\ \Lambda^{(2)}(\omega) &= 2\overline{\mathscr{G}_\omega\mathscr{F}^+_\omega}[1 + \Lambda^{(1)}(\omega)] - [\overline{\mathscr{G}^2_\omega} + \overline{(\mathscr{F}^+_\omega)^2}]\Lambda^{(2)}(\omega),\end{aligned} \tag{39.27}$$

where

$$\mathbf{p}\overline{\mathscr{G}^2_\omega} = \frac{n}{(2\pi)^3}\int |u(\mathbf{p} - \mathbf{p}')|^2\mathscr{G}^2(p')\mathbf{p}'\,d\mathbf{p}' = \frac{\mathbf{p}\Delta^2}{4\tau_1\eta_\omega(\sqrt{\omega^2 + \Delta^2})^3},$$
$$\mathbf{p}\overline{(\mathscr{F}^+_\omega)^2} = \frac{n}{(2\pi)^3}\int |u(\mathbf{p} - \mathbf{p}')|^2[\mathscr{F}^+(p')]^2\mathbf{p}'\,d\mathbf{p}' = \frac{\mathbf{p}\Delta^2}{4\tau_1\eta_\omega(\sqrt{\omega^2 + \Delta^2})^3},$$
$$\mathbf{p}\overline{\mathscr{F}^+_\omega\mathscr{G}_\omega} = \frac{n}{(2\pi)^3}\int |u(\mathbf{p} - \mathbf{p}')|^2\mathscr{F}^+(p')\mathscr{G}(p')\mathbf{p}'\,d\mathbf{p}' = \frac{i\mathbf{p}\omega\Delta}{4\tau_1\eta_\omega(\sqrt{\omega^2 + \Delta^2})^3}.$$

Solving (39.27), we find that

$$\begin{aligned}\Lambda^{(1)}(\omega) &= \frac{\Delta^2}{2\tau_1(\omega^2 + \Delta^2)\left(\sqrt{\omega^2 + \Delta^2} + \dfrac{1}{2\tau_{tr}}\right)},\\ \Lambda^{(2)}(\omega) &= \frac{i\omega\Delta}{2\tau_1(\omega^2 + \Delta^2)\left(\sqrt{\omega^2 + \Delta^2} + \dfrac{1}{2\tau_{tr}}\right)}.\end{aligned} \tag{39.28}$$

Substituting (39.28) into (39.26), and then substituting $\boldsymbol{\Pi}^{(1)}(p, p)$ into (39.24), we obtain the expression

$$\begin{aligned}Q &\equiv Q(k)\\ &= \frac{Ne^2}{m}\Bigg\{1 + T\sum_\omega\int d\xi\Bigg[(\xi^2 - \omega^2\eta^2_\omega + \Delta^2\eta^2_\omega)\left(1 + \frac{\Delta^2}{2\tau_1(\omega^2 + \Delta^2)^{3/2}\eta_{\omega,\,tr}}\right)\\ &\quad + \frac{2\Delta^2\omega^2\eta^2_\omega}{2\tau_1(\omega^2 + \Delta^2)^{3/2}\eta_{\omega,\,tr}}\Bigg]\frac{1}{[\xi^2 + (\omega^2 + \Delta^2)\eta^2_\omega]^2}\Bigg\}\end{aligned} \tag{39.29}$$

for the kernel $Q(\mathbf{k})$, where we have written

$$\eta_{\omega,\,\mathrm{tr}} = 1 + \frac{1}{2\tau_{\mathrm{tr}}\sqrt{\omega^2 + \Delta^2}}.$$

Here, as in Sec. 37.1, we again encounter a formally divergent integral, and for the same reasons, we have to first carry out the sum over the frequencies. Writing the first term in parentheses in the right-hand side of (39.29) as

$$\xi^2 + (\omega^2 + \Delta^2)\eta_\omega^2 - 2\omega^2\eta_\omega^2,$$

we apply the Abel transformation

$$\sum_{n=1}^{k} (B_n - B_{n-1})u_n = B_k u_k - B_0 u_1 - \sum_{n=1}^{k-1} (u_{n+1} - u_n)B_n, \tag{39.30}$$

which is the generalization to the case of series of the principle of integration by parts. Applying (39.30) to the series

$$2\pi T \sum_{n=1}^{\infty} \frac{1}{\xi^2 + (\Delta^2 + \omega^2)\eta_\omega^2},$$

where

$$B_n = (2n + 1)\pi T = \omega,$$
$$B_n - B_{n-1} = 2\pi T,$$

we are able to formally cancel the divergent terms in (39.29). This gives the formula

$$Q = \frac{Ne^2}{m} 2\pi T\Delta^2 \sum_{n=1}^{\infty} \frac{1}{(\omega^2 + \Delta^2)\left(\sqrt{\omega^2 + \Delta^2} + \dfrac{1}{2\tau_{\mathrm{tr}}}\right)}. \tag{39.31}$$

which involves only the transport time between the collisions. For the penetration depth, we have

$$\delta = \frac{1}{\sqrt{4\pi Q}}.$$

As $1/\tau \to 0$, this formula goes into the usual London formula

$$\delta = \sqrt{\frac{m}{4\pi N_s e^2}},$$

where N_s is the number of "superconducting" electrons. In the opposite limiting case $l \ll \xi_0$, we can neglect the square root in (39.31), and the remaining series can easily be summed. As a result, we find that the penetration depth for "dirty" alloys is

$$\delta = \frac{1}{2\pi}\sqrt{\frac{1}{\Delta\sigma \tanh(\Delta/2T)}},$$

where σ is the conductivity of the normal metal.

BIBLIOGRAPHY[1]

A1 Abrikosov, A. A., *Contribution to the theory of highly compressed matter, II*, Sov. Phys. JETP, **14**, 408 (1962).

A2 Abrikosov, A. A. and I. E. Dzyaloshinski, *Spin waves in a ferromagnetic metal*, Sov. Phys. JETP, **8**, 535 (1959).

A3 Abrikosov, A. A. and L. P. Gorkov, *On the theory of superconducting alloys, I. The electrodynamics of alloys at absolute zero*, Sov. Phys. JETP, **8**, 1090 (1959).

A4 Abrikosov, A. A. and L. P. Gorkov, *Superconducting alloys at finite temperatures*, Sov. Phys. JETP, **9**, 220 (1959).

A5 Abrikosov, A. A., L. P. Gorkov and I. E. Dzyaloshinski, *On the application of quantum field theory methods to problems of quantum statistics at finite temperatures*, Sov. Phys. JETP, **9**, 636 (1959).

A6 Abrikosov, A. A., L. P. Gorkov and I. M. Khalatnikov, *A superconductor in a high-frequency field*, Sov. Phys. JETP, **8**, 182 (1959).

A7 Abrikosov, A. A., L. P. Gorkov and I. M. Khalatnikov, *Analysis of experimental data relating to the surface impedance of superconductors*, Sov. Phys. JETP, **10**, 132 (1960).

A8 Abrikosov, A. A. and I. M. Khalatnikov, *On a model for a non-ideal Fermi gas*, Sov. Phys. JETP, **6**, 888 (1958).

A9 Akhiezer, A. I. and V. B. Berestetski, *Quantum Electrodynamics*, second revised edition, translated by G. M. Volkoff, Interscience Publishers, New York (1963).

B1 Bardeen, J., L. N. Cooper and J. R. Schrieffer, *Theory of superconductivity*, Phys. Rev., **108**, 1175 (1957).

[1] The titles of all papers from Russian journals are given in English. The translations of titles given in the journal *Soviet Physics, JETP* (a translation of the *Journal of Experimental and Theoretical Physics of the Academy of Sciences of the USSR*, published by the American Institute of Physics, New York, N.Y.) have been preserved, except for a few corrections of style and spelling. The Russian original of this journal is abbreviated as *Zh. Eksp. Teor. Fiz.* The papers B1, B3, B4, B7 (part I), C3, G1, G2, G3, G6, H4, L3, L4, L6 and M6 are reprinted in D. Pines (editor), *The Many-Body Problem*, W. A. Benjamin, Inc., New York (1961). (*Translator*)

B2 Belyaev, S. T., *Application of the methods of quantum field theory to a system of bosons*, Sov. Phys. JETP, **7**, 289 (1958).

B3 Belyaev, S. T., *Energy spectrum of a non-ideal Bose gas*, Sov. Phys. JETP, **7**, 299 (1958).

B4 Belyakov, V. A., *The momentum distribution of particles in a dilute Fermi gas*, Sov. Phys. JETP, **13**, 850 (1961).

B5 Bogoliubov, N. N., *On the theory of superfluidity*, Izv. Akad. Nauk SSSR, Ser. Fiz., **11**, 77 (1947).

B6 Bogoliubov, N. N., *A new method in the theory of superconductivity, I*, Sov. Phys. JETP, **7**, 41 (1958); *Part III*, ibid., p. 51.

B7 Bogoliubov, N. N., V. V. Tolmachev and D. V. Shirkov, *A New Method in the Theory of Superconductivity*, translated from the Russian, Consultants Bureau, Inc., New York (1959).

B8 Bogoliubov, N. N. and S. V. Tyablikov, *Retarded and advanced Green's functions in statistical physics*, Dokl. Akad. Nauk SSSR, **126**, 53 (1959).

B9 Brewer, D. F., J. G. Daunt and A. K. Sreedhar, *Low-temperature specific heat of liquid He*^{3} *near the saturated vapor pressure and at higher pressures*, Phys. Rev., **115**, 836 (1959).

B10 Brueckner, K. A. and K. Sawada, *Bose-Einstein gas with repulsive interactions. General theory*, Phys. Rev., **106**, 1117 (1957).

C1 Casimir, H. B. G. and D. Polder, *The influence of retardation on the London-van der Waals forces*, Phys. Rev., **73**, 360 (1948).

C2 Cohen, M. and R. P. Feynman, *Theory of inelastic scattering of cold neutrons from liquid helium*, Phys. Rev., **107**, 13 (1957).

C3 Cooper, L. N., *Bound electron pairs in a degenerate Fermi gas*, Phys. Rev., **104**, 1189 (1956).

D1 Dzyaloshinski, I. E., E. M. Lifshitz and L. P. Pitayevski, *Van der Waals forces in liquid films*, Sov. Phys. JETP, **10**, 161 (1960).

D2 Dzyaloshinski, I. E., E. M. Lifshitz and L. P. Pitayevski, *The general theory of Van der Waals forces*, Advances in Physics, **10**, 165 (1961).

D3 Dzyaloshinski, I. E. and L. P. Pitayevski, *Van der Waals forces in an inhomogeneous dielectric*, Sov. Phys. JETP, **9**, 1282 (1959).

E1 Edwards, S. F., *A new method for the evaluation of electric conductivity in metals*, Phil. Mag., series 8, **3**, 1020 (1958).

E2 Eisenschitz, R. and F. London, *Über das Verhältnis der van der Waalsschen Kräfte zu den homöopolaren Bindungskräften*, Z. für Physik, **60**, 491 (1930).

E3 Eliashberg, G. M., *Interaction between electrons and lattice vibrations in a superconductor*, Sov. Phys. JETP, **11**, 696 (1960).

E4 Eliashberg, G. M., *Temperature Green's function for electrons in a superconductor*, Sov. Phys. JETP, **12**, 1000 (1961).

E5 Eliashberg, G. M., *The low-temperature specific heat of metals*, Sov. Phys. JETP, **16**, 780 (1963).

F1 Feynman, R. P., *Atomic theory of the two-fluid model of liquid helium*, Phys. Rev., **94**, 262 (1954).

F2 Fradkin, E. S., *The method of Green's functions in quantum statistics*, Sov. Phys. JETP, **9**, 912 (1959).

F3 Fröhlich, H., *Theory of the superconducting state, I. The ground state at the absolute zero of temperature*, Phys. Rev., **79**, 845 (1950).

G1 Galitski, V. M., *The energy spectrum of a non-ideal Fermi gas*, Sov. Phys. JETP, **7**, 104 (1958).

G2 Galitski, V. M. and A. B. Migdal, *Application of quantum field theory methods to the many-body problem*, Sov. Phys. JETP, **7**, 96 (1958).

G3 Gell-Mann, M. and K. A. Brueckner, *Correlation energy of an electron gas at high density*, Phys. Rev., **106**, 364 (1957).

G4 Gell-Mann, M. and F. Low, *Bound states in quantum field theory*, Phys. Rev., **84**, 350 (1951).

G5 Ginzburg, V. L. and L. D. Landau, *On the theory of superconductivity*, Zh. Eksp. Teor. Fiz., **20**, 1064 (1950).

G6 Gorkov, L. P., *On the energy spectrum of superconductors*, Sov. Phys. JETP, **7**, 505 (1958).

G7 Gorkov, L. P., *Microscopic derivation of the Ginzburg-Landau equations in the theory of superconductivity*, Sov. Phys. JETP, **9**, 1364 (1959).

H1 Henshaw, D. G., A. D. B. Woods and B. N. Brockhouse, *Dispersion curve in liquid helium by inelastic scattering of neutrons*, Bul. Am. Phys. Soc., series II, vol. 5, no. 1, part 1, p. 12 (1960).

H2 Huang, K. and C. N. Yang, *Quantum-mechanical many-body problem with hard-sphere interaction*, Phys. Rev., **105**, 767 (1957).

H3 Huang, K., C. N. Yang and J. M. Luttinger, *Imperfect Bose gas with hard-sphere interaction*, Phys. Rev., **105**, 776 (1957).

H4 Hugenholtz, N. M. and D. Pines, *Ground-state energy and excitation spectrum of a system of interacting bosons*, Phys. Rev., **116**, 489 (1959).

K1 Keesom, W. H., *Helium*, Elsevier Publishing Co., New York (1942).

K2 Kerr, E. C., *Orthobaric densities of* He^3, 1.3° *K to* 3.2° *K*, Phys. Rev., **96**, 551 (1954).

K3 Khalatnikov, I. M., *Theory of transport phenomena in He II*, Uspekhi Fiz. Nauk, **59**, 673 (1956).

K4 Khalatnikov, I. M., *The hydrodynamics of He II*, Uspekhi Fiz. Nauk, **60**, 69 (1956).

K5 Khalatnikov, I. M. and A. A. Abrikosov, *The modern theory of superconductivity*, Advances in Physics, **8**, 45 (1959).

L1 Landau, L. D., *The theory of superfluidity of helium II*, Zh. Eksp. Teor. Fiz., **11**, 592 (1941).

L2 Landau, L. D., *On the theory of superfluidity of helium II*, J. Phys. USSR, **11**, 91 (1947).

L3 Landau, L. D., *The theory of a Fermi liquid*, Sov. Phys. JETP, **3**, 920 (1956).

L4 Landau, L. D., *Oscillations in a Fermi liquid*, Sov. Phys. JETP, **5**, 101 (1957).

L5 Landau, L. D., *The properties of the Green's function for particles in statistics*, Sov. Phys. JETP, **7**, 182 (1958).

L6 Landau, L. D., *On the theory of the Fermi liquid*, Sov. Phys. JETP, **8**, 70 (1959).

L7 Landau, L. D. and E. M. Lifshitz, *Quantum Mechanics, Non-Relativistic Theory*, translated by J. B. Sykes and J. S. Bell, Addison-Wesley Publishing Co., Inc., Reading, Mass. (1958).

L8 Landau, L. D. and E. M. Lifshitz, *Statistical Physics*, translated by E. Peierls and R. F. Peierls, Addison-Wesley Publishing Co., Inc., Reading, Mass. (1958).

L9 Landau, L. D. and E. M. Lifshitz, *Electrodynamics of Continuous Media*, translated by J. B. Sykes and J. S. Bell, Addison-Wesley Publishing Co., Inc., Reading, Mass. (1960).

L10 Lee, T. D. and C. N. Yang, *Many-body problem in quantum mechanics and quantum statistical mechanics*, Phys. Rev., **105**, 1119 (1957).

L11 Lehmann, H., *Über Eigenschaften von Ausbreitungsfunktionen und Renormierungskonstanten quantisierter Felder*, Nuovo Cimento, **11**, 342 (1954).

L12 Lifshitz, E. M., *Superfluidity (Theory)*, Supplement to Russian translation of Reference K1, Izd. Inostr. Lit., Moscow (1949). English translation available as first chapter of *A Supplement to "Helium,"* by E. M. Lifshitz and E. L. Andronikashvili, Consultants Bureau, Inc., New York (1959).

L13 Lifshitz, E. M., *The theory of molecular attraction forces between solid bodies*, Zh. Eksp. Teor. Fiz., **29**, 94 (1955).

L14 London, F. and H. London, *The electromagnetic equations of the supraconductor*, Proc. Roy. Soc. London, Ser. A, **149**, 71 (1935).

L15 London, F. and H. London, *Supraleitung und Diamagnetismus*, Physica, **2**, 341 (1935).

L16 Luttinger, J. M., *Fermi surface and some simple equilibrium properties of a system of interacting fermions*, Phys. Rev., **119**, 1153 (1960).

L17 Luttinger, J. M. and J. C. Ward, *Ground-state energy of a many-fermion system, II*, Phys. Rev., **118**, 1417 (1960).

M1 Matsubara, T., *A new approach to quantum-statistical mechanics*, Prog. Theor. Phys., **14**, 351 (1955).

M2 Mattis, D. C. and J. Bardeen, *Theory of the anomalous skin effect in normal and superconducting metals*, Phys. Rev., **111**, 412 (1958).

M3 Maxwell, E., *Isotope effect in the superconductivity of mercury*, Phys. Rev., **78**, 477 (1950).

M4 Maxwell, E., *Superconductivity of Sn^{124}*, Phys. Rev., **79**, 173 (1950).

M5 Migdal, A. B., *The momentum distribution of interacting Fermi particles*, Sov. Phys. JETP, **5**, 333 (1957).

M6 Migdal, A. B., *Interaction between electrons and lattice vibrations in a normal metal*, Sov. Phys. JETP, **7**, 996 (1958).

N1 Nozières, P. and D. Pines, *Electron interaction in solids. General formulation*, Phys. Rev., **109**, 741 (1958).

P1 Peierls, R. E., *Quantum Theory of Solids*, Oxford University Press, New York (1955).

P2 Pippard, A. B., *An experimental and theoretical study of the relation between magnetic field and current in a superconductor*, Proc. Roy. Soc. London, Ser. A, **216**, 547 (1953).

P3 Pitayevski, L. P., *On the derivation of a formula for the energy spectrum of liquid He^4*, Sov. Phys. JETP, **4**, 439 (1957).

P4 Pitayevski, L. P., *Properties of the spectrum of elementary excitations near the decay threshold of the excitations*, Sov. Phys. JETP, **9**, 830 (1959).

P5 Pitayevski, L. P., *Attraction of small particles suspended in a liquid at large distances*, Sov. Phys. JETP, **10**, 408 (1960).

P6 Pitayevski, L. P., *On the superfluidity of liquid He^3*, Sov. Phys. JETP, **10**, 1267 (1960).

P7 Pitayevski, L. P., *The problem of the form of the spectrum of elementary excitations of liquid helium II*, Sov. Phys. JETP, **12**, 155 (1961).

R1 Reynolds, C. A., B. Serin, W. H. Wright and L. B. Nesbitt, *Superconductivity of isotopes of mercury*, Phys. Rev., **78**, 487 (1950).

R2 Ryazanov, M. I., *Phenomenological study of the effect of a nonconducting medium in quantum electrodynamics*, Sov. Phys. JETP, **5**, 1013 (1957).

T1 Tsuneto, T., *On dirty superconductors*, Technical Report of the Institute of Solid State Physics, University of Tokyo, series A, no. 47 (1962).

V1 Vedenov, A. A., *Thermodynamic properties of a degenerate plasma*, Sov. Phys. JETP, **9**, 446 (1959).

Y1 Yarnell, J. L., G. P. Arnold, P. J. Bendt and E. C. Kerr, *Excitations in liquid helium. Neutron scattering measurements*, Phys. Rev., **113**, 1379 (1959).

Z1 Zubarev, D. N., *Double time Green's functions in statistical physics*, Uspekhi Fiz. Nauk, **71**, 71 (1960); English translation, Sov. Phys. Uspekhi, **3**, 320 (1960).

NAME INDEX

SUBJECT INDEX

E

F

G

H

T

V

W

Z

A CATALOG OF SELECTED

DOVER BOOKS

IN SCIENCE AND MATHEMATICS

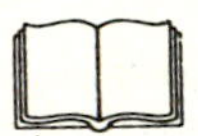

NUMERICAL METHODS FOR SCIENTISTS AND ENGINEERS, Richard Hamming. Classic text stresses frequency approach in coverage of algorithms, polynomial approximation, Fourier approximation, exponential approximation, other topics. Revised and enlarged 2nd edition. 721pp. 5⅜ × 8½.
65241-6 Pa. $14.95

THEORETICAL SOLID STATE PHYSICS, Vol. I: Perfect Lattices in Equilibrium; Vol. II: Non-Equilibrium and Disorder, William Jones and Norman H. March. Monumental reference work covers fundamental theory of equilibrium properties of perfect crystalline solids, non-equilibrium properties, defects and disordered systems. Appendices. Problems. Preface. Diagrams. Index. Bibliography. Total of 1,301pp. 5⅜ × 8½. Two volumes. Vol. I 65015-4 Pa. $14.95
Vol. II 65016-2 Pa. $14.95

OPTIMIZATION THEORY WITH APPLICATIONS, Donald A. Pierre. Broad-spectrum approach to important topic. Classical theory of minima and maxima, calculus of variations, simplex technique and linear programming, more. Many problems, examples. 640pp. 5⅜ × 8½. 65205-X Pa. $14.95

THE MODERN THEORY OF SOLIDS, Frederick Seitz. First inexpensive edition of classic work on theory of ionic crystals, free-electron theory of metals and semiconductors, molecular binding, much more. 736pp. 5⅜ × 8½.
65482-6 Pa. $15.95

ESSAYS ON THE THEORY OF NUMBERS, Richard Dedekind. Two classic essays by great German mathematician: on the theory of irrational numbers; and on transfinite numbers and properties of natural numbers. 115pp. 5⅜ × 8½.
21010-3 Pa. $4.95

THE FUNCTIONS OF MATHEMATICAL PHYSICS, Harry Hochstadt. Comprehensive treatment of orthogonal polynomials, hypergeometric functions, Hill's equation, much more. Bibliography. Index. 322pp. 5⅜ × 8½. 65214-9 Pa. $9.95

NUMBER THEORY AND ITS HISTORY, Oystein Ore. Unusually clear, accessible introduction covers counting, properties of numbers, prime numbers, much more. Bibliography. 380pp. 5⅜ × 8½. 65620-9 Pa. $9.95

THE VARIATIONAL PRINCIPLES OF MECHANICS, Cornelius Lanczos. Graduate level coverage of calculus of variations, equations of motion, relativistic mechanics, more. First inexpensive paperbound edition of classic treatise. Index. Bibliography. 418pp. 5⅜ × 8½. 65067-7 Pa. $11.95

MATHEMATICAL TABLES AND FORMULAS, Robert D. Carmichael and Edwin R. Smith. Logarithms, sines, tangents, trig functions, powers, roots, reciprocals, exponential and hyperbolic functions, formulas and theorems. 269pp. 5⅜ × 8½. 60111-0 Pa. $6.95

THEORETICAL PHYSICS, Georg Joos, with Ira M. Freeman. Classic overview covers essential math, mechanics, electromagnetic theory, thermodynamics, quantum mechanics, nuclear physics, other topics. First paperback edition. xxiii + 885pp. 5⅜ × 8½. 65227-0 Pa. $19.95

SPECIAL FUNCTIONS, N.N. Lebedev. Translated by Richard Silverman. Famous Russian work treating more important special functions, with applications to specific problems of physics and engineering. 38 figures. 308pp. 5⅜ × 8½. 60624-4 Pa. $8.95

OBSERVATIONAL ASTRONOMY FOR AMATEURS, J.B. Sidgwick. Mine of useful data for observation of sun, moon, planets, asteroids, aurorae, meteors, comets, variables, binaries, etc. 39 illustrations. 384pp. 5⅜ × 8¼. (Available in U.S. only) 24033-9 Pa. $8.95

INTEGRAL EQUATIONS, F.G. Tricomi. Authoritative, well-written treatment of extremely useful mathematical tool with wide applications. Volterra Equations, Fredholm Equations, much more. Advanced undergraduate to graduate level. Exercises. Bibliography. 238pp. 5⅜ × 8½. 64828-1 Pa. $7.95

POPULAR LECTURES ON MATHEMATICAL LOGIC, Hao Wang. Noted logician's lucid treatment of historical developments, set theory, model theory, recursion theory and constructivism, proof theory, more. 3 appendixes. Bibliography. 1981 edition. ix + 283pp. 5⅜ × 8½. 67632-3 Pa. $8.95

MODERN NONLINEAR EQUATIONS, Thomas L. Saaty. Emphasizes practical solution of problems; covers seven types of equations. ". . . a welcome contribution to the existing literature. . . ."—*Math Reviews*. 490pp. 5⅜ × 8½. 64232-1 Pa. $11.95

FUNDAMENTALS OF ASTRODYNAMICS, Roger Bate et al. Modern approach developed by U.S. Air Force Academy. Designed as a first course. Problems, exercises. Numerous illustrations. 455pp. 5⅜ × 8½. 60061-0 Pa. $9.95

INTRODUCTION TO LINEAR ALGEBRA AND DIFFERENTIAL EQUATIONS, John W. Dettman. Excellent text covers complex numbers, determinants, orthonormal bases, Laplace transforms, much more. Exercises with solutions. Undergraduate level. 416pp. 5⅜ × 8½. 65191-6 Pa. $9.95

INCOMPRESSIBLE AERODYNAMICS, edited by Bryan Thwaites. Covers theoretical and experimental treatment of the uniform flow of air and viscous fluids past two-dimensional aerofoils and three-dimensional wings; many other topics. 654pp. 5⅜ × 8½. 65465-6 Pa. $16.95

INTRODUCTION TO DIFFERENCE EQUATIONS, Samuel Goldberg. Exceptionally clear exposition of important discipline with applications to sociology, psychology, economics. Many illustrative examples; over 250 problems. 260pp. 5⅜ × 8½. 65084-7 Pa. $7.95

LAMINAR BOUNDARY LAYERS, edited by L. Rosenhead. Engineering classic covers steady boundary layers in two- and three-dimensional flow, unsteady boundary layers, stability, observational techniques, much more. 708pp. 5⅜ × 8½. 65646-2 Pa. $18.95

LECTURES ON CLASSICAL DIFFERENTIAL GEOMETRY, Second Edition, Dirk J. Struik. Excellent brief introduction covers curves, theory of surfaces, fundamental equations, geometry on a surface, conformal mapping, other topics. Problems. 240pp. 5⅜ × 8½. 65609-8 Pa. $7.95

THE FOUR-COLOR PROBLEM: Assaults and Conquest, Thomas L. Saaty and Paul G. Kainen. Engrossing, comprehensive account of the century-old combinatorial topological problem, its history and solution. Bibliographies. Index. 110 figures. 228pp. 5⅜ × 8½. 65092-8 Pa. $6.95

CATALYSIS IN CHEMISTRY AND ENZYMOLOGY, William P. Jencks. Exceptionally clear coverage of mechanisms for catalysis, forces in aqueous solution, carbonyl- and acyl-group reactions, practical kinetics, more. 864pp. 5⅜ × 8½. 65460-5 Pa. $19.95

PROBABILITY: An Introduction, Samuel Goldberg. Excellent basic text covers set theory, probability theory for finite sample spaces, binomial theorem, much more. 360 problems. Bibliographies. 322pp. 5⅜ × 8½. 65252-1 Pa. $8.95

LIGHTNING, Martin A. Uman. Revised, updated edition of classic work on the physics of lightning. Phenomena, terminology, measurement, photography, spectroscopy, thunder, more. Reviews recent research. Bibliography. Indices. 320pp. 5⅜ × 8¼. 64575-4 Pa. $8.95

PROBABILITY THEORY: A Concise Course, Y.A. Rozanov. Highly readable, self-contained introduction covers combination of events, dependent events, Bernoulli trials, etc. Translation by Richard Silverman. 148pp. 5⅜ × 8¼. 63544-9 Pa. $5.95

AN INTRODUCTION TO HAMILTONIAN OPTICS, H. A. Buchdahl. Detailed account of the Hamiltonian treatment of aberration theory in geometrical optics. Many classes of optical systems defined in terms of the symmetries they possess. Problems with detailed solutions. 1970 edition. xv + 360pp. 5⅜ × 8½. 67597-1 Pa. $10.95

STATISTICS MANUAL, Edwin L. Crow, et al. Comprehensive, practical collection of classical and modern methods prepared by U.S. Naval Ordnance Test Station. Stress on use. Basics of statistics assumed. 288pp. 5⅜ × 8½. 60599-X Pa. $6.95

DICTIONARY/OUTLINE OF BASIC STATISTICS, John E. Freund and Frank J. Williams. A clear concise dictionary of over 1,000 statistical terms and an outline of statistical formulas covering probability, nonparametric tests, much more. 208pp. 5⅜ × 8½. 66796-0 Pa. $6.95

STATISTICAL METHOD FROM THE VIEWPOINT OF QUALITY CONTROL, Walter A. Shewhart. Important text explains regulation of variables, uses of statistical control to achieve quality control in industry, agriculture, other areas. 192pp. 5⅜ × 8½. 65232-7 Pa. $7.95

THE INTERPRETATION OF GEOLOGICAL PHASE DIAGRAMS, Ernest G. Ehlers. Clear, concise text emphasizes diagrams of systems under fluid or containing pressure; also coverage of complex binary systems, hydrothermal melting, more. 288pp. 6½ × 9¼. 65389-7 Pa. $10.95

STATISTICAL ADJUSTMENT OF DATA, W. Edwards Deming. Introduction to basic concepts of statistics, curve fitting, least squares solution, conditions without parameter, conditions containing parameters. 26 exercises worked out. 271pp. 5⅜ × 8½. 64685-8 Pa. $8.95